STRUCTURAL FIREFIGHTING
STRATEGY AND TACTICS

Bernard "Ben" J. Klaene

Thomas C. Lakamp

JONES & BARTLETT
LEARNING

World Headquarters
Jones & Bartlett Learning
25 Mall Road
Burlington, MA 01803
978-443-5000
info@jblearning.com
www.jblearning.com
www.psglearning.com

National Fire Protection Association
1 Batterymarch Park
Quincy, MA 02169-7471
www.NFPA.org

Jones & Bartlett Learning books and products are available through most bookstores and online booksellers. To contact the Jones & Bartlett Learning Public Safety Group directly, call 800-832-0034, fax 978-443-8000, or visit our website, www.psglearning.com.

> Substantial discounts on bulk quantities of Jones & Bartlett Learning publications are available to corporations, professional associations, and other qualified organizations. For details and specific discount information, contact the special sales department at Jones & Bartlett Learning via the above contact information or send an email to specialsales@jblearning.com.

Copyright © 2021 by Jones & Bartlett Learning, LLC, an Ascend Learning Company.

All rights reserved. No part of the material protected by this copyright may be reproduced or utilized in any form, electronic or mechanical, including photocopying, recording, or by any information storage and retrieval system, without written permission from the copyright owner.

The content, statements, views, and opinions herein are the sole expression of the respective authors and not that of Jones & Bartlett Learning, LLC. Reference herein to any specific commercial product, process, or service by trade name, trademark, manufacturer, or otherwise does not constitute or imply its endorsement or recommendation by Jones & Bartlett Learning, LLC and such reference shall not be used for advertising or product endorsement purposes. All trademarks displayed are the trademarks of the parties noted herein. *Structural Firefighting: Strategy and Tactics, Fourth Edition* is an independent publication and has not been authorized, sponsored, or otherwise approved by the owners of the trademarks or service marks referenced in this product.

There may be images in this book that feature models; these models do not necessarily endorse, represent, or participate in the activities represented in the images. Any screenshots in this product are for educational and instructive purposes only. Any individuals and scenarios featured in the case studies throughout this product may be real or fictitious, but are used for instructional purposes only.

The NFPA and the publisher have made every effort to ensure that contributors to *Structural Firefighting: Strategy and Tactics, Fourth Edition* materials are knowledgeable authorities in their fields. Readers are nevertheless advised that the statements and opinions are provided as guidelines and should not be construed as official NFPA policy. The recommendations in this publication or the accompanying resources do not indicate an exclusive course of action. Variations, taking into account the individual circumstances and local protocols, may be appropriate. The NFPA and the publisher disclaim any liability or responsibility for the consequences of any action taken in reliance on these statements or opinions.

20375-2

Production Credits
Public Safety Group:
Vice President of Sales, Professional Education: Phil Charland
VP, Product Development: Christine Emerton
Director of Product Management - Fire: Bill Larkin
Director of Content Management, Professional Education: Donna Gridley
Senior Editor: Carol B. Guerrero
Editorial Assistant: Alex Belloli
Manager, Project Management: Kristen Rogers
Senior Project Specialist: Alex Schab
Senior Digital Project Specialist: Angela Dooley

Digital Project Specialist: Rachel DiMaggio
Marketing Manager: Jessica Cicciu
Production Services Manager: Colleen Lamy
VP, Manufacturing and Inventory Control: Therese Connell
Composition: S4Carlisle Publishing Services
Cover Design: Scott Moden
Text Design: Scott Moden
Media Development Editor: Faith Brosnan
Rights Specialist: Liz Kincaid
Cover Image (Title Page, Front Matter): © Courtesy of David J. Jones, Cincinanti, Ohio.
Printing and Binding: LSC Communications

Library of Congress Cataloging-in-Publication Data

Names: Klaene, Bernard J., author. | Lakamp, Thomas C., author.
Title: Structural firefighting : strategy and tactics / Bernard "Ben" J. Klaene, Thomas C. Lakamp.
Description: Fourth edition. | Burlington, Massachusetts : Jones & Bartlett Learning, 2020. | Includes bibliographical references and index. | Summary: "Safe and effective structural firefighting requires a complex thought process. It is not a simple matter of "how to." Decisions depend on many factors, from the type of building, to the likelihood of occupancy, to the water supply. The fourth edition of Structural Firefighting: Strategy and Tactics leads readers through all phases of planning, evaluation and implementation to enable them to effectively manage structure fire incidents safe and effective manner, regardless of size or complexity"-- Provided by publisher.
Identifiers: LCCN 2020025781 | ISBN 9781284203752 (paperback)
Subjects: LCSH: Fire extinction. | Fire prevention. | Fire departments--Equipment and supplies.
Classification: LCC TH9310.5 .K535 2020 | DDC 628.9/25--dc23
LC record available at https://lccn.loc.gov/2020025781

6048

Printed in the United States of America
27 26 25 24 23   10 9 8 7 6 5 4 3 2

Brief Contents

CHAPTER **1** Organizing, Coordinating, and Commanding Emergency Incidents — 2

CHAPTER **2** Procedures, Preincident Planning, and Size-Up — 44

CHAPTER **3** Developing an Incident Action Plan — 96

CHAPTER **4** Company Operations — 122

CHAPTER **5** Fire Fighter Safety — 136

CHAPTER **6** Life Safety — 202

CHAPTER **7** Fire Protection Systems — 228

CHAPTER **8** Offensive Operations — 254

CHAPTER **9** Defensive Operations — 306

CHAPTER **10** Property Conservation — 328

CHAPTER **11** The Role of Occupancy — 338

CHAPTER **12** High-Rise Buildings — 384

APPENDIX **A** FESHE Correlation Guide — 426

GLOSSARY — 427

INDEX — 432

Contents

About the Authors xi
Dedication xii
Acknowledgments xiii
Preface xv

CHAPTER 1
Organizing, Coordinating, and Commanding Emergency Incidents — 2

Introduction 6
Evolution of the National Incident Management System 6
Command 7
  National Incident Management System 7
  Unified Command 7
  Initial Command 8
  Command by a Chief Officer 10
  Transfer of Command 11
  Delegation 12
  Command Post 12
  Span of Control 13
  Calling for Additional Resources 14
  Demobilization 15
  Staging 15
NIMS Organization and Positions 16
  Modular Organization 17
  Command Staff 17
  General Staff 19
Naming Geographic Areas of a Building 28
Communications 31
  Face-to-Face Communications 31
  Messengers 31
  Cell Phones 31
  Satellite Phones 32
  Hard-Wire Communications Systems 32
  Building Communications Systems 32
  Public Address Systems 32
  Computer Systems 32
  Advances in Communications 33
  Radio Designation for the IC 33
  Communications Network 33
Summary 35
Wrap UP 36
  Chapter Summary 36
  Key Terms 38
  Suggested Activities 39
  References 43

CHAPTER 2
Procedures, Preincident Planning, and Size-Up — 44

Introduction 47
Developing Standard Operating Procedures 47
  Purpose of Standard Operating Procedures 48
  Relationship of Standard Operating Procedures to Training and Equipment 49
  The Standard Operating Guidelines Controversy 50
Evaluating Response District Resources and Challenges 50
  Response Time 51
  Water Supply 51
  Construction Methods 54

Evaluating a Specific Property	55
Security Concerns	55
Preincident Plans	58
What Structures Are Preplanned?	67
Modifying Standard Operating Procedures	67
Estimating Life-Safety Needs	68
Estimating Extinguishment Needs	68
Estimating Property Conservation Needs	68
Relationship of Preplanning to Size-Up	68
Analyzing the Situation Through Size-Up	68
Life Safety/Fire Fighter Safety	70
Structure	78
Extinguishment	83
Property Conservation	86
General Factors	87
Size-Up Chronology	90
Standard Operating Procedures and the Preincident Plan	91
Shift/Day/Time	91
Alarm Information	91
While Responding	91
Visual Observations at the Scene	91
Information Gained During Continuing Operations	91
Overhaul	91
Wrap UP	92
Chapter Summary	92
Key Terms	92
Suggested Activities	94
References	94

CHAPTER 3
Developing an Incident Action Plan   96

Introduction	99
Determining Life-Safety Needs	99
Evaluating Structural Conditions	99
Estimating Resource Capability and Evaluating Resource Requirements	100
Developing an Offensive or a Defensive Incident Action Plan	101
Formulating an Incident Action Plan	101
Deployment	101
Scenario 1: Single-Family Detached Dwelling	102
Risk-Versus-Benefit Analysis	103
Incident Action Plan	105
Deployment	105
Scenario 2: Fourteen-Unit Apartment Building	105
Risk-Versus-Benefit Analysis	106
Incident Action Plan	108
Deployment	108
Scenario 3: Supreme Meat Packing Company—Processing Plant Complex	108
Risk-Versus-Benefit Analysis	110
Incident Action Plan	110
Deployment	110
Scenario 4: Church Fire	111
Risk-Versus-Benefit Analysis	113
Incident Action Plan	113
Deployment	113
Scenario 5: High-Rise Apartment Building	114
Risk-Versus-Benefit Analysis	115
Incident Action Plan	116
Deployment	116
Wrap UP	117
Chapter Summary	117
Key Terms	117
Suggested Activities	117
References	121

CHAPTER 4
Company Operations   122

Introduction	124
Engine Company Tasks	124
Truck Company Tasks	126
A Note about Staffing	126
Coordinating Company Operations	126
Quint and Quad Companies	128
Ventilation	129
Apparatus Positioning	130

Wrap UP	132
Chapter Summary	132
Key Terms	132
Suggested Activities	133
References	135

CHAPTER 5
Fire Fighter Safety — 136

Introduction	139
Incident Safety Officer	139
Fire Fighter Injuries and Fatalities	140
Risk Management Applied to the Fire Ground	148
Fire Intensity	150
Fuel Load	150
Building Design Loads	152
Structural Stability	152
Construction Methods and Materials	153
Roof Operations	157
Green Construction	159
Floor Construction	159
Basement Fires	160
Prefire Conditions and Fire Conditions	165
Prefire Conditions	165
Fire Conditions Leading to Structural Collapse	166
Fire Extension	166
Hazard Control Zone	168
Electrical Hazards on the Fire Ground	172
Shock Hazard Warning Precautions	172
Relationship of Time, Fire Intensity, and Structural Stability	174
Time: Ignition to Effective Actions	177
Detection/Transmission	177
Dispatch Time	177
Turnout Time	177
Travel Time	177
Setup Time	178
Adequate Number of Personnel	178
Tactical Reserve	180

Correlation between Elapsed Time and Progression to Flashover	180
Fire-Ground Operations	182
Communications	182
Command and Control	182
Accountability	183
Safety Officer	184
Alternative Egress	184
Rapid Intervention Crews	185
Company Safety Responsibilities	188
Declaring a Mayday	188
Opposing Fire Streams	189
Personal Protective Clothing	190
Overhaul	192
Rehabilitation	194
Wrap UP	196
Chapter Summary	196
Key Terms	196
Suggested Activities	197
References	200

CHAPTER 6
Life Safety — 202

Introduction	204
Evaluating the Probability of Extinguishment	204
Rescue versus Fire Attack	205
Assessing the Ventilation Profile	206
Analyzing the Available Rescue Options	207
Interior Stairs	208
Fire Escapes	208
Aerial Ladders and Elevated Platforms	209
Ground Ladders	209
Elevator Rescues	209
Rope Rescues	209
Helicopter Rescues	210
Classifying Evacuation Status	210
Estimating the Number of People Needing Assistance	213
Surveying Floor Layout and Size	214
Prioritizing Rescues by Location/Proximity	216

Evaluating the Medical Status of Victims	217
Evaluating Victims in Mass-Casualty Incidents	218
Triage, Prioritizing, and Transport	218
Evaluating the Need for Shelter	220
Estimating Life-Safety Staffing Requirements	220
Summary	221
Wrap UP	222
Chapter Summary	222
Key Terms	222
Suggested Activities	223
Reference	227

CHAPTER 7
Fire Protection Systems — 228

Introduction	231
Sprinkler Systems	231
Working at a Sprinkler-Protected Building with No Signs of Fire	235
Gaining Entry	235
Checking the Main Control Valve	236
Checking the Fire Pumps	236
Checking the Building for Fire and/or Sprinkler Operation	237
Supplying the Fire Department Connection	237
Working at a Sprinkler-Protected Building with Evidence of Fire Showing from the Exterior	238
Gaining Entry	238
Checking the Main Control Valve and Fire Pump	238
Supplying the Fire Department Connection	238
Letting the System Do Its Job	239
Backing Up the System	239
Ventilating	240
Performing Property Conservation Tasks	240
Placing the System Back in Service	240
Working at a Property Protected by a Deluge System	240
Checking the Control Valve and Fire Pump	240
Operating the Deluge Valve	240
Checking Interlocks	241
Letting the System Do Its Job	241
Backing Up the System	241

Working at a Building Equipped with a Standpipe System	241
Checking Fire Pumps and Main Control Valves	242
Supplying Fire Department Connections	242
Providing Standpipe Equipment	243
Connecting to the Standpipe Discharge	246
Nonwater-Based Extinguishing Systems	246
Foam Systems	246
Carbon Dioxide Systems	247
Halon and Other Clean Agents	247
Dry and Wet Chemical Systems	248
Working in Areas Protected by Total Flooding Carbon Dioxide or Clean Agent Systems	248
Letting the System Do Its Job	248
Final Extinguishment and Rescue	249
Manual Activation	249
Checking Interlocks	249
Checking Agent Supply	249
System Restoration	249
Working in Areas Protected by Local-Application Carbon Dioxide, Clean Agents, Dry Chemical, or Other Special Extinguishing Agents	249
Letting the System Do Its Job	249
Checking the Interlocks	249
Manual Activation	249
Backing Up the System	249
System Restoration	250
Responses to Building Fire Alarm Systems	250
Wrap UP	251
Chapter Summary	251
Key Terms	251
Suggested Activities	252
References	253

CHAPTER 8
Offensive Operations — 254

Introduction	257
Calculating Rate of Flow	257
The Royer/Nelson Formula	258
The National Fire Academy Formula	259
Sprinkler Rate of Flow Calculations	260
Estimating the Size of the Largest Area	261

Estimating the Percentage of Area on Fire	263
Comparing Rate of Flow Calculations	265
Which Rate of Flow Calculation Is Best?	267
First Floor	268
Second Floor	268
Third Floor	268
Selecting the Attack Hose Size	268
Selecting the Nozzle Type	276
Selecting the Method of Attack	276
Indirect Attack	277
Direct Attack	278
Estimating the Number of Attack Hose Lines	282
Evaluating Exposures	282
Estimating Backup Needs	284
Estimating the Number of Hose Lines Needed Above the Fire	284
Evaluating Other Hose Lines Needed	285
Estimating Water Supply Needs	285
Estimating Ventilation Needs	287
Calculating Staffing Needs	289
Engine 1 (First-Arriving Engine Company)	290
Engine 2 (Second-Arriving Engine Company)	291
Truck 1 (First-Arriving Truck Company)	292
Chief Officer	292
Engine 3 (Third-Arriving Engine Company)	294
Engine 4	294
Heavy Rescue, Truck 2, and Engine 5	294
Determining Apparatus Needs	296
Wrap UP	298
Chapter Summary	298
Key Terms	298
Suggested Activities	299
References	305

CHAPTER 9
Defensive Operations 306

Introduction	309
Classifying as a Defensive Attack	309
Establishing a Collapse Zone	310
Evaluating Exposures	312
Evaluating a Direct Attack	312
Estimating the Number and Type of Master Streams Needed	313
Estimating Water Supply Needs	314
Calculating Staffing Needs	315
Determining Apparatus Needs	315
Conflagrations and Group Fires	316
Classifying as a Nonattack	321
Wrap UP	322
Chapter Summary	322
Key Terms	322
Suggested Activities	322
References	327

CHAPTER 10
Property Conservation 328

Introduction	330
Classifying as an Offensive Attack	330
Estimating Indirect Damage	331
Evaluating Water Damage	331
Evaluating Smoke Damage	333
Calculating Staffing Needs	333
Evaluating Property Conservation Needs	333
Evaluating Overhaul Needs	334
A Word about Fire Investigation	335
Wrap UP	336
Chapter Summary	336
Key Terms	336
Suggested Activities	336
Reference	337

CHAPTER 11
The Role of Occupancy 338

Introduction	342
Classifying the Occupancy Type	342
Assembly Occupancies	343
Places of Worship	344
Eating and Drinking Establishments	345

Sports Arenas	347
Convention Centers	348
Theaters	349
Educational Occupancies	350
Elementary Schools	351
Middle, Junior High, and High Schools	353
Colleges and Universities	353
Healthcare Occupancies	353
Hospitals	354
Nursing Homes	355
Limited Care Facilities	356
Ambulatory Care Facilities	357
Residential Board and Care Occupancies	357
Detention and Correctional Occupancies	359
Residential Occupancies	360
One- and Two-Family Dwellings	362
Apartment Buildings	363
Dormitories	364
Hotels and Motels	366
Mercantile Occupancies	366
Shopping Centers	367
Enclosed Shopping Malls	367
Lifestyle Centers	368
Big-Box Stores	368
Multilevel Department Stores	369
Business Occupancies	369
Storage Occupancies	370
Industrial Occupancies	372
Multiple (Mixed or Separated) Occupancy Buildings	374
Buildings Under Construction, Renovation, or Demolition	374
Buildings Under Construction	374
Renovated Buildings	376
General Considerations for Special Occupancy Fires	376
Estimating the Number of Potential Victims	377
Wrap UP	378
Chapter Summary	378
Key Terms	379
Suggested Activities	379
References	382

CHAPTER 12
High-Rise Buildings 384

Introduction	387
Developing and Revising High-Rise Standard Operating Procedures	388
Fire Fighter Safety	388
Fire Fighter Use of Elevators	388
Ground Support	392
Life Safety	393
Rescuing and Evacuating Occupants	393
Helicopter Rescues	393
Partial or Sequential Evacuation	394
Emergency Voice/Alarm Communications System	394
Extinguishment	395
Command Post Location	396
Developing Building-Specific High-Rise Preincident Plans	397
Analyzing the Situation Through Size-Up	399
Smoke Movement	400
Heat of the Fire	401
Stack Effect	401
Wind	401
Developing and Implementing an Incident Action Plan	404
Use of Stairways	405
Ventilation	406
Interior Exposures	406
Property Conservation	406
Lead Time	406
Establishing a Wide Fire Zone	407
Applying the National Incident Management System to a High-Rise Fire	408
Communications	408
Tactical Worksheets	409
Base (Exterior Staging for High-Rise Fires)	410
Staging (Interior)	411
Lobby Control	412
High-Rise Case Histories	412
The One Meridian Plaza Fire	412

The First Interstate Bank Building Fire	413	Suggested Activities	419
Comparing the One Meridian Plaza and First Interstate Bank Building Fires	413	References	425
Terrorist Attacks at the World Trade Center	415	APPENDIX A **FESHE Correlation Guide**	**426**
The MGM Grand Fire	417	Glossary	427
Wrap UP	418	Index	432
Chapter Summary	418		
Key Terms	419		

About the Authors

Ben Klaene

District Chief Ben Klaene retired from the Cincinnati Fire Department after 30 years of service. Klaene completed basic fire fighter training and was assigned to a highly industrialized area.

His first promotion was to Cincinnati's heavy rescue unit as a driver. The heavy rescue company responded to a wide variety of emergencies, including fires in the downtown district, all extra alarms, entrapments, hazardous materials, water, and high-angle rescue calls throughout the city. All heavy rescue members were certified paramedics, scuba divers, and hazardous materials technicians.

Klaene advanced through the ranks from recruit to district chief, where he was assigned as training and safety chief. As safety officer, he responded to more than 50 multiple-alarm fires, the largest incident being the BASF Chemical Company Fire and explosions that severely damaged the chemical plant covering a city block, killing two, and injuring 83. He also served on a three-person committee writing department standard operating procedures (SOPs) and responded to all three-alarm or greater fires as the safety officer.

Klaene attended classes at the University of Cincinnati, earning an associate degree in fire science and technology, a bachelor of science degree in fire and industrial safety, and a master's degree in education. He is also a graduate of the U.S. National Fire Academy Executive Fire Officer program. Klaene served as a subject matter expert during the NIST high-rise fire-ground field experiments.

Using his education to the advantage of fire students, Klaene taught fire science courses at the University of Cincinnati for 27 years. He served as an adjunct instructor and subject matter expert at the National Fire Academy where he also assisted in revising courses in the Degrees at a Distance program. Klaene used his skills and education as a fire and safety consultant, training industrial fire brigades, assessing fire protection in Ohio and Michigan cities, as well as in a variety of businesses.

Thomas C. Lakamp

Assistant Fire Chief Thomas Lakamp is a 32-year veteran of the Cincinnati Fire Department. Lakamp is responsible for the operations division, which oversees all fire, special operations, and emergency medical services delivery for the City of Cincinnati.

Prior to assignment as the operations division assistant chief, Lakamp was assigned to the human resources division where he managed the training, risk management, and disciplinary issues for the department. As a district chief, Lakamp served as the department's first special operations chief, managing the department's response to technical rescue, hazardous materials, bomb squad, and the city's municipal airport response. Lakamp has also served as a district chief in Fire District 4 on the city's east side for 4 years and the training chief for 7 years.

Lakamp holds an associate degree in fire science technology and a bachelor of science degree from the University of Cincinnati, where he is an adjunct instructor in fire tactics. Lakamp also holds a master's degree in homeland security from the Naval Postgraduate School in Monterey, California. Lakamp is a graduate of the National Fire Academy Executive Fire Officer program and was formerly a task force leader for FEMA Ohio Task Force 1—Urban Search and Rescue team. He is currently the commissioner for the Hamilton County, Ohio—Region Six USAR team.

Dedication

© Courtesy of David J. Jones, Cincinanti, Ohio.

This book is dedicated to our colleagues who gave their lives serving others. Hopefully, by sharing our many years of fire service training and experience, future line-of-duty deaths can be prevented.

Courtesy of Bill Strite, Cincinnati, Ohio.

Courtesy of Bill Strite, Cincinnati, Ohio.

Acknowledgments

Reviewers

Vince Ashcraft
Assistant Chief (ret.)
Clark State Community College
Springfield, Ohio

Gregory Barton
Fire Chief
Beverly Hills, California

David Blair, AD Architecture
Paramedic, Rescue Tech, Haz-Mat Tech, Fire Instructor, Peer Support Member
Lieutenant, Columbus Ohio Division of Fire
Cleveland State University, Center for Emergency Preparedness
Career and Technology Education Center of Licking County
Columbus, Ohio

Robert J. Colameta, Jr.
Public Safety Education Network
Everett, Massachusetts

Joey Davis
Mill Spring Fire Department
Mill Spring, North Carolina

Timothy Dorsey, BS, AAS, EMT-P
Division Chief/Commission Chairman
Lake Ozark Fire Protection District
Missouri State Fire Education/Advisory Commission Chairman
Lake Ozark, Missouri

Chris Farrell
Senior Emergency Services Specialist
NFPA
Quincy, Massachusetts

Curt Floyd
Senior Specialist, Technical Lead, Engineering Services
NFPA
Quincy, Massachusetts

Brian S. Gettemeier
Captain
Cottleville Fire Protection District
St. Charles, Missouri

Christopher L. Gilbert, PhD
Training Officer, Tactical Medic, Hazardous Materials Officer
Alachua County
Gainesville, Florida

Todd Gilgren
Battalion Chief
Arvada Fire Protection District
Arvada, Colorado

Bryan Goustos
Firefighter and Fire Service Instructor
Brampton Fire and Emergency Services
Brampton, Ontario, Canada

Perry Hall
Battalion Chief
High Point Fire Department
High Point, North Carolina

Brandon Harrill
Fire Chief
Rutherfordton Fire and Rescue
Rutherfordton, North Carolina

Jordan T. Hood
City of Charlotte Fire Department
Charlotte, North Carolina

Robert Lindstedt, MSEd
Assistant Professor
Southern Maine Community College
South Portland, Maine

Tony Mecham
Fire Chief
CAL FIRE
San Diego County Fire
San Diego, California

Nick Morgan
Captain
St. Louis Fire Department
St. Louis, Missouri

Douglas Rohn
Verona Fire Department
Verona, Wisconsin

Chad Smith
Battalion Chief
Georgetown Fire Department
Georgetown, Kentucky

Curt Smith, CFO, EFO
Assistant Chief
Hastings Fire and Rescue
Hastings, Nebraska

Robert E. Swiger
Garner, North Carolina

Terry Wattenbarger
Chief
Laurel County Fire Department
London, Kentucky

William Wren, BS, FS1-2, FO-1, ISO
NYS Fire Instructor—Adjunct Instructor
NYSAFS—New Hartford Fire Department
New Hartford, New York

Christopher Yoch, MS, FO
Keystone Fire Company No. 1 of Shillington
Shillington, Pennsylvania

Preface

The goal of this text is to explain proven tactics and strategies used at structure fires. The guiding objective throughout this text is to prepare the fire officer to apply appropriate tactics and strategies at structure fires, fully using available resources in a safe and effective manner.

It is assumed that the reader has a basic understanding of firefighting and associated tasks as outlined in *Fundamentals of Fire Fighter Skills* and such National Fire Protection Association (NFPA) documents as NFPA 1001, *Standard for Fire Fighter Professional Qualifications*; NFPA 1002, *Standard for Fire Apparatus Drive/Operator Professional Qualifications*; NFPA 1021, *Standard for Fire Officer Professional Qualifications*; or other similar firefighting manuals and standards. Qualified fire fighters with basic training and some level of experience who are seeking positions within a fire department as company and/or chief officers are the intended audience for this text.

Many fire departments adopted the past editions of *Structural Firefighting* as part of their reading list for promotional examinations. To ensure consistency throughout a fire department, it is best when all department members use the same strategy and tactics text. The lessons learned in *Structural Firefighting* apply to all members of the department, and it is an excellent text for providing consistent, department-wide strategy and tactics training.

The basis for an incident action plan that leads to a safe and effective fire-ground operation is an operational priority list. This text is designed around a three-point list:

1. Life safety
2. Extinguishment
3. Property conservation

Several other priority lists exist, but each list basically prioritizes life safety, followed by a logical sequence of activities aimed at reducing property loss. It must be stated that crossover exists among the listed priorities, and quite often, a lower-priority item provides the means to achieve a higher priority. For example, when the fire is extinguished, occupants and fire fighters are in less danger. Thus, extinguishment becomes an essential part of the life-safety priority. The importance of extinguishment cannot be overemphasized and is a recurring theme throughout this text.

An important goal of any fire strategy and tactics text, including this one, is to give the reader the necessary tools to achieve maximum productivity under adverse fire-ground conditions. While fire-ground experience is a prerequisite to using the tools presented in this text, this text will enable readers to learn fire-ground procedures at an accelerated pace, thus reducing the cost in lives and property associated with learning by experience only. In addition to updating materials, answers to frequently asked questions are embedded in this fourth edition. Furthermore, explanations of many of the topics covered in the prior editions have been expanded to include information from several recent studies by the National Institute of Standards and Technology (NIST), Underwriters Laboratories (UL), and others.

This is a dynamic and exciting time in the fire service. Many research studies with direct application to structural firefighting have been conducted since the publication of the past editions of this text. Staffing studies by NIST prove the need for adequate staffing as a result of quantifying the effects of understaffing. Positive-pressure ventilation studies increase our understanding of this tactical method. The UL ventilation and fuel-load studies indicate a need to change fire-ground tactics. Future studies are planned to further promote scientific knowledge in ways that directly affect the way fire fighters "go to work" at structure fires. Of particular interest to these authors is a proposed study investigating rate-of-flow requirements. Some old myths, including some held by these authors, have been discredited. The information gained through these various studies should improve safety and effectiveness at the fire scene. However, it is essential to closely study the parameters of research studies. To take maximum advantage of the knowledge gained through research, we must share experiences.

The development of an effective fire officer or incident commander involves several steps over time. Training, education, and experience all play a role in the developmental process. A successful fire officer's background should include the following:

- Basic fire fighter training (as outlined in NFPA 1001, *Standard for Fire Fighter Professional Qualifications*)
- Experience as a fire fighter
- Advanced fire fighter training
- Company officer training (NFPA 1021, *Standard for Fire Officer Professional Qualifications*)
- Education in tactics, building construction, command, and organization

This list is not all inclusive, and the mix of experience, education, and training will vary in the developmental process. Experience is especially difficult to categorize. The needed experience may not be provided by serving several years with a department or company that has few structure fires. In turn, the assignment of a fire fighter or officer to a busy unit does not guarantee diverse learning. For example, a fire fighter on a busy company may always be in the attack position with a 1¾-inch (44-mm) hose line. This fire fighter will be proficient at advancing hose lines but may not be knowledgeable regarding other aspects of firefighting, such as fires requiring master streams, truck company operations, and other seldom-encountered fires.

While it is not possible (nor desirable!) to totally eliminate experience from the process of becoming an effective company officer or incident commander, training and education do reduce the experience necessary to become proficient. Of particular value is live fire training and realistic simulations. Some level of experience is necessary to "read the fire ground" and to understand the capabilities and limitations of resources and the impact of lead time. Ongoing training and education play a key role in the development of even the most experienced fire officer.

By applying the principles described in this text, a company officer or incident commander can use his or her experiences more effectively at the scene of structure fires. The fire officer who relies entirely on experiential learning essentially places the lives and property of others at a greater risk while he or she learns the trade. Sadly, many departments do not provide adequate structural firefighting training. After the attack on the World Trade Center on September 11, 2001, the NFPA conducted a fire department needs assessment for the U.S. Fire Administration. The study found, in part, that more than half of the fire departments (55%) surveyed did not provide structural firefighting training to all members. A follow-up report, *Four Years Later—A Second Needs Assessment of the U.S. Fire Service*, found that 53% of departments still failed to provide structural firefighting training to all members. A third needs assessment, conducted by the NFPA in 2010 showed a slight improvement, indicating that 46% of the departments responsible for structural firefighting have not formally trained all of their personnel. With nearly half of all departments lacking adequate training, there is a definite need to better train and educate members of the fire service in strategy and tactics. This text provides a readily available tool to improve this unacceptable statistic.

Fire is an integral part of our society. We depend on it for warmth and convenience. Fire allowed humans to progress from cavemen to the computer age. Yet uncontrolled fire is feared, and rightfully so, as it continues to destroy life and property. The modern fire department uses public safety education and code enforcement to help prevent and reduce the impact of fire. When these proactive efforts fail, fire fighters are forced into a reactive, suppression mode. Even where considerable effort is expended to prevent fires, they do occur. In this text, fire is viewed as a formidable adversary, and understanding "the enemy" is crucial if we are to be successful. Firefighting is as much an art as a science, but applying the available science can aid us in our war against fire.

Many have compared fire-ground management to commanding a military battle or coaching a football game. These analogies are correct and useful, but with one significant difference: The military commander and football coach face an unpredictable enemy; in contrast, the fire officer, if he or she has done effective preincident planning, has a predictable enemy. Fire is bound by the laws of chemistry and physics. The surprises that may occur are a result of incomplete information or failing to understand basic natural laws governing the fire, building construction, and the impact of the fire on the structure.

This text is by no means the only one a fire fighter endeavoring to be a proficient fire officer should read. The modern fire department may respond to hazardous materials releases, medical emergencies, technical rescues, wildland fires, and a wide variety of other emergencies. The scope of this text is limited to structural firefighting. The incident commander should be a qualified safety officer and be well versed on fire fighter safety laws and standards. The NFPA, the Occupational Safety and Health Administration (OSHA), and the National Institute of Occupational Safety and Health (NIOSH) have developed standards related to fire fighters' health and safety that should be familiar to the fire officer.

This text includes numerous case studies, incident summaries, and references to other cases discussed in fire service periodicals, NFPA Fire Investigations, the NIOSH Firefighter Fatality Investigation and Prevention Program, and the Technical Report Series from the U.S. Fire Administration. Because some types of fires may occur only once in a lifetime (conflagration, large area fire, major high-rise fire), fire officers relying strictly on experience may be at a loss when confronted with these rare incidents. Collectively, the fire service has learned many lessons about strategies and tactics that were applied at these major events. Only by reviewing case studies and ongoing education can most fire officers learn about these most challenging incidents.

Preincident planning and standard operating procedures (SOPs) are essential elements of successful operations. Preincident planning is particularly important in large and complex structures. This text reiterates the importance of preincident planning to successful operations. Advanced knowledge of some of the factors leading to strategic decisions before an incident occurs will reduce the information processing needs of the fire officer in the crucial early moments of an operation. Predesignating initial operations using SOPs allows the incident commander to process information while a predictable course of action takes place.

Although a great deal of science is associated with fighting fires in structures, the application of this science requires a less-than-scientific approach. As a matter of fact, the rigid application of scientific principles sometimes leads to theories that are technically correct but empirically fallacious. As you progress through this text, you will find several places where a fallacy is introduced but then disputed. Some commonly held fallacies are described for the purpose of disproving misconceptions. "Fallacy/Fact" boxes alert you to commonly held beliefs that are fallacious.

This text is designed to be used in a classroom environment or in a distance-learning format, with an instructor providing additional insight and practice materials. "Suggested Activities" assist the student and instructor in applying the information contained in each chapter. Instructor materials include slide presentations, lecture outlines, and additional activities. Given the distance-learning design, the text is ideal for self-study.

The authors have many people to thank. First and foremost is Russ Sanders, coauthor of the first and second editions. Russ not only contributed mightily to the text, he remains a mentor and information source. Although he was not involved in the actual writing of this fourth edition, his experience as chief of department in Louisville, Kentucky, and as an NFPA staff member are evident on every page. Photographers David J. Jones, Bill Strite, Nick Morgan, Denny Baker, Mike Carey, Bernard Erwin, J.B. Forbes, David Mullis, and Jeff Neal provided live-action fire photographs.

Dr. Rita Fahy, manager of applied research at the NFPA, researched and provided much of the statistical information contained in this text. Chris Farrell and Curt Floyd provided background information related to fire investigations, as well as pointing us in the right direction when searching for "real-life" fire experiences related to specific topics.

Ben Klaene
Thomas C. Lakamp

Courtesy of Bill Strite, Cincinnati, Ohio.

CHAPTER 1

Organizing, Coordinating, and Commanding Emergency Incidents

LEARNING OBJECTIVES

- Given a fire scenario, deploy units and organize the operation.
- Identify and define the main functions within the National Incident Management System (NIMS).
- Given different scenarios, organize an operation using the NIMS.
- Identify the minimum fire company staffing for initial response per NFPA 1710.
- Discuss and contrast fire-ground management compared to administrative management.
- Discuss the history and evolution of incident management systems including the development of the NIMS.
- Define unified and single command, listing the advantages and disadvantages of each.
- Compare command modes available to the first-arriving officer, determining situations where each mode would be appropriate.
- Develop an initial on-scene report.
- Explain the importance of and develop a status report.
- Analyze the command transfer process, discussing when and how command should be transferred.
- List the dangers related to freelancing.
- List the attributes of a good command post.
- Define and explain the importance of maintaining a reasonable span of control.
- Describe and enumerate the importance of staging.

- Compare a staged company to a parked apparatus.
- Define and explain the duties of the incident commander (IC).
- Identify, define, and place command staff positions on a NIMS organization chart.
- Identify, define, and place the five sections on a NIMS organization chart.
- Describe the position and function of a chief's aide.
- Explain the advantages of assigning a chief's aide.
- Define the functions of branches, divisions, groups, task forces, and strike teams.
- Explain the two-in/two-out rule.
- Organize an operation using geographic and functional assignments, and describe when each should be used.
- Given a fire situation, apply an intuitive naming system for various tactical-level management units.
- Recognize and articulate the importance of fireground communications.
- List general rules for incident scene communications.
- Define and explain unity of command.
- List and compare various means of communication that could be used at the incident scene.
- Explain the advantages of various kinds of fireground communication (face-to-face, radio, messenger).
- Develop a communications network that supports a NIMS organization.
- Explain methods that can be used to reduce radio communications to and from the IC.
- Explain the chain of command from company officer to the IC at a major incident with all levels of the incident command system in use.
- List types of essential information to be communicated to the IC when functioning as a division commander.
- Given an incident command system organizational chart for a major incident, develop an effective communications network using face-to-face, radio, and predeveloped signals using vehicle air horns and other alert systems.
- List the advantages of a mobile communications vehicle and other equipment that is typically available in a mobile command van.
- Explain the importance of always having someone in command at the incident scene.
- List important information that should be communicated in an initial on-scene report.
- Given a fire scenario where you are the first-arriving company officer at the scene of a structure fire and temporarily in command until a chief officer arrives, develop a status report to be communicated to the higher-ranking officer assuming command.
- Define the attributes of a good command post.
- Define *span of control* as used in the fire service.

Note: NFPA standards listed as quoted in this publication were the latest edition at the time of publication and provide a general outline of NFPA standard requirements. It is highly advised that the latest version of the standard be reviewed when applying NFPA standards to operating procedures or legal issues.

Courtesy of Julie Miklaszewicz.

Case Study

The following table lists the responding fire companies, fire officers, and emergency medical resources available in a particular area. There is a reliable and adequate county water supply with high-volume hydrants spaced every 1000 ft in this area of town. Using this information, answer the following questions.

	Engine Companies	Truck Companies	Other
Initial Response	Your Town Eng. 1 Any Town Eng. 2 Any Town Eng. 3	Any Town Truck 1	Your Town Chief Your Assistant Chief 1 Your Department EMT Ambulance
2nd Alarm	Next Town Eng. 4 Next Town Eng. 5	Next Town Truck 2	Any Town Chief of Dept. Next Town Chief of Dept. Next Town Heavy Rescue Next Town Paramedic Ambulance 1
3rd Alarm	Far Town Eng. 6 Far Town Eng. 7	Far Town Quint 1	Far Town Chief of Dept. Jones Town Paramedic Ambulance

© Jones & Bartlett Learning.

1. Describe the actions you would take as the officer on the first-arriving engine company (Your Town Engine 1) (e.g., secure a water supply from nearest fire hydrant and begin defensive attack on side "Charlie" using a deck-mounted master stream).
2. Assume the role of the officer of Any Town Truck Company 1. (You arrive 5 minutes after Your Town Engine Co. 1.) Explain your operation.
3. Assume the role of "Your Town Chief of Department" arriving at the scene. The companies on scene are working as described in the answers to the previous questions.
   a. Would you reassign companies working at the scene?
   b. Would you order additional fire companies?
   c. If additional companies are ordered, explain their assignments.
   d. Develop an incident action plan (IAP).
   e. Create a National Incident Management System (NIMS) chart showing all on-scene units.
4. The Any Town Engine 2 is arriving at the scene awaiting your orders. Assign this mutual aid engine.
5. You order a second alarm response.
   a. Assign responding units and call for additional alarms as needed.
   b. Develop an organization chart showing all responding units.
6. Download the National Institute for Occupational Safety and Health (NIOSH) investigation report at: www.cdc.gov/niosh/fire/reports/face201208.html
   a. Develop a list of problems with the operation as described in the NIOSH Report 2012-08.
   b. Compare your proposed deployment to the actual operation at this fire as described in the NIOSH Firefighter Fatality Investigation and Prevention Program, 2012-08 report.

Data from: NIOSH FIRE FIGHTER FATALITY INVESTIGATION AND PREVENTION PROGRAM, 2012-08. Volunteer Lieutenant Killed and Two Fire Fighters Injured Following Bowstring Roof Collapse at Theatre Fire Wisconsin.

Introduction

This chapter discusses the importance of adopting and implementing the National Incident Management System (NIMS) for use at all structure fires. Use of the NIMS is essential to enable the incident commander (IC) to safely and effectively manage incident resources. NIMS is instrumental in ensuring that fire fighter safety, as well as the three operational priorities of *life safety*, *extinguishment*, and *property conservation*, are addressed at the fire scene. Because even the best organization is at a distinct disadvantage without good communications and this is probably the most cited deficiency at actual emergencies and exercise scenarios, communications is also addressed in this chapter.

Decision making at the incident scene is different than day-to-day administrative decision making. The IC must decide on the proper course of action with limited information available in a relatively short period of time. This process is referred to as recognition-primed decision making (RPD). Administrative decision making is less time sensitive; therefore, input should be included from diverse sources, and careful analyses of all options should be considered. This is referred to as rational decision making (RDM). The time constraints of the fire ground require RPD. Proficient ICs make sound decisions within the time and information constraints of the fire ground by taking advantage of previous training, education, preincident planning, and matching the situation at hand to similar experiences at previous fires. Building types, occupancies, and potential fire scenarios are nearly endless, making it virtually impossible to gain experience in handling every possible fire scenario. Well-constructed fire simulation training is highly recommended to increase experiential learning.

Adoption of the NIMS, meeting the criteria outlined in NFPA 1561, *Standard on Emergency Services Incident Management System and Command Safety*, is a must (NFPA 1561, 2020). Such a system will allow for the expansion of forces while maintaining a reasonable span of control. NIMS is often referred to as IMS (incident management system). The use of NIMS is highly recommended. Call it what you like, but use it at every emergency incident regardless of size.

Fire-ground operations have often been compared to military operations. A military commander uses a battle plan much as an IC uses a preincident plan. Standard military maneuvers (the equivalent of fire department standard operating procedures, or SOPs) are developed in advance. The military has an effective organizational structure that limits the span of control to a relatively small number. A general trying to give orders to each of the thousands of soldiers under their command would soon be overwhelmed and lose control. This general may have success with skirmishes in which few soldiers are actually involved in battle, but could not hope to win a major campaign.

On the fire ground, it is a mistake for a chief officer to attempt to control the entire incident. Like the general, the IC may be successful in handling a skirmish (e.g., a one-alarm fire involving only four or five companies); however, a chief who tries to apply a micromanaging style to a major incident will quickly become overwhelmed.

NFPA 1500, *Standard on Fire Department Occupational Safety, Health, and Wellness Program*, requires the use of NIMS, as do current Occupational Safety and Health Administration (OSHA) regulations dealing with hazardous materials response (NFPA 1500, 2018; Superfund Amendments and Reauthorization Act, 1986).

> **NOTE**
>
> The terms *NIMS* and *IMS* are often used interchangeably; however, National Incident Management System (NIMS) is the official name of the system mandated by Congress (HSPD-5). NIMS provides a consistent, nationwide approach that can be used by federal, state, and tribal governments; the private sector; and nongovernmental organizations. NIMS enables groups to work together efficiently as they prepare for, respond to, and recover from domestic incidents, regardless of cause, size, or complexity.
>
> NIMS is a system that defines the roles and responsibilities to be assumed by responders as well as the standard operating procedures to be used in managing and directing an emergency incident and other functions.
>
> In this text the term *NIMS* is used exclusively, as it is the official name of this management system.

Evolution of the National Incident Management System

A significant amount of work has gone into developing systems to manage emergency incidents of all types. Forest fires in California in 1970 prompted the development of the FIRESCOPE Incident Command System (NICS; National Interagency Incident

Management System, 1981). FIRESCOPE greatly improved operations at large wildland fires, and its use by fire departments of all sizes in California provided the testing ground that led to the urbanization of the FIRESCOPE Incident Command System. As use of the system expanded, improvements were made based on the experience gained at actual incidents. The result was an extremely useful tool that could be used by the fire service and others involved in any emergency response. This is the system that is used to coordinate resources within the NIMS.

Command

National Incident Management System

It is essential that the entire response community, not just the fire service, be familiar with and trained in the use of NIMS. Agencies such as police, health departments, and local disaster agencies, as well as mutual aid departments, should all work under the same system. The use of NIMS provides common terminology and operational assignments for all agencies at the incident scene. This uniformity can greatly facilitate the successful conclusion of an incident. Homeland Security Presidential Directive (HSPD)-5 establishes NIMS as the national system, requiring its use by federal agencies and making its adoption mandatory when the federal government is providing preparedness assistance.

NIMS should be used for all incidents, regardless of their size. If personnel are using NIMS for smaller, routine incidents, they will be better prepared to use it for large-scale incidents. It is a mistake to wait until incident grows to a certain size before implementing NIMS; by then, it is too late to start trying to put together the organizational segments required to support the incident. It is much better to let the NIMS grow with the incident and integrate other agencies and jurisdictions as needed. The command system should be regional in scope and capable of handling large numbers of resources from beyond the local jurisdiction. NIMS is national in scope, and because of its flexibility and adaptability, it is capable of handling the largest imaginable structure fire including fires involving multiple buildings. The focus of this textbook is on structure fires, but NIMS is an all-hazard system capable of effectively managing any emergency situation that the IC may encounter.

The Oklahoma City bombing of the Alfred P. Murrah Federal building occurred some 30 years after large California wildfires emphasized the need for an incident management system capable of managing incidents with large numbers of personnel spread over a huge geographic area. The Federal Emergency Management Agency (FEMA) combined efforts with the National Fire Protection Association (NFPA) and others to develop a practical command system for use by emergency responders. This system led to the development of the NIMS.

The explosion at the nine-story Oklahoma City Federal building eventually resulted in the response of 43 fire departments as well as many police departments, federal agencies, and others. There is no doubt that this incident, which resulted in 168 fatalities and 475 injuries and spread over 48 square blocks, required an effective command system.

Unified Command

NIMS addresses situations in which more than one jurisdiction or agency has responsibility by establishing a **unified command**. Unified command provides an invaluable method for controlling large incidents in which multiple agencies or jurisdictions are involved. In a unified command scenario, different jurisdictions and/or agencies share responsibility for developing the incident action plan.

A word of caution about unified command is in order. Whenever possible, a **single command** (one person designated as the overall IC) is preferred to a shared, or unified, command. However, some situations dictate the use of a unified command system, such as when the incident objectives are not clearly defined as being one of fire or law enforcement.

Structure fires are clearly under the jurisdiction of the local fire department unless the structure or structures span more than one jurisdiction. Much has been written about the difficulty in communications and coordination between fire and police at the World Trade Center on September 11, 2001. Some feel this was a case for a unified command. The police department played a major role, but the emergency phase of the operation clearly should have been under the jurisdiction of the fire department. It is not unusual for police to assist in evacuating or isolating the area during a structure fire; they also control vehicular and pedestrian traffic around the incident scene. There could be a large contingent of police officers at the fire scene, but the fire department should retain control, with a police department supervisor coordinating police activities from the fire department incident command post. The police department supervisor (law branch director or law group supervisor) should be equipped with a radio capable of communicating with all police personnel, enabling the IC to communicate through the police supervisor. This police supervisor could also be assigned to the operations section. Although the World Trade Center events on September 11, 2001, may have been a case for unified command, seldom would a unified command be warranted for a structure fire.

> **NOTE**
> Tactical-level management unit (groups, divisions, and branches) names should indicate the area or function being managed. Tactical-level management units that manage police personnel are named as Law Branch or Law Group in this book. Naming them as Police Branch or Police Group would also be correct.

A unified command can also evolve after the emergency has been mitigated and the incident has moved into the clean-up phase. For example, once the life-safety issues have been resolved, several agencies might become engaged in a fire involving hazardous materials.

Problems can arise when various agencies with different priorities are charged with developing and implementing an incident action plan. When using the unified command structure, plan development is shared, but NIMS requires that one operations chief direct all field units.

Fallacy	Fact
Whenever more than one agency responds or the fire spreads over more than one jurisdiction, always establish a unified command.	Whenever possible, establish a single command with one person assuming the role of incident commander.

Initial Command

Major incidents typically begin with one or two fire companies arriving at the scene while other fire companies and the IC are responding. It is critically important that the person serving as the initial IC get the operation off to a good start by establishing command and following department procedures. It is very difficult to recover once an operation becomes chaotic.

It is certainly desirable for the first-arriving company officer to establish a formal command post outside of the structure and concentrate on directing

> **NOTE**
> The ultimate success or failure of a fire operation is quite often dependent on the actions or inactions of the first-arriving company.

operations. However, few fire departments have sufficient staff to allow the company officer to establish a stationary command post on arrival. It is important to recognize that the officer of a first-in unit, who is performing task-level functions (as well as commanding the operation), is at a distinct disadvantage over the chief officer, who is at a stationary command post with a single function. If the first-in officer delays taking effective task-level actions in order to establish a stationary command post, the fire may make substantial progress during this delay.

It is essential that someone always be in command. Typically, departments respond to structure fires as fire companies, with a company officer or member in charge of the company. Most departments permit three command options for this first-arriving company officer:

- **Investigation:** Generally, nothing showing on arrival, no need to take immediate action to save lives or protect property (command post not established)
- **Fast attack:** Obvious signs of fire or some other dangerous conditions that require immediate action where the company officer's direct involvement will be beneficial to the outcome; command option considered a mobile command as the first-arriving company officer is in command while taking action with their company (command post not established)
- **Command:** A situation of such a magnitude or complexity that the company officer must assume a command position and not become directly involved in the operation so that the first-arriving officer can coordinate the actions of personnel and other incoming units (command post established)

> **NOTE**
> On occasion there are circumstances when a company officer is not available. However, there must always be an officer or appointed acting officer assigned to every fire company.

When fire companies are staffed by four or more personnel, the investigative or fast attack options allow the operation to continue while the company officer is busy performing initial tasks as part of the entry team. Some departments assign the first-arriving officer to establish an exterior command post with two or more fire fighters making entry to conduct an offensive fire attack. This procedure has the advantage of providing a better command function, but in many cases results in less experienced personnel conducting the initial interior attack with no apparent leader. For these reasons the preferred fast attack mode places the company officer as part of the interior attack crew. There are some situations where an individual fire fighter or officer arrives at the scene prior to the arrival of a fire company. In this case, this person is the initial IC. Company officers must be reminded of the importance of establishing command at fires and other emergencies. To ensure fire fighter safety, there should always be a strong command presence. Establishing command during training does much to remind the company officer of this important concept. Postincident critiques and tactics training are also necessary if company officers are expected to establish a stationary command post when circumstances permit or require it.

Most departments adhere to the strict interpretation of establishing command by announcing command. Others are satisfied with carrying out the duties, without an announcement of command, when the officer is in the investigation or fast attack mode. Announcing command reinforces the fact that this first-arriving officer is in command, and this action is therefore recommended. Others believe that the command announcement is unnecessary, as command is assigned to the first-in officer per SOPs, but this could lead to confusion when two or more companies arrive simultaneously. Regardless, there must always be someone in command of every incident, no matter how large or small. The first-arriving officer must report conditions upon arrival at the scene. The expected initial report must be included in the department SOPs. Some of the common items included in an initial report are as follows:

- Confirm command
- Confirm address
- Announce command mode (investigation, fast attack, or command)
- Brief description of building or specific name of well-known structures
- Occupancy type
- Conditions (e.g., heavy smoke with occupants at windows)
- Actions being taken
- Resources needed

For example, an initial report could include the following:

> "Engine 1 is on the scene at 1234 Main Street.
> Heavy fire and smoke are visible from the second floor
> of a three-story, wood-frame apartment building;
> one victim can be seen at a second-floor window at the front of the building.
> Engine 1 has a water supply and we are advancing a one-and-three-quarter-inch hose line to the second floor.
> Respond a second alarm assignment to this location.
> Engine 1 is Main Street Command."

Whenever feasible, the first-arriving company officer should attempt to view all sides of the building. The officer can often view three sides of a building by having the driver pull just past the building involved. The officer then only needs to get a view of one remaining side. In some cases, due to the size or complexity of the building or area, having first-arriving units view all four sides of a building would result in a considerable delay in initiating operations. When this is the case, it may be best to begin operations immediately. The IC should then assign later-arriving personnel to check the rear and other sides of the building. Sometimes the best route to the rear of the building is through the building.

For example, the building shown in FIGURE 1-1 is a large multistory residential building. Walking completely around a large building would result in a

FIGURE 1-1 Multistory residential building.
Courtesy of Ben Klaene.

considerable delay in the initial attack if a complete walk-around were carried out prior to commencing operations. The rear and connecting walls of these buildings could be checked by walking through the fire building. In some departments the roof crew is assigned to check the rear of the building. This is a good procedure, providing a roof crew is actually needed and the roof is safe.

The information required in the initial report should match department needs and resources. For example, it may be necessary to include information regarding water supply if water supply requirements are not specific in the SOPs or when parts of the jurisdiction do not have water mains and hydrants.

Use good communications techniques when using the radio:

- Take a deep breath.
- Think about what you are going to say before transmitting message.
- Use clear text, no **ten codes**.
- Key the radio, followed by a very short delay. *This is very important; with some radios, part of your transmission will be lost if you start speaking immediately.*
- Speak slowly and distinctly

It is imperative that all members who could possibly be in charge of the first-arriving unit practice transmitting an initial report. Suggested activities at the end of this chapter provide an opportunity to do just that.

Command by a Chief Officer

The three command mode options (investigation, fast attack, and command) apply only to company-level operations. When a chief officer assumes command, it is critical that a stationary command post be established.

As higher-ranking chief officers arrive, they have the option of accepting the command role and reassigning the previous command officer to a position that may be mobile (e.g., safety officer), to an interior tactical-level management unit, or to another general and/or command staff position. This should occur only after a formal transfer of command.

The chief officer assuming command must be provided with a status report by the previous IC. As soon as the chief officer taking command has an opportunity, a progress report should be transmitted to dispatch. Part of the size-up and status report process involves reconnaissance from units at the scene. This is most often done using the radio system to request a progress report from operating units, for example:

> "Engine 1 to command, we have the fire knocked down on the second floor."
> "Truck 1 to command, one victim has been rescued and the primary search completed for the entire structure."

The IC would then communicate a status report:

> "Main Street Command to dispatch, we have the fire under control, one victim rescued, and the primary search is complete."

The IC will also need to use the communications network to relay orders to tactical-level management units or individual companies depending on the size and complexity of the incident. It is important to first identify the unit you are calling (e.g., "*Command to Engine 1*"). When Engine 1 replies, issue brief, specific, and clear orders. The unit being assigned should repeat the order. ICs must remember that completing a task will take time, and on occasion the unit assigned to complete the task will be unsuccessful. When an order is issued, the IC will assume that the objective is being successfully completed unless the company given the directive advises otherwise. However, the IC should request a status report from the company assigned the task after a reasonable time has elapsed to complete the assignment. When available, a chief's aide or planning section chief can estimate the time needed for task completion and track elapsed time. Once the IC gives an order, the companies assigned the task(s) must complete the objective(s) and notify the IC that the assignment is complete or advise the IC that they are unable to complete the assignment and reasons why.

On occasion, a crew could be redirected by another officer who is not in their **chain of command**. This practice of working outside the chain of command can have severe negative consequences and should be avoided.

> **NOTE**
>
> In some jurisdictions, chief officers are not on duty after normal business hours; therefore, response by a chief officer could be delayed. Some departments have collaborated with neighboring departments to form regional incident management teams to provide a command-level officer and support staff when necessary.

NFPA 1561 specifically addresses this issue (NFPA 1561, 2020):

5.8.8.1 Where conflicting orders are received at any level of the incident management system, the individual receiving the conflicting order shall inform the individual giving the order that a conflict exists.

A.5.8.8.1 The guideline for clarifying conflicting orders shall not apply to imminent hazard situations where immediate action is needed to avoid a dangerous situation.

5.8.8.2 If the conflicting order is to be carried out, the individual giving the conflicting order shall so inform the individual who provided the initial order.

Transfer of Command

Command transfer must be addressed in department SOPs. In larger departments where there is a multi-level rank structure, it is essential that later-arriving company officers know whether they are required to assume command or if they have other options. The first-arriving company officer may not be at a command post. When command is transferred to another officer, it should be formalized at a stationary command post.

Unless there is a compelling reason, command should not be transferred between company officers. The situations where command transfer between company officers is permitted should be clearly outlined in department SOPs. Multiple command transfers in the initial stages of the operation generally results in confusion and could reduce the number of people engaged in hands-on life-safety and extinguishment activities. Later-arriving company officers should reevaluate conditions upon arrival. An officer who finds an unsafe operation in progress upon arrival must immediately take command. In this case, a formal command post must be established and immediate corrective actions taken.

When command is transferred, there is the possibility of information being lost in the process. It is essential that the person assuming command communicate with the previous IC to determine both situation and resource status of the operation in progress. If a tactical worksheet is in use it should be passed to the new IC. While this communication is taking place, arriving units should be awaiting orders. The longer it takes to transfer command, the greater the chance of freelancing. A strong command presence and efficient transfer process will ensure a smooth transition and eliminate independent actions. Keeping everyone focused on achieving the tactical objectives outlined in the incident action plan is imperative.

The IC assuming command must evaluate the safety and effectiveness of the operation in progress. If the operation is unsafe or is not focused on accomplishing the incident priorities of life safety, extinguishment, and property conservation, the IC assuming command must reorganize the operation and reassign operating units as needed. It is very difficult to reorganize and redirect operations in the heat of battle, but it is sometimes necessary. If the current operation is being conducted safely and effectively, the task of assuming command is much easier. The person assuming command should improve the operation by updating the size-up and augmenting operations in need of additional resources. Information available to the IC improves as companies relay critical information from the interior or remote positions and more time is available to review preincident plans and other information sources.

If department procedures give a more senior or higher-ranking officer the option of taking command or allowing the present IC to continue serving in the position, numerous and unnecessary command changes may be eliminated. However, the higher-ranking officer, especially chief officers, face a dilemma in allowing a lower-ranking officer to retain command: If there is a major problem in resolving the incident, the highest-ranking chief officer present onsite will be held accountable, even if they did not assume command. Remember, you can delegate authority, but you cannot delegate responsibility! Thus, the department and its officers must balance the possible efficiency gained in retaining command with the fact that higher-ranking officers will be held responsible and accountable. An alternative is to assign the lower-ranking officer as the operations section chief. Delegating the operations section allows the IC to retain overall command, but provides tactical-level experience to the lower-ranking officer.

Fallacy	Fact
Responsibility can be delegated.	Responsibility cannot be delegated; only authority can be delegated.

Command transfers tend to disrupt the continuity of operations, but department culture or rules may require command changes, particularly between higher-ranking chief officers.

On rare occasions, command of a structure fire could be transferred to another agency. Fires and explosions are used as weapons of terror. This could be a case where law enforcement would assume command from the fire department after the fire situation is controlled. The February 26, 1993, explosion and fire at the World Trade Center would be an example of a structure fire where command might eventually be transferred to police authorities (Isner and Klem, 1993).

Fires involving hazardous materials can also result in transferring command to another agency. The accidental fire involving a large quantity of solvents at the BASF plant in Cincinnati is an example. Once the fire was brought under control and all victims removed, the Ohio Environmental Protection Agency (EPA) assumed command and maintained a presence at the scene for 5 months during clean-up operations. The Ohio EPA retained command during this 5-month period even though fire units were at the scene whenever clean-up operations were in progress.

Information passed on during a transfer of command should include the following as a minimum:

- Progress report
- Company location and assignments
- Current NICS organization structure/assignments
- Known hazards such as hazardous materials, downed power lines, building hazards, terrain issues, etc.
- Additional resources requested but not assigned

Delegation

Establishing control over available resources entails the delegation of authority. The IC develops the strategy, while division/group supervisors or branch directors develop tactics within the overall strategy by assigning and coordinating tasks for units that are working under their supervision. Information exchange through good communications is the key to managing the fire ground.

Command Post

The first-arriving company officer will be in command but generally will be working inside the building during an offensive fire attack. However, fires that are obviously defensive operations from the beginning are normally handled by the first-arriving officer establishing a stationary command post and assigning other first alarm companies.

A good command post will have the following characteristics:

- Be established in a location that is known and easily found
- Be outside the hot zone, in an area where personal protective clothing is not required (cold zone)
- Provide a view of the two most important sides of the building
- Never hinder apparatus movement

In most cases, companies and agencies will report to the command post for instructions and information. Therefore, the command post should be easily found and accessible. When establishing a command post, the IC should communicate the location using the street name for an exterior command post (e.g., First Street Command) **FIGURE 1-2** or the building name when using a command center within a building (e.g., Scripts Command) **FIGURE 1-3**.

FIGURE 1-2 An exterior command post can be named for its street location, for example, "Main Street Command."
Courtesy of David J. Jones, Cincinnati, Ohio.

FIGURE 1-3 Fire command room.
Courtesy of Ben Klaene.

For outside command post locations, the IC should identify the side of the building where the command post is located (e.g., Third Street Command). There are times when some degree of isolation is required. Companies can be assigned to report directly to group and division supervisors or branch directors. Outside agencies should report to a liaison officer, and the media should be managed by the public information officer.

Positioning the command post so that two sides of the building can be viewed is usually good practice. There is a distinct advantage in being able to see the effects of tactical decisions. However, there are times when seeing the scene can be a distraction, causing the IC to focus on the visible while failing to deal with higher priorities that cannot be seen. Generally, the larger and the more complex the situation, the farther away and more isolated the command post should be. Isolation from distractions at the scene can be an important factor in developing good incident action plans. Most often the command post will be located at or near the scene of a structure fire. When confronted with a large-scale operation involving very large buildings or multiple buildings, it is sometimes a good idea to place the command post in a nearby building with good lighting and communications. For a catastrophic incident, such as the September 11, 2001, attack on the World Trade Center, a command post should be established at a location well away from the incident. Most city and county governments have preplanned emergency operations centers (EOCs) that could be used for this purpose. The command post is where the IC is located; even if an EOC is functioning for a structure fire, when the IC is at the incident scene, then so is the command post.

> **NOTE**
>
> Wherever the incident commander is located—that is the command post.

Managing an incident requires the IC's undivided attention. The command post should be in a location that supports command, control, and coordination of all incident activities. Good communication is critically important to the command function.

Most fires are managed by an IC with a few command staff, sections, or tactical-level management units. ICs at these everyday incidents generally use an automobile or sport utility vehicle as a command post. Some ICs prefer to establish command at an outside position away from their vehicle, or outside at the rear

FIGURE 1-4 An example of an exterior command post.
Courtesy of David J. Jones, Cincinanti, Ohio.

of their vehicle **FIGURE 1-4**. Many command vehicles have status boards and other command-post equipment in the rear vehicle compartment or trunk. Other departments require or strongly suggest that the IC stay inside the vehicle and place command equipment in the front seat area to support this policy. Sport utility vehicles are particularly well suited for placing command post materials and equipment in the front seat or rear compartment. There is also an advantage in using four-wheeled drive vehicles with a higher road clearance in areas where snow is expected or where response may require off-road travel. Working from the vehicle generally provides better communications, as vehicle radios are usually more powerful and the vehicle engine maintains a constant charge. Working inside the vehicle also affords a measure of security and safety for the IC, as well as providing climate control. Commanding from an exterior position in high winds, heavy snow, torrential rains, or other types of inclement weather can be distracting or even dangerous. The larger and more complex the incident, the greater the need for command post facilities and space.

Span of Control

Safety factors, as well as sound management planning, both influence and dictate span-of-control considerations. The span of control is the number of people reporting to a supervisor. For any individual with emergency management responsibility, the span of control should range from three to no more than seven, with five being the average.

The type of incident, the nature of the task, hazard and safety factors, and the distance separating tactical units will influence span-of-control decisions. An important consideration in span of control is to anticipate change rather than react to it. This is especially true during rapid build-up of the command structure, when good management is made difficult because of numerous reporting units.

Large-scale operations where the span of control exceeds manageable limits become chaotic and unsafe. Span-of-control limitations are generally in the area of five subordinates reporting to a single supervisor under emergency fire conditions. NIMS recommends a span of control ranging from three to seven people reporting to a supervisor. A one-to-one span of control serves no useful purpose, making the NIMS organizational structure more complex than necessary. The larger and more complex the NIMS organization, the more difficult it is to control. An important rule to remember is "keep it simple." Form tactical-level management components as necessary to maintain a reasonable span of control and to coordinate specific geographic and functional operations. However, remember that each organizational layer places the IC further away from having direct contact with operating units. In some large, complex incidents, this is necessary and desirable; however, at most incidents it is not.

A question sometimes arises regarding a two-to-one span of control. If there is a need to sector an operation to maintain a reasonable span of control, and only two units are needed in a geographic area or to perform a specific function, a two-to-one span of control may be acceptable FIGURE 1-5.

Calling for Additional Resources

Sometimes a piecemeal approach might be recommended when requesting help, to avoid being overwhelmed when the responding units report for assignments. A better approach is to stage arriving resources until assignment needs are identified. Companies or crews can be sent directly from staging to a group, division, or other assignment. Remember, it is best to call for additional help before it is needed. If you call for additional resources after the need is obvious, they may arrive too late. The need for additional units must be anticipated and requested well in advance of when the units should actually be in position. Consider how traffic, weather conditions, distance, and other factors will affect response times.

Smaller departments do not have the benefit of the large number and variety of resources that are typically part of a larger department. In these situations, smaller departments enter into mutual aid agreements, which will ensure that resources needed for larger events, including a tactical reserve, are

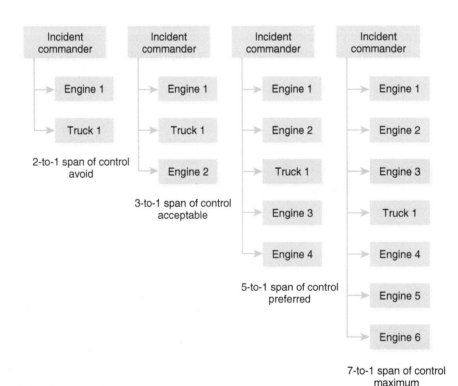

FIGURE 1-5 Span of control.
© Jones & Bartlett Learning.

available when requested. Mutual aid resources can be dispatched at the time the alarm is received from the public (automatic mutual aid) or dispatched only when requested by the responding department. Automatic mutual aid is gaining popularity, as many departments do not have the resources needed to combat a working structure fire. It is recommended that mutual aid agreements be written and signed by community leaders to alleviate any misunderstandings that may arise. Agreements should spell out what type of aid will be available, staffing levels, financial obligations, liability such as injuries and equipment loss, and how to decide who will be in charge, to name a few. It is further recommended that mutual aid departments periodically train together. Training together can reduce costs, uncover incompatible equipment (i.e., hose threads, hydrant connections), increase comfort levels of working with fire fighters in other departments, and provide an opportunity to become familiar with each other's SOPs.

Demobilization

As the incident is brought under control and essential tasks are completed, it is time to start returning units to stations and "available" status. There is a natural tendency to return later-arriving units first. Company apparatus that arrived first are being used to pump water, provide access via ladders, or are otherwise committed to needed operations. However, the first-arriving units are more likely to be fatigued or totally exhausted and more prone to injury, whereas later-arriving units may have been held in reserve and be fresh.

A word of caution is in order regarding the release of resources; maintaining an adequate reserve and rapid intervention crew (RIC) should continue as long as companies remain on the scene. Many injuries occur after emergency-phase operations appear to be complete. Always expect the unexpected! For example, on June 17, 1972, Boston Fire Department lost nine members when the Hotel Vendome collapsed during overhaul operations.

Staging

Staging areas should be established to locate resources that are immediately available for an assigned task. There are two types of staging. The most common type of staging places unassigned, fully staffed units at an exterior location near the fire scene ready for immediate assignment. Some departments also

Incident Summary

Hotel Vendome

On June 17, 1972, at 2:35 PM the Boston Fire Department responded to a fire alarm at the 101-year-old Hotel Vendome. A few minutes later, a working fire was called in with subsequent alarms sounded. A total of 24 fire companies responded to the fire and were able to bring it under control by 4:30 PM. Three companies remained on the scene to complete the overhaul process. At 5:28 PM, suddenly and without warning, a large section of the five-story building collapsed on top of a ladder truck and 17 fire fighters. Nine fire fighters were killed.

stage later-arriving initial response companies while the first-in company completes their size-up or investigation. This is sometimes referred to as Level 1 staging. Level 2 staging is usually implemented when additional alarms are requested and the apparatus and personnel are staged at a distance from the emergency scene.

The second-due engine is often assigned to stage for level 1 at a fire hydrant. When staged at a hydrant, it is good practice to check the hydrant. In areas where hydrants are not available, the water source may be a static source such as a cistern, lake, pond, or swimming pool. When assigned to stage at a static water supply, ensure that water is available. Static water sources are known to dry up during droughts. A pool or cistern could be drained for periodic maintenance. In any case, ensure the water supply is viable when staged at a source of water.

SOPs must address staging, including staging at high-rise buildings. This is addressed in detail in Chapter 12, *High-Rise Buildings*.

Level 2 staging allows the IC to control access to an incident scene, while deploying resources in a safe and effective manner. A level 2 staging area can be located anywhere mobile equipment can be temporarily parked while awaiting assignment. It is best to stage units two or more blocks from the actual incident to avoid the temptation to freelance into action and to ensure that staged units are in a safe area. When establishing a level 2 staging area, locate it far enough away to avoid obstructing or slowing access to the incident scene, but close enough to allow companies to arrive quickly once summoned. When weather conditions are extreme, such as freezing temperatures, shelter should be provided for companies that could be staged for a long period of time.

The best time to identify good staging locations is during preincident planning—not in the middle of the night during the heat of battle. Predetermining the location of the staging area is especially important for large, complex properties. Depending on factors such as property size, access issues, and weather conditions, the staging area could be an open-area parking lot, a nearby firehouse, or a predetermined location that would provide some measure of security and protection from the elements.

When a chief officer arrives on the scene, there is an immediate need to reevaluate the situation and complete the incident action planning process. During these early stages of an operation, the IC can be easily distracted by information overload from several sources, including radio communications from both on-scene and incoming units. If the IC has not completed the incident action planning process, including the determination of how responding fire forces are to be deployed to carry out the plan, incoming units will have a tendency to "freelance" into action. Furthermore, an IC who is already overwhelmed with radio traffic is often reluctant to call for additional assistance, even when the need is obvious. Establishing a staging area early on, and directing all responding companies to that location unless otherwise ordered, allows the IC to better manage on-scene units, establish a ready tactical reserve, and eliminate potentially dangerous freelancing. Establishing a tactical reserve is an important command consideration. If all units at the scene are committed to the operation, a tactical reserve is needed to cover unanticipated problems and provide relief to operating crews. Staging is an important tactical tool that is underutilized by many departments.

The level 2 staging area can also be used as a parking area for out-of-service apparatus. During a large-scale offensive operation, later-arriving units often are only needed to deliver additional fire fighters. The apparatus pumps, ladders, and tools are not needed; therefore, the unneeded apparatus are parked out of the way in the staging area, or base at a high-rise fire when a base is established. It is important to distinguish between an out-of-service apparatus parked in staging and a staged unit that is part of the tactical reserve. A staged unit is a fully staffed unit, such as an engine company, truck company, or EMS unit that is ready to respond immediately to the incident scene. Apparatus without adequate staffing are classified as out of service, rather than staged.

> **NOTE**
> Staging is an important management tool that increases incident scene safety and efficiency by providing a means to immediately deploy forces to achieve identified tactical objectives while preventing dangerous freelancing.

SOPs should outline how the staging area is managed, the duties of the staging manager and of the company officers whose units are assigned to staging, as well as the physical characteristics of an effective staging location. During large-scale incidents, a staging manager is assigned to manage staging and a more formal staging area is established in an area that provides adequate space and access to the incident scene. The staging manager is responsible for managing all incoming resources and dispatching resources at the request of the IC or operations section chief. Equipment and staffing in staging areas must be ready for immediate response. Crews should remain intact and available for immediate deployment. Span of control is not a problem in staging. The staging manager can manage numerous companies because of their inactive status.

Interior staging is typically used for a fire on an upper floor of a high-rise building. High-rise staging procedures are addressed in Chapter 12, *High-Rise Buildings*.

NIMS Organization and Positions

As the need for resources increases, there is a need for a larger and more complex organization. An incident management system is a tool used to organize an operation. The implementation of the incident management system is not a tactical objective, but a means to command and control an incident to achieve incident action plan objectives. With this in mind, the NIMS organization should be as simple as possible.

Modular Organization

The NIMS organizational structure develops in a modular fashion based on the type and size of the incident. First and foremost, there must always be an IC. Other line and staff positions are assigned by the IC according to incident priorities.

The specific organizational structure that is established for any given incident will be based on the management needs of the incident. If one individual (the IC) can simultaneously manage all of the major geographic and functional areas, no further organization is required. However, if one or more of the areas require independent management, then a qualified individual should be assigned to manage each of the necessary areas. If the IC does not specifically delegate the geographic areas or incident functions, then the IC retains responsibility for all that are not delegated. The NIMS positions should be thought of as job descriptions used by the IC. The IC decides whether to do the task or to have someone else perform that function. NIMS positions are divided into command staff and general staff (sometimes referred to as line officers).

> **NOTE**
>
> The only NIMS position that must be established at every incident is the IC.

Command Staff

All command staff positions report directly to the IC. Command staff positions are established to assume responsibility for key activities that are not included in the line organization. NIMS identifies the following three command staff positions:

1. Incident safety officer
2. Liaison officer
3. Public information officer

These positions are generally placed in "staff" position on the NIMS organizational chart as shown in **FIGURE 1-6**.

Incident Safety Officer

The **incident safety officer** is one of the key positions on the fire ground and is important to fire fighter safety, particularly at large-scale or complex incidents. This position should be staffed by an experienced officer who meets the requirements outlined in NFPA 1521, *Standard for Fire Department Safety Officer*

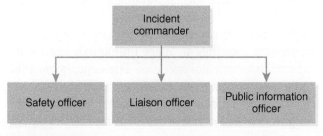

FIGURE 1-6 NIMS showing command staff positions.
© Jones & Bartlett Learning.

Professional Qualifications (NFPA 1521, 2020). This standard outlines situations when the position should be formally established on the fire ground:

> **5.1.1** The fire department incident safety officer (ISO) shall meet the requirements of Fire Officer, Level 1 as specified in NFPA 1021 and the job performance requirements (JPRs) defined in sections 5.2 to 5.7.

Incident safety officer is the command staff position that should be staffed most often. The IC should be at the command post, but the incident safety officer must monitor all areas where fire fighters are conducting operations. Therefore, it is virtually impossible for one person to effectively handle both functions (IC and incident safety officer) at a large or long-term incident. Some large-scale incidents may require assistant incident safety officers. When more than one incident safety officer is assigned, assistant incident safety officers are generally appointed to monitor specific areas, with all assistant incident safety officers reporting to the incident safety officer responsible for monitoring and advising the IC of overall incident safety. Seldom is more than one incident safety officer assigned at a structure fire. Appointing an incident safety officer does not relieve the IC of responsibility for safe operations. The IC has overall responsibility for scene safety, tactical-level management supervisors are responsible for the safety of personnel assigned to them, the company officer is responsible for the safety of their crew, and each member has a safety responsibility to themselves and other members of their crew.

It is imperative that the incident safety officer focus on the overall operation and major risks to fire fighters. There is a tendency for incident safety officers to spend too much time concentrating on minor safety infractions while failing to recognize potentially deadly hazards. For example, the incident safety officer checking the proper donning of protective clothing could fail to evaluate structural stability.

The incident safety officer is charged with monitoring all areas where fire fighters are working at the incident. This provides the incident safety officer with a unique overview of the entire operation that should be shared with the IC as the incident safety officer completes a tour of the incident scene. This also helps the incident safety officer review the incident action plan in regard to safety issues, including checking the NIMS organizational chart.

Liaison Officer

Liaison is the point of contact for agencies that are not assigned to operations functions. If more than one person or unit from an agency responds to the incident, one person from that agency should be appointed to communicate with the liaison officer (or the IC if liaison is not staffed). This person (the agency representative) represents the agency they are from and is responsible for coordinating other people and units from the agency that are working to resolve the incident. The agency representative must have the authority to speak on behalf of their agency, although sometimes the agency representative will consult with their agency prior to making a commitment. In most cases, the police department reports to the liaison officer if this command staff position is staffed. However, if law enforcement personnel are playing a major role in meeting incident action plan objectives, they would then report to a general staff position within the operations section or directly to the IC.

If the IC does not staff the liaison officer position, representatives from all responding agencies will communicate directly with the IC. Most structure fires do not require the separate staffing of the liaison officer position. However, when many agencies are assisting and all of them are reporting to the IC, it will become extremely difficult for the IC to effectively communicate with emergency resources operating at the scene. When this is the case, the IC should staff the liaison position.

Public Information Officer

The public information officer (PIO) disseminates information to the public, usually via the media. This officer provides critical information to the community regarding protective actions that need to be taken as well as information of general interest. There should only be one PIO per incident to assure continuity in information being released to the public. An efficient PIO will monitor communications at the command post and keep abreast of the current status of the operation. It is essential that all communications to the public concerning emergency operations related to the incident be from the IC, or from the PIO in consultation with the IC. All information relayed to the public must be cleared by the IC. This is particularly critical when there are injuries or fatalities, or when there is suspicion that a crime has been committed.

If there is a high level of public interest in the incident, the media will require a substantial amount of time, thereby making it difficult for the IC to maintain communications with operating units and the media. In this case, a PIO should be appointed or other general NIMS positions handed off to allow the IC time to communicate with the media. Designating a media area **FIGURE 1-7** that provides a view of the

FIGURE 1-7 Designated media area.
Courtesy of Captain Nick Morgan, St. Louis Fire Department.

fire or operating equipment assists in developing good relations with the media and discourages them from seeking out potentially dangerous locations on their own. The IC or PIO must prepare to provide factual information during news briefings while not disclosing information that should not be made public. The media may turn to unreliable sources for information if the fire department fails to provide the information they need for their reporting. Many departments now take advantage of social media sites to better address the public's need to know. The media may also have information useful to the IC, such as fire reported in a nearby community.

General Command Staff Considerations

Some question arises as to whether the command staff positions should be counted in the span of control for the IC. A valid argument can be made for not counting these positions against the recommended three to seven reporting positions in the IC's span of control, as they serve in roles similar to those of command aides or adjuncts.

Some departments preassign staff officers to command staff positions. For example, the training chief may function as the safety officer at the scene. These individuals receive special training and often develop useful contacts prior to the incident. For example, the liaison officer might attend local emergency planning committee meetings to meet individuals representing nongovernmental agencies. The department PIO could develop a rapport with media personnel.

There is value in preassignment. However, preassignments may delay the staffing of important positions. To ensure that positions such as safety and liaison are filled, it is often necessary to make temporary assignments until the preassigned personnel arrive at the scene. Whoever is assigned to fill one of these positions must be thoroughly trained and qualified to carry out the responsibilities.

Large fire events often require the staffing of all three command staff positions. In most cases, a person can be assigned to each position. Only one person should be assigned to each command staff position. If the safety, liaison, or public information officer position involves more work than can be effectively completed by a single person, assistant safety, liaison, and public information officers can be assigned, but they are subordinate to the incident safety, liaison, and public information officer. At times it is possible to combine assignments. One person may be able to manage both liaison officer and public information officer, provided that the workload is not too great. However, when delegated, the safety officer should always be a separate assignment. Command staff officers and section chiefs are frequently located at the command post, whereas personnel who are assigned as safety officers should be mobile and able to monitor the entire operation.

General Staff

General staff positions provide a pyramid-structured hierarchy capable of coordinating and controlling the incident. The IC is at the top of the pyramid with five possible organizational layers between the IC and individual responders FIGURE 1-8. Rarely would it be necessary to use all NIMS hierarchical levels and all subordinate units at a structure fire. Positions are staffed when incident conditions require separate management of geographic areas or functions.

Sections

Section chiefs report directly to the IC FIGURE 1-9. The five main sections of the NIMS organization under the direction of the IC are as follows:

- Finance/administration section
- Logistics section

FIGURE 1-8 NIMS hierarchy.
© Jones & Bartlett Learning.

- Operations section
- Planning section
- Information/intelligence section

It is important to note that these positions are *sections*, not sectors.

As priorities are established, the five separate sections can be assigned. Each section can have several subordinate units. Only in the most extreme circumstances would all section positions be staffed at a structure fire. When established, section chiefs can further delegate management authority for their area of responsibility as required. If necessary, functional units within the section may be established by the section chief. Similarly, each functional unit leader can assign individual tasks within the unit as needed. Units subordinate to each section are staffed depending on incident requirements. When the section leader requires assistance in handling section duties, a deputy section chief is assigned.

Finance/Administration. The finance/administration section is unlikely to be separately staffed at the scene of a structure fire. Finance/administration,

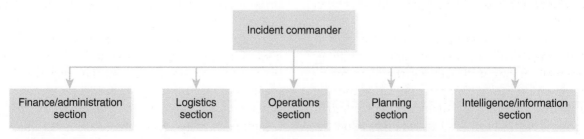

FIGURE 1-9 Section positions.
© Jones & Bartlett Learning.

as the name indicates, manages financial matters and/or provides administrative services; the functions provided by this section are shown in FIGURE 1-10.

There are some cases where the fire department is able to recover expenses, such as during hazardous materials–related incidents. For instance, the Cincinnati Fire Department was able to recover expenses after the previously referenced BASF Chemical Company Fire. However, it is important to remember that, even at smaller incidents where the finance/administration section is not staffed, these duties are important and remain the IC's responsibility. In these instances, finance/administration duties might be limited to completing the National Fire Incident Reporting System form that could be later used by the property owner and/or insurance company. However, at larger, more complex incidents, detailed documentation might be necessary to secure reimbursement for the following:

- Damaged equipment
- Payroll for responding or back-fill personnel
- Compensation for injuries to personnel
- Other incurred expenses

At such incidents, the finance/administration position should be staffed.

Logistics Section. The logistics section locates and provides the materials, equipment, supplies, and facilities required to support incident operations. The logistics section chief can be thought of as the supply sergeant or quartermaster. Most of the resources needed to support incident operations are already on the scene or provided by additional fire companies; therefore, the logistics section is seldom staffed at minor structure fires. However, there are circumstances where supplies or facilities are needed. For example, a fire situation where companies are on the scene for an extended period of time would require rehabilitation with food, water, sheltered areas, and possibly toilet facilities. The logistics section chief would be responsible for bringing these resources to the scene. Some departments have a rehab vehicle available that can supply all of these needs.

One of the most important logistics section units is the communications unit. This unit assists in setting up the communications network, as well as providing and maintaining communications equipment. Some departments have a communications specialist respond to the scene of major or long-term emergencies with spare radio batteries and chargers. The need to staff the communications unit increases as the communications network becomes more complex.

The ground support unit would provide items such as fuel at the scene of a structure fire. Logistics section subordinate units are shown in FIGURE 1-11. Some

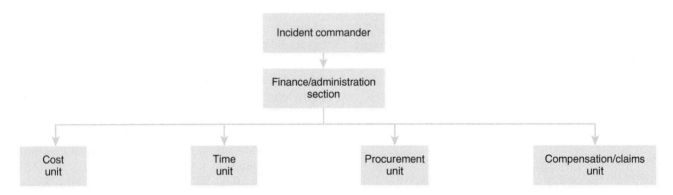

FIGURE 1-10 Finance/administration section subordinate units.
© Jones & Bartlett Learning.

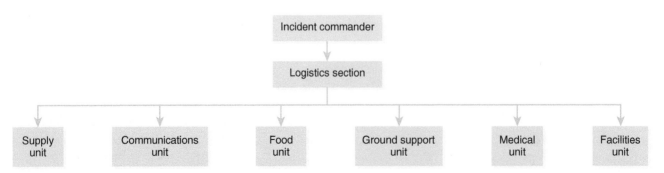

FIGURE 1-11 Logistics section subordinate units.
© Jones & Bartlett Learning.

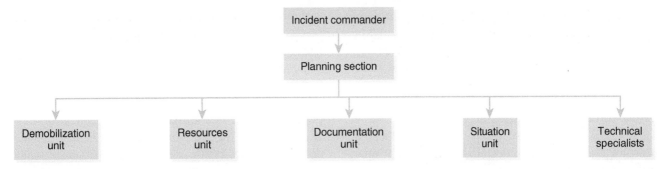

FIGURE 1-12 Planning section subordinate units.
© Jones & Bartlett Learning.

charts further subdivide the logistics section into service and support branches.

Planning Section. The planning section chief is the information manager. Planning should be the most frequently implemented section position. The planning chief gathers information about the incident, tracks resources, and assists the IC in developing the incident action plan and contingency plans. Planning should be one of the first sections staffed during a major incident. When transferring command, consider assigning the previous IC as the planning section chief, as the previous IC should know the status of the operation as well as the number and location of incident resources. A major part of the planning section chief's responsibilities is tracking and documenting incident status and on-scene resources, oftentimes referred to as "SIT-STAT" (situation status) and "RE-STAT" (resource status), as shown in **FIGURE 1-12**.

The demobilization unit prepares and implements a plan to return personnel and resources to service as an incident is being brought under control. This is an important planning section function when large areas have been stripped of resources.

The documentation unit collects incident information, which is important to a formal incident evaluation. The documentation unit becomes more important when there will be a formal investigation or study following the incident and the finance/administration section is not staffed.

Most technical specialists report to the planning section, but they could report to other sections, the command staff, or the IC as needed.

Chief's Aide. Closely related to planning is the position of chief's aide. At one time most large city fire departments provided chief officers with a chief's aide, but due to budget constraints, many cities eliminated this position. Some departments called the chief's aide a driver, or chauffeur, which made it more difficult to defend retaining the position. As administrators examined fire departments in search of ways to cut the budget, the chauffeur position was a likely candidate for elimination.

The chief's aide drives the chief's vehicle. However, driving is a small but important part of the duties assigned to an aide. Having an aide drive the chief officer to the fire scene allows the chief officer to concentrate on communications as well as accessing preincident plans and other sources of information to start the size-up process. At the scene of an incident, the chief's aide can manage command tasks so the chief officer can concentrate on developing an incident action plan and deployment of forces. The chief's aide can also be assigned to do a complete walk-around of the fire building, handle communications, manage the accountability system, maintain the NIMS organizational chart, and research information sources.

The aide's duties could be described as the planning section or as a planning section subordinate unit, and can include providing resource status or situation status updates (RE-STAT and SIT-STAT). Departments who have chief's aides strongly defend keeping the position and usually refer to the chief's aide by a name more suitable to the duties they perform, such as field incident technician, staff aide, or command adjunct.

The Phoenix Fire Department assigns each battalion chief a support officer. This program not only provides the battalion chief assistance at the command post, but also provides invaluable command post experience for future battalion chiefs.

A chief's aide can assist the IC in organizing and coordinating a safe and effective operation. Departments who do not have members assigned as chief's aides often assign responding personnel to these duties at the incident scene. This temporary assignment is highly recommended as an alternative. However, this temporary assignment is not equal, as command assistance will be delayed and lacks the team concept developed between a chief officer and a regularly assigned aide. To overcome some of the shortcomings of using a member from a responding company, it is recommended that these selected members attend additional training as a chief's aide. It does not make much sense to assign a fire fighter to assist the IC if the IC has to stop and train the fire fighter for this role while the fire is in progress. Also, this is definitely not a position for the newest member of the crew. Additional training might include maintaining an ICS chart. Attendance at a formal planning section chief class may fulfill this requirement.

Knowledge of communications procedures and the ability to communicate effectively on the radio are important attributes for a chief's aide. The member should have a good working knowledge of firefighting tactics to provide the IC with accurate information that can be applied to the strategy selected. The chief's aide must know the procedures and methods used to communicate with other agencies and/or the media to allow the aide to serve as a liaison or PIO at emergencies. A list of qualifications for chief's aide should be maintained by the department.

Information/Intelligence Section. The intelligence/information section provides information related to the prevention of criminal activity including terrorism or the apprehension of perpetrators of criminal activity. In a fire situation, this section could be used to determine the fire cause or lead to the apprehension of an arsonist. Depending on the situation, the information/intelligence function could be located in one of four places on a NIMS organizational chart. It could be part of the general staff. When the information gathered is vital to the safety and welfare of the community and responders will play a significant role in the investigation of a terrorist or criminal event, or when response involves specialized and technical information, the fifth section chief would be assigned. When the information gathered will affect firefighting tactics, this function could be located within the operations section. In situations where the information is neither critical, sensitive, nor has little tactical benefit, but provides some real-time information to the IC, this function could be part of the command staff. However, in most situations the appropriate location of the information/intelligence function is within the planning section.

Operations Section. The operations section manages all tactical operations such as search and rescue, extinguishment, and providing medical care to victims. When the IC hands off the operations section, the operations chief will make all tactical assignments and control all resources working to resolve the incident. Delegating the operations section allows the IC to focus on the overall strategy and other command functions. However, it is important to note that staffing the operations section means the IC no longer has direct control over operating companies, so there is another layer of organization between the IC and operating units. Typically, the IC delegates authority for operations only at large-scale incidents. Senior-ranking officers sometimes assign the person they are relieving of command to the operations section to provide the lower-ranking officer with experience in managing larger operations, while the senior officer maintains overall control as the IC.

The operations section generally has the most resources and the most complex organization, which may require several hierarchical levels to effectively maintain a reasonable span of control. The hierarchical levels within the operations section are shown in **FIGURE 1-13**.

The operations section is capable of controlling as many as 625 companies or crews with a five-to-one span of control **FIGURE 1-14**. At the maximum recommended span of control (7 to 1), a total of 2,401 companies or crews can be effectively controlled. Numerous large-scale disasters, including conflagrations, hurricanes, wildland fires, and earthquakes, have demonstrated that when properly implemented, NIMS includes all of the tools necessary to effectively

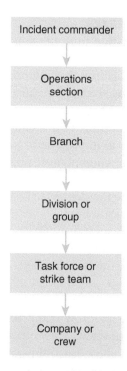

FIGURE 1-13 Operations section hierarchy.
© Jones & Bartlett Learning.

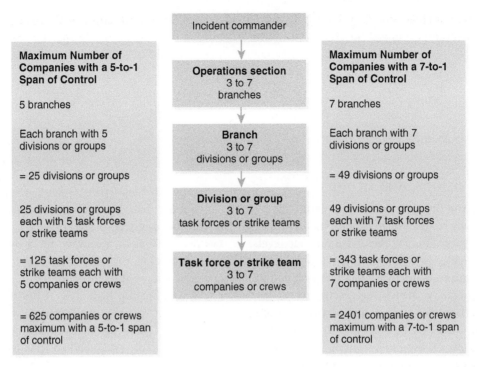

FIGURE 1-14 Operations section span-of-control capabilities.
© Jones & Bartlett Learning.

manage the largest and most complex operations. Obviously, NIMS has the capacity to manage the largest imaginable structure fire.

The sections that the IC should hand off depend on the incident. Some officers may be reluctant to hand off the operations section, as they lose direct contact with tactical operations. Some departments require the highest-ranking officer at the scene to assume command. Other departments give a higher-ranking officer the option to assume command of the incident, to fall into the command structure under the direction of the present IC, or to serve as a senior advisor.

Allowing a lower-ranking officer to retain command shows confidence in the officer and provides experience in commanding a large-scale situation. However, in the end, the senior officer will be responsible for the outcome of the incident, regardless of whether he or she formally assumed command. By handing off operations to the lower-ranking officer, for example, the senior officer can assume command and direct the general strategy while the lower-ranking officer gains experience in handling the dynamic tactical assignments.

Command staff and section chiefs make up the **incident management team**, which provides the necessary staff and line functions for an incident. Incident management teams for most structure fires are made up of members from the responding department or from departments that normally provide mutual aid assistance.

However, FEMA encourages the development of "all-hazard" incident management teams on a regional and state level. Regional, state, or national incident management teams would include members certified to function in the various command staff and section positions, much like the "Red-Card" system used by the forest service when combating large-scale wildland fires. The U.S. Fire Administration provides NIMS organization and other incident management team training to facilitate certification as an all-hazard member of an incident management team. Some regional teams establish training and certification standards identified within the region. Regional teams are particularly valuable when the responding department does not have personnel specifically qualified to supervise a command staff or section position. For example, members of regional departments could attend training and educational sections to qualify as safety officers. A qualified safety officer could then be dispatched to any report of a working structure fire within the region.

A more detailed description of each section's responsibilities and subordinate units is provided in the U.S. Department of Homeland Security publication, *National Incident Management System* (DHS, 2017).

Throughout an incident, from inception to conclusion, the duties of each section are the ultimate responsibility of the IC. The IC chooses whether to retain the duties or delegate sections that will assist in managing the incident depending on the incident and the resource requirements.

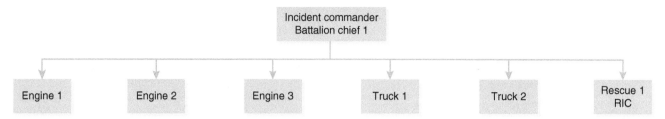

FIGURE 1-15 NIMS chart for a small incident.
© Jones & Bartlett Learning.

Branches, Divisions, and Groups

In using NIMS, the first management assignments by the IC will normally be to hand off geographic and functional areas of responsibility at the division/group level. A **division** is in charge of units working in a geographic area, whereas a **group** is in charge of units assigned to complete a function (e.g., search and rescue group).

As the fire progresses from a one-alarm response to extra alarms, there may be a need to divide the operation to maintain a reasonable span of control or to better manage specific functions. For example, suppose a fire department responds to an apartment fire with three engine companies, two truck companies, and a heavy rescue company. The command structure would be as shown in **FIGURE 1-15**.

Engine 1 represents the first-arriving engine company, which arrives nearly simultaneously with Truck 1 and District 1 Chief from the same fire station. Given the fact that an interior operation is indicated, two members remain outside in order to fulfill the two-out function. The pump operator and the IC satisfy the two-out requirement.

Notice that a rapid intervention crew (RIC) and five fire companies are reporting to the IC in Figure 1-15. This small incident exceeds a five-to-one span of control and is nearing the recommended maximum span of control of seven to one.

Many fire departments send a second chief officer to every fire. In many instances this second chief officer is assigned to supervise the fire floor. Using **FIGURE 1-16** as an example, for a fire on the second floor, a truck company is assigned to search and rescue, and two engine companies are assigned to lay an attack and backup hose line. A division supervisor is now in direct control of the critical second floor operation with a three-to-one span of control, and the IC's span of control is reduced to four to one, as seen in Figure 1-16.

Another example is a five-story apartment building with a fire on the first floor extending into the hallway and the floor above. Rescue operations are needed on all five floors of the building. The IC would normally assign forces working in the same geographic area to a division to reduce the span of control. For example, suppose the apartment building has a fire on the first floor with three engine companies and a truck company working that area. An engine company and a truck company are working on the second floor, and an engine company is assigned to each of the remaining three floors. An additional truck company is assigned to ventilate the building. The incident could be managed in several different ways, but with 11 companies working at the scene, there is a definite need to assign tactical-level management units. There

> **NOTE**
>
> **Two-in/two-out rule:** Fire fighters working inside the hazard area must work in crews of at least two people (two-in). During the initial stages of an operation the two people working in the hazard area must be backed up by at least two people outside the hazard area (two-out) who are properly equipped and immediately available to come to the aid of the inside crew. The two-out rule does not apply to subsequent arriving units. The two-out team is often referred to as the initial rapid intervention crew, or IRIC, which is replaced by the rapid intervention crew (RIC) as additional units arrive.

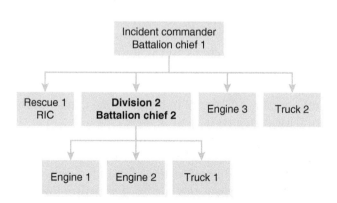

FIGURE 1-16 NIMS chart for a small incident with a division supervisor assigned to the fire floor.
© Jones & Bartlett Learning.

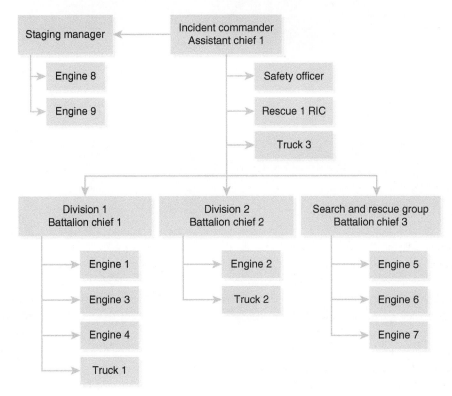

FIGURE 1-17 NIMS chart for a five-story apartment building using divisions and groups.
© Jones & Bartlett Learning.

are many correct ways to organize the incident; one of many correct methods is depicted in the organizational chart in **FIGURE 1-17**.

It is recommended that all companies working on the first floor be under the supervision of a single branch or division. Some companies are performing search and rescue tasks and therefore could be assigned to the search and rescue group.

However, having units in the same general area reporting to different supervisors is not a good practice. In most cases, geographic rather than functional assignments are preferred. Most ICs share this preference because communications are improved and the status of the assignment can be personally monitored by the division supervisor, thus eliminating the need for status reports within the division.

The three companies working on the upper floors in the apartment building example could report directly to command, but functional assignments also assist the IC in managing this important component of the incident action plan. The supervisor in charge of the search and rescue group will manage search and rescue assignments on all three floors, reassigning or requesting assistance as needed to complete the search and rescue on upper floors. The IC's incident action plan calls for removing the occupants. Therefore, the search and rescue group supervisor applies the tactics necessary to achieve command's objectives and issues the necessary task-level assignments to companies within the search and rescue group. For example, the company assigned to the third floor finds very little smoke and can quickly complete the primary search on that floor. However, the company assigned to the top floor may find occupants who need assistance, and the primary search may be difficult due to heavy smoke conditions. The search and rescue group supervisor could then reassign the company from the third floor to assist personnel assigned to the top floor. After the primary search is complete, the group supervisor could assign companies to the secondary search, avoiding repeat searches by the same companies.

All tactical operations within this group are managed by a single supervisor, who organizes and coordinates all activities related to the search and rescue on the designated floors. The IC is regularly informed of progress being made, but does not assign or reassign companies, as these responsibilities are assigned to the group or division supervisors. Truck 3 is assigned to ventilate as a single resource reporting directly to the IC.

Branches can be used in place of divisions or groups, but this is not the recommended way to initially reduce the span of control. Branches should be reserved for operations beyond the span of control of a single division or group, or when a contingent of units from another agency is working together, such as a law

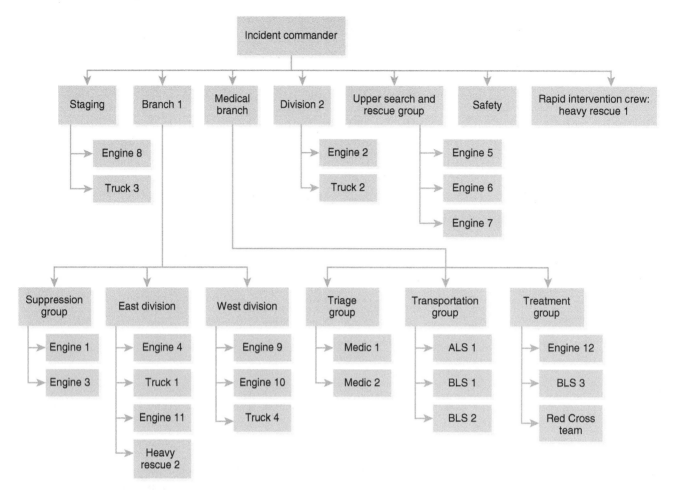

FIGURE 1-18 NIMS using branches.
© Jones & Bartlett Learning.

branch for police personnel. For example, if nine companies were assigned to the first floor in the apartment building example, division 1 would become branch 1 with a search and rescue group or groups and a suppression group reporting to the branch director.

A large medical or police contingent could also dictate the establishment of branches. If medical services are provided by a separate agency, a branch assignment is even more likely for the medical component of the operation.

Nine companies working on the first floor under a single branch, as well as the entire medical operation under a medical branch, are shown in the organizational chart in **FIGURE 1-18**.

In 1990 when the BASF fire occurred (see Incident Summary: BASF Chemical Plant), NICS terminology included the term "finance section" instead of the present "finance/administration section." The IC began by dividing the BASF operation geographically, but there were 15 companies working on the Dana Avenue side of the fire. The IC recognized that the recommended span of control was being exceeded; therefore, the Dana Avenue side was further subdivided into the Dana Division East and Dana Division West. Even with this subdivision, there were eight companies reporting to Dana Division East, which exceeds the recommended seven-to-one maximum span of control. To reduce the span of control, the Dana Avenue side could have been assigned to a branch with three or more subordinate tactical-level management units. Also worth noting is that not everyone reporting to a branch needs to be at the division/group level. Individual companies, task forces, or strike teams could also report to a branch. A revised version of the organization chart for the Dana Avenue side of this fire is shown in **FIGURE 1-19**. There are many correct ways to organize this operation, but the geographic sectoring allowed division supervisors, many of whom were company officers, to manage the operation in their area of responsibility using face-to-face communications. Note that a rapid intervention crew is not shown on the chart. At the time of this fire, rapid intervention crews were not in widespread use. If this fire were to occur today, there should be at least two rapid intervention crews (RICs) with one stationed at the front and one at the rear of the complex.

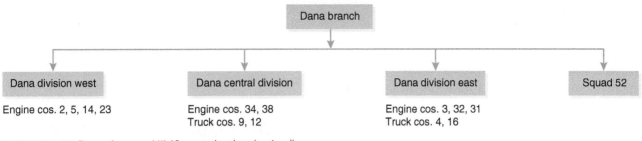

FIGURE 1-19 Dana Avenue NIMS organization (revised).
© Jones & Bartlett Learning.

Incident Summary

BASF Chemical Plant

On the afternoon of July 19, 1990, an explosion and fire at the BASF chemical plant in Cincinnati, Ohio, required the services of 25 fire companies, EMS units, police, and other agencies. The NICS was used to coordinate and control activities at the scene **FIGURE IS1-2**. The rear of the plant bordered Norwood, Ohio. As companies arrived, they found severely injured employees, a massive fire, and continuing explosions. Two plant employees were killed, and 88 people were injured. The 16 buildings on the BASF complex were heavily damaged or destroyed; 161 non-BASF buildings were damaged. Command was transferred to the Ohio EPA on the evening of July 20, 1990. Clean-up efforts with fire companies in a standby mode continued for nearly 5 months.

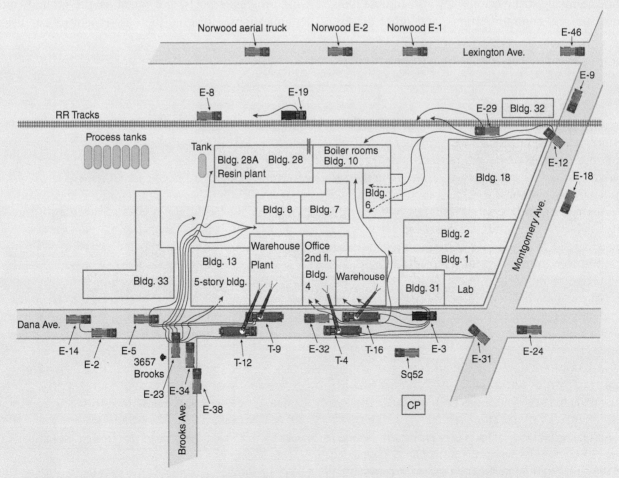

FIGURE IS1-2 BASF chemical plant operations after explosion.
© Jones & Bartlett Learning.

A Word about Sectors

Prior to the formal implementation of NIMS, the term *sector* was commonly used to identify geographic or functional tactical-level management units at the division/group level. The current editions of NIMS (2017) and NFPA 1561 (2020) do not recognize the term sector (NIIMS, 1981; DHS, 2017; NFPA 1561, 2020).

Task Force and Strike Team

The use of task forces and strike teams is an additional way to reduce the span of control, thus reducing the communications load at an incident **FIGURE 1-20**. A **task force** can be any combination of resources, whereas a **strike team** must be resources of the same type. Strike teams are a common way to organize large numbers of mutual aid companies or private sector resources. Some states define strike teams as a specific number of companies to be used as a statewide resource. A strike team could be any number of units of the same type, but the jurisdiction defining the strike team must predetermine the number of units. For example, some state and local governments assign five engine companies and a strike team leader to an engine company strike team. They also address minimum staffing, communications, and other factors.

Task forces are much more flexible and are more likely to be formed at the time of the incident. A common application of the task force concept would be two engine companies and a truck company working as a team.

Naming Geographic Areas of a Building

In Figure 1-18, Branch 1 is in charge of operations on the first floor. The upper search and rescue group is distinguished from the east and west groups assigned to search and rescue on the first floor. Both of these assignments are inside the building.

When naming areas outside the building, the normal designation is to name the sides of the building using an alphabetical designation where side A (Alpha) is designated as the front of the building, then continuing clockwise around the building, side B (Bravo) is the left side of the building, side C (Charlie) is the rear of the building, and side D (Delta) is the right side of the building, as shown in **FIGURE 1-21**. This commonly used system is useful when assigning locations, providing progress reports, and for other fire-ground communications. However, there is sometimes confusion

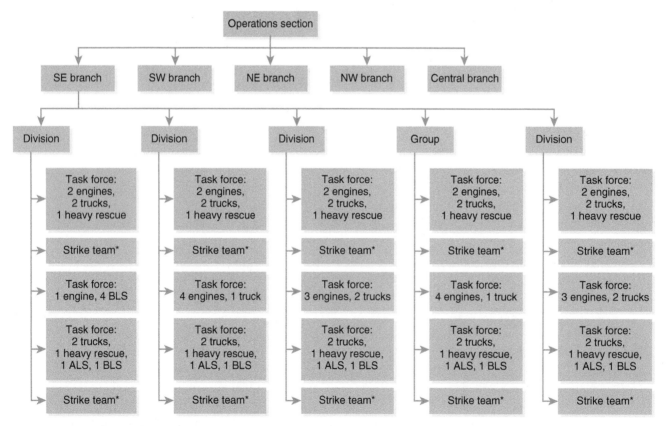

*The strike team in this jurisdiction is five engine companies.

FIGURE 1-20 Utilization of task forces and strike teams to increase organizational capabilities.
© Jones & Bartlett Learning.

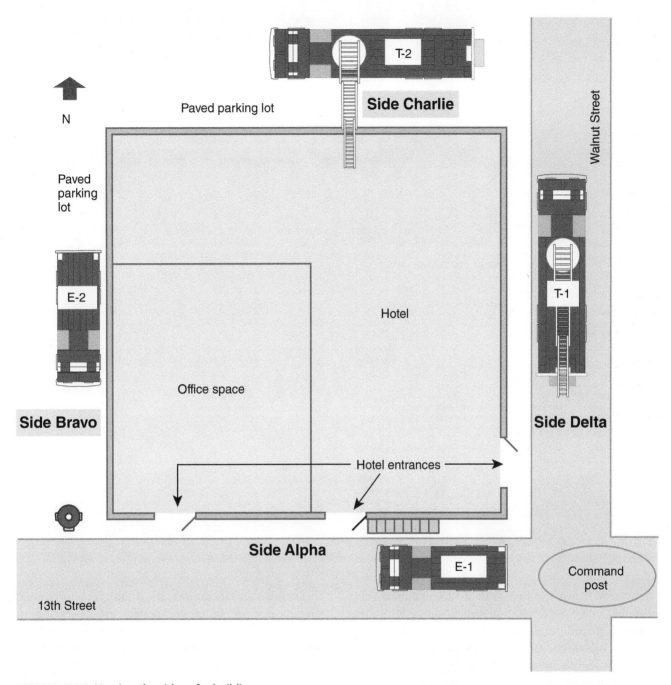

FIGURE 1-21 Naming the sides of a building.
© Jones & Bartlett Learning.

regarding which side to name as side Alpha. As shown in Figure 1-21, there are two entrances to the hotel portion of the building, and the command post is at a good location for a fire in the hotel viewing both sides A and D. Some departments name the side of the building where the command post is located as side A. Using that criteria, in Figure 1-21, either the Walnut Street or 13th Street side could be named side A. Confusion regarding the location of fire companies or when providing progress reports, maydays, etc. can negatively affect operations, resulting in unsafe and ineffective tactics. Naming the address side of the building as side A eliminates much of the confusion. In Figure 1-21, the office building address is 1210 Thirteenth Street and the hotel's address is 1220 Thirteenth Street. Therefore, side A of either building is the 13th Street side regardless of the location of the command post. The two attached buildings in Figure 1-21 could have addresses on different streets. The hotel could possibly have a Walnut St. address. If operations were being conducted in both the office and hotel buildings, there could be confusion as to which side of the building is side A.

FIGURE 1-22 shows the ICS chart that was used for the BASF chemical plant incident. Geographic areas of responsibility are designated using street names. The drawing in the BASF Chemical Company incident

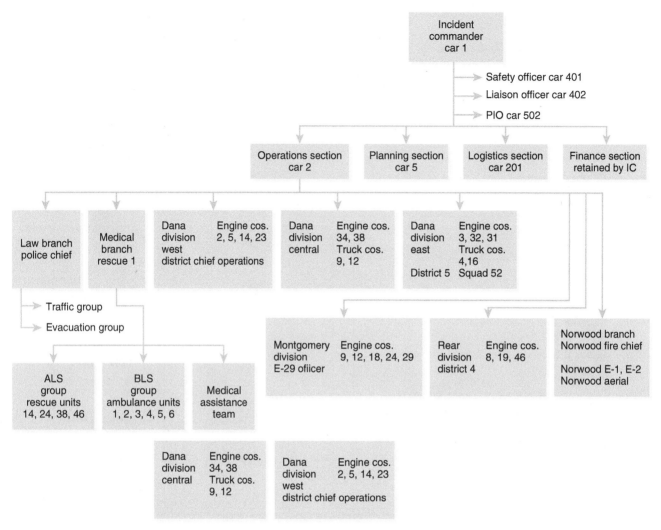

FIGURE 1-22 BASF ICS chart.
© Jones & Bartlett Learning.

summary shows the deployment of companies for a fire that involves several buildings and operations occurring inside the complex between buildings. The normal alphabetic naming system would be difficult to apply in this situation where sides Alpha, Bravo, Charlie, and Delta are not well defined.

> **NOTE**
>
> When identifying the side of a building using an alphabetical naming system, for example: A = front, B = left side, C = rear, D = right side, some letters are difficult to distinguish such as "B" and "D." Therefore, the addition of a phonetic equivalent is recommended such as: "Alpha" = A, "Bravo" = B, "Charlie" = C, and "Delta" = D.

Floors within a building are designated by floor level, thus Division 21 would be supervising multiple companies assigned to the 21st floor of a high-rise building. Naming tactical-level management units by floor number is intuitive and highly recommended. When one or more companies are operating on a floor level, they retain their company name if a tactical-level management unit is not assigned. Division 21 indicates one person is managing all operations on the 21st floor. When working with mutual aid departments, everyone should agree on a common system, include it in the SOPs, and use it during joint training exercises and tactical simulations as well as at the emergency scene. Whatever system you select, use it consistently and at every incident to make sure everyone is comfortable with the labeling/identifying system.

Fallacy	Fact
A floor level is always named as a division.	A floor level is named as a division only if a tactical-level management unit supervisor is assigned to direct multiple units on that level.

Communications

Communication is the lifeblood of any command system. It is impossible to coordinate and control a successful operation without effective communications.

During major emergencies it is highly recommended that a chief officer with command experience and who is familiar with day-to-day operations at the dispatch center be assigned to manage operations and make resource allocation decisions at the dispatch center, such as taking the following actions:

- Reallocating companies to vacated stations
- Relocating and staffing extra apparatus
- Requesting mutual aid and outside agency assistance
- Recalling off-duty members
- Converting companies (e.g., truck companies to engine companies)
- Suggesting the dispatch of special equipment

Some general rules for communication at the incident scene include the following:

- Use face-to-face communication whenever possible.
- Provide mobile communication to units that are remote from the command post.
- Ensure that all operating units have some form of communication that ultimately relays information to the command post.
- Place representatives of agencies on different frequencies at the command post to handle communications within their agency.
- Follow the command organization structure, facilitating unity of command.
- Keep the number of radio channels used by any supervisor to an absolute minimum, preferably no more than two. If more than two channels are in use, each channel should be monitored by an aide or another fire fighter at the command post.
- Do not clutter radio channels with unnecessary transmissions.
- Use standard terminology.
- Use clear text; do not use ten-codes.

Within the NIMS structure is a communications unit that reports to the logistics section. At large-scale incidents, establishing a communications unit is critical. This unit is responsible for establishing a communications plan and installing, procuring, and maintaining the communications equipment at the scene. At incidents where a formal communications unit is not established, preplanned resources should be used.

All radio communications at the incident should be in plain English. Codes should be avoided because they can easily be misunderstood. Furthermore, not all personnel operating at large-scale incidents from different jurisdictions or agencies will be familiar with the codes.

All communications should be confined to essential messages. It is important to avoid clogging the airways with nonessential messages and to keep the frequency available for vital transmissions. Each transmission should be brief, not a long narrative of what is transpiring within the area.

Radios are the most common mobile communications tool for emergency operations, but other communications devices offer advantages and should be considered during long-term or communications-intensive situations. In addition to radios, the following means of communication are often used at the scene:

- Face-to-face communication
- Messengers
- Telephones (cell, satellite, and hard-wire)
- Computers/mobile data terminals/mobile data computers/tablet computers
- Public address systems

Face-to-Face Communications

The most effective form of communication is face to face; however, running an entire large-scale incident using only face-to-face communications would be all but impossible and is limited to those within earshot of communicator. Face-to-face communication may be the best option when the information is sensitive and confidential, such as the fire cause or when a fire fighter or occupant has been killed or seriously injured.

Messengers

In some cases, during radio failure, messengers can be used to communicate orders and directions to units or tactical-level management unit supervisors who are in remote locations, but this will result in delayed communication. However, this can be an efficient way to relay messages between the command post and nearby section chiefs or to nearby units working under a division supervisor.

Cell Phones

Cell phones are widely available, are often used to contact outside resources and are used by many departments as part of the fire-ground communications network. In some cases, cell phones have been found to be more reliable in large and high-rise buildings than radios. One problem with cell phones, however, is that

> **Incident Summary**
>
> **Oklahoma City Bombing: Cells on Wheels**
>
> After the Oklahoma City bombing, the local cell phone providers were able to provide additional capacity in the area of the Alfred P. Murrah Federal Building by bringing in portable cell transmitters called "cells on wheels." They also provided the many agencies operating at the scene with cell phones at no cost. These were programmed with priority numbers that could grab a channel. Several locations were established where telephone personnel were available around the clock to repair phones and provide charged batteries. Use of the cell system was instrumental in coordinating the overwhelming logistics that this incident required.
>
> *Data from* Ed Comeau and Stephen Foley, Oklahoma City, April 19, 1995, NFPA Journal, July/August, 1995.

the system can be overwhelmed or damaged during a major disaster. In some cases, media representatives and others at the scene tie up the available cell phone channels, locking in open lines by keeping their initial call active. During an extended incident, the cell phone companies might be able to provide additional channel capacity and resources, or provide priority service for first responders. A "cells on wheels" (COW) can be brought into the area to provide additional network coverage and capacity. A fire department representative should communicate with wireless service providers to determine their emergency capabilities as part of the emergency planning process.

Satellite Phones

Satellite phones are also finding their way into the fire service. Satellite service overcomes some of the shortcomings of local cell towers and repeaters. With satellite systems, signals are transmitted between the sender and receiver through a satellite, as opposed to a cell tower. Cell towers can be damaged due to catastrophic weather conditions and are subject to terrorists. Satellites are less vulnerable. However, satellite phones can be more complicated, and satellite service can be interrupted during storms when a clear satellite signal is not possible.

Hard-Wire Communications Systems

One of the best examples of a hard-wired system is a building emergency communication system. Over the years this type of system has evolved from handsets, similar to a landline telephone, which can be carried by the firefighter and plugged into phone jacks located throughout the building. Some newer systems have permanently mounted "fire phones" located in selected areas such as stairwells, elevator lobbies, etc. These systems are designed to be used primarily by fire fighters; however, they can also be part of occupant emergency evacuation plan.

Building Communications Systems

Security and maintenance personnel often have radio systems that are designed to work within their building. Larger complexes may even have a cache of radios reserved for fire department use that are kept in the building's fire command center. In addition, a group assigned to work in different areas or floors can use installed landline telephones as a means of communication. For example, the supervisor of a search and rescue group conducting searches on multiple floors could be stationed at a telephone or telephone bank in a safe location, and units can report in and receive their next assignment via telephone, reducing radio traffic.

With the approval of the authority having jurisdiction, bidirectional antennas are finding their way into new high-rise buildings in lieu of hard-wired systems. These radio systems are designed to improve radio reception inside the building. It is recommended that local fire departments test this type of system to ensure it works as designed before approving the system. Also consider the possibility of future radio replacements; will new radios communicate effectively using the installed bidirectional antennas?

Public Address Systems

Public address systems can be used for mass notification within a building. For example, in a high-rise incident, the public address system can be used to direct occupants to remain in place or to exit via a specific stairwell. Although rarely used, public address systems can also be used as a means of communicating instructions to fire fighters.

Computer Systems

Computers can provide nonverbal communications that free "air time," as well as provide additional information, especially in graphic form. Mobile data terminals, status message terminals, and tablet computers are all systems that are commonly used to relay digital information. These terminal-based systems are dependent on a data network. Mobile data computers (MDCs) and other available devices combine dispatch communications with onboard data storage, thus reducing dependence on a wireless network and the dispatch center. Use of on-scene computer systems in providing preincident plans will be discussed further in Chapter 2, *Procedures, Preincident Planning, and Size Up*.

Advances in Communications

Many departments now have access to multiple forms of communications. General communication improvements in the private sector are now being adapted to fire service use. Computers are commonplace on fire apparatus and in some departments; laptop computers are being replaced by tablet computers. Many departments now have access to, or own, field communications units that are often equipped with the latest innovations in communication **FIGURE 1-23**.

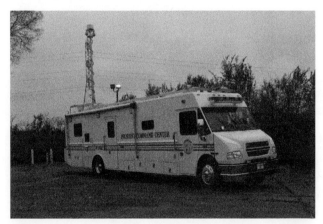

A

B

FIGURE 1-23 Mobile command vehicle. **A.** Exterior. **B.** Interior.
Courtesy of Thomas Lakamp, Cincinnati, Ohio.

Radio Designation for the IC

For purposes of radio communications, the IC is simply referred to as "command," a designation that should be reserved solely for the IC. Some departments refer to each division, group, or branch using the term "command," such as "rear command" or "interior command." This is strongly discouraged as the use of multiple command designations results in confusion regarding who is in overall command of the incident.

The IC is designated as "command" whether this person is the chief of the department or a fire fighter working as an acting officer. Department SOPs must define the term "command" and specify how it is to be used and by whom. By using well-defined terminology, it is possible to eliminate confusion during command transfers. Operating units, divisions, and branches might not remember that Chief Jones assumed command, but they do recognize the importance of orders being issued by command and the need to communicate through channels to command.

Communications Network

It is crucial that the communications network support **unity of command** within the NIMS organization. An example communications network following the NIMS structure is shown in **FIGURE 1-24**. It is important to keep the communications network as simple as possible. It is not necessary to designate scores of radio channels even if you have the required equipment to do it. However, there are sometimes good reasons to assign multiple channels to reduce interference among various units involved in the incident.

In Figure 1-24, the IC has retained operations and is using Channel 2 to communicate with all field units. Notice that the IC is using only two radio channels: one to communicate with dispatch and the other for field communications. The IC is communicating with the law branch by cell phone.

With the many divisions and fire companies using Channel 2, radio discipline would be imperative, and this single channel would be very crowded. If multiple channels were available, the communications network could be expanded, with each division using a separate channel to communicate with units under their supervision and Channel 2 or another channel to communicate with the Interior Branch. For example, command could communicate with the Interior Branch on Channel 2, and the Interior Branch could use Channel 4 to communicate with Divisions 1, 2, 3,

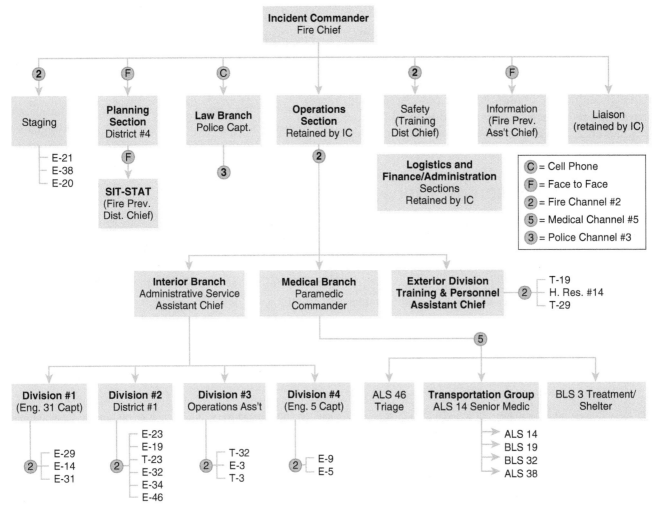

FIGURE 1-24 Communications network following NIMS organization.
© Jones & Bartlett Learning.

and 4, as shown in **FIGURE 1-25**. Each division could then be assigned a different channel. More air time would then be available for units communicating with the Interior Branch; however, the communications network would be more complex, and there would be more communications layers between the IC and companies reporting to the divisions within the Interior Branch. Units subordinate to the Interior Branch could no longer listen to messages being transmitted by command.

Interoperability is a concern at incidents involving multiple agencies or more than one jurisdiction. Technology provides a means of communicating on many radio channels. With proper planning and equipment, it is possible to communicate with all agencies at the scene, as well as agencies providing support from off-site. Even if a department has the capability of communicating with all responding agencies, communicating via multiple radio channels is problematic. It is essential that the IC and all units that report to the operations section limit the number of radio channels used to reduce the possibility of missing critical messages from operating companies or crews. Although it is possible to monitor several radio channels using radios as scanners, it is very difficult to conduct two-way communications on more than two channels. This problem can be solved in several ways:

- Assign a fire fighter or command aide to monitor each radio channel at the command post and relay pertinent information to the IC.
- Place a representative of the assisting agency with their communications equipment at the command post. The agency representative manages and communicates with responders from their agency.
- Assign a liaison officer to communicate with other agencies.
- Assign the logistics section to communicate with agencies that are providing equipment and supplies.

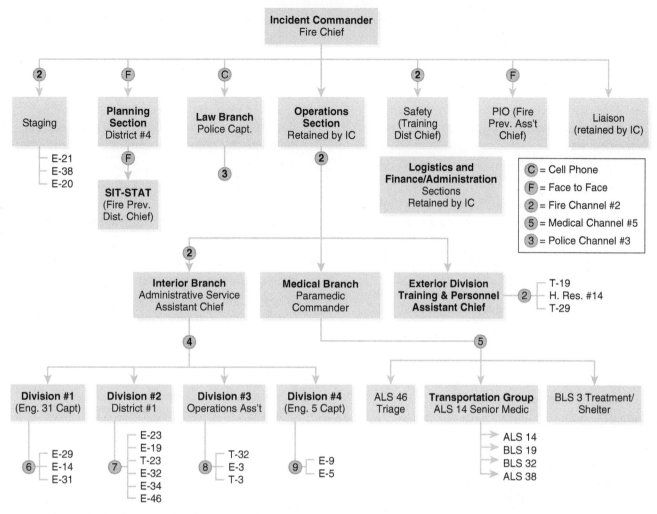

FIGURE 1-25 Revised communications network.
© Jones & Bartlett Learning.

- Direct communications to technicians who retransmit critical messages to the IC. This could be part of the communications unit within the logistics section.

Also consider using alternative communications methods such as computers, cell phones, or satellite phones when communicating with other agencies.

Common terminology is the cornerstone of effective interoperability. It is essential that common terminology be established for any management system, especially one that will be used in joint operations by numerous and diverse participants under emergency conditions. This commonality of terms is particularly important in the areas of organizational functions, resources, and facilities.

Positions within the NIMS organizational framework are named, and the same names are used for the function regardless of the type of emergency. Line and staff positions remain the same, even though incident situations and objectives may differ significantly.

Summary

The fire ground can present complex challenges. A tremendous amount of information must be processed rapidly and accurately to ensure the safety of fire fighters and victims. The only safe and effective way to manage fire-ground information and make the proper command decisions is to use NIMS from the beginning of the incident to its conclusion. This allows the IC to maintain a proper span of control and ensure accountability while safely and efficiently accomplishing the objectives listed in the incident action plan.

Wrap UP

CHAPTER SUMMARY

- The incident commander (IC) must manage resources, ensure fire fighter safety, and target priorities.
- Good communications are essential to decision making.
- The National Incident Management System (NIMS) enables the IC to delegate tasks and make decisions at both major and minor incidents. It should be used at all incidents, regardless of size. NIMS is the required incident management system for federal agencies and for departments using federal preparedness funding.
- NIMS should be used by the entire response community, not just fire departments. NIMS provides common terminology and operational assignments for all agencies.
- A unified command can be used when more than one jurisdiction or agency has responsibility. Although a single command is preferable, some situations require a unified command. Unified command is seldom warranted for the emergency phase of a structure fire, but may evolve during the clean-up phase.
- When using a unified command structure, plan development is shared, but one operations chief should direct all field units.
- The first-arriving company officer is in command and should follow department procedures, but in most cases will adopt a fast attack or investigative command mode.
- Three command options are available as follows:
    - Investigation: No obvious hazard is evident on arrival; no obvious need for immediate action to save lives or protect property.
    - Fast attack: Obvious signs of fire or another dangerous condition that require immediate action.
    - Command: Large or complex incident requiring that the company officer coordinate the actions of personnel and other incoming units.
- Standard operating procedures (SOPs) allow the operation to continue while a chief officer develops and implements an incident action plan.
- Strong command presence is critical to fire fighter safety.
- The initial report should confirm incident location, type, and conditions.
- Good communications and clear initial reporting are essential to command.
- When a chief officer assumes command, a stationary command post must be established.
- Officers assuming command must provide an updated status report based on observation and reconnaissance from units at the scene.
- Communications between command and field units must be established through the communications network.
- Command transfer must be addressed in department SOPs so company and chief officers know whether they are required to assume command.
- When command is transferred to another officer, even another company officer, it should be formalized at a stationary command post. To avoid confusion, command should not be transferred between company officers unless there is a compelling reason to do so.
- An officer who finds an unsafe operation in progress upon arrival must immediately take command, establish a formal command post, and take immediate corrective action.
- The person assuming command must communicate with the previous IC to determine both situation and resource status of the operation in progress.
- Command may be transferred to another agency in specific circumstances (terrorist attack, hazardous waste spill).
- The highest-ranking officer at the scene is held responsible and accountable for incident operations.
- The IC develops the incident strategy.

- Branch directors, division, and group supervisors develop tactics within the overall strategy.
- Information exchange and good communications are the keys to a safe and effective operation.
- A good command post is located in a known or easily found place, within the cold zone, in a location that supports command, control, and coordination of all incident activities. The IC is located at the command post.
- In larger, more complex incidents, the command post should be isolated from the scene.
- The span of control is the number of people reporting to a supervisor, and should include from three to no more than seven people reporting to a single supervisor.
- The type of incident, the nature of the task, hazard and safety factors, and the distance separating tactical units will influence span-of-control decisions.
- Tactical-level management components should maintain a reasonable span of control and coordinate specific geographic and functional operations.
- Anticipate the need for additional units and request them well in advance.
- Staging areas should be established for resources not immediately assigned to a task. Staging allows the IC to control access to an incident scene while deploying resources in a safe and effective manner.
- Staging areas should be far enough away to avoid impeding access to the scene or freelancing into action, but close enough to allow companies to arrive quickly once summoned.
- Early establishment of a staging area allows better management of on-scene units, establishes a tactical reserve, and eliminates freelancing. Management of staging areas should be part of SOPs.
- NIMS is a tool used to organize an operation—a means of providing command and control at an incident to achieve incident action plan objectives. NIMS organization should be as simple as possible.
- NIMS organizational structure develops in a modular fashion based on the type and size of the incident. The IC is the one position that must always be staffed within a NIMS organization. Some NIMS positions may be combined at smaller incidents.
- The IC determines organizational structure based on the management needs of the incident.
- Three command staff positions in NIMS include the safety officer, liaison officer, and public information officer. Command staff and line positions are staffed when incident conditions require separate management of the functions of that position.
- NIMS includes five possible sections, each with its own section chief: finance/administration, logistics, operations, planning, and intelligence/information. The intelligence/information section is rarely needed or implemented for a structure fire.
- A division is in charge of a geographic area; a group is in charge of a functional area. Use geographic rather than functional assignments whenever possible to improve communications and eliminate the need for status reports within the division.
- Branches are for operations beyond the span of control of a single division or group, or when a contingent of units from other agencies are working together.
- Task forces and strike teams can be used to reduce the span of control and the communications load at an incident.
- Strike teams combine a set number of resources of the same type. Strike teams can be used to organize large numbers of mutual aid companies or private sector resources.
- Task forces combine different resources. Task forces are more flexible than strike teams and more likely to be formed at the time of the incident.
- Areas of responsibility are designated by using an alphanumeric system whenever possible.
- The naming system used should be addressed in department and regional SOPs.
- Standard rules for maintaining good communications should be observed.
- A communications unit reporting to the logistics section chief is important at large or complex incidents.

- All radio communications should be in plain English, not codes, and confined to essential messages.
- Make use of other available forms of communication in addition to the radio to facilitate communications and planning.
- The communications system must support unity of command.
- When multiple agencies are responding, coordination of communications should be streamlined to avoid information conflicts.

KEY TERMS

automatic mutual aid Resources dispatched on the initial alarm without special request from the authority having jurisdiction, and based on preexisting agreements between agencies.

branches Immediately subordinate to section in NIMS hierarchy; these units are subordinate to the logistics, finance/administration, and planning sections. They are used to reduce the span of control at very large operations or to manage a particular function/agency.

chain of command Organizational structure establishing a line of authority and responsibility along which orders and instructions are passed (e.g., IC to operations section, to branch director, to division supervisor, to company officer, to fire fighter).

division Tactical-level management unit in charge of a geographic area.

finance/administration section The section that tracks and provides financial and administrative services required to compensate personnel or organizations providing goods and services at the incident scene.

freelancing Performing tasks outside the incident organization structure.

group The tactical-level management unit in charge of a function.

incident commander (IC) The person responsible for all incident activities, including the development of strategies and tactics and the ordering and release of resources.

incident management team (IMT) A team consisting of the incident commander and appropriate command and general staff personnel assisting the IC in managing an incident. May also consist of specially trained and credentialed members who, at the request of another jurisdiction, are called in to assist or manage an incident.

incident safety officer The command staff position assigned to monitor the scene for hazards or unsafe operations; enforces safety practices, establishes a safety plan, and coordinates with representatives from cooperating and assisting agencies.

intelligence/information section The section that provides information related to the prevention of criminal activity, including terrorism or the apprehension of perpetrators of criminal activity.

liaison officer A member of the command staff who is the point of contact for agencies that are not assigned to operations functions.

logistics section The section that obtains needed supplies, equipment, and facilities.

National Incident Management System (NIMS) A U.S. Department of Homeland Security system designed to enable federal, state, and local governments and private-sector and nongovernmental organizations to effectively and efficiently prepare for, prevent, respond to, and recover from domestic incidents, regardless of the cause, size, or complexity, including acts of catastrophic terrorism.

operations section The section that manages all tactical units deployed at an incident scene.

planning section The section that gathers and evaluates information, assists the incident commander in developing the incident action plan, and tracks progress; also tracks resource status.

public information officer (PIO) A member of the command staff responsible for interfacing with the public and media or with other agencies with incident-related information requirements.

rational decision making (RDM) A form of decision making in which input is obtained from diverse sources and careful analyses of all options is considered.

recognition-primed decision making (RPD) A form of decision making in which the incident commander must decide on the proper course of action with limited information available in a relatively short period of time.

single command A command structure in which one person is designated as the incident commander. This person is responsible for the development and implementation of the incident action plan. The incident commander can delegate staff and command positions as needed to assist in command and control functions.

span of control The number of people reporting to a supervisor. The span of control should not exceed seven people reporting to a single supervisor under emergency conditions. For example, a fire captain supervising an apparatus operator and three fire fighters would have a four-to-one span of control.

strike team A set number of resources of the same kind and type that have an established minimum number of personnel (e.g., five engine companies). These teams always have a leader (usually in a separate vehicle) and have common communications among resource elements.

task force Any combination of resources that can be temporarily assembled for a specific mission. These teams should be established to meet specific tactical needs and should be demobilized as single resources. All resource elements within this team must have common communications and a leader.

ten-codes Numeric codes used to communicate predefined situations or conditions. For example, "10-4" generally means "finished communicating." Their use is discouraged.

two-in/two-out rule A rule mandating that fire fighters working inside the hazard area must work in crews of at least two people; these two people must be backed up by at least two people outside the hazard area who are properly equipped and immediately available to come to the aid of the inside crew. This rule applies to the first-arriving unit.

unified command An application of the National Incident Management System that allows all agencies with jurisdiction responsibility for an incident or planned event, either geographic or functional, to manage an emergency incident or planned event by establishing a common set of incident objectives and strategies.

unity of command A pyramidal command system ensuring that no one reports to more than one supervisor.

SUGGESTED ACTIVITIES

1. Develop an organizational chart and communications network for a room-and-contents fire in a two-story, single-family residential building. The fire is on the first floor. Assume that all first-alarm units are on the scene performing the following tasks:

   - Engine 1, per SOP, has secured a water supply via a forward lay using 5-in. (127-mm) hose. The Engine 1 crew (officer and fire fighter) is attacking the fire on the first floor with a 1¾-in. (44-mm) hose line. The Engine 1 hydrant fire fighter joins the officer and other fire fighter on the interior and the apparatus operator continues operating the pump.

   - Engine 2 secures a 5-in. (127-mm) water supply and stretches a 1¾-in. (44-mm) backup hose line behind Engine 1. If Engine 1 has controlled the fire, Engine 2 will advance their hose line to the second floor for fire extension.

   - Engine 3 is assigned as the rapid intervention crew (RIC).

   - Truck 1 divides into two crews: Crew 1 is conducting a primary search of the first floor fire area; Crew 2 conducts a primary search of the second floor.

   - Battalion Chief 1 assumes command at a command post on the A–D corner of the building after driving past the A–B side.

2. Develop a communications network and NIMS organizational chart showing all operating units for a large fire on the second floor of a two-story warehouse. It is 2:00 AM. The building is tightly secured and unoccupied. Preincident plans indicate that the second floor is undivided with dimensions of 100 ft (45.7 m) wide by 50 ft (15.2 m) deep with a 25 ft (7.6 m) high ceiling.

   - Engine 1, per SOP and preincident plan instructions, has secured a water supply via a forward lay using 5-in. (127-mm) hose and split into two crews. Crew 1 is the initial rapid intervention crew (IRIC) stationed outside the structure and consists of the hydrant fire fighter and apparatus operator. The officer and other fire fighter make up the entry crew and advance to the second-floor fire with a 2½-in. (64-mm) hose line. There is smoke in the hallway. They encounter a heavy volume of fire as they force entry into the second floor and are unable to advance through the door. They report conditions to the IC, close the door, and back down the stairs waiting for additional assistance.

- Engine 2 relieves the IRIC with all four members forming the RIC.
- The Engine 1 hydrant fire fighter joins the officer and other fire fighter on the interior. The Engine 1 apparatus operator continues operating the pump.
- Engine 3 secures a second source of water. The apparatus operator remains with the apparatus while the three remaining members advance a second 2-½-in. (64-mm) hose line toward the second floor. The Engine 1 and Engine 3 interior crews open the door and attack the fire with two 2½-in. (64-mm) hose lines.
- Engine 4 secures a third water source. The apparatus operator remains with the apparatus while the two remaining fire fighters and officer connect a 2½-in. (64-mm) hose line to the Engine 1 pumper and the three Engine 4 members operate a third 2½-in. (64-mm) hose line on the second floor. All three companies are able to make progress on the fire.
- Truck 1 is assisting units on the second floor by venting horizontally from the exterior by opening windows opposite the entry stairs.
- Truck 2 is assigned to evaluate the roof conditions and prepare to ventilate from their aerial ladder if necessary.
- Truck 3 gains access to the first floor and is checking the first floor for occupants and fire extension.
- Battalion Chief 1 assumes command at a command post on the A–B corner of the building after driving past the A–D side and orders a second alarm assignment (four additional engine companies, a truck company, and a deputy chief).
- Engine 5 connects a 2½-in. (64-mm) hose line to the Engine 3s pumper and the four-member company is advancing a 2½-in. (64-mm) hose line into the first floor and assists Truck 3 in checking for fire extension.
- Engine 6 connects a 2½-in. (64-mm) hose line to the Engine 1 pumper and the four-member company is standing by with a 2½-in. (64-mm) backup hose line.
- Engines 7 and 8 are staged.
- Truck 4 is staged.
- ALS 1 is managing the medical function with two primary responsibilities: providing rehabilitation and medical treatment for fire fighters and providing medical transportation as needed.
- ALS 2 is assigned to rehabilitation, which is being set up in a nearby building
- BLS 1 and BLS 2 are assigned to transportation.
- The deputy chief assumes command and assigns Battalion Chief 1 as the planning section chief.
- The department safety officer is assigned as safety officer.

3. A large six-story warehouse is fully involved in fire, and the IC has decided on a defensive attack for the warehouse. There are occupied exposure buildings on each side of the warehouse, as shown in **FIGURE 1-26**. The IC has assigned companies to protect and evacuate the exposures on both sides of the warehouse. Develop a NIMS organizational chart for this operation as it is presently being conducted.

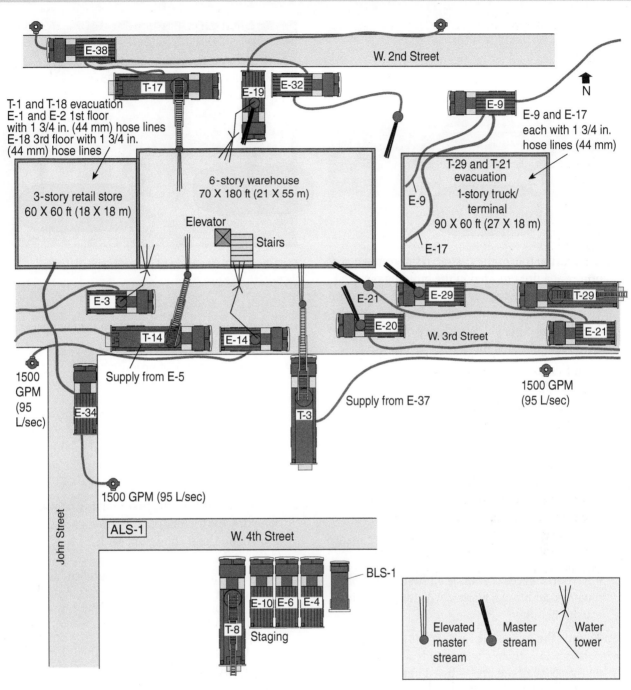

FIGURE 1-26 Warehouse fire and exposures.
© Jones & Bartlett Learning.

FIGURE 1-27 Cinema at 111 West Third Street.
Courtesy of Ben Klaene.

4. Assume the role of the first-arriving company officer at a fire reported in the building pictured in **FIGURE 1-27**. The fire was reported at 11 West Third Street at the Empire Theater. However, the correct address is 111 West Third Street. Develop a report to dispatch including the following:

    - Address confirmation
    - Command confirmation
    - Command mode (offensive, defensive, investigative, etc.)
    - Brief explanation of the building (or the name of the building if it is well known)
    - Occupancy (if not obvious in the building description)
    - Conditions on arrival (working fire, occupants at windows, nothing showing, etc.)
    - Resource needs (continue all first alarm companies, additional alarms, will advise, Engine 1 can handle)

5. Assume the role of the first-arriving chief officer (District 1) with an engine and truck company on the scene of the fire pictured in **FIGURE 1-28**. The fire was reported as 1234 Main Street, which is the correct address. Develop a status report to dispatch, including the following:

    - Address confirmation
    - Command confirmation
    - Brief description of building (or name of well-known building)
    - Occupancy (if not obvious in the building description)

FIGURE 1-28 Fire at 1234 Main Street.
Courtesy of Bill Strite, Cincinnati, Ohio.

- Conditions on arrival (nothing showing, occupants at windows, etc.)
- Progress (or lack of progress)
- Resource needs

6. Develop an incident action plan for the fire shown in Figure 1-28. A truck company is on scene, but fire has not been found on the second floor. Develop a message assigning the next arriving truck to perform this task.

7. Assume the role of the first-arriving engine company for the fire pictured in Figure 1-28. You are on the second floor with a 1¾-in. (44-mm) hose line encountering heavy smoke conditions, but you have not found the fire. District 1 requests a status report. "Command to Engine 1, what is your status?" Develop a status report including the following:

    - Unit number (Engine 1 to command, wait for command reply)
    - Your location
    - Progress/conditions
    - Resources
    - Any safety issues

8. Analyze incident communications by listening to audio media from actual incidents. Most dispatch centers have the capability of duplicating audio media, and fire scene communications are often posted on Internet sites.

REFERENCES

Federal Emergency Management Agency, U.S. Department of Homeland Security (DHS). 2017. *National Incident Management System*, Third edition. https://www.fema.gov/media-library-data/1508151197225-ced8c60378c3936adb92c1a3ee6f6564/FINAL_NIMS_2017.pdf.

Isner, Michael S., and Thomas J. Klem. February 26, 1993. *NFPA Fire Investigation Report: World Trade Center Explosion and Fire*. New York: NFPA.

National Fire Protection Association. 2020. *NFPA 1561: Standard on Emergency Services Incident Management System and Command Safety*. Quincy, MA: NFPA.

National Fire Protection Association. 2020. *NFPA 1521: Standard for Fire Department Safety Officer Professional Qualifications*. Quincy, MA: NFPA.

National Fire Protection Association. 2018. *NFPA 1500: Standard on Fire Department Occupational Safety, Health, and Wellness Program*. Quincy, MA: NFPA.

National Interagency Incident Management System (NIIMS). 1981. *Incident Command System: Operational System Description* (NICS 120-1). Stanford, CA: FIRESCOPE Program.

Superfund Amendments and Reauthorization Act of 1986. 1986. "Section 1910.120: Hazardous Waste Operations and Emergency Response." *Code of Federal Regulations*, title 29. Last amended May 14, 2019. https://www.ecfr.gov/cgi-bin/text-idx?SID=d9021eba51d4feb7753b89d8bc776312&mc=true&node=se29.5.1910_1120&rgn=div8.

CHAPTER 2

Procedures, Preincident Planning, and Size-Up

LEARNING OBJECTIVES

- List the kinds of operations that should be covered by standard operating procedures (SOPs).
- List the operations, occupancies (or structures), and special building systems that should be covered by SOPs.
- Explain the importance of preincident planning as it relates to on-scene size-up.
- Discuss the relationship between SOPs, preincident plans, and size-up.
- Examine the relationship between SOPs, equipment, and training.
- Explain the cycle used to maximize the value of newly introduced equipment.

- Compare standard operating *procedures* to standard operating *guidelines*, explaining the role of a "reasonable person" clause.
- Evaluate response district resources and challenges as part of the preincident planning process.
- List alternative water supplies that could be used when no piped public hydrants are available.
- Identify the kinds of underground flammable and/or hazardous material pipelines used and the potential hazards associated with these pipelines.
- List types of security systems and the potential problem they represent to building occupants as well as measures that can be taken to gain entry and assist building occupants in escaping a fire.

- Describe how enhanced security precautions affect response time and tactics.
- Articulate the main components of preincident planning, identifying a step-by-step process used to gather, review, store, and retrieve preincident plan information.
- List the major steps taken during size-up, and identify the order in which they will take place at an incident.
- Recognize the relationship between preplanning and construction characteristics common to a community.
- List new construction materials and methods introduced over the past 50 years and the effect they have on fire fighter and occupant safety.
- Explain the process of analyzing construction methods during everyday responses and while surveying buildings under construction and demolition.
- Preincident plans must be accurate and up-to-date to be of value; name ways preincident plans can be kept current.
- Explain a step-by-step process used to keep preincident plans up-to-date.
- Explain the usefulness of computers when recording and recalling preincident plans.
- Contrast life hazards in a large multi-story apartment building housing senior citizens compared to the same building housing younger families.
- Construct a priority chart of buildings to be preplanned by occupancy type.
- List factors to be considered during size-up, and briefly define and explain the significance of each factor.
- Demonstrate your knowledge of fire behavior and the chemistry of fire.
- Recall the basics of building construction and how they interrelate to preincident planning and size-up.
- Describe the role played by occupancy type in developing a preincident plan and in on-scene size-up.
- Define and explain the difference between *occupancy*, *occupant*, and *occupied*.
- Develop a list of occupancies that should be preincident planned.
- Explain the size-up process related to the chronological order in which information is received.
- Evaluate a specific fire department's SOPs.
- Develop a standard operating procedure (SOP) that could be used to develop preincident plans.
- List the occupancies that are preincident planned in your jurisdiction or residence.
- Prioritize occupancies to undergo preincident planning in a specific jurisdiction.
- Develop a preincident plan for a nursing home or hospital in your response area.
- Create a preincident plan drawing and narrative.
- Perform an initial size-up based on limited information.
- Apply size-up factors to a fire situation, and categorize factors as primary or secondary.
- Develop a pre-incident plan for a large and complex building with high value contents; explain how you would safeguard building information, while providing firefighters easy access to the pre-incident plan, keys, etc. while securing the information so nonauthorized personnel could not gain access to the information.
- Calculate and compare the rate of flow required in two 50 ft × 100 ft (15.2 m × 30.5 m) open office occupancies, one with a 10-ft (3 m) high ceiling, the other with a 15-ft (4.6 m) high ceiling. Analyze whether your standard preconnected hose line would be adequate for both offices.
- Use the Size-Up/Preplan Checklist to develop tactical needs, given a working fire that can safely be attacked offensively.

Courtesy of David J. Jones, Cincinanti, Ohio.

Case Study

This four-story nonsprinklered building with attic was constructed as a hospital in 1889. The hospital closed in 1974 and was later renovated to become a 161-unit apartment building to house low-income senior citizens over age 62 and other tenants with disabilities. The renovation created many dead-end areas within the building that are only accessible via a single stairway. There are multiple stairways within the structure, however only the stairways at the ends of the building are equipped with a standpipe. Apartment configuration is different from floor to floor. The building contains many concealed spaces including several open elevator and dumbwaiter shafts, which have been covered with gypsum board. The ground floor is public space for residents while floors two through four are all private apartments.

At the present time there is a major construction project in progress on the street in front of the building. The road in front of the building has divided lanes. Eastbound lanes are 30 ft (9.1 m) from the westbound lanes in front of the building. The eastbound and westbound lanes are each three lanes wide. At times, the construction workers completely shut down one side of the road, causing massive traffic jams.

A fire is reported on the third floor in Apartment 316. While en route, the first-due engine company reports "heavy smoke in the area; east and westbound lanes are both down to one open lane; there is a major traffic tie-up."

1. How would a preincident plan affect the initial size-up?

2. How does prior knowledge of the building and information gained through the preincident plan affect response effectiveness?

3. What provisions should be made to provide accessibility to the building?

Introduction

This chapter will discuss the importance of **preincident plans**, **standard operating procedures (SOPs)**, and their relationship to size-up. Every strategist recognizes the importance of gaining intelligence. Military officers go to great lengths to understand the factors that are involved in all operations, as well as the particulars of the specific battleground and enemy. In the fire service, factors that apply to fire-ground operations can and should be outlined in SOPs. The battle particulars are a result of preincident planning and analysis. As with military operations, the better the preincident plan, plan analysis, and SOPs, the fewer decisions that will need to be made in the heat of battle, which allows the incident commander (IC) to focus on important incident-related factors when developing an incident action plan.

SOPs, preincident plans, and incident-specific information are interrelated and are all-important components of the size-up. The IC who is faced with a large or complex building and no preincident plan or applicable SOPs is at a great disadvantage. SOPs and a good size-up are necessary prerequisites in the development of an incident action plan. For larger and more complex properties, preplanning is essential.

Important information about specific buildings can be obtained and analyzed in advance through preincident planning. This information is critical to the fire officer when making strategic and tactical decisions at the incident scene. Of course some things will only be known after the incident occurs, as is addressed in this chapter (see the section "Analyzing the Situation Through Size-Up"). Prefire information and incident-specific information are evaluated in terms of factors related to the operational priorities of life safety, extinguishment, and property conservation. Subtopics are structural conditions, resource needs, contents, occupancy type, access limitations, water supply, and special challenges.

The whole complex process of evaluating an incident and developing an incident action plan must take place in a few minutes. To accomplish this, it is necessary to focus on the major (primary) factors during the initial plan development process. As more information becomes available and the IC has time to reevaluate, the size-up information and the incident action plan should improve.

Developing Standard Operating Procedures

SOPs are general guidelines to be used at all structure fires or fires in similar occupancies. SOPs recognize similarities, covering areas such as those listed in **TABLE 2-1**.

SOPs address any operation that can be handled using a standard approach. SOPs will vary among

TABLE 2-1 Areas Covered in SOPs

Area	Related Elements
Command	- Implementation of the incident management system - Designation of person in charge - Establishment of command post - Command transfer - Communications - Strategy development - Develop an offensive or defensive strategy - Development of an incident action plan - Rapid intervention crew - Accountability
Staging	- Exterior staging - Interior staging
Water supply	- Adequate water supply - Attack pumper - Forward or reverse hose lays - Pumper/hose wagon - Tender shuttle - Water relay
Special occupancy operations	- Places of assembly - Educational occupancies - Healthcare occupancies - Board and care occupancies - Hotels - Dormitories - Detention facilities - Mercantile - Business - Storage - Industrial - High-rise - Large-area buildings

(continues)

TABLE 2-1 Areas Covered in SOPs (Continued)

Area	Related Elements
Company operations	- Truck company - Engine company - Quint/quad company - Rescue company - Medical unit - Special functions (e.g., collapse, communications, hazardous materials, foam)
Special operations	- Technical rescues - Hazardous materials - Natural disasters - Electrical fires - Active shooter - Civil disturbances - Terrorism
Private fire protection	- Standpipe - Sprinkler - Nonwater-based fire protection systems - Alarm systems

© Jones & Bartlett Learning.

departments. The types of property to be protected, resources available, equipment, and training, among other factors, guide the promulgation of SOPs. This variability among departments does not mean fire departments should not share procedures. Instead, it is necessary to learn all that can be learned from the experiences of others. However, the final procedure must reflect department-specific needs and resources. For example, a high-rise SOP would be of no value to a small rural department that does not respond to high-rise buildings within their jurisdiction or mutual aid response area. On the other hand, the SOP for an automobile fire would be similar in both a rural and a large urban fire department.

It is necessary to revise SOPs regularly. An automobile fire SOP written prior to the widespread use of alternate fuels, electric, and hybrid vehicles would be wholly inadequate. SOPs should be modified to include tactics related to recent research regarding fire behavior, venting, **flow path**, and staffing.

Large urban departments can quickly summon hundreds of fire fighters to the scene of a large structure fire. A smaller rural jurisdiction would, by necessity, handle a fire in the same size and type of structure differently, by calling on surrounding communities through mutual or automatic aid agreements. SOPs for handling the same type of fire may vary from department to department.

Although SOPs must be written specifically for the department, there is a need for regional planning when writing procedures. It makes little sense to have specific procedures for the incident management or accountability systems limited to just one department. There is a good probability that these systems will be critical during large-scale incidents when other regional departments are on the scene providing mutual aid. Chapter 4 of NFPA 1720, *Standard for the Organization and Deployment of Fire Suppression Operations, Emergency Medical Operations, and Special Operations to the Public by Volunteer Fire Departments*, requires procedures to ensure that an effective firefighting force can communicate with each other and conduct emergencies in a uniform manner (NFPA 1720, 2020).

Purpose of Standard Operating Procedures

Good department SOPs and preincident plans take the guesswork out of those first few precious moments on the fire ground. Company officers have predesignated assignments that allow them to take immediate action within the overall plan or, in some cases, to stage, awaiting orders from command.

In the absence of SOPs and preincident plans, the IC would be so busy assigning companies that little time would be left to collect the information necessary for the development of an incident action plan. Freelancing might also occur while the IC gathers the intelligence necessary for the formulation of an incident action plan, including the development of strategy and tactics. Either case results in inefficient and possibly dangerous operations.

With this in mind, it is best to provide specific SOPs for the first-arriving engine company. For example, some departments require the first-arriving engine company to secure a source of water and advance a hose line to attack the fire. The SOP for the second-arriving unit could be specific or general in nature, depending on whether the first-arriving unit is directed to secure a source of water. If SOPs do not require the first-arriving engine company to secure a water supply, the second-arriving engine company is generally assigned to supply water. If a water supply is not needed immediately, departments may preassign the second-arriving engine to Level 1 staging, act as the rapid intervention crew (RIC), or provide general guidelines if the second-arriving company is not assigned specific tasks. The first-arriving truck company will generally choose from a menu of possible tasks. This is further enumerated in Chapter 4, *Company Operations*. Truck company operations must be coordinated with the engine company attacking the fire.

During large-scale operations, the IC is faced with numerous, complex decisions. SOPs provide a structure for the decision-making process, including answering the questions of who makes what decisions, at what level of command, and from where? The National Incident Management System (NIMS) does an excellent job of establishing a command structure and describing the roles of various players at the incident scene and must be included in SOPs.

Relationship of Standard Operating Procedures to Training and Equipment

Regarding SOPs, NFPA 1500, *Standard on Fire Department Occupational Safety, Health, and Wellness Program*, states (NFPA 1500, 2018):

> **4.1.2** The fire department shall prepare and maintain written policies and standard operating procedures that document the organization structure, membership, roles and responsibilities, expected functions, and training requirements, including the following:
>
> (1) The types of standard evolutions that are expected to be performed and the evolutions that must be performed simultaneously or in sequence for different types of situations
>
> (2) The minimum number of members who are required to perform each function or evolution and the manner in which the function is to be performed
>
> (3) The number and types of apparatus and the number of personnel that will be dispatched to different types of incidents
>
> (4) The procedures that will be employed to initiate and manage operations at the scene of an emergency incident
>
> **5.1.2** The fire department shall provide training, education, and professional development for all department members commensurate with the duties and functions that they are expected to perform.
>
> **5.1.6** The fire department shall provide all members with training and education on the fire department's written procedures.
>
> **5.1.10** Training programs for all members engaged in emergency operations shall include procedures for the safe exit and accountability of members during rapid evacuation, equipment failure, or other dangerous situations and events.

NFPA 1500 makes a definite statement that training must be commensurate with SOPs. If new equipment is placed in service, the availability of the equipment makes a statement that fire fighters are expected to be able to use the equipment safely. SOPs must outline how and under what circumstances the equipment is to be used.

When a new tactic or SOP is introduced, the equipment necessary to perform the new tasks must be procured and fire fighters must be trained to safely implement the tactic. Seldom-used equipment and tactics require frequent training, as fire fighters will lack field experience. For example, if a fire department decides to purchase wind-control devices to counteract the effect of wind-driven fires in high-rise buildings, as discussed in the high-rise ventilation study conducted by the National Institute of Standards and Technology (NIST; Averill and Moore-Merrell, 2013), it would be necessary to make inquiries regarding available wind-control devices, purchase the selected equipment, and then train all members who could conceivably be ordered to deploy the wind-control device. Training should include the conditions under which the wind-control device should be used along with actual hands-on training in deploying the device.

Underwriters Laboratories (UL) published research related to vertical ventilation (Kerber, 2013). The International Association of Fire Chiefs (IAFC) Safety, Health, and Survival Section and the International Association of Fire Service Instructors (IAFSI) issued suggested guidelines to be implemented based on the findings of this and other research (IAFC and ISFSI, 2013). One of their recommendations states, "Water should be applied to a fire as soon as possible and from the safest location." This tactic is often referred to as transitional attack or **softening the target**.

In this case, it is not necessary to purchase new equipment; you can employ the suggested tactics using existing nozzles and ventilation equipment. However, the suggested fire attack is much different from methods currently used by most fire departments. In many departments, this represents a radical departure from current fire attack methods. There could be considerable resistance to implementing these tactics, and there is a definite need for training that not only demonstrates the new method, but also a need to explain why the change is being made. Any department taking this advice to "attack the fire as soon as possible and from the safest location" should carefully study the associated research and note limitations prior to implementing these new tactics. After training members on the proper techniques, it is best to objectively evaluate fires where the new procedures are implemented, both within the department and in other jurisdictions.

Consideration should be given to SOPs when developing new apparatus and equipment specifications. For example, if department SOPs state that the first-arriving pumper should pull past the main entrance to allow room for the first-arriving truck company, preconnected hose lines should be specified in locations that permit efficient deployment to the rear of the apparatus.

There is a relationship between SOPs, equipment, and training, as illustrated in **FIGURE 2-1**. Any time new equipment is introduced or a new procedure is written, the entire cycle must be completed.

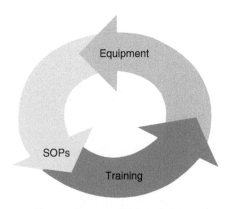

FIGURE 2-1 The equipment–SOP–training cycle.
© Jones & Bartlett Learning.

The Standard Operating Guidelines Controversy

Within the fire service, there is some discussion about whether to refer to standard operating *procedures* as standard operating *guidelines*, *general operating guidelines*, or another, similar name. Semantics can be important, but not in this case. It is far more important to have written procedures or guidelines than it is to argue about the name. Many in the fire service believe that calling procedures "standard operating guidelines" or "general operating guidelines" somehow changes their strength, in that a guideline may be considered less stringent than a procedure.

In reality, procedures are guidelines, and guidelines become procedures through practice. If a task is always performed the same way, that becomes the procedure for carrying out that task. Likewise, most procedures should be taken as general guidelines. Situations will arise in which the SOP does not fit the circumstance, and the fire officer would be expected to modify actions accordingly. For example, an SOP that requires the first-arriving truck company to ventilate would not be appropriate if an occupant requiring a ladder rescue were visible or the fire had self-vented, an attack hose line was not in position to control the fire and flow path, or there was an obvious defensive operation in progress. Some SOPs are always in effect, such as an SOP stating that fire fighters *must* wear self-contained breathing apparatus (SCBA) in contaminated atmospheres. Essentially, a statement is made that procedures are to be followed but that fire fighters should follow a reasonable course of action when confronted with a situation in which modification of the procedure is appropriate. When a decision is made to modify procedures, the noncompliant unit or member is required to communicate the variance and must be prepared to justify the modification.

Whether you call them procedures or guidelines, write them down, train to them, and use them consistently.

Evaluating Response District Resources and Challenges

The most valued fire suppression resource in any community is the fire fighter. A community with the best apparatus, equipment, and water supply is underprotected when the fire department is understaffed. Volunteer departments expend a good deal of time and effort in recruiting and retaining an adequate fire force. Paid departments struggle with budget issues that often result in understaffing. Determining the staffing level is based on many parameters. Over the years, many studies have been conducted in an attempt to establish the staffing needed for a structure fire. NFPA 1710, *Standard for the Organization and Deployment of Fire Suppression Operations, Emergency Medical Operations, and Special Operations to the Public by Career Fire Departments*, specifically addresses staffing at residential fires (NFPA 1710, 2020). A study conducted by NIST verified the residential staffing requirements (NIST, 2010). A more detailed discussion of staffing levels established in the NFPA 1710 standard can be found later in this chapter. (See also Chapter 5, *Fire Fighter Safety*.).

Apparatus requirements are also important when evaluating community fire protection. Apparatus needs vary depending on the water supply, height of buildings in the area, number of fire fighters per rig, and other factors.

Response Time

The ability to save lives and property is directly related to response time. Distance is but one factor affecting response time. A major highway with a higher speed limit will allow apparatus to travel faster compared to a winding two-lane road with numerous uncontrolled traffic signals. However, interstate highways may become extremely congested during rush hour. Likewise, flat terrain is easier to traverse with heavy apparatus than hilly topography. Railroad crossings, construction areas, weather, and other factors can also adversely affect response time. Response time will be discussed at length in Chapter 5, *Fire Fighter Safety*.

The street layout, numbering system, and addresses are sometimes confusing. Natural and man-made barriers often break up street continuity. For example, a street that previously continued for a considerable distance may be divided into different sections due to the construction of a limited-access highway. A hillside, river, or other obstruction may also result in a street being divided into different sections and still retaining the same name. When the street is divided into different sections, the address usually continues in the next street segment. Street addresses may end in the 1900 range at the bottom of a hill and continue in the 2100 range at the top of the hill. Continuous streets sometimes change names as they progress through different areas or jurisdictions. Likewise, streets laid out in a square pattern may have angular streets intersecting at various points. The street pattern in Washington, DC, has many such angular streets in what is basically a square street layout. A rapid response requires that fire fighters know the street layout and addresses in their response area. Having good street maps and/or a global positioning system (GPS) available is essential. The response information can be directly populated from the computer-aided dispatch (CAD) into the mobile data terminal (MDT) to provide directions to the incident. However, this data must be continually updated and verified as correct, and it often may not provide the most efficient response route. Modern technology is a helpful tool, but it should not be relied upon to replace response district familiarity and knowledge. Therefore, it is critical that training programs include knowledge of street addresses and response routes, as well as alternate routes. Fire stations in communities with confusing numbering systems or complicated street layouts often place large area maps in the fire station alarm room or apparatus floor to provide a quick reference.

Water Supply

Water supply is a critical part of community fire protection. SOPs and tactics must be modified to match the water supply. Fire fighters should know the capabilities of the water supply system available in their response area. In addition, it is necessary to have an alternate water supply strategy. Even the best water systems can be damaged due to an accident or natural disaster. Earthquakes have resulted in the loss of major sections of municipal water supplies, such as the damage in San Francisco during the Loma Prieta earthquake in 1989. Many areas are subject to flooding, resulting in (1) flooded areas that are not inaccessible via standard response routes, and (2) inaccessible fire hydrants **FIGURE 2-2**. Snow or floods can make hydrants inaccessible. Even if a fire hydrant can be accessed, many water utilities shut down water supplies to flooded areas to prevent contamination of the water supply and major loss of water if fire hydrants are damaged.

FIGURE 2-2 Residential area with flooding.
Courtesy of Thomas Lakamp, Cincinnati, Ohio.

During extremely cold weather, water mains have been known to freeze. Static sources of water can freeze, dry up, or be inaccessible. As water supplies improve, there is a tendency to abandon water tender or drafting operations even in areas with plentiful access to static sources of water. During Hurricane Sandy, pumpers drafted water from flooded streets to combat the Breezy Point conflagration. Preincident planning involves considering normal operations as well as the potential for disruption of normal resources.

A well-maintained water distribution system with closely spaced fire hydrants and large flow capacity provides the ideal water supply. Commercial facilities that are remote from a reliable water supply may be required to store firefighting water onsite. These onsite water supplies may be in the form of aboveground storage tanks or reservoirs. In most cases, these properties will be required to store an adequate water supply based on the hazard.

When all or part of the community is not serviced by a water distribution system with closely spaced fire hydrants, special provisions are necessary, such as the following:

1. Working from the apparatus tank
2. Setting up water relays
3. Implementing water tender shuttles

Evaluate these options during preincident planning by first determining the rate of flow (see Chapter 8, *Offensive Operations*, for rate of flow calculations). Once the rate of flow has been determined, the next step is to evaluate various sources of water and their accessibility and reliability.

Of the three options listed above, supplying water from the apparatus tank is the fastest and least labor intensive. The preferred strategy is an offensive attack whenever it can be carried out safely and effectively. Setting up water relays and shuttles requires considerable time and sometimes results in a delayed attack. Conducting an offensive attack when the risk-versus-benefit analysis indicates otherwise is dangerous, and it places fire fighters in unnecessary jeopardy. Most residential fires are within the capacity of a 1¾-in. (44-mm) hose line and, if properly applied, can be extinguished, or at least controlled, with the water carried on the pumper.

Using the onboard water supply to attack the fire is sometimes referred to as an "attack pumper" tactic. Some departments apply this tactic at all residential fires. When an attack pumper deployment is used, the second-arriving pumper usually supplies the first-due pumper with water from the hydrant system or other source. This multiphase approach may be the best tactic when water supplies are located far away from a small structure. When an attack pumper deployment is used, fire fighters must be aware of the limited water supply provided by the apparatus tank and limit the time of attack to allow enough water to protect their retreat if necessary. This process should be addressed in the department training program. Whether the first-arriving pumper uses an attack pumper tactic or not, make provisions to ensure a reliable, adequate, and continuous water supply.

When the distance from the fire to the pumper that is supplying hose lines to attack the fire exceeds pump, hydraulic, or hose limitations, an alternative water supply layout is needed. Alternative layouts are more complicated and time consuming. If the water supply is from a static source, a drafting operation will also be required.

Relay pumping is one way to provide a continuous water supply when the water supply is a considerable distance from the fire. Relay pumping involves moving water from the source (static or hydrant) through multiple pumpers to the apparatus operating at the fire scene. This can be very effective in providing 1000 gallons per minute (GPM) (3785 liters per minute [L/min]) or more over a long distance when large-diameter hose or multiple relay hose lines are used. **FIGURE 2-3** shows a relay from a static source.

Nearly every community has static sources of water, and the fire department should provide training and equipment necessary to set up a relay from a static water supply.

After evaluating the water supply during preincident planning and deciding on a water relay, many

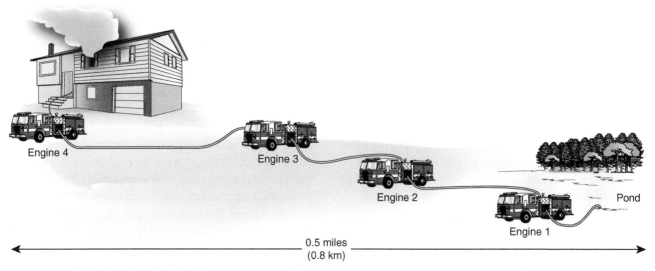

FIGURE 2-3 Water relay from a static source.
© Jones & Bartlett Learning.

FIGURE 2-4 Water relay sign.
Courtesy of Ben Klaene.

departments place permanent signs on the access road showing the position for each pumper in the relay **FIGURE 2-4**.

Communities sometimes rely on shuttling water via **tenders** (also referred to as a tanker or mobile water supply apparatus). High flow rates are also possible when using tenders, provided the water source can meet the demand, the source is relatively close to the fire scene, and several large-capacity tenders are available. However, moving the same amount of water the same distance, as in the water relay example shown in Figure 2-3, is generally more difficult when using tenders. NFPA 1142, *Standard on Water Supplies for Suburban and Rural Fire Fighting*, provides the following formula to determine the amount of water a tender can supply in GPM (L/min) (NFPA 1142, 2017):

$$Q = \frac{V}{A + T_1 + T_2 + B} \times k$$

where:

- Q = maximum continuous flow capabilities (GPM [L/min])
- V = tank volume of the mobile supply apparatus in gallons (or liters)
- A = time in minutes for the mobile water supply apparatus to drive 200 ft (61 m), dump water into the drop tank, and return 200 ft (61 m) to the starting point
- T_1 = time in minutes for the mobile water supply apparatus to travel from the fire to the water source
- T_2 = time in minutes for the same mobile water supply apparatus to travel from the water source back to the fire
- B = time in minutes for the mobile water supply apparatus to drive 200 ft (61 m), fill the mobile water supply apparatus at the water source, and return 200 ft (61 m) to the starting point
- k = 1 for pressure/vacuum water supply apparatus and 0.9 for other water supply apparatus

Setting up a tender operation using a single 3000-gallon (13,638-L) tender with a pressure/vacuum system transporting water 1 mile (1.6 kilometers [km]) from the source to the fire traveling at an average speed of 35 miles per hour (56 km/hr) yields:

$$V = 3000 \text{ gallons } (13{,}638 \text{ L})$$
$$A = 3 \text{ minutes}$$
$$T_1 \text{ and } T_2 = 2.35 \text{ minutes each}$$
$$B = 4$$
$$k = 1$$
$$Q = 256 \text{ GPM } (1164 \text{ L/min})$$

Determine flow requirements and the best method to meet the calculated flow during preincident planning. Obviously, the calculations necessary to determine the flow rate from each tender and additional calculations to determine how many tenders are needed would be very difficult to carry out at the incident scene. It is best to try various tactics prior to the fire, as well as determining the time required to fill and empty available tenders. The A, B, V, and k variables in the formula can be known in advance. The T_1 and T_2 components can also be measured or calculated prior to the incident.

If a water shuttle operation is implemented, the IC should establish a water supply group. The water supply group supervisor would have the responsibility of determining the number of tenders needed to meet the flow requirement established by the IC. Adding more tenders will increase the continuous flow capacity. However, other variables, such as congestion at fill and dump sites, will affect the total flow. Available fill and dump sites, as well as routes to and from these sites, should be preplanned. If not preplanned, the water supply group supervisor will need to determine routes and dump and fill sites at the incident scene. When determining routes, it is generally best to set up separate pathways to and from the fire, taking traffic, road surfaces, bridge and road weight limits, and other factors into consideration.

Many serious accidents have occurred when vehicles not specifically designed as water tenders were used to transport water at the fire scene. Tank trucks that are not designed to transport water present unique safety hazards and should be inspected and approved before being pressed into service. This inspection should include, but not be limited to, adequate construction, vents, and discharges large enough to quickly dump and fill water, and the presence of baffle plates. The previous contents of the tank should be determined and the tank interior checked for residual product. An added consideration is the weight capacity of the tank truck. Water, at 8.3 pounds per gallon (1.0 kg/L), is heavier than many other liquids that are transported via tank truck. A full load of water could easily overload a tank truck that normally transports products lighter than water. Another important safety consideration is the vehicle's center of gravity.

Construction Methods

Most fire districts have a predominance of buildings of similar construction. A great deal can be learned about properties in your jurisdiction during emergency medical services (EMS) or other responses. This is particularly true for buildings that are not formally preincident planned such as single-family residential buildings. In many cases, buildings in the same area were built as a subdivision and share common traits such as having inground basements, walk-out basements, cellars, or no basements. For example, the 93 houses in the subdivision shown in **FIGURE 2-5** were built in the 1970s. Every house on the four streets that make up this subdivision has a basement, but none of the basements has direct access to the outside. Every house has a garage and lightweight wood truss roof construction, but none were originally constructed using lightweight wood truss or wooden I-beam floor support systems. At the time these homes were constructed, there was a shortage of natural gas, so all houses in the subdivision are "total electric," with no gas service. All electric service is provided via overhead wires. An adjoining subdivision has underground electric service. Ironically, underground natural gas and petroleum pipelines traversed both subdivisions.

Pipelines have a good safety record but on occasion they are damaged and present a significant fire hazard. Petroleum pipelines tend to present a greater

A

B

FIGURE 2-5 A subdivision in which the homes all have similar construction features.
Courtesy of Ben Klaene.

hazard than natural gas pipelines as the products in petroleum pipelines are generally liquid and tend to run off, emitting vapors that are often heavier than air. These products often run off into streets, sewers, and buildings, creating a significant risk to occupants and fire fighters. Natural gas is lighter than air therefore it tends to rise. Of course, either petroleum products or natural gas leaking into a building presents a significant risk to occupants and fire fighters. The department preincident planning process should include identifying pipeline locations.

Buildings constructed during a given period and by the same builder tend to be very similar unless the building has undergone major renovation or repair. Frame buildings built prior to 1940 may be **balloon-frame construction**; whereas frame buildings built later will probably be **platform-frame construction**. Similarly, frame structures built prior to 1940 typically will have solid beam roof structures rather than lightweight truss roofs.

Fire companies should survey their district on a regular basis, paying particular attention to buildings being demolished and under construction. Buildings being demolished will show the common construction characteristics for an era and provide a look inside the building's skeleton. When a building is under construction, the interior framework and other features can be seen **FIGURE 2-6**. Fire fighters often discover new construction methods and materials when examining a building under construction. This is a good time to take photographs and share discoveries with the other crews. Studying your response area by subdivisions and/or eras can reveal a lot about the residential and small commercial buildings within your jurisdiction.

FIGURE 2-6 This is formerly a one-story, ordinary commercial building that is being renovated into a three-story residential multiunit dwelling. Notice the lightweight parallel trusses in the foreground.
Courtesy of Thomas Lakamp, Cincinnati, Ohio.

Evaluating a Specific Property

During an official visit to a facility, it makes sense to gather information about all kinds of problems that may occur at a property. If the facility falls under the requirements of Title III of the **Superfund Amendments and Reauthorization Act (SARA)**, then hazardous materials planning is mandated by law.

Establishing SOPs is step 1 in the size-up. Preincident planning is step 2. After SOPs are developed, individual properties should be examined for specific hazards and characteristics. Preincident plans are a natural extension of SOPs. For example, a fire department should have a written procedure for operations conducted in buildings protected by sprinkler systems. These procedures would not be repeated in the preincident plan, but any deviations from the normal operation would be outlined. For example, if the water supply at a property is reliable and no off-site water supply exists, the preincident plan might modify the typical SOP at this sprinkler-protected building by not automatically requiring that the sprinkler system be supplied at the fire department connection.

Security Concerns

Security concerns are much greater now, resulting in building owners and management increasing security measures. Armed security personnel are commonplace, especially at government facilities. The World Trade Center attack of 2001 substantially increased awareness of the threat of terrorism. Schools and other properties are now taking precautions to protect occupants from violence. Recent "active shooter" incidents have resulted in extra security measures. The NFPA developed NFPA 3000, *Standard for an Active Shooter/Hostile Event Response (ASHER) Program* (NFPA 3000, 2018). The 2018 edition was developed as a provisional standard to ensure prompt dissemination.

Security measures often run contrary to fire safety. Locked doors, limited access, energized electrical fences, fences with razor or barbed wire, and other security measures can greatly affect fire department operations. Security barriers often delay fire department entry onto the property and hinder occupant evacuation **FIGURE 2-7**. It is necessary to cooperate with building management and note any specific security concerns on the preincident plan.

Lockboxes

In the past, many fire departments carried keys to buildings or gates leading to complexes on fire

FIGURE 2-7 Security features at an industrial complex.
Courtesy of Ben Klaene.

apparatus. These keys were often misplaced or lost. Sometimes keys could no longer be identified or the property owner would change locks, failing to notify the fire department. This system resulted in a security breach for the property owner and in many cases did not provide the fire department access to the building or property. Many communities now require a lockbox system at specific commercial properties. Lockbox systems are usually placed at properties with the following features:

- An alarm system is tied to a central station
- Chemicals are stored in reportable quantities according to the SARA
- Properties that require immediate access by the fire department

The local authority having jurisdiction decides which properties are required to have lockboxes and selects a standard key lock system that will be used throughout the community. A dedicated key opens all lockboxes in that community. The property owner procures the lockbox and contacts the local fire department to secure the key in the box after ensuring that the key placed in the lockbox will provide access to areas and buildings on the property. All properties requiring lockboxes according to the local code use the same dedicated key. This dedicated key is secured on fire apparatus and staff vehicles. It is important to secure the dedicated key that provides access to many major properties within the jurisdiction. Some departments use a key code system to access the apparatus lockbox that tracks when the dedicated key box was opened and by whom.

Lockboxes come in various sizes that can hold sets of keys, preincident plan information, or safety data sheets. In some cases, the dedicated key can also be used to unlock fire department connections to sprinkler or standpipe systems. The use of a key not only allows easy access to the building or facility, but it can also eliminate the need for forcible entry, thus reducing property damage. NFPA 1, *Fire Code*, now requires that a standard elevator key be used throughout the jurisdiction (NFPA 1, 2018). If a standard key cannot be used, property owners are required to place a nonstandard elevator key in a lockbox located near the elevator.

Safeguarding Preincident Plan Information

The information gathered while preincident planning could be invaluable to the would-be terrorist or thief. Historically, fire departments have been liberal in sharing information and somewhat lax in securing SOPs and preincident plans. Many times, SOPs are posted on the Internet and, thus, are available to the general public. Preincident plans are often carried in three-ring binders inside unsecured vehicles. Lockbox and other keys are sometimes left in plain view within a command vehicle or apparatus.

It is recommended that fire departments take a more aggressive approach in protecting and securing preincident plans, department SOPs, and keys, including developing procedures for the disposal of computers, SOPs, preplans, and the medium on which they are stored. Procedures should be established for releasing SOP and preincident plan information.

Under the current climate of terrorism, the fire department would be open to criticism if they fail to be good stewards of information in their possession or neglect to take steps necessary to protect it.

Collecting Information and Preparing Preincident Plans

Preincident planning is vital to the safety of fire fighters and the efficient delivery of community fire protection. Therefore, it is imperative that preincident planning be approached in a positive manner. The information collected is essential to conducting safe and effective operations at large, complex facilities or high-hazard properties.

Personnel assigned to preincident planning must be properly trained and familiar with the preincident planning process used in their department. Some departments assign fire prevention staff to collect and store preincident plan information. Fire prevention personnel can often incorporate preincident planning into their inspection schedule and are generally well qualified. However, if possible, line personnel who are most likely to respond to an incident at the property should conduct the preincident plan survey. Whoever conducts the preincident survey should share their findings with other members who routinely respond to the premises, preferably in the form of an instructional program. There is a training effect in actually surveying the property and also in sharing important findings with other responders. Once the information about a property is gathered, it may be best to have specially trained personnel enter the information into the department storage/retrieval system to maintain uniformity and avoid entry errors.

Preincident planning is a time-intensive activity that requires extensive data collection and data entry. A building's occupancy, layout, and contents may change frequently.

Preincident plans must be kept up to date. The entry gate shown in Figure 2-7 is blocked by concrete barriers. Formerly, this was the main entrance at this large industrial plant. A fire company arriving at the previous main entrance would not be able to proceed beyond the concrete barriers, and may not know where the new main entrance to this large industrial complex is located. Access to the plant via the new main entrance requires reentering the highway and proceeding approximately one-quarter mile (0.4 km) west, then reentering the complex by way of a newly constructed road. The old main entrance is only sporadically staffed by security personnel. At times there is no one at the gate to provide directions to the new entrance.

Changes to the interior of a building can greatly affect safety and tactics. In some cases, a building that contained a fairly light fuel load may change occupants, with a subsequent change in the life hazard and/or fuel load. For example, a storage warehouse that previously stored noncombustible commodities could change occupants, with the new tenant storing rubber tires in open racks, greatly increasing the fuel load. If this storage warehouse is protected by a sprinkler system, the new fuel load could exceed the design limitations of the sprinkler system.

Keeping preincident plans current is at least as important as creating the initial preincident plan. Bad information can be worse than no information. The revision process is also time intensive, especially if plans must be completely redrawn and the information must be reentered. The revision process is when the computer can be used to great advantage; minor details can be changed and the information printed out and/or stored electronically with little effort. If computers are available at the incident scene, the process of finding the right preincident plan and the information needed to develop an incident action plan can be greatly simplified.

Imagine having hundreds of preincident plans, with thousands of pages and accompanying information (e.g., safety data sheets, hydrant maps), filed in three-ring binders. Finding the right property and safety data sheet could involve considerable time and effort. If a department has relatively few preincident plans, a three-ring binder or card system may be sufficient. However, if the department protects large numbers of properties requiring a preincident plan, then a computerized

FIGURE 2-8 Onboard computerized preincident plan.
© Jones & Bartlett Learning.

system, such as the one shown in **FIGURE 2-8**, is highly recommended.

A word of caution: The use of onboard computers is the preferred method of storing large numbers of preincident plans. However, before investing a great deal of time and effort in computerizing preincident plans, ensure that your computer system can handle all the required data, photographs, and drawings. It is essential that stored preincident planning and response information can be quickly retrieved. Loading a CD or flash drive, waiting for the computer to boot, or downloading your preincident plan can cause a considerable time delay. Preincident plan information that is not immediately available will be of limited use to the IC or company officers during initial operations.

The ultimate onboard computer system would remain in a ready mode and automatically load the incident address from dispatch into the vehicle computer, while a global positioning system (GPS) would select the best route to the fire location, and a geographic information system (GIS) would visually display the location of nearby fire hydrants, water mains, flood zones, underground pipelines, and other valuable information.

Some applications can display aerial photographs or plot plans of buildings. If a preincident plan were available for the address, this ideal system would load the preincident plan, making it immediately available by touch screen or cursor navigation. In addition, the preincident plan drawing would be displayed on top with additional information available by accessing the narrative, via **hovering** over a drawing element or by **drilling down** for more text-based or graphic information. This "ultimate system" has one major drawback—it relies on information being received from a remote source (e.g., dispatch, telecommunications contractor). The reliability of the network must be verified via testing prior to making a final decision on system components and programs. For additional precaution, preincident plan information should also be stored onboard fire vehicles as a backup when the communications network or terminal fails.

Many computer programs are available that allow easy navigation from page to page or provide information when the cursor is placed over the item. Preplan software programs have been developed specifically for fire service use, or common software programs such as PowerPoint can also be used. Easy navigation provides the means to quickly access needed information.

Probably the most difficult to construct and most important part of the preincident plan is the drawing(s). Most architectural drawings produced today are created electronically.

When considering new preplan software, it is best to purchase a program that can import and modify computer-generated drawings. If the property being preplanned has architectural drawings, you may be able to simply load the drawings and add your own symbols. Most architectural drawing programs are done in layers, allowing the user to eliminate unwanted drawing information and add layers with information important to the fire department. Some preincident plan drawings cannot be downloaded, or drawings may not be available. A powerful user-friendly drawing software program is an indispensable tool when developing and updating preincident plan drawings.

As an alternative to computerization or as an interim measure while preincident plans are being converted to onboard computer files, it may be a good idea to place printouts in the facility lockbox, provided the lockbox is large enough to accommodate the printout.

Preincident Plans

Formal preincident plans include both a narrative and drawings. Drawings are the most important part of the preincident plan. The IC and company officers do not have time to read pages of text-based information. However, some narrative information is needed, such as the address, occupants, emergency contacts, telephone numbers, and other general information.

Narratives are best written in outline form with extremely important information highlighted, color coded, or otherwise identified to draw attention to critical information. Much of the text-based information can be embedded in graphic features, allowing the user to drill down or hover over the graphic feature to access additional graphic or text-based information. For example, a symbol representing a standpipe can be used to show the location of the standpipe inside the building, with additional information such as pressure, size, flow rate, pressure regulator operation, and other pertinent details revealed when the cursor is placed over the standpipe symbol.

Preincident plans can take various forms. A preincident plan that includes both a narrative and a drawing would be a formal preincident plan. On the other end of the spectrum is a building where a simple notation is made about a particular problem, such as holes in the floor from a previous fire.

> **NOTE**
>
> There should be an SOP describing the preincident planning system.

NFPA 1620, *Standard for Pre-Incident Planning*, outlines the steps involved in developing, maintaining, and using a preincident plan (NFPA 1620, 2020). Various occupancies are also defined in NFPA 1620, including the following:

- Assembly
- Educational
- Health care
- Detention and correctional
- Apartment buildings
- Dormitories
- Hotel
- Lodging and rooming house
- Residential board and care
- Mercantile
- Business
- Industrial
- Warehouse and storage

Annex B of NFPA 1620 contains case histories of several fires where preincident planning led to success or where a lack of preincident planning led to failure. For example, NFPA 1620, B.11 discusses a fire in a nursing facility where preincident planning resulted in a safer and more effective operation, as described here. On the morning of March 29, 2008, fire units responded to an elder care/skilled nursing facility less than 1 mile from Station #2. The first-due company, Engine 2 had done a preincident plan for the four-story building. The fire was reported to be in the laundry room area, which was in a difficult to access area of the facility. Using the preincident plan information, the company officer was able to direct the other responding units to the most appropriate location and establish the best course of action.

Incident Summary

Paint Manufacturing Facility

Fire units responded to a fire alarm at a paint manufacturing facility. The department's hazardous materials team had just conducted a preplan at that facility the previous week. Shortly after arriving on scene, the IC requested additional resources, including the hazardous materials team, due to an explosion at the facility. The product involved was nitrocellulose, a DOT Class 4 flammable solid.

The first-arriving companies attempted to gather information by looking the product up in the emergency response guide but were frustrated by the fact that there were six different entries for nitrocellulose. Due to the knowledge gained during the preplanning process, the hazardous materials unit was able to tell the companies exactly which entry to use until the hazardous materials team arrived. The hazardous materials team also knew the exact location of the product in the facility and the location of the deluge valve to turn off the suppression system. Because of the preplanning process, responding personnel had information on the product, its location, and the location of the pertinent fire protection systems without having to send a team into the hazard zone and without having to wait for property representatives to arrive on scene. Company officials were impressed with the fire department's knowledge base and the speed with which the incident was mitigated. While the incident could have been mitigated without a preplan, it certainly made the operation much faster and, more importantly, much safer.

Types of Preincident Plans

There are many ways to describe and categorize preincident plans. This text considers the following three levels or types of preincident plans:

1. **Complex preincident plan.** This type of plan is used when a property has more than three buildings or when it is necessary to show the layout of the premises and the relationship between buildings on the site. Complex preincident plans are used to identify building and fire protection features as well as hazards for each building. Fire companies and the IC cannot be expected to know the location of every building within a large complex. A street-level

view of a large industrial complex is shown in **FIGURE 2-9**.

Complex preincident plan drawings provide an overview of the complex to assist in locating buildings or areas within the complex. A preincident plan of this type may have several drawings of specific floor areas but also includes a drawing showing an overview of the complex and the position of various structures by name, number, letter, or other naming system, as shown in **FIGURE 2-10**. If "Building I" is given as the fire location, fire fighters and the IC use the overview drawing to locate Building I.

FIGURE 2-9 Industrial complex.
Courtesy of David J. Jones, Cincinanti, Ohio.

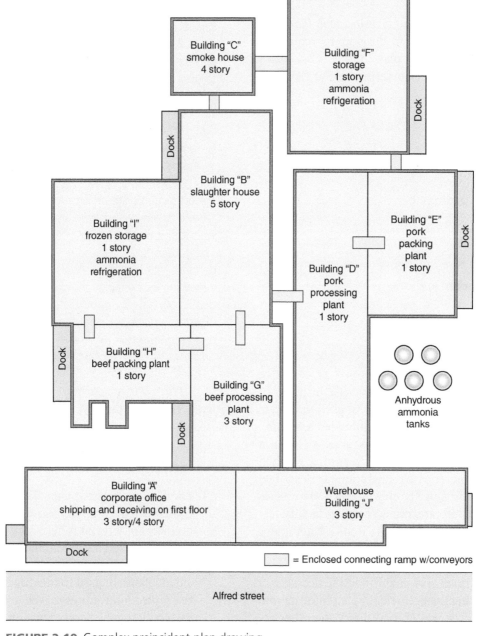

FIGURE 2-10 Complex preincident plan drawing.
© Jones & Bartlett Learning.

Responding units use the formal preincident plan for Buildings "H" and "I" to assist in sizing up the situation and formulating the incident action plan. Often, buildings are named according to their function or when they were constructed, rather than using a rational pattern that is apparent to the casual observer.

2. **Formal preincident plan.** A property with a substantial risk to life and/or property should be the subject of a formal preincident plan. This formal preincident plan would include a drawing of the property, specific floor layouts, and narrative describing important features. Several formal preincident plans could be included within a preincident plan for a single building within a complex preincident plan. One large, complex property could easily fill a three-ring binder with drawings of specific areas as well as the relationship between buildings.

The formal preincident plan drawing shows the interior layout for two attached buildings **FIGURE 2-11**. These buildings are located

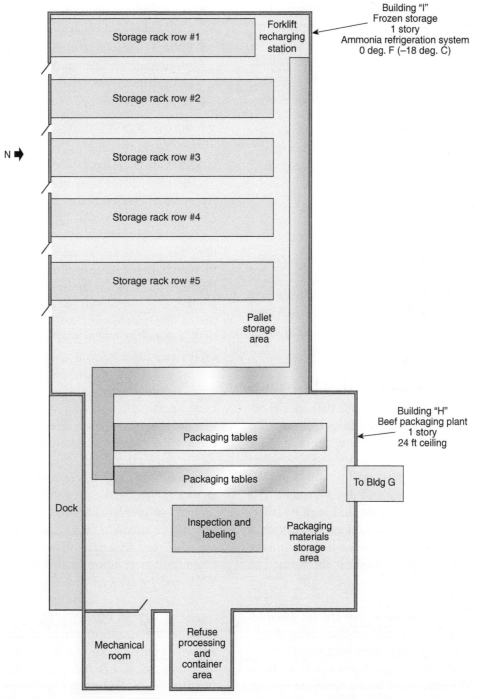

FIGURE 2-11 Formal preincident plan drawing.
© Jones & Bartlett Learning.

within the complex shown in Figures 2-9 and 2-10. Only items of importance to the IC are shown on the formal preincident plan. The narrative portion of this preincident plan would further describe the anhydrous ammonia refrigeration system, storage height, and any other significant features.

3. **Notation.** A simple notation may be made about the premises, such as building damage from a previous fire. Computer-aided dispatch systems usually have provisions for making a special notation for specific addresses. The dispatch center may be used to store information that is transmitted to the IC and companies at the incident scene. If communications applications are part of the computerized preincident plan system, this information may be shown on the computer screen as part of the graphic information available for the address.

Marking the exterior of dangerous buildings, such as painting a large "X" at the top level of a building that should not be entered because of structural problems, would be another form of notation preplan.

Preincident Planning by Occupancy

A fourth category of preincident planning by occupancy could be described as either a training issue or the topic of an SOP. For example, church fires have common traits that should be understood when conducting the size-up. See Chapter 11, *The Role of Occupancy*, for more details about occupancies.

Preincident Plan Checklist and Drawings

Preincident planning SOPs should identify special occupancies or specific types of buildings that are to be included in the preincident planning process. Some departments prefer to gather and distribute information using a standard preincident plan form. The use of forms has advantages and disadvantages. The advantages include being able to find specific information in a predictable location on each preincident plan and having defined categories, which ensure the gathering of data for items listed on the form. The disadvantages include having large amounts of "not applicable" space on the form and possibly failing to gather information about items that are not listed on the form. Preformatted preincident plans using checklists tend to be several pages long because of generic formats, and much of the information may be irrelevant. If a building does not have a sprinkler system, there is really no need to state that it is not sprinkler protected.

It is recommended that a detailed format, based on the checklist titled Site Data Collection Card in Annex D of NFPA 1620, be used. However, there could be other important data that is not listed.

A partial list of categories listed in NFPA 1620, Annex D include the following:

- Site data: street building name, address, and occupancy type
- Exposures on all sides
- Water supply
- Fire protection and alarm systems
- Life safety: occupant load, means of egress, roof access, disabled occupants
- Building data: construction materials; heating, ventilation, and air conditioning (HVAC); stairs; elevators; utilities
- Hazardous materials
- Emergency contact information

The Memory Lane list in **FIGURE 2-12** also provides a good checklist.

The preincident plan narrative for Memory Lane Apartments lists items that can be known in advance. This information would typically be gathered using a checklist. To yield a more "user friendly" preincident plan, the information should then be formatted to match the size-up/preincident plan checklist in Figure 2-12. When a factor is assumed or covered in SOPs, it is not included. For example, the factor "additional hose lines needed" does not need to be addressed in this preincident plan narrative, as "nothing special" (SOPs call for a backup and hose line on the floor above the fire) should be adequate. Factors that do not apply to the property should not be listed. Memory Lane Apartments is not equipped with a standpipe or sprinkler system; therefore, no mention is made of a sprinkler or standpipe system in the narrative.

The size-up/preplan checklist includes 76 items (not counting titles where subtitles fully address the factor). If you closely examine the Memory Lane Apartment preplan narrative and drawing, you can see that 27 of these factors are directly addressed, 5 are addressed in a general way, and 2 do not apply. Property conservation in this instance would follow SOPs and be typical; therefore, the 10 factors listed under property conservation are also addressed in a general way. In other words, information about 44 of 76 of the size-up/preincident plan factors can be known or partially known in advance. In some buildings, many or few factors may be known in advance. Incident-related factors, such as time and weather conditions, will not be known until the day of the fire or arrival at the fire scene.

In addition to addressing the size-up/preplan factors, the building name and address and the owner's/manager's/agent's name, with telephone numbers and emergency contact information, should be included in the preincident plan narrative.

The IC and responding units do not have time to read an encyclopedia about the property. They should be able to quickly find and easily read the needed information.

Notice the Maltese cross at the top of the Memory Lane Apartments preincident plan narrative. This is meant to be a quick first reference. This symbol could also be included in the drawing on a formal preincident plan. The Maltese cross marking is based on NFPA 1, *Fire Code*, Annex E, which describes the Fire Fighter Safety Building Marking System (NFPA 1, 2018). **TABLE 2-2** provides an overview of the marking system. This simple icon provides basic information regarding construction type, hazards of contents, sprinkler and standpipe systems, and occupancy or life hazards. The center of the cross is reserved for special hazards such as hazardous materials.

For example, on the Maltese cross for the Memory Lane Apartments, the top of the cross is used to designate the construction method: "ORD" for ordinary construction. The right side addresses built-in fire protection features. The "N" signifies no protection; this building is not protected by a sprinkler or standpipe system. The bottom of the Maltese cross identifies the relative life hazard, which is rated as "H" (high) due to the number of occupants and their immobility. The left side is used to indicate the content hazard. The "M" indicates a moderate fire hazard regarding contents in this building.

The center of the Maltese cross is reserved for special hazards. Typically, this is used to identify hazardous chemicals using NFPA 704, *Standard System for the Identification of the Hazards of Materials for Emergency Response* (NFPA 704, 2017). When hazardous materials are present and the department is using an electronic retrieval system, this or any other space on the Maltese cross can be used to navigate to additional information. In the case of hazardous materials, the NFPA 704 symbol could be linked to the chemical list, which could then be linked to another layer containing safety data sheets for each hazardous chemical. The center of the Maltese cross can also be used for other important information. The center is left blank on the Memory Lane Apartments preincident plan. NFPA 1 proposes using the Fire Fighter Safety Building Marking System on the exterior of a building. The authority having jurisdiction would need to pass a regulation, law, or ordinance to enforce the use of this marking system. However, using the system on preincident plans is at the discretion of the fire department and highly recommended.

It is said that a picture is worth a thousand words, and nowhere is this more true than at an incident scene. Including a drawing in a preincident plan is extremely useful. **FIGURE 2-13** is a sample of a preincident plan drawing.

Using intuitive drawing symbols on preincident plans that do not require a legend to understand is highly recommended. NFPA 170, *Standard for Fire Safety and Emergency Symbols* (NFPA 170, 2018) should be consulted when developing a symbol system for use in preincident planning. The symbols in Chapter 5 of NFPA 170 (Symbols for use by the Fire Service) are intuitive and particularly well suited to preincident plan drawings. The architectural and engineering drawings in Chapter 6 of the NFPA 170 standard use markings and symbols that are commonly used in professional drawings (NFPA 170, 2018). Knowing these markings and symbols will assist in reading building drawings made available by building management. If using a computer to retrieve and view drawings, note that many programs have the ability to describe the symbol or marking when the cursor is placed over (hovers over) the drawing.

Drawings should include the following:

- Entry and exit doors
- Fences, gates, and other exterior security features
- Areas of safe refuge
- General floor layout
- Building and area dimensions
- Stairways
- Fire escapes
- Elevators
- Fire separations
- Utility shutoffs
- Alarm panels
- Lockbox
- Fire protection system and sectional control valves
- Fire protection system intakes
- Fire protection system manual actuation
- Fire protection system agent supply
- Hose outlets for standpipe systems
- Fire pumps
- Fire hydrants
- Potential staging areas
- Hazardous materials locations
- Emergency vents and controls
- Other information important to the IC

Memory Lane Apartments
26 Memory Lane

Life Safety/Fire Fighter Safety
Occupancy Type
- 89-unit apartment building
- Most occupants are elderly, some with severe handicaps
- Seven small retail stores on first floor facing Main St:
    - Video Rental Store (50 ft × 45 ft [15.2 m × 13.7 m])
    - Joe's Barber Shop (24 ft × 45 ft [7.3 m × 13.7 m])
    - Hair by Gloria Beauty Parlor (24 ft × 45 ft [7.3 m × 13.7 m])
    - Old Tyme Antiques (12 ft × 45 ft [3.6 m × 13.7 m])
    - Nail Boutique Manicure Salon (12 ft × 45 ft [3.6 m × 13.7 m])
    - Song Shop CDs (24 ft × 45 ft [7.3 m × 13.7 m])
    - Thrift Shop (24 ft × 45 ft [7.3 m × 13.7 m])
- Basement
- Bingo Hall (180 ft × 45 ft [55 m × 17.3 m])
    - Bingo hall posted for 500 maximum occupants
    - Exits for Bingo hall stairs to Memory Lane and Main St. sides

Estimated Number of Occupants
- Bingo hall could have 500 on Tuesday and Thursday evenings
- Occasionally used as a party hall
- Most apartments occupied by a single resident
- Shops vary, seldom more than 10 per shop

Primary and Alternative Egress Routes
- Three stairways and fire escape (see drawing)
- One-story roof to rear could provide emergency or secondary egress from second floor

Access to Building Exterior
- Access streets to front (Memory Lane) 50 ft (15 m) setback
- Right side (Main St)
- Driveway on left side
    - Narrow, but apparatus accessible
    - Too close for aerial use
    - U-shaped driveway to front (may not support apparatus without damage)
    - Rear: one-story roof; good access to second floor

Construction Type
- Ordinary construction (brick exterior wooden floor/ceiling assembly)

Roof Construction
- Built-up flat tar roof supported by $2^2 \times 12^2$

Condition
- Well maintained, no noted damage

Live and Dead Loads
- Average load for residential property

Enclosures and Fire Separations
- Only open areas in basement and stores
- Metal fire doors from apartments to hallways
- Hallway not separated, open from end to end

Extension Probability
- Stairways are not enclosed
- Utility openings penetrate floors

Concealed Spaces
- Inaccessible crawl space between top floor and roof

Age
- Built in 1932

FIGURE 2-12 Sample preincident plan narrative.
© Jones & Bartlett Learning.

Height and Area
- Four-story building
- Basement to rear of building above grade level
- 220 ft × 180 ft (67 m × 55 m) U-shaped (see drawing)

Complexity and Layout
- U-shaped hallway with center hallway (see drawing)
- Access to basement down stairs
- Stores all face Main St. with direct access to street

Extinguishment

Probability of Extinguishment
- Basement only significant problem

External Exposures
- College of Design on left side (Side B) 15 ft (4.6 m)

Internal Exposures
- Johnny's Restaurant attached to rear (Side C)
- Open stairways and floor to floor via walls/ceilings

Fuel Load
- Average residential
- Stores and bingo hall: average

Calculated Rate of Flow Requirement
- Rate of flow for apartments and stores within the capacity of standard preconnect and backup hose line
- Bingo hall 810 GPM (15 L/sec) required

Number and Size Hose Lines Needed for Extinguishment
- Standard preconnect except bingo hall where 2½-in. (64-mm) hose lines may be required

Additional Hose Lines Needed
- Nothing special (SOP calls for backup hose line and hose line on floor above the fire)

Water Supply
- Hydrant located to front of building and on Main St.
- Additional hydrants on Main St.
- Hydrant flow on Memory Lane 1000 GPM (63 L/sec) per hydrant
- Hydrant flow on Main St. 1500 GPM (95 L/sec) per hydrant
- Water supply on grid system 8-in. (20-cm) main on Memory Lane; 20-in. (51-cm) main on Main St.
- System flow in area is approximately 12,000 GPM (757 L/sec)
- System can be cross-tied to two other supply systems for a total 30,000 GPM (1892 L/sec)

Property Conservation
- Ordinary household items in apartments
- Stores have property of moderate value

General Factors
- First alarm = 3 engines, 2 trucks, 1 heavy rescue, 2 district chiefs, 1 RIT truck
- Second Alarm = 3 engines, 2 trucks, 1 heavy rescue, district chief, advanced life support (ALS) unit, deputy chief
- Third alarm = 3 engines, 2 trucks, all staff, fire chief
- Fourth Alarm = 3 engines, training chief, fire prevention chief

Additional Company and Command Staff Available
- 11 engine companies
- 8 truck companies
- 1 heavy rescue company
- 4 ALS, 6 basic life support (BLS) units
- Mutual aid

Staging/Tactical Reserve
- Suggested staging area: parking lot off Main St. and Taft Road

Utilities: Water, Gas, Electric
- Gas, water, and electric shut-offs in basement utility room just left of Bingo hall

FIGURE 2-12 *(Continued)*

TABLE 2-2 Maltese Cross: Fire Fighter Safety Building Marking System

Top of Cross Construction Type	Right Side Fire Protection	Bottom of Cross Life Hazard	Left Side Content Hazard	Center of Cross Special Designation
FR = Fire Resistive	A = Full Sprinkler	L = Low	L = Low	At discretion of authority having jurisdiction
NC = Noncombustible	P = Partial Sprinkler	M = Moderate	M = Moderate	Hazardous Materials NFPA 704
ORD = Ordinary	S = Standpipe	H = High	H = High	
HT = Heavy Timber	N = None			
C = Combustible				

© Jones & Bartlett Learning.

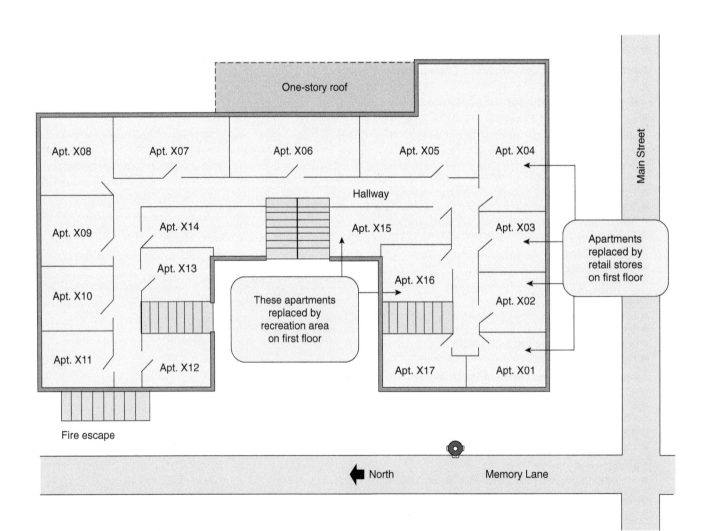

FIGURE 2-13 Sample preincident plan drawing.
© Jones & Bartlett Learning.

An important feature of a good preincident plan is its usefulness in the field. A hard-copy preincident plan consisting of more than two double-spaced printed pages in outline form plus a drawing page tends to be of little value during initial operations. The use of computerized preincident plans allows much more information to be accessed without cluttering the top layer of the drawing and narrative.

Simulating a fire during the preplanning process provides good training in preparing for fires in a building, but rarely should information about a specific fire scenario be part of the preincident plan. SOPs ensure that a company will supply the sprinkler intakes and that the incident management system will be implemented. Repeating SOP items unnecessarily or including too many minor details on drawings or the narrative weakens the preincident plan. The IC does not always care about the location of bathrooms, plumbing, and other details included in architectural drawings. However, there are occasions when this information is useful. For example, to remove water from the upper floors of a building, crews can remove the toilets from the floors, creating a 4-in. (10-cm) drain. If a computerized preincident plan is properly designed, this information could be located on a level accessible by drilling down.

When preparing a preincident plan, one of the most common errors is preassigning companies to respond to specific locations. This is an acceptable practice up to a point. For example, in the absence of visual signs of fire, a company could be assigned to check the alarm system with other companies standing by, or a second truck company could be assigned to the rear of the building. However, it is usually poor practice to preselect a specific entry point or specific tactics or to make company assignments in advance. If the actual fire occurs in a location other than the assumed area, companies will be in the wrong locations, and the preincident plan would then have made the situation worse. For this reason, including specific tactical assignments on the preincident plan should be avoided. Preassigning companies to specific tasks could result in vital tasks being unassigned or totally missed during the heat of battle.

What Structures Are Preplanned?

Just as some SOPs are department-specific, a decision regarding what properties to preincident plan depends on the jurisdiction being protected. Few communities have the resources necessary to preplan all buildings, or even all commercial buildings. The operational priority list (life safety, extinguishment, property conservation) provides direction regarding which buildings may need to have preincident plans. If there is a high life hazard (including fire fighter safety issues), a particularly difficult extinguishment problem, or high-value property, then there is a need to prepare a preincident plan. Properties with high life hazards should be the highest priority. Preincident planning deals more with potential demand (properties with the potential for a large loss of life or property) than it does with realized demand (the actual number of fires in a building or type of building).

Buildings that present an extraordinary challenge in terms of life safety, extinguishment, and salvage should be preplanned. Other buildings that are generally preplanned, whether or not they present a high life-safety, extinguishment, or property conservation problem, include the following:

- Buildings protected by fire protection systems (with the exception of small restaurants with a hood system)
- High-rise buildings
- Complexes with a large number of buildings

Modifying Standard Operating Procedures

SOPs describe a standardized method for addressing predictable operational circumstances. Preplanning addresses what is different or unusual. Preincident plans are building-specific, whereas SOPs are general. If the specific property fits the general category of operations, the SOP is assumed to be in effect; however, there may be special circumstances in which the SOP is not the most effective course of action.

Fallacy	Fact
All SOPs are always in effect.	SOPs are not etched in stone. They can be modified during preplanning or whenever situations dictate a different course of action.

For example, department SOPs may require the first-arriving engine company to secure a source of water. The specific building being preplanned is small and located a considerable distance from the nearest hydrant. Based on information obtained while preplanning the building, a recommendation is made to make an exception to the department water supply procedure. It is determined that the extra time involved in supplying water would result in complete destruction of the structure, and the rate of flow indicates that the apparatus tank has more than sufficient water capacity to conduct an offensive attack.

Because the preplanned tactical consideration is different from the department SOP, it is important that this recommended tactic be approved for this property and then shared with all other responding fire companies and command staff.

Estimating Life-Safety Needs

As mentioned earlier, any building that poses an unusually high risk to fire fighters or occupants must be included in the preplanning process. Occupancies with long-span roofs present a hazard to fire fighters but could also be preplanned due to the large rate of flow required to extinguish a fire. Heavy roof loading or unusual building features that could lead to partial or total collapse should be noted. A notation-type preincident plan may be enough for buildings that have previously been damaged by fire or weather.

Nursing homes, hospitals, places of assembly, schools, churches, and other places holding large numbers of people or a significant number of disabled people should also be preplanned.

Estimating Extinguishment Needs

Compartmentation plays a major role in determining extinguishment needs. Most residential buildings are compartmented into a series of small rooms requiring much less water than in a large undivided warehouse. A building with compartments requiring more than two standard preconnected hose lines, as calculated by using the volume of the fire compartment divided by 100, should be preplanned. (Rate of flow will be covered in more detail in Chapter 8, *Offensive Operations*.) If highly combustible or hazardous materials are present in quantity, such as flammable liquids, explosive or toxic materials, or materials presenting an extraordinary fuel load such as plastics, rubber tires, and idle wood pallets, smaller areas may also require a preincident plan.

The estimated rate of flow should be included on the preincident plan when applicable. Where large areas with heavy fuel loads exist, a decision may be made in advance of the fire regarding defensive or non-attack situations for fires beyond a certain percentage of involvement.

Estimating Property Conservation Needs

Any number of items could be considered high-value contents. Jewelry, electronic equipment, and many other items could justify a preplanning effort. Identifying the location of floor drains is generally beyond the scope of a preincident plan. However, if information of this type would be of value to responding units and the IC, then it should be included in the preincident plan. When using drawing software, it is often possible to include information such as floor drains on a separate layer. The user can then turn on or drill down to the floor drain layer if this information is needed. Some property is more susceptible to damage, and the factors increasing its vulnerability should become part of the preincident plan. Stock that is piled on floors, rather than in racks or on pallets, is more likely to sustain water damage. Determining where the water from sprinkler activation or firefighting streams will go can be important.

For example, a damaged sprinkler pipe in a high-rise building resulted in water flowing into a computer room that processed credit card transactions for the entire east coast of the United States. Immediate action by employees and fire fighters was able to stop the flow of water, which prevented interruption of this vital service.

Relationship of Preplanning to Size-Up

Size-up is a natural extension of the SOP and preplanning process. SOPs get the operation off to a predictable start, whereas preplans provide specific information about the building in advance of the fire. Whenever possible, it is best to gather as much information in advance as possible; therefore, the list of size-up factors is also used as a preincident plan list. Use the size-up list in FIGURE 2-14, and the description of each factor, as a guideline when preplanning.

A brief explanation of what is meant by each of the size-up/preincident plan factors is provided in this chapter. Many of these factors are further explained in other chapters of this text.

Analyzing the Situation Through Size-Up

Size-up factors, which may change from incident to incident, are difficult to categorize in terms of relative importance. Incident conditions determine which size-up factors are most important. Factors related to life safety are most likely to be critical. These most important factors are known as primary factors. The initial size-up analysis is limited to evaluating primary factors. Less important factors are categorized as secondary factors.

Size-Up/Preplan Checklist

Life Safety/Fire Fighter Safety
- ☐ Smoke and fire conditions
  - ☐ Fire location
  - ☐ Direction of travel
- ☐ Ventilation status
- ☐ Occupancy type*
- ☐ Occupant status
  - ☐ Estimated number of occupants*
  - ☐ Evacuation status
  - ☐ Occupant proximity to fire
  - ☐ Awareness of occupants*
  - ☐ Mobility of occupants*
  - ☐ Occupant familiarity with building*
  - ☐ Primary and alternative egress routes*
  - ☐ Medical status of occupants*
- ☐ Operational Status
  - ☐ Adherence to SOPs
  - ☐ Fire zone/perimeter
  - ☐ Accountability*
  - ☐ Rapid intervention
  - ☐ Organization and coordination*
  - ☐ Rescue options*
  - ☐ Staffing needed to conduct primary search
  - ☐ Staffing needed to conduct secondary search
  - ☐ Staffing needed to assist in interior rescue/evacuation
  - ☐ Staffing needed for exterior rescue/evacuation
  - ☐ Apparatus and equipment needed for evacuation
  - ☐ Access to building exterior*
  - ☐ Access to building interior* (forcible entry)
- ☐ Structure
  - ☐ Signs of collapse
  - ☐ Collapse zone*
  - ☐ Construction type*
  - ☐ Roof construction*
  - ☐ Condition*
  - ☐ Live and dead loads*
  - ☐ Water load
  - ☐ Enclosures and fire separations*
  - ☐ Extension probability*
  - ☐ Concealed spaces*
  - ☐ Age*
  - ☐ Height and area*
  - ☐ Complexity and layout*

Extinguishment
- ☐ Probability of extinguishment*
- ☐ Offensive/defensive/non-attack
- ☐ Ventilation status
- ☐ External exposures*
- ☐ Internal exposures*
- ☐ Manual extinguishment
  - ☐ Fuel load*
  - ☐ Calculated rate of flow requirement*
  - ☐ Number and size hose lines needed for extinguishment*
  - ☐ Additional hose lines needed*
  - ☐ Staffing needed for hose lines*
  - ☐ Water supply*
  - ☐ Apparatus pump capacity*
  - ☐ Manual fire suppression system*
- ☐ Automatic fire suppression equipment*

Property Conservation
- ☐ Salvageable property*
- ☐ Location of salvageable property*
- ☐ Water damage
  - ☐ Probability of water damage*
  - ☐ Susceptibility of contents to water damage*
  - ☐ Water pathways to salvageable property*
  - ☐ Water removal methods available*
  - ☐ Water protective methods available*
- ☐ Smoke damage
  - ☐ Ventilation status
  - ☐ Probability of smoke damage*
- ☐ Susceptibility of contents to smoke damage*
- ☐ Damage from forcible entry and ventilation*

General Factors
- ☐ Total staffing available versus staffing needed*
- ☐ Total apparatus available versus apparatus needed*
- ☐ Staging/tactical reserve*
- ☐ Utilities (water, gas, electric)*
- ☐ Special resource needs*
- ☐ Time
  - ☐ Time of day
  - ☐ Day of week
  - ☐ Time of year
  - ☐ Special (e.g., holiday season)
- ☐ Weather
  - ☐ Temperature
  - ☐ Humidity
  - ☐ Precipitation
  - ☐ Winds

* This factor can be at least partially known in advance of the fire through preplanning.

FIGURE 2-14 Size-up/preplan checklist.

Incident conditions will dictate which conditions are primary. For example, weather can be categorized as a critically important (primary) factor when at the extreme and as a less important (secondary) factor during moderate conditions. The fire officer must first sort through information and evaluate available data related to primary factors.

Size-up actually begins before the incident with the development of SOPs and preincident planning. When an alarm occurs, fire officers and the IC consider what is already known about the specific property and about the type of property and area in general. Important factors such as weather conditions, time of the alarm, and day of the week will be known as the response begins. Other incident-related information begins with the dispatcher's information about the location of the fire, method of alarm, and other conditions reported to dispatch. Fire service personnel should maintain awareness about what is going on in their community. Are there any scheduled activities during their shift, such as protests, political activities, or religious holidays? Have there been any recent trends such as frequent arson fires?

In many situations, at least one fire company will arrive before the dispatched chief officer(s). Communications from the scene should add to or confirm dispatch information. Once on the scene, the responding chief officer will add to what is known through personal observation, communications with fire companies and building personnel, and reconnaissance. Once the on-scene information is processed, the IC must quickly evaluate the action that is taking place, fire stage, area of involvement, and condition of the building.

The amount and quality of information will improve with time, and the IC should have time to better evaluate the situation once initial assignments are made. Size-up continues throughout the incident and should continue through the overhaul phase.

Life Safety/Fire Fighter Safety

Smoke and Fire Conditions

Smoke and fire conditions are directly related to occupant survival as well as fire fighter safety, and these are primary factors at a structure fire. Heavy, dark, pressurized smoke and visible fire conditions may necessitate a defensive attack, whereas light smoke with no fire evident indicates a high probability of saving occupants who are still inside the building. It must be remembered that reading smoke and fire is not an exact science, and a hidden or ventilation-limited fire can rapidly increase when it breaks out of containment or when it receives additional air. Research conducted by UL (Kerber, 2010), NIST, and others indicates that most fires in occupied residential buildings are oxygen limited and go through a decay phase where a decrease in temperature and smoke production could result in no visible signs of smoke on the exterior. A ventilation opening, including opening a door to gain access, could create a flow path that increases the oxygen concentration, resulting in a dramatic increase in visible smoke and flames. **Flashover** is a critical indicator at a structure fire. Occupants inside post-flashover compartments (rooms) have a very low probability of survival. Smoke and fire conditions can also provide a warning of an impending **backdraft**.

Experienced fire officers learn to evaluate pressure, smoke characteristics, and other factors in determining the intensity of the fire **FIGURE 2-15**. However, interior reconnaissance is generally the best way to realistically determine fire intensity when an interior attack is possible.

FIGURE 2-15 Smoke and fire conditions.
Courtesy of Captain Nick Morgan, St. Louis Fire Department.

Smoke volume, velocity, density, and color provide clues about the possibility of flashover, location of the fire, and survivability inside the burning building. Smoke and fire conditions will not be known until the time of the fire, but the compartment (room) size and tightness, as well as fuel content, provide prefire indicators of the relative fire and smoke potential.

Fire Location. The fire's location determines the method and direction of attack. Extinguishment is the primary life-safety tactic and an operational priority. In most cases, company officers will be able to determine the location of the fire on arrival. The information received from dispatch, alarm systems, occupants, and visual clues assist in finding the fire location. The first-in crews will typically be attacking or setting up to attack the fire when the chief officer arrives. There are times when the fire location is not obvious (as is shown in Figure 2-15) or when dispatch information is wrong.

When information is incomplete or incorrect, finding the fire can be difficult, especially in a multistory building. High-rise buildings can hide a fairly large interior fire, and smoke alarms may be sounding far away from the actual fire location. A fairly small fire located at a window may appear to be much larger, whereas a large fire toward the core of the building may not be visible from the exterior. Locating the fire and its extension paths is critical to the development of an effective incident action plan.

Open flaming is a definite indicator of fire location, but the fire could have originated at another location. The old adage "where there's smoke, there's fire" is true, but the primary goal here is to find the seat of the fire. Knowing the basic chemistry and physics of fire is helpful in learning to read smoke. All matter, including smoke and heat, naturally try to reach equilibrium. When openings are available, smoke will try to reach the same temperature as the temperature on the other side of the opening. In most cases, smoke will go to the nearest and largest opening unless wind conditions reverse the vent path. Likewise, all gases, including fire gases, are governed by the physical properties of gases, which state that a given volume of gas will expand as the temperature of the gas rises. If the gas is contained under conditions in which expansion is not possible, the heated gas will undergo an increase in pressure. This increase can be calculated using the ideal gas law; however, the fire ground is no place for precise calculations. It is enough to understand that smoke rapidly flowing from an opening is a clue that the fire may be nearby **FIGURE 2-16**. Fires develop flow paths that are based on air providing the oxygen necessary to sustain combustion flowing into

FIGURE 2-16 Building on fire that some would evaluate as an offensive attack; however, there are also indications that a defensive attack is correct.
Courtesy of Bernard Erwin.

the area and hot fire gases, smoke, and fire flowing out. Flow paths can be controlled by applying water and/or ventilation. It must be remembered that ventilation can occur either by fire fighters intentionally making ventilation openings or by the fire causing unintentional and uncontrolled venting.

Direction of Travel. Knowing where the fire is most likely to spread is important to life safety and extinguishment tactics. Search teams with hose lines need to protect and evacuate occupants in areas where the fire is likely to extend. Basic chemistry and physics of fire determine the direction of fire travel. The products of combustion travel upward until reaching a barrier. They will then travel horizontally to fill the top of the compartment and, if contained, will travel downward. Fire, heat, and smoke will therefore travel upward via the path of least resistance, before traveling horizontally or downward. Fire will most likely travel up stairways and other available openings to the floors above the fire. The upward path could also be through concealed wall areas (balloon-frame construction is particularly noted for this).

Fire and smoke entering a wall will travel to the attic unless it meets a solid barrier. Nearly all buildings have vertical paths to the top of the building, although some are much better protected than others. Fire in an open wall, such as in balloon-frame construction, will travel unimpeded to the attic. In platform-frame construction and other types of construction with fire barriers, the path may not be as direct, but smaller pathways via utility openings for drains, water pipes, electrical wiring, and others are generally present.

Although it is not possible to know the exact means of fire travel during preincident planning, it is possible to know the probability of fire travel. Rated and nonrated fire compartments tend to limit fire spread. Construction methods and alterations that affect fire spread are factors that can be known in advance.

Ventilation Status

Ventilation, whether intentional or unintentional, may change the direction of smoke and fire travel. Opening a door creates an air intake opening, supplying oxygen to the fire and a low-pressure opening allowing smoke and fire gases to escape. Water application or other conditions also influence ventilation. For example, wind-driven fires have been known to reverse venting and supply a superabundance of oxygen. Vent openings generally increase the available oxygen and often result in a rapid increase in fire intensity and size. Improper venting can spread and accelerate fire spread, change the ventilation flow path, and precipitate a backdraft or flashover, all of which increase the danger to occupants and fire fighters.

Notice that ventilation status is listed under all three operational priorities (life safety, extinguishment, and property conservation). Ventilation is a key factor during all phases of the operation. Venting for life safety involves moving the fire away from occupants and fire fighters. Understanding smoke and fire movement, as discussed previously under other factors, is essential to good venting. As mentioned earlier, the fire will follow the path of least resistance—upward, then horizontally, and finally downward. The ventilation opening should direct the flow path and fire away from occupants and fire fighters.

A common life-safety venting tactic, included in the SOPs for many departments, is to remove a scuttle or other cover over a built-in opening at the top of a stairway in a multistory building. The intended purpose of this tactic is to clear the stairway of smoke. However, this tactic can result in the unintended consequence of creating a flow path that extends fire, heat, and smoke into the stairway being vented. **FIGURE 2-17**, **FIGURE 2-18**, and **FIGURE 2-19** illustrate this unintended consequence. In this scenario, the first-arriving engine company is advancing a hose line to the fire floor, while the first truck company is split into two crews. One truck crew is gaining access to the roof, and the other is in the stairway making entry

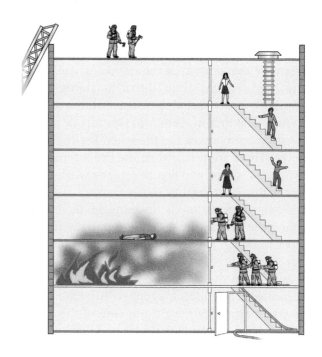

FIGURE 2-17 Preparing to vent via scuttle.
© Jones & Bartlett Learning.

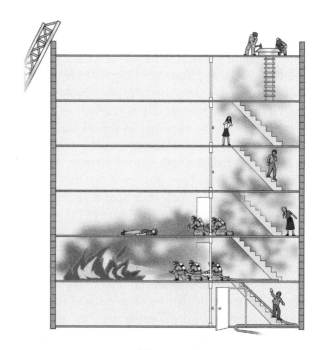

FIGURE 2-18 Unvented fire attack.
© Jones & Bartlett Learning.

FIGURE 2-19 Fire vented at top of stairway, changing direction of travel. Fire is entering the stairway at the second floor, causing heavy smoke above.
© Jones & Bartlett Learning.

FIGURE 2-20 Fixed roof vents.
Courtesy of Ben Klaene.

into the floor above the fire to check for extension and begin the primary search as well as facilitating evacuation (Figure 2-17).

When the engine company opens the door to attack the fire, an unvented fire will be released from confinement and will enter the stairway (Figure 2-18). This will tend to make smoke conditions in the stairway worse, even if there is no vent opening at the top of the stairway. If an opening is made at the top of the stairway, fire and smoke could be pulled into this "chimney-like" flow path of least resistance, making conditions much worse in the stairway (Figure 2-19). Opening the scuttle at the top of the stairs can be a good tactic if the fire is controlled but can cause serious injury or death if done under the wrong circumstances. Partially closing the door at the base of the stairs could reduce the unintended consequences of venting the stairs via the roof scuttle.

Ventilation possibilities can be known through preincident planning. Roof openings such as scuttles and vents may be present. The roof vents shown in **FIGURE 2-20** could be opened to vent the area below.

Areas where aerial apparatus can be positioned to reach the roof can be known in advance. Department SOPs may determine what types of roofs are to be vented, as well as how and when positive-pressure ventilation is to be implemented. When venting an area, it is essential that hose lines be in place to attack the fire.

Depending on the fire scenario and staffing availability, it may be advisable to "soften the target" by applying water prior to making entry. Earlier laboratory research conducted by UL was verified in cooperation with NIST and the New York City Fire Department (FDNY) on Governors Island, New York, during a series of live burns in residential buildings. This research further verified that limiting air intake and applying water to soften the target prior to making entry reduced temperatures at nearly every level of every room (UL Firefighter Safety Research Institute, 2012).

Occupancy Type

The building's use will determine how likely the building is to be occupied at the time of the fire, number of occupants, fuel load, fuel type, value of the contents, and many other essential facts. Major occupancies can and should be identified in advance of the fire through preincident planning. Smaller occupancies can often be readily identified (e.g., a single-family detached residence). However, some buildings are used for purposes other than their original intended use (e.g., single-family residence being used as a real estate office). See Chapter 11, *The Role of Occupancy*, for a complete discussion of various occupancies and tactics related to occupancy.

> **NOTE**
>
> Occupancy: Building's primary use (e.g., school, hospital, nursing home)
> Occupant: Person who could be in the building
> Occupied: Presence of occupant(s) in the building

Estimated Number of Occupants. Estimating the number of occupants in a large building is difficult at best. However, the building's intended use (occupancy) provides evidence of relative occupant density. NFPA 101, *Life Safety Code* (NFPA 101, 2018) establishes the maximum occupant load per square foot by occupancy type (e.g., one person per 100 ft^2 [9 m^2] in an office building [business]). In places of public assembly, the maximum number of occupants should be posted. In hotels and apartment buildings, the number of available rooms/apartments provides an indicator of maximum occupant load. All of these details provide a starting point for estimating the number of people that could be in the building. Time factors, such as hours of operation, combined with the maximum occupant load and evacuation status help determine the number of people endangered by fire and the possible need for some level of assistance. Part of knowing your response district is being familiar with the location of local bars, restaurants, and other places that are most likely to be overcrowded. Some departments conduct special "life-safety" inspections in places of assembly during times when overcrowding is most likely to occur (e.g., a bar is most likely to be overcrowded on a weekend night).

Evacuation Status

Size-up of evacuation status considers the following factors:

- Occupant proximity to fire
- Awareness of occupants
- Mobility of occupants
- Occupant familiarity with building
- Primary and alternative egress routes
- Medical status of occupants

Estimating the number of people still inside the building is the next logical step after estimating how many people may have occupied the building upon discovery of the fire. Some occupancies (e.g., primary and secondary schools) are required to routinely conduct fire drills and account for occupants when a fire alarm sounds. Recently, this requirement has received scrutiny due to the threat of an active shooter. As a measure of precaution, schools should vary the times and days of the week when drills are conducted.

The NFPA recommends that each home have an evacuation plan following the Exit Drills in the Home (EDITH) educational program. Many times, family members account for other family members; unfortunately this is not always the case, and information from occupants is not always accurate. Most buildings do not have an occupant accountability system.

Consider a fire in a large apartment building. Some of the residents probably would not be home when the fire occurred, others would evacuate and leave the scene, and some occupants may have guests. For all these reasons and more, the only way to be sure the building has been evacuated is to conduct a primary search, and even then there is a need to verify the search with a secondary search. Information from occupants who have escaped would be less reliable in a larger building without a formal accountability system than in a school or single-family home.

If a preplanned building has an occupant accountability system, note the specifics of the plan on the preincident plan. Of particular interest is who will report to the fire department and where this person will be located (including an alternate site). Also note where occupants will be assembled. During preincident planning, be sure that occupant assembly areas do not conflict with expected fire department response routes or on-scene operations.

Occupant Proximity to Fire. Obviously, being close to a well-involved fire places occupants in imminent danger **FIGURE 2-21**. Chapter 6, *Life Safety*, explains the life safety priorities for occupants on various levels of a multistory building. Knowing the location of occupied areas through preincident planning can be very useful during the size-up process.

Awareness of Occupants. People who are awake and alert are more likely to hear an alarm or sense the products of combustion and take action to evacuate the building. Occupants who are asleep or mentally incapacitated may not be aware of the fire. Awareness is related to occupancy and will be further explained in Chapter 11, *The Role of Occupancy*.

Identifying the occupancy type during preincident planning provides an indication of occupant awareness (e.g., a nursing home fire at night where occupants are likely to be sleeping and possibly under the influence of medications).

Mobility of Occupants. Most large buildings will contain people who are unable to fully evacuate on their own. As with other factors, knowing the occupancy of the building can be helpful in determining if occupants are likely to successfully escape on their own.

Occupant Familiarity with Building. In most cases, the people inside a place of public assembly are not familiar with the building layout or alternate exit facilities. Many large loss-of-life fire reports

FIGURE 2-21 Fire fighters outside rear of building with an occupant hanging out the third floor window.
Courtesy of Bill Strite, Cincinnati, Ohio.

in places of assembly address unfamiliarity as a major problem. People who live or work in a building would be expected to know the building layout and location of exits. However, people will be most familiar with their regular ingress path. For example, a person working on an upper floor of a high-rise office building likely enters the lobby and takes an elevator to their office every workday. Unless the building management regularly conducts fire drills, many of the employees on upper floors will not use the stairway or even know its location.

Primary and Alternative Egress Routes. This factor is related to the mobility factor inasmuch as most occupants will escape unassisted if there is sufficient egress. Codes specify egress facilities, and preplan drawings should show the location of all exits. Some occupancies, such as hospitals, designate a defend-in-place strategy for occupants.

Many public buildings have an **area of refuge** where immobile occupants wait for assistance. It is essential that preincident plans address these special facilities. Fire fighters must know the location, how many occupants could be at the area of refuge, and what the occupant and building managers expect to happen during an emergency. Often, fellow occupants or security personnel are designated to assist people in these locations. The fire department must always check areas of refuge during a fire emergency.

Medical Status of Occupants. This factor requires a twofold analysis: (1) What is the normal medical status of people occupying the building, and (2) what effect is the fire having on building occupants? Whenever people are still in the building upon arrival, EMS units need to be summoned to the scene. A medical branch is needed if there are large numbers of potential victims.

Operational Status

Most of the operational status factors apply to the scenario in which an IC arrives after the first-due companies have begun operations, and the IC assumes command of an operation that is already in progress.

SOPs provide a standard method of going to work at a fire scene. This allows the operation to get started, following a predictable plan. However, there are situations that require a different approach. The IC must continually evaluate the safety and effectiveness of the standard operation and determine whether a nonstandard attack would be safer and/or more effective **FIGURE 2-22**.

Safety should be the IC's most important consideration. If the building is in danger of early collapse or the situation necessitates a transition from an offensive to a defensive operation for other reasons, the IC must order all personnel out of the building.

The effectiveness of the operation must constantly be evaluated. The IC must consider the following questions:

- Is the search being conducted in a systematic manner?
- Are the occupants who are in the most danger being rescued?
- Has the fire been properly vented to control fire spread?
- Is progress being made in controlling the fire?
- Are rate of flow requirements being met?
- Has salvage been considered?

It is critical that the IC continually reevaluate the situation in terms of risk management. The question that should regularly be asked is whether the desired benefit is worth placing fire fighters at risk.

Adherence to SOPs. This factor is evaluated when command is transferred. Are on-scene units following

FIGURE 2-22 Fire showing at roof with aerial ladder extended to roof.
Courtesy of David J. Jones, Cincinnati, Ohio.

SOPs? Is the operation progressing as expected? Safety is a major part of this analysis. Is the operation being conducted safely? Are SCBAs and turnouts being used as prescribed? Is there a need to change from an offensive to a defensive strategy? This is one of the few size-up factors about which little can be known in advance of the fire through preincident planning. However, SOPs that are developed in preparation for fire emergencies provide an advance indicator.

Hazard Control Zones and Perimeters. Fire zones and perimeters will be discussed in Chapter 5, *Fire Fighter Safety*. Generally, the fire zone refers to an area where a specified level of protective clothing is required or possibly a safe area where no protective clothing is needed. The fire perimeter is set up to keep nonresponse people out of the area. The IC who is assuming command should evaluate the present zones/perimeter to determine if they are sufficient. General rules for fire zones/perimeters, including zones established during an active shooter incident, should be part of department SOPs. In addition, preincident plans for large, secured properties may designate specific perimeters.

Accountability. The primary fire fighter accountability system is utilizing the incident management system or NIMS. A properly organized incident provides the first level of fire fighter accountability. NIMS is more than a means of organizing an operation; it is also a safety system. It is important that freelancing be eliminated and that units working in the hazard area work as a unit (usually a fire company at a structure fire) within the incident command system. A formal accountability system is required by NFPA 1500 (NFPA 1500, 2018). The IC must ensure that the local system is being used. Chapter 5, *Fire Fighter Safety*, includes a detailed description of accountability systems.

> **NOTE**
>
> The fire department accountability system accounts for all fire fighters and their location at the incident scene.
> It does not account for building occupants.

Rapid Intervention. NFPA 1500, NFPA 1710, and NFPA 1720 require a **rapid intervention crew (RIC)**. There are many other terms for RIC, such as rapid intervention team (RIT), rapid assistance team (RAT), and firefighter assist and search team (FAST). A team of at least two fire fighters must be immediately available to rescue fellow fire fighters who need assistance. Although not required by regulations and standards, good practice dictates that at least one officer and three fire fighters be assigned to the RIC as soon as possible. Chapter 5, *Fire Fighter Safety*, also includes a detailed description of RICs.

Organization and Coordination. NIMS, as described in Chapter 1, *Organizing, Coordinating, and Commanding Emergency Incidents*, is the acceptable method of organizing an incident. All units on the scene must be included in the organizational structure. Activities of units must also be coordinated. The first-arriving company begins this process by organizing the initial attack and communicating orders to incoming units. When command is transferred, the person assuming command must be sure all units are working toward common tactical objectives within the overall strategy.

Rescue Options. The IC first evaluates the ways occupants can be removed from the building and then selects the safest and most efficient option. The preincident plan can identify the primary means of egress and locations of the preferred evacuation pathways. It should also address alternatives such as fire escapes and ladder positions. See Chapter 6, *Life Safety*, for a more detailed discussion of rescue options.

Estimate the search and rescue personnel needed to conduct each of the following segments of a search and rescue operation **FIGURE 2-23**. Conduct the following segments:

- Primary search
- Secondary search
- Building exterior check
- Occupant rescue

Staffing Needed to Conduct the Primary Search. The size of the area to be searched, smoke conditions, rescue methods available, and the condition of the occupants determine how many crews should be assigned to the primary search. In a single-family dwelling, a single company or possibly one company per floor conducts the primary search. In a large area with obstacles and poor visibility, several companies may be required to complete the primary search of a single floor. If victims must be physically removed, there will be a need for additional staffing.

FIGURE 2-23 Fire showing at roof with smoke at second floor and fire fighters at front of building.
Courtesy of David J. Jones, Cincinanti, Ohio.

Staffing Needed to Conduct the Secondary Search. Often the same teams that perform the primary search will conduct a secondary search in an area other than the one they originally searched. It is recommended that a different company perform the secondary search, as they may look in areas that the first team overlooked. In most cases, the secondary search will not require increased staffing.

Staffing Needed for Interior Rescue/Evacuation. This task could involve the same staffing required for the primary search, or additional teams may be assigned to remove victims who are found during the primary search. In many cases, occupants attempt to evacuate on their own; staffing may be needed to assist and direct them to safety.

Staffing Needed for Exterior Rescue/Evacuation. If ladder or other exterior rescue methods are warranted, additional staffing will be needed. Most exterior rescues require more staffing per rescue than removal via the interior stairs.

Apparatus and Equipment. Needed for Exterior Rescue/Evacuation. In most situations, an aerial apparatus or ground ladder will be needed for an exterior rescue. Other rescue equipment is rarely used (e.g., rope rescue equipment).

Access to the Building's Exterior. When preincident planning a building, it is important to note apparatus entry points for secured properties. Special arrangements are often made with building security for these properties. Street or road access around the structure should be included on the preincident plan drawing. Also note aerial access points and roadways that are not safe or not accessible to fire apparatus. When buildings are set back a considerable distance from the public roadway, fire companies may need to position apparatus on a private driveway to carry out rescues via aerials or to be within range of their attack hose.

Access to the Building's Interior (Forcible Entry). In response to the September 11, 2001, attacks and the increase in active shooter incidents, there has been an increase in security on the exterior and interior of buildings. Buildings that were formerly open to the public are now blockaded. School lockdowns often place students and staff in concealed, secure locations. Be aware of changes to buildings and security measures in your jurisdiction. The need for forcible entry can significantly delay search and rescue operations as well as the initial fire attack.

Note the times when the building is secured or when special security features such as internal locks and gates impede entry. While performing preincident plan surveys, check the accessibility and security of doors while the building is open. Fire codes state that no special knowledge or key should be required to exit through a required exit.

Structure

A thorough risk-versus-benefit analysis determines whether the operation will be offensive or defensive. A major risk-versus-benefit consideration involves evaluating structural conditions. If the structure is in imminent danger of collapse, no expected benefit is worth the risk posed by entry into the structure. If the building is being damaged but adequate time remains before a potential collapse, an offensive attack posture to save lives could be warranted. A decision may be made to contain the fire and rescue occupants, with a switch to a defensive attack once the primary search and rescue are complete. The effect of the initial operations should be reevaluated once occupants are removed from the structure. The necessary risk-versus-benefit analysis is somewhat subjective. However, experience and training can greatly improve accuracy.

Signs of Collapse

Collapse can be catastrophic (failure of a supporting structure with complete or nearly complete destruction of the building). However, many partial collapses can injure and kill, such as a floor collapse where fire fighters are struck by the floor/ceiling assembly above

---

Incident Summary

Hamlet, North Carolina, Chicken Processing Plant

On September 3, 1991, a fire claimed the lives of 25 employees at a chicken processing plant in Hamlet, North Carolina. Locked exits were reported as the reason workers could not exit. These same locked exits could prevent fire fighters from exiting the building or delay the rapid intervention crew in gaining access.

U.S. Fire Administration/Technical Report Series. Chicken Processing Plant Fires: Hamlet, North Carolina and North Little Rock, Arkansas. USFA-TR-057/June/September 1991. https://www.usfa.fema.gov/downloads/pdf/publications/tr-057.pdf. Accessed May 6, 2020.

or fall into a burning lower level. See Chapter 5, *Fire Fighter Safety*, for a list of possible collapse cues. Always be aware that many collapses occur with no perceptible warning, especially in buildings constructed using lightweight materials such as trusses.

Collapse Zone

A collapse zone encompassing the height of the structure plus an allowance for debris scatter is recommended (see Chapter 5, *Fire Fighter Safety*). In most cases, this translates to a collapse zone that is one and a half times the height of the building. It is difficult to estimate the height of the building looking up from the street level. It is much easier to measure or estimate building height during preincident planning. Other pragmatic decisions can be made during the preplanning tour, such as determining whether the width of the street is less than the predicted collapse zone. In such a case, a decision could be made to avoid placing apparatus in the street in front of the building when there is a collapse hazard.

All fire department personnel should know the strengths and limitations of tools and equipment used on the fire ground. Fire streams, including master streams, have limited flow and reach. Knowing the potential collapse zone distance is important when the use of fire streams, including master streams, is being considered.

Determining a collapse zone provides a good example of how the same size-up factor can be primary or secondary. When confronted with a well-involved fire in a noncombustible structure, determining the collapse zone is a primary factor **FIGURE 2-24**. In contrast, it would be a secondary factor for a minor fire in a fire-resistive building.

Construction Type

It is highly recommend that all members of fire departments become familiar with building types and common building problems. Few fire fighters would be able to conduct an engineering analysis to determine structural stability, but they can recognize potential structural problems during preincident planning and by studying past fires in different types of construction. *Brannigan's Building Construction for the Fire Service* (Corbett and Brannigan, 2021) is a "must read" for anyone who might take command at a structure fire, including the first-arriving officer.

During preincident planning, the building should be classified by construction type:

- Type I construction: Fire-resistive
- Type II construction: Noncombustible

FIGURE 2-24 Building in danger of imminent collapse.
Courtesy of Bill Strite, Cincinnati, Ohio.

- Type III construction: Ordinary
- Type IV construction: Heavy timber
- Type V construction: Wood frame

See NFPA 220, *Standard on Types of Building Construction* (NFPA 220, 2018), for specific fire-resistive requirements for each type of construction. Some buildings defy classification, particularly buildings that have been renovated. The original construction might be classified as fire-resistive construction, but removal of protective coatings on steel structural members could reduce the building to noncombustible status. NFPA 5000, *Building Construction and Safety Code* (NFPA 5000, 2018), would classify a building so modified as the weakest form of construction (noncombustible in this example). There can also be different types of construction in various areas of a building. An older section of a building may be constructed using heavy wooden beams while a newer section may have wood or steel truss construction.

Code requirements change. Typically, the code enforced at the time of construction is the code that will apply to a building. For example, in some jurisdictions, sprinklers were not required in high-rise buildings until 1972. Many older buildings remain unsprinklered, although many cities now require high-rise buildings to be retrofitted with sprinklers, regardless of when they were built.

The fire-resistive (Type I) building, as the name indicates, is superior to all other building types in regard to structural stability under fire conditions. Some might argue whether wood frame or noncombustible is most likely to collapse. Wood frame is generally of limited size, whereas noncombustible construction is used on large buildings. Most large noncombustible buildings are sprinkler protected. Chapter 5, *Fire Fighter Safety*, discusses the relative safety of the five major types of construction.

Roof Construction

The roof covering can be important, particularly if it is combustible, such as wood shingle. However, the structure that supports the roof is generally what is most important to fire fighters in this era when nearly all roof coverings are noncombustible. Roof collapse is a killing mechanism in its own right, but also a precursor to catastrophic collapse. The roof tends to tie in the walls of the building. When walls damaged by fire lose roof support, they often fail. Of particular concern are truss and other lightweight engineered supporting structures. Most modern buildings with large open areas will have a lightweight engineered roof support system. These long spans are particularly dangerous.

The roof structure should be identified and noted during preincident planning. If a truss roof is present, especially a long-span truss roof, it should be highlighted on the preplan narrative and a symbol added to the drawing indicating the truss roof. Truss roofs have been responsible for many fire fighter fatalities. The truss roof collapse mechanism and safety issues will be further discussed in Chapter 5, *Fire Fighter Safety*.

Condition

A building in poor repair or one that has been previously damaged presents an extra hazard for fire fighters. Abandoned buildings often have structural damage caused by weather, previous fires, or unauthorized occupants. Even if the building does not warrant formal preincident planning, a notation should warn responding fire fighters of structural damage or hazards such as holes in the floor. Some departments place a placard directly on the building to indicate structural damage or other structural problems. Even when a warning placard is affixed to the building, a backup notation system is recommended. The fire officer must consider serious prefire damage as a critical factor when deciding whether to take an offensive or a defensive approach **FIGURE 2-25**.

FIGURE 2-25 Vacant/dilapidated building with fire visible on top floor and roof.
Courtesy of Davis Mullis.

Live and Dead Loads

The **dead load** is the load imposed on structural members by the building and permanent attachments. The type of construction will be a major factor in determining the overall dead load. Of special consideration are heavy roof loads such as roof-mounted equipment, particularly if the roof is supported by unprotected truss construction. Heavy roof loads should be noted on the preincident plan.

Live loads are those loads produced by the building contents. The live load will vary from very light to extra hazard. Warehouse buildings tend to carry heavy live loads, with a possible added impact load due to materials-handling equipment. If the materials being stored are combustible, they also add to the fuel load. Extraordinary live loads should be noted on preincident plans. Knowing a building's live load is important when developing incident-specific tactics.

Fire Suppression Water Load

The obvious water load is the weight of the water that fire fighters discharge into the building during fire operations. Each gallon of water weighs 8.33 pounds (3.8 kg). Therefore, a 1000-GPM (63-L/sec) master stream operating into the building is adding 8330 pounds (3778 kg) to the building for each minute of operation. Given the fact that defensive operations often require several master streams operating over a prolonged period of time, it is easy to see how this water load could affect building integrity. Fortunately, most buildings will hold only a small percentage of the water from these master streams, as the water will run down vertical openings to the basement or outside. It is difficult, if not impossible, to accurately determine how much water is being retained in any area of the building. Here again, preplan information can be invaluable. If stored materials are stacked directly on the floor and are absorbent, a large percentage of water will be retained. In contrast, storage on skids tends to allow water to run off. NFPA 1, *34.4.1.3*, addresses the hazard created by the weight of water (NFPA 1, 2018):

> **34.4.1.3** Where storing water-absorbent commodities, normal floor loads shall be reduced to take into account the added weight of water that can be absorbed during firefighting operations.

Water-absorbent materials are sometimes stacked against, or near, walls. As these materials absorb water from sprinklers or fire streams, they expand and push on the walls, sometimes resulting in an unexpected collapse. NFPA 1, requires that water-absorbent materials be stored no closer than 24 in. (0.6 m) from a wall (NFPA 1, 2018). Other factors can also affect runoff. When a building is severely damaged, structural members, wall materials, and contents can fall on the floor, forming dams that retard water runoff. When pumping large quantities of water into a building, observe the runoff and consider the possibility that the weight of the water could facilitate structural failure.

Enclosures and Fire Separations

The type of construction and occupancy, both of which can be known in advance, will be major clues in determining extension probability. Fire-resistive construction contains fire-rated compartments. However, the size of the compartments will vary by occupancy. A business occupancy such as the World Trade Center may have large, open office spaces, which will allow the fire to freely spread within the compartment. Residential high-rise buildings may have a few large common areas but will mainly consist of small rooms, which will impede fire extension.

Noncombustible (Type II) construction may also have large open areas, but the fire-resistive qualities between areas will generally be inferior to fire-resistive barriers.

The balloon-frame building is noted for a lack of compartmentation, having no rated fire assemblies to prevent fire spread. However, even nonrated doors and walls will retard fire spread for a period of time.

In a multistory building, the floor/ceiling assembly provides a barrier between floors. This barrier will be more effective in fire-resistive construction than in other types of construction. Floor/ceiling assemblies are seldom perfect barriers due to penetrations made for wiring, plumbing, air movement equipment, and other building services.

Extension Probability

Extension probability is directly related to the enclosures and fire separation factor discussed previously and the concealed space factor. Fire spreads through concealed spaces and from area to area when enclosures are missing or weak. During size-up, the fire officer must consider how the fire spread will endanger fire fighters and occupants.

Concealed Spaces

Most buildings (except Type IV heavy-timber construction) contain multiple concealed spaces. Of special note are attics, particularly when several buildings or units share a common attic. Fire separations are often damaged or removed in buildings that were originally constructed with attic fire separations.

Sometimes the only separation between the attic and useable space below is a suspended ceiling. Many ICs have been surprised by a fire that breaks through the roof or extends to a distant apartment or store. If the concealed space involved in fire contains a truss roof or floor assembly, expect rapid collapse. Immediately communicate this finding to the IC.

In multistory buildings, suspended ceilings create a false space between the actual floor or roof assembly above and the ceiling tiles below. Flow paths created in this false space can result in fire cutting off fire fighter egress. As fire fighters make entry, they should use a thermal imaging camera (TIC) to determine whether the fire has penetrated the ceiling and made its way into the false ceiling area. Under certain conditions, TICs are inaccurate; therefore, it is sometimes necessary to push up ceiling tiles to see if fire has extended into this space.

> **NOTE**
>
> The thermal imaging camera is not actually a camera; therefore, NFPA documents, such as NFPA 1801, *Standard on Thermal Imagers for the Fire Service*, use the term *thermal imager* (NFPA 1801, 2018). However, because it is in widespread use within the fire service, this text uses the term *thermal imaging camera*, or its abbreviation, TIC.

Fire fighters must use extreme caution when gaining access to this false space. Opening the ceiling could provide air to a ventilation-limited fire. A hose stream should always be immediately available before lifting ceiling tiles. The time to discover concealed spaces and truss roofs/floors is during preincident planning. Smoke conditions may not allow an examination during the fire.

Age

The age of a building can have positive and negative effects. Most older buildings will have heavier, more fire-resistive construction. However, buildings that have been renovated may contain new lightweight construction in the renovated areas. Not everyone who remodels secures a building permit, but renovations can be discovered as you tour your response area. It is more likely that a building permit will be secured when larger commercial buildings are remodeled.

The fire department should be part of the permit approval process, or at the very least be notified when such construction is taking place. Renovated buildings using lighter construction can be an extra hazard because fire fighters expect heavy construction methods in older buildings. The negative side of the building's age is that the building may be getting weaker as it ages. This depends on the construction materials and building maintenance. The year the building was constructed should be noted on the preincident plan. In many counties, the auditor maintains a website that lists all properties in the county, including the date the building was constructed. Other useful information may also be available at the auditor's site, such as drawings or aerial photographs of properties.

Height and Area

It is impossible to have a large fire in a small building; thus the building's size will at least partially dictate the total volume of fire. Rate of flow formulas are based on the size of the fire compartment. Rate of flow is thoroughly discussed in Chapter 8, *Offensive Operations*. The location and size of large undivided areas of buildings should be noted on preincident plans.

The height of the building determines the effectiveness of ground-based fire apparatus. When preincident planning a multistory building, determine the maximum reach of available aerial apparatus in terms of floors (e.g., aerial will reach the sixth floor). There may be areas around the building where aerial apparatus cannot be used due to obstructions or lack of access. Aerial access points should be identified in advance of the fire.

The height of the building will also affect the number of possible occupants, the collapse zone, the type of construction, fuel load, and other factors. The height of the building should be included in the preincident plan. All high-rise buildings should be preplanned.

Complexity and Layout

The preincident plan should include a general floor layout for the building along with any information that might affect firefighting operations or fire fighter safety. The layout of some buildings is mazelike, particularly large buildings with multiple additions. **FIGURE 2-26** shows a large medical complex that has undergone multiple additions and demolitions. When visiting a large, complex building, take a moment to orient yourself. Try to determine your location in relation to the stairway, street, standpipe, and other critical landmarks. Could you find your way out under heavy smoke conditions? Storage occupancies may have aisles and cross-aisles that could be confusing under low-visibility conditions. Large open areas in business occupancies may have cubicles separating work areas. The layout of these cubicles can be extremely confusing and difficult to search under fire conditions.

FIGURE 2-26 Large, complex building.
Courtesy of Ben Klaene.

Lightweight structural members and energy-efficient windows are considered green. The problems associated with lightweight construction under fire conditions are well known. Energy-efficient windows delay self-venting, which can be an advantage or disadvantage, depending on circumstances.

Extinguishment

Probability of Extinguishment

The probability of extinguishment is also important to life safety. When the fire is extinguished, the probability of death and injury are greatly diminished. Extinguishment is nearly always a primary factor at a working structure fire. Rate of flow is discussed later in this chapter, as well as in Chapter 8, *Offensive Operations*. Flow requirements and the extent of fire are key factors in determining whether a rapid, life-saving extinguishment operation is possible.

Offensive/Defensive/Nonattack

Deciding the overall attack strategy is critical. The entire operation hinges on this early decision. Whenever command is transferred, the person assuming command must consider whether the present strategy is correct. Critical factors change as tactical objectives are achieved. The most important objectives are those related to life safety. Once the building is cleared of occupants, the reason for continuing an offensive attack is reduced; therefore, a new risk-versus-benefit analysis should be conducted to determine if an offensive attack should continue.

Ventilation Status

Finding the fire location is much easier when the fire has self-vented in a building that is still safe to enter. However, a vented fire will generally progress toward full involvement much faster than an oxygen starved fire (ventilation-controlled fire). If the fire is unvented, it is generally best to place hose streams in an attack position prior to venting.

Venting for extinguishment involves making openings to create a flow path that directs heat and smoke away from the hose crew to allow a quick and efficient

Green Buildings

The increased emphasis on energy conservation has resulted in an increase in "green energy technology." Many of the "green" concepts, such as solar panels, have been in use for decades. Solar panels have been placed on roofs for many years, but their use is now more commonplace. Standards regulating the installation of solar panels are evolving. Many existing solar panels do not meet current standards. Green roofs designed to grow grass and other vegetation on the roof are a fairly new phenomenon. Both solar panels and green roofs add to the roof load and hinder vertical ventilation. Solar panels have proven difficult to deenergize (Fire Protection Research Foundation, 2014), and the wiring system is complex. Solar panels collect energy in the form of direct current electricity, which is then transformed into alternating current using an inverter. Disrupting power to the building at the inverter panel may not affect the energy level being emitted from the roof panels, which is usually 600 volts.

Fire departments should survey their response area to determine the location of green roofs, massive arrays of solar panels, roof-mounted wind turbines, and other green construction. If such structures are found, contact the building department and request an analysis of the structural stability of the roof support system. The preincident plan should address whether roof operations should be conducted along with precautions to be taken.

fire attack. Part of the preincident planning process involves identifying available built-in vents and evaluating the roof structure for roof operations.

External Exposures

Are nearby structures (not connected), vehicles, and other property threatened by the fire? Protecting exposures can be done in several ways. These are discussed further in Chapter 8, *Offensive Operations*, and Chapter 9, *Defensive Operations*. The preincident plan drawing should show the location of nearby exposures. The narrative part of the preplan should identify the occupancy and any special considerations regarding exposures. Many times, large liquefied petroleum gas tanks or tanks containing various hazardous materials are located at the rear of a building. A liquid oxygen tank can be seen in Figure 2-26.

Check all sides of the building when noting the location of exposures.

Internal Exposures

Other parts of the fire building or connected structures are categorized as internal exposures. Identifying fire pathways is best done prior to the fire. At the time of the fire, the IC evaluates whether the fire is likely to follow one of the preidentified pathways. Tactics for protecting internal exposures are also discussed in the Chapter 8, *Offensive Operations*.

Manual Extinguishment

If there is no automatic fire suppression system or the system does not control the fire, manual extinguishment will be necessary. Chapter 8, *Offensive Operations*, and Chapter 9, *Defensive Operations*, both address this topic.

Fuel Load. Fuel load varies as to quantity, type of fuel, geometric orientation, and other factors. Fuels identified as "extra hazard" in sprinkler calculations would generally require more water to extinguish than fires in ordinary combustibles. While fuel loads tend to change, in some buildings, the maximum and average fuel load can be evaluated during preincident planning. The UL study, "Impact of Ventilation on Fire Behavior in Legacy and Contemporary Residential Construction" (Kerber, 2010), indicates that residential fuel loads found today are higher and burn faster than in the past.

Calculated Rate of Flow. The big question in calculating rate of flow is whether the standard attack hose line with a backup hose line can extinguish expected fires. It is impractical to calculate rate of flow at the time of the fire. If the fire compartment is larger than two standard preconnects can extinguish, or if the fuel load will create a fire situation beyond the control of two standard attack hose lines, the rate of flow should be precalculated for these areas and noted in the preincident plan. Chapter 8, *Offensive Operations*, explains the formulas used to calculate rate of flow.

Number of Hose Lines Needed for Extinguishment. The number of hose lines needed for extinguishment is an extension of rate of flow. Once the rate of flow and the flow provided by department hose lines are known, the required number of hose lines needed to extinguish an advanced fire can be estimated.

Additional Hose Lines/Master Streams Needed. Extinguishment is the primary purpose for using handheld hose lines and master streams, but there is also a need to cover internal and external exposures, provide a backup hose line, and protect critical egress routes.

Staffing Needed for Hose Lines. In most cases, a full fire company should be assigned to each hose line, except when the apparatus operator is needed at the pump panel (two members are assigned as the initial RIC). The importance of deploying the initial attack hose line on the fire cannot be overstated. It may require the first two companies to work together to get the hose line to the fire. Departments should operate nozzles and hose lines under simulated fire conditions during training. This training should include advancing the hose up stairways, through doorways, and around obstacles to determine the actual staffing needs for each hose and nozzle combination. Keeping company members together also results in better accountability.

Water Supply. Once a fire occurs, it is too late to evaluate the water supply capability. Some areas are fortunate to be protected by a high-flow and reliable water supply. Most areas have weaknesses in the water supply system and others have no public water supply.

Growing communities often outstrip their water supply, resulting in reliability issues. These communities are likely to experience lower pressures and lower flows during business hours as demand for domestic water increases on a marginal water system. Even well-established high-volume water supplies will contain problem areas if hydrants are on dead-end or small mains. If the system pressure is supplied by gravity and the protected area is hilly, pressures and flows will be different at various locations, depending on the difference in height between the hydrant and the water supply tank as well as the distance from the water supply (friction loss).

Some public and nearly all private water systems have a limited supply. These limitations must be understood when deploying large-flow appliances. Private hydrants are likely to have a separate supply that may be connected to automatic fire suppression systems. Sometimes connecting to these private hydrants will have a negative effect on the automatic fire suppression system.

Large municipal water systems are often supplied by multiple water systems. These multiple systems can usually be cross-tied to supply water from adjoining systems, thus increasing the total flow available at the scene of a major emergency **FIGURE 2-27**. If the water system has this capability, members should know how to open valves (or how to get assistance in opening valves) connecting water systems. Many water utilities have the capability to increase water pressure when large flows are needed.

Some communities or sections of communities do not have a public water system and rely on other sources such as ponds, rivers, and pools as a water supply. As with the municipal water system, the reliability of the water source must be known in advance. Questions that should be asked regarding these water sources include the following:

- Will the source freeze during cold weather?
- Is there access for apparatus assigned to draft water?
- Are dry hydrants or other water supply equipment available?
- Are water sources dry during part of the year?
- When tender operations are anticipated, how many tenders will be required to maintain the required fire flow?

FIGURE 2-27 Water supply from ladder to roof of a two-story residential structure.
Courtesy David J. Jones, Cincinnati, Ohio.

Even departments with a strong water supply should plan for situations when the water supply could be reduced or unavailable. After Hurricane Katrina, water supplies were unavailable, requiring large municipal departments to use alternative water supplies. Some departments with reliable water supplies have removed hard-suction hose from their pumpers and no longer train pump operators to draft water from static sources. As mentioned previously, FDNY found it necessary to draft water from flooded streets to suppress a conflagration during Hurricane Sandy. All departments should consider alternative sources of water as part of their disaster preparedness.

Apparatus Pump Capacity. Apparatus specifications should consider available water supply and fire hazards when pumps are specified. Individual apparatus pump capacity is a known factor. However, the total flow needed for a given situation is incident dependent. Seldom is the apparatus pump capacity challenged during an offensive attack. During defensive operations, the need for water could be greater than the apparatus pump capacity. In many cases, total pump capacity exceeds the available water supply.

Manual Fire Suppression Systems. In most cases, manual fire suppression system refers to a standpipe system. The fire officer must make a decision whether the system is to be used in a one-story or low-rise building or in the lower floor of a high-rise building. Usually, it is best to use the standpipe because its use reduces the work required to advance a hose line. However, some standpipe systems are supplied by the same water supply as the sprinkler system, or they may be designed for occupant use, thus having a minimal flow. There are three types of standpipe systems: Class 1, Class 2, and Class 3. Class 1 and 3 systems are designed to be used by fire fighters and have a required minimum flow rate. Sprinkler and standpipe systems will be discussed in more detail in the Chapter 7, *Fire Protection Systems*. The type of sprinkler and standpipe as well as the flow capacity should be noted in the preincident plan.

When there is a chance that the sprinkler system will be ineffective because hose lines are using water needed to supply the automatic sprinkler system, it is best to avoid using the standpipe. The water pressure and flow are preestablished for standpipe systems, although pressure and flow can usually be increased by pumping into the fire department connection. If the standpipe has a severely limited flow or low pressures, it may be best to use hose taken directly from the apparatus to attack a fire. The reliance on the standpipe is directly related to the height of the fire floor; the higher in the building the fire, the greater the reliance on the standpipe.

Preincident plans should show the location of control valves (main and sectional), pumps, fire department connections, and hose outlets. Some systems have pressure or flow-reducing valves. The preincident plan should explain the use of these valves, whether they can be adjusted, and what equipment is needed to adjust the pressure/volume. SOPs and preincident plans should specify hose, nozzles, and equipment needed during standpipe operations. Standpipe systems are discussed in Chapter 7, *Fire Protection Systems,* and in Chapters 12, *High-Rise Buildings*.

Automatic Fire Suppression Equipment

There may be more than one sprinkler system in a large building. These systems may or may not be interconnected. Just as with the standpipe system, the location of control valves (main and sectional), pumps, fire department connections, and hose outlets should be shown on the preincident plan for sprinkler-protected buildings. When a building is equipped with an automatic sprinkler system, the primary tactic involves letting the system do its job while fire fighters support the system and move in for final extinguishment. The sprinkler system becomes the first line of defense. It is essential that fire protection systems be properly maintained and that maintenance records be consulted when preincident planning a sprinkler-protected building.

Other automatic systems may be found in buildings; they must be addressed in the preincident plan. See Chapter 7, *Fire Protection Systems,* for more information on automatic fire suppression systems and tactics to be used at protected properties.

Property Conservation

Property conservation is the third operational priority. Seldom will factors related to property conservation take on the urgency associated with life safety and extinguishment. Information about property value and location within the building can usually be known in advance; preplanning property conservation measures can significantly reduce the overall loss. See Chapter 10, *Property Conservation,* for specific property conservation tactics.

Salvageable Property

Nearly every property has some form of salvageable property. During preincident planning, it is important to determine what there is to salvage and the value of the property. Remember that property might also have nonmonetary value. Most fires occur in residential occupancies that contain personal property, such as heirlooms and photographs that are considered invaluable by the occupants. Being careful and protecting these personal items can improve the public's perception of the fire department.

Location of Salvageable Property. In a residential occupancy, salvageable property can be found throughout the building. In other occupancies there may be concentrations of high-value property, such as computer rooms, in a school or business occupancy. Preincident plans should identify these locations and possibly provide guidance in handling the salvage effort (e.g., the location of accounts payable records, which are generally vital to business continuation). Make a note on preplans where critical files are kept or stored, and recommend to the owner that they protect these files.

Water Damage

The primary extinguishment tactic involves applying water to the burning material, thus there will be some quantity of water in the building. Water will migrate through various openings and possibly damage property on floors where water is discharged and below.

Probability of Water Damage. Some buildings tend to hold water while others have drains or pathways that allow water to harmlessly drain off. Property in water pathways below the fire have a high probability of water damage.

Susceptibility of Contents to Water Damage. Susceptibility of contents to water damage has to do with how easily the property is damaged. Some property is highly vulnerable to water damage (e.g., paper files and electronic equipment).

Water Pathways to Salvageable Property. Water will flow downward through paths of least resistance. If the floor contains openings or holes, water will flow through the openings or holes to the floor below. Otherwise, it will flow to stairways, elevators, drains, or other routes.

Water Removal Methods Available. Preferred methods involve using built-in features that drain the water out of the building. Otherwise, water removal will involve pushing or directing water to channels leading out of the building, or possibly removing it with water vacuums.

Protective Measures Available. When water cannot be directed away from salvageable property, the most common way to protect property from water damage is to place covers over the exposed contents, starting with the property that is in the water pathway and most susceptible to water damage. At times it is possible to move contents above floor level (e.g., placing furniture on skids or placing more valuable property on top of less valuable property). Water removal, as mentioned previously, is also a means of protecting property from flowing water.

Smoke Damage

Smoke can infiltrate the entire building, especially through ventilation systems. However, the most common pathway for smoke is upward; therefore, most smoke damage occurs on the fire floor and above.

Ventilation. The best way to reduce smoke damage is to ventilate the building. Some ventilation methods that are ill advised prior to full fire containment are acceptable after fire control or extinguishment. For example, venting the stairway, as mentioned under the section "Ventilation Status," could place fire fighters and occupants in danger. Later in the operation, this may be a good option for reducing smoke damage.

Property that is susceptible to smoke damage on the fire floor and above is most likely to be damaged. Materials that absorb smoke are more susceptible to damage as is sensitive electronic equipment.

Forcible entry and ventilation are often necessary to achieve tactical objectives, but they can be overdone. If a building is locked and no key is available, forcible entry will nearly always result in some damage, as will ventilating. Damage should be limited to what is needed to achieve tactical objectives.

General Factors

Total Staffing Available Versus Staffing Needed

The implementation of an incident action plan will require resources, and until the actual plan is developed, the exact number of resources needed is unknown. However, the IC needs to make approximations during the early size-up. Staffing is generally the most important and difficult resource to obtain when initiating an offensive attack.

As the size and complexity of the property increase, the required staffing will also increase, as demonstrated in the NIST high-rise report (Averill

FIGURE 2-28 A. High-rise building located at 223 23rd Street, Crystal City, Virginia. **B.** Line-of-fire apparatus in staging.
A: NIST Report on High-Rise Fireground Field Experiments. https://nvlpubs.nist.gov/nistpubs/TechnicalNotes/NIST.TN.1797.pdf. Figure 1, pg. 30, left-hand photo; **B:** NIST Report on High-Rise Fireground Field Experiments. https://nvlpubs.nist.gov/nistpubs/TechnicalNotes/NIST.TN.1797.pdf. Figure 1, pg. 30, right hand photo.

and Moore-Merrell, 2013) **FIGURE 2-28**. If the calculated fire flow requires multiple 2½-in. (64-mm) attack hose lines, more staffing will be needed to meet the rate of flow. Staffing requirements enumerated in NFPA 1710 will be further discussed in Chapter 12, *High-Rise Buildings*, as will findings of the NIST high-rise staffing report.

Searching large areas, physically removing victims, meeting larger rate of flow requirements, or responding to areas beyond a fixed water supply will require additional staffing.

Available water resources become a high-priority issue in areas without hydrants or when the required rate of flow is large. Fire departments working in rural areas can sometimes summon large numbers of fire fighters to the scene, but the immediately available water supply is usually limited.

Offensive operations are typically labor intensive. Conversely, defensive operations require fewer fire fighters operating apparatus and equipment that require large-volume water supplies. In any case, the IC needs to estimate the staffing resources required.

Once an estimate of staffing requirements is made, the IC must consider how the responding resources match incident requirements. Augmenting resources could take the form of requesting additional alarms or mutual aid. If resource capabilities cannot match the needs of an offensive attack, attempting an offensive operation will place fire fighters in extreme danger. If the needs of neither an offensive nor a defensive attack can be met, the IC's only safe option may be to assign available resources to exposure protection.

The IC must also include resources necessary to provide an RIC when considering an offensive attack. At least four fire fighters should be assembled to function as the RIC to provide for initial fire fighter rescue on the fire ground. An actual rescue (locating the fire fighter and removal) may require multiple companies to be successful.

Total Apparatus Available Versus Apparatus Needed

Offensive operations are staffing intensive; the initial response generally provides an adequate number of apparatus. Defensive operations are apparatus intensive and require more apparatus to supply water for master streams, as well as master stream equipment requirements such as aerial apparatus and apparatus-mounted master streams. Each master stream will generally require a minimum of one apparatus. If the apparatus does not have pumps, such as an aerial truck without pumps, an additional apparatus will be needed to supply water.

Staging/Tactical Reserve

SOPs should address staging in a general way. Some properties, especially large complexes, may have specific staging opportunities that should be identified when preincident planning.

When all staffing or all apparatus are being used and the incident is not resolved, a tactical reserve is needed. For a small-scale incident, this reserve is usually at least one engine company and one truck company.

Larger and more complex incidents require a larger tactical reserve. When confronted with a working fire, it is good practice to maintain a tactical reserve in staging, so forces can be rapidly deployed from a staging area. Waiting for resources to respond from distant fire stations can result in additional loss of life and property.

> **NOTE**
>
> A parked apparatus is not a staged fire company.
>
> A staged company is a standby crew in full fire fighter gear with or without an apparatus.

Utilities (Water, Gas, Electricity, Other)

Some departments routinely shut down the electrical supply in residential property. This can provide a safer environment and reduce chances of reignition. However, doing so can also be a dangerous operation, even in a residence. Fire fighters should not attempt to disrupt the power supply in larger properties with high-voltage service. If the department routinely interrupts the power supply at residential properties, SOPs and preincident plans should specifically state that fire fighters should not disconnect power supplies when dealing with properties with high-energy service.

Residential gas or fuel supplies can be shut down if they are causing a problem or are leaking. Utility company personnel have the experience, training, and equipment to shut down electrical and piped gas supplies. Preincident plans should show the location of gas and electric shut-offs.

> **NOTE**
>
> Beware: some properties have electric generators or other emergency power sources that could self-activate when power is interrupted.

Special Resource Needs

Although not always considered a special resource, police and EMS are often needed at the scene. The IC should assess the need for these resources as part of the size-up. In many areas, police and EMS routinely respond to a reported structure fire. Other resources could also be required, such as air trucks, hazardous materials teams, utility companies, Red Cross, and others. Local emergency management plans should include a list of resources available to the community with emergency contact information. The fire department should have access to this information.

Time

By examining the size-up/preincident plan factors, much can be determined in advance of the fire through preincident planning. Time is one factor that will not be known until a fire occurs. Consider the following time factors:

- **Time of day:** Time of day can determine the likelihood of people being in the building and their degree of awareness. An apartment building fire at noon generally presents a much different scenario than a fire in the same building at midnight. A fire in a Christian church on Sunday morning could involve an extreme life hazard. A fire in the same church late on a Monday night might gain considerable headway before being discovered, but probably would not involve an extreme life hazard. Some properties (e.g., nursing homes) are continuously occupied.
- **Day of week:** As with time of day, occupancies are generally fully occupied on different days of the week. A school would normally be occupied during the week during the school year and minimally occupied at other times, except when a special event is taking place, such as an evening basketball game.
- **Time of year:** The time of year should also be considered. In the northern United States and Canada, winter is a time when people tend to stay inside more and use heating systems, which can serve as additional sources of ignition.
- **Special times:** Finally, special times that should be considered would include holidays. An office building early on Christmas day would contain few occupants, whereas this same building could house thousands of people on a normal workday. Likewise, a Christian church would be expected to contain a maximum number of occupants early on Christmas day.

Weather

The IC must consider the effect of the present weather conditions on the operation. The variety of weather conditions is nearly endless, but the IC should concentrate on extremes. High heat and humidity take a toll on fire fighters. Extreme cold also affects fire fighters and their equipment as well as victims who are being removed from a structure **FIGURE 2-29**. High wind affects fire spread and ventilation, especially in high-rise structures or at large fuel fires, such as those at lumberyards. NIST has done extensive research on the subject of wind-driven fires, especially in high-rise buildings. However, wind-driven fires can be problematic in any building, as was experienced in

FIGURE 2-29 Fire fighters fighting fire on an icy street.
Courtesy of David J. Jones, Cincinnati, Ohio.

a one-story residence in Houston, Texas, that resulted in two fire fighter on-duty fatalities (NIOSH, 2010).

Fire departments in communities that experience little or no wind on a typical day can be caught off guard during rapidly changing weather conditions, such as thunderstorms, that can temporarily produce hurricane-force winds and change the dynamics of a fire. Improper ventilation tactics or sudden failure of windows can put fire fighters in peril before they realize what has happened. A vigilant IC, working with a planning and safety officer, should notice or anticipate changes in weather and adjust tactics accordingly.

The general kinds of weather that an area experiences can be known in advance, but the weather at the time of the incident is unknown until the alarm is transmitted. Fire fighters in southern Florida expect to deal with high temperatures and humidity for many months of the year, just as fire fighters in Alaska expect to deal with cold weather extremes; of course, a fire could occur in Alaska during the summer, with heat being a potential problem for fire fighters who are unaccustomed to working in warmer temperatures.

Extremely cold or hot weather conditions may require occupant shelter and fire fighter rehabilitation. Very cold weather can also result in frozen fire hydrants, frozen pumps on apparatus, and damaged apparatus equipment. During very cold or very hot weather, most buildings will be tightly sealed to maintain internal temperatures, increasing the likelihood of ventilation-limited fires. Fires occurring during moderate weather conditions are more likely to be well ventilated via open windows or other openings.

Humidity is particularly important during hot weather extremes because it increases the fatigue factor. Humidity can also affect smoke movement. On a hot, humid day, smoke may stratify, creating heavy smoke conditions outside the building; this heavy smoke may require apparatus to be positioned farther away or require apparatus operators to don SCBA. High humidity and heat conditions can also cause smoke stratification within high-rise buildings.

Precipitation in the form of rain is less problematic than frozen precipitation, which slows response and makes it more dangerous, as well as creating hazardous on-scene operating conditions.

High winds can push the fire toward exposures or affect venting. High winds are particularly problematic during high-rise fire operations.

A list of size-up/preplan factors is included in Figure 2-14. An attempt has been made to include major factors affecting size-up, but invariably, one or more factors will be omitted from such a list. The three priorities (life safety, extinguishment, and property conservation) are the basis of the size-up/preincident plan checklist. All of these factors are interrelated at some level. For example, weather conditions have an effect on fire fighter and occupant safety. Weather also affects ventilation and thus extinguishment. Structural stability affects the attack method and almost every other tactic at the incident scene.

Size-Up Chronology

Size-up follows a chronological sequence:

1. SOPs
2. Preincident plan
3. Shift/day/time
4. Alarm information
5. En route
6. Visual observations at the scene
7. Information gained during continuing operations
8. Overhaul

Size-up is a continuing and dynamic process. The on-scene evaluation should improve as observations and reports from working units provide additional and improved information.

Standard Operating Procedures and the Preincident Plan

SOPs and preincident planning activities are done well in advance of an alarm but play a significant role in the size-up when developing an incident action plan. They provide the IC with a head start in the size-up process and actually become part of the incident size-up.

Shift/Day/Time

As the fire officer reports for a shift, unusual weather conditions, day of the week, and related factors are considered. On occasion, special measures are taken, such as adding an extra company to certain locations because of snow-covered streets. The time of the alarm may also be very important. The life hazard in a high-rise office building will be much greater at 2:00 PM on a workday than at 2:00 AM on a weekend.

Special information may change from day to day. The water system may be out of service, a parade may be scheduled, or the area might have been experiencing extremely hot, dry weather for a prolonged period. These are all important factors to consider.

Alarm Information

When the alarm is received, the time of day factor is immediately processed along with dispatch information. Dispatch information can include the following:

- Building location/address
- Fire location
- Fire intensity (numerous calls usually translate into big fires)
- Occupant status (e.g., report of an occupant trapped on the second floor)

Additional information may also be known about the area or the specific building, or the dispatch file may contain a preincident plan notation.

While Responding

The dispatcher may be able to provide additional information to the responding units. An onboard computer could also provide specific preincident plan information. Companies arriving on the scene should provide status reports. For example, there may be visual indications of a working fire, such as visible smoke or flames. During the response, all previous information is reconsidered, such as what effect high winds will have on the operation. Initial information received from units on the scene is critical in the formulation of an incident action plan.

Visual Observations at the Scene

Many questions are answered as the IC gets the first look at the situation. When conditions permit, it is a good idea to get a view from several angles. By driving up to the building and going past the front prior to setting up a command post, the IC may be able to view three or even four sides of the building. However, this must be done quickly so that an incident action plan can be developed and implemented.

The combination of visual information and reconnaissance from companies working on the interior or other unseen locations provides the basis for the initial incident action plan. This may be the first opportunity to evaluate structural conditions and the effectiveness of action being taken. The crucial question is: Will the current incident action plan accomplish the desired objectives?

The quality and quantity of information will greatly increase over time. The IC goes from a dearth of information to what may be information overload. At this point, it is important to prioritize information as primary and secondary and spend time reevaluating the incident action plan on the basis of information received.

Information Gained During Continuing Operations

If the operation is successful, good news such as "the primary search is complete and the fire is extinguished" will reach the IC. If on-scene resources are unable to accomplish tactical objectives, either bad news or no news will be communicated to the IC. Bad news requires an adjustment in plans. On occasion, the entire action plan is changed, such as shifting from an offensive attack to a defensive attack. After allowing time for companies to make entry and begin operations, no news is unacceptable. The IC must call for a status report when none is forthcoming.

Overhaul

Overhaul should be planned and deliberate. The emergency phase is over, and extreme caution should be taken to avoid injuring fire fighters. At this point, there is no justification for rushing to make decisions that could put fire fighters at risk.

Wrap UP

CHAPTER SUMMARY

- Standard operating procedures (SOPs), preincident plans, and incident-specific information are interrelated and are important components of the size-up.
- When preincident plans are available for a property, their use in evaluating the situation is crucial.
- A proper size-up is crucial to developing effective firefighting strategies and the incident action plan.
- Due to time constraints, the fire officer must focus on primary size-up factors when developing an initial incident action plan. Incident conditions will determine which size-up factors are more important (primary) or less important (secondary).
- The incident action plan should adapt to changes in circumstances or to additional critical information that indicates a need to change tactics.
- Structural firefighting SOPs are guidelines to be used at all structure fires except when circumstances indicate the need for a different approach.
- SOPs must address department-specific training and equipment but should also reflect regional procedures to facilitate operations with other jurisdictions and agencies.
- SOPs allow first-arriving units to take immediate action or stage while awaiting orders from command.
- SOPs specify action to be taken but should be flexible enough to allow fire fighters to follow a reasonable course of action when confronted with a situation in which modification of the procedure is appropriate.
- Preincident plans must be kept up to date.
- Preincident plans are needed when there is a high life hazard, a difficult extinguishment problem, or high-value property.
- Preincident plans should estimate the required rate of flow for large undivided areas, note any hazardous materials, and evaluate fuel loads.
- Building materials, roof construction, age and overall condition, fuel load, live loads, dead loads, and layout are key factors in determining likelihood of fire extension and building collapse.

KEY TERMS

area of refuge A floor area with at least two rooms separated by smoke-resisting partitions in a building protected by a sprinkler system, or a space located in an egress path that is separated from other building spaces.

backdraft A fire condition that occurs when oxygen (air) is introduced into a superheated, oxygen-deficient compartment charged with smoke and pyrolytic emissions, resulting in an explosive ignition.

balloon-frame construction An older type of wood-frame construction in which the wall studs extend vertically from the basement of the structure to the roof.

compartmentation Subdividing of a building into small areas (rooms) capable of limiting the spread of fire and the products of combustion.

complex preincident plan A plan used when a property has more than three buildings or when it is necessary to show the layout of the premises and relationship between buildings on the site.

dead load The weight of a building; consists of the weight of all materials of construction incorporated into a building, including but not limited to floors, roofs, ceilings, stairways, built-in partitions, finishes, cladding, and other similarly incorporated architectural and structural items, as well as fixed service equipment.

drilling down Using a computer to navigate by pointing and clicking through a series of drop-down menus in a graphical user interface.

flashover An oxygen-sufficient condition in which room temperatures reach the ignition temperature of the suspended pyrolytic emissions, causing all combustible contents to suddenly ignite.

flow path The movement of heat and smoke from the higher pressure fire area toward lower pressure areas on the interior and exterior of the structure.

formal preincident plan A plan for a property with a substantial risk to life and/or property; includes a drawing of the property, specific floor layouts, and a narrative describing important features.

fuel load Fuels provided by a building's contents and combustible building materials; also called fire load.

hovering Accessing text or graphic items on a computer screen by placing the computer cursor over a specific area or by touching the screen in a touch screen environment.

live loads The weight of the building's contents, people, or anything that is not part of or permanently attached to the structure.

notation A piece of information about the premises, such as damage to the building from a previous fire. This information may accompany a preincident plan or may be available when the building does not have a preincident plan.

platform-frame construction A construction technique using separate components to build the frame of a structure (one floor at a time). Each floor has a top and bottom plate that act as fire-stops.

preincident plans Written documents resulting from the gathering of general and detailed information to be used by public emergency response agencies and private industry for determining the response to reasonable anticipated emergency incidents at a specific facility.

primary factors The most important factors, assessed during size-up, which change from incident to incident and depend on specific incident conditions.

rapid intervention crew (RIC) A minimum of two fully equipped personnel onsite, for immediate rescue of injured or trapped fire fighters.

rate of flow The minimum water application rate required for extinguishment

secondary factors Less important factors at an incident, which change from incident to incident and depend on specific incident conditions.

softening the target Cooling hot fire gases using rapid, short-duration application (usually 15 to 60 seconds) from a straight or smooth-bore fire stream aimed at a steep angle toward the ceiling from the safest effective location. When using this tactic, water often is applied from the exterior to the interior of a building.

standard operating procedures (SOPs) Written rules, policies, regulations, and procedures intended to organize operations in a predictable manner.

Superfund Amendments and Reauthorization Act (SARA) A federal law enacted in 1986 and also known as the Emergency Planning and Community Right to Know Act. Title III of this law requires businesses that handle or store hazardous chemicals in quantities above specific limits to report the location, quantity, and hazards of those chemicals to the State Emergency Response Commission (SERC), Local Emergency Planning Committee, and local fire department. Some of this information has been classified since the attack on the World Trade Center on September 11, 2001.

tenders NIMS term for vehicles that transport water from a water source to the fire scene; also referred to as a tanker or mobile water supply apparatus.

type I construction Construction method in which the structural members, including walls, columns, beams, girders, trusses, arches, floors, and roofs are of approved noncombustible or limited-combustible materials and have the highest level of fire-resistance ratings.

type II construction Construction method in which the structural members, including walls, columns, beams, girders, trusses, arches, floors, and roofs are of approved noncombustible or limited-combustible materials but the fire resistance rating does not meet the requirements for Type I construction.

type III construction Construction method in which exterior walls and structural members that are portions of exterior walls are of approved noncombustible or limited-combustible materials, and interior structural members, including walls, columns, beams, girders, trusses, arches, floors, and roofs, are entirely or partially of wood of smaller dimensions than those required for Type IV construction or of approved noncombustible, limited-combustible, or other approved combustible materials.

type IV construction Construction method in which exterior walls and structural members that are portions of exterior walls are of approved noncombustible or limited-combustible materials. Other interior structural members, including walls, columns, beams, girders, trusses, arches, floors, and roofs are of solid or laminated wood without concealed spaces. Wood columns supporting floor loads are not less than 8 in. (20 cm) in any dimension; wood columns supporting roof loads only are not less than 6 in. (15 cm) in the smallest dimension and not less than 8 in. (20 cm) in depth. Wood beams and girders supporting floor loads are not less than 6 in. (15 cm) in width and not less than 10 in. (25 cm) in depth. Wood beams and girders and other roof framing supporting roof loads only are not less than 4 in. (10 cm) in width and not less than 6 in. (15 cm) in depth. Specifics for other structural members are required to be large-dimension lumber as well.

type V construction Construction method in which exterior walls, bearing walls, columns, beams, girders, trusses, arches, floors, and roofs are entirely or partially of wood or other approved combustible material smaller than material required for Type IV construction.

wood truss An assembly made up of small-dimension lumber joined in a triangular configuration that can be used to support either roofs or floors.

SUGGESTED ACTIVITIES

1. Review your department's SOPs:
   A. Are they up to date? When was the last revision?
   B. Does the department have SOPs addressing:
      - Use of NIMS, including command transfer, establishing a command post, and specifying the IC?
      - Apparatus management—staging, positioning, collapse zones?
      - Structural firefighting, including water supply, initial attack, and company duties?
      - Operating in buildings with built-in fire equipment?
      - Special hazard operations, such as hazardous materials and technical rescues?
      - High-rise, large-area buildings and other special occupancies?
2. Select a department SOP that is out of date or otherwise needs revision. Revise the SOP.

REFERENCES

Averill, Jason D., Lori Moore-Merrell, Raymond T. Ranellone, Jr., et al. 2013. *Report on High-Rise Fireground Field Experiments*. Gaithersburg, MD: National Institute of Standards and Technology. https://nvlpubs.nist.gov/nistpubs/TechnicalNotes/NIST.TN.1797.pdf.

Corbett, Glenn P., and Francis L. Brannigan. 2021. *Brannigan's Building Construction for the Fire Service*, Sixth edition. Burlington, MA: Jones & Bartlett Learning.

Fire Protection Research Foundation. June 2014. "Forum on PV Panel Fire Risks." Las Vegas, NV: NPRF.

International Association of Fire Chiefs (IAFC), Safety, Health and Survival Section and International Society of Fire Service Instructors (ISFSI). 2013. "Firefighter Safety Call to Action: New Research Informs Need for Updated Procedures, Policies." http://www.iafc.org/Media/articlePR.

Kerber, Steve. 2010. *Impact of Ventilation on Fire Behavior in Legacy and Contemporary Residential Construction*. Northbrook, IL: Underwriters Laboratories Firefighter Safety Research Institute.

Kerber, Steve. 2013. *Study of the Effectiveness of Fire Service Vertical Ventilation and Suppression Tactics in Single Family Homes*. Northbrook, IL: Underwriters Laboratories Firefighter Safety Research Institute.

National Fire Protection Association. 2017. *NFPA 704: Standard System for the Identification of the Hazards of Materials for Emergency Response*. Quincy, MA: NFPA.

National Fire Protection Association. 2017. *NFPA 1142: Standard on Water Supplies for Suburban and Rural Fire Fighting*. Quincy, MA: NFPA

National Fire Protection Association. 2018. *NFPA 1: Fire Code*. Quincy, MA: NFPA.

National Fire Protection Association. 2018. *NFPA 101: Life Safety Code*. Quincy, MA: NFPA.

National Fire Protection Association. 2018. *NFPA 170: Standard for Fire Safety and Emergency Symbols*. Quincy, MA: NFPA.

National Fire Protection Association. 2018. *NFPA 220: Standard on Types of Building Construction*. Quincy, MA: NFPA.

National Fire Protection Association. 2018. *NFPA 1500: Standard on Fire Department Occupational Safety, Health, and Wellness Program*. Quincy, MA: NFPA.

National Fire Protection Association, 2018. *NFPA 1801: Standard on Thermal Imagers for the Fire Service*. Quincy, MA: NFPA.

National Fire Protection Association. 2018. *NFPA 3000: Standard for an Active Shooter/Hostile Event Response (ASHER) Program*. Quincy, MA: NFPA.

National Fire Protection Association. 2018. *NFPA 5000: Building Construction and Safety Code*. Quincy, MA: NFPA.

National Fire Protection Association. 2020. *NFPA 1620: Standard for Pre-incident Planning*. Quincy, MA: NFPA.

National Fire Protection Association. 2020. *NFPA 1710: Standard for the Organization and Deployment of Fire Suppression Operations, Emergency Medical Operations, and Special Operations to the Public by Career Fire Departments*. Quincy, MA: NFPA.

National Fire Protection Association, 2020. *NFPA 1720: Standard for the Organization and Deployment of Fire Suppression Operations, Emergency Medical Operations and Special Operations to the Public by Volunteer Fire Departments*. Quincy, MA: NFPA.

National Institute for Occupational Safety and Health (NIOSH). 2010. "Career Probationary Fire Fighter and Captain Die as a Result of Rapid Fire Progression in a Wind-Driven Residential Structure Fire—Texas. Latest revision December 8, 2014. http://www.cdc.gov/niosh/fire/reports/face200911.html.

National Institute of Standards and Technology. 2010. *Report on Residential Fireground Field Experiments*. Gaithersburg, MD: NIST.

Underwriters Laboratories Firefighter Safety Research Institute, National Institute of Standards and Technology, and New York City Fire Department. 2013. *Governors Island Experiments*. Northbrook, IL: UL FSRI.

CHAPTER 3

Developing an Incident Action Plan

LEARNING OBJECTIVES

- Explain the importance of deciding on an overall strategy in the development of an incident action plan.
- Describe how extinguishment is both an operational priority and tactical objective with an emphasis on the relationship between life safety and extinguishment.
- Explain the relationship between the risk-versus-benefit analysis and the incident action plan.
- Compare and explain the differences between an offensive and defensive attack.
- Evaluate conditions leading to an offensive or defensive operation.
- Compare probability of occupant survival to fire and building conditions.
- List situations when a written incident action plan is needed.
- Explain the role of an engine or truck company officer arriving at the scene of a working fire prior to the arrival of a chief officer in relationship to developing the incident action plan.
- Use a case study or actual fire to develop an incident action plan based on a risk-versus-benefit analysis.
- Develop a deployment and incident action plan tailored to specific occupancy types.
- Develop a tactical worksheet.

CHAPTER 3 Developing an Incident Action Plan

Courtesy of Bill Strite, Cincinnati, Ohio.

Case Study

The following table lists the responding fire companies, fire officers, and emergency medical services (EMS) resources available in a particular area. All engine, truck, and heavy rescue companies are staffed with an officer, driver, and two fire fighters. EMS units are staffed with two paramedics. The only chief officer with a chief's aide is the fire chief.

Responding Fire Department Fire Companies, Fire Officers, and EMS Resources			
	Engine Companies	**Truck Companies All Equipped with 100-ft (30.5-m) Aerial Ladder**	**Other Units**
1st Alarm	Engine 1 Engine 2 Engine 3	Truck 1 Truck 2	District Chief 1 Paramedic 1
2nd Alarm	Engine 4 Engine 5 Engine 6	Truck 3 Truck 4	District Chief 2 Operations assistant chief Safety officer Paramedic 2 Heavy Rescue 1
3rd Alarm	Engine 7 Engine 8 Engine 9	Truck 5	Fire chief Operations assistant chief District Chief 2 Paramedic 2 Heavy Rescue 1

© Jones & Bartlett Learning.

1. Engine 1 and Truck 1 are located in a fire station approximately one-half mile from this incident.
   - Assume the role of Engine Company 1.
   - You arrive just in advance of Truck 1. A man standing at the front of the building (side Alpha) on the west side (side Bravo) is frantically waving for help as you arrive. You notice fire issuing from the roof of a two-story apartment building on the opposite side of the street. The driver stops at the fire hydrant and establishes a water supply.
   - You order the two Engine 1 fire fighters to advance the preconnected 250-ft (76.2-m) 1¾-in. (44-mm) hose line to the side door at the Alpha–Bravo corner of the building. You notice that the man waving from across the street has severely burned hands and chest. He meets you at the side door leading to a stairway at the front of the building.
   - There is one apartment on each floor. The basement serves as a utility room with a furnace/air conditioning unit, washer/dryer, and a storage area for each of the two apartments.
   - The injured man informs you that he occupies the second-floor apartment with his wife and 6-year-old daughter.
   - The injured man explains that he removed the gas tank from his car and was emptying it into a large bucket so he could repair a leak in the gas tank. Suddenly a huge fireball filled the basement. His 6-year-old daughter was with him at the time. When asked if she escaped he replies: *"I don't know if she got out or is still in there; we live on the second floor."*
   - You encounter a large volume of fire as you approach the basement from the stairway.
   - You report conditions to dispatch: *"There is a working fire in the basement and west side stairway. The roof has self-vented."*
   - Explain your action plan.

2. Now assume the position of the Truck Company 1 Officer.
   - You hear the report from Engine 1 and raise your aerial ladder and ground ladder to the second-floor windows.
   - You request and receive a progress report from Engine 1. Engine 1 reports: *"We are making some progress on the basement fire, but we have not checked the remainder of the building for occupants; however, there are at least two second-floor residents who may still be in the building in the second-floor apartment and a severely burned man on the sidewalk at the front of the building."*
   - Visual information can be seen on the photo provided for this Case Study.

3. Now assume the position of District 1 as the incident commander.
   - Formulate an incident action plan.
   - If necessary, reassign Engine Company 1 and/or Truck Company 1.
   - List the tasks remaining to be assigned to the responding first-alarm companies. (Two additional engine companies, one additional truck company, and Paramedic 1.)
   - Is a one-alarm response sufficient? If not, what additional assistance would you request?
   - Assign all units requested!

4. Explain the basic differences between an offensive and a defensive incident action plan.

5. Would you classify your incident action plan as offensive or defensive?

6. List the risks/dangers to the occupants and fire fighters at this incident.

7. What tactics should be included in an offensive incident action plan when exposures are threatened?

Introduction

An **incident action plan** provides the central focus for operations. An offensive/defensive strategy decision based on a risk-versus-benefit analysis is the first step in developing an incident action plan. All tactical decisions and task assignments are based on the overall strategy. Everyone on the fire ground must be aware of the strategic mode. The offensive/defensive decision and overall incident action plan are derived from an analytical approach to information gained through preincident planning and size-up. The result should be a straightforward, easy-to-understand plan outlining the major tactical objective. All tactics should lead to completion of the major objectives identified in the incident action plan.

Because conditions constantly change, size-up is a continuous process; therefore, the incident action plan must remain flexible. Often, as conditions change, the tactics necessary to accomplish a strategy are modified. Incident commanders (ICs) should develop alternative incident action plans that anticipate possible changes and be ready to react as necessary (i.e., change from an offensive to defensive strategy). A chief's aide or planning section chief can be of great assistance in developing a "plan B," allowing the IC to focus on the current operation.

The primary strategic considerations are always life safety, extinguishment, and property conservation. However, the three priorities are not mutually exclusive. When resources are limited—and they usually are in the beginning stages of an operation—the fire officer must take action related to life safety first, extinguishment second, and property conservation third. However, in most cases, extinguishment and life safety are closely related. If the fire is extinguished, rescue often takes care of itself, and the overall operation is much safer. Extinguishment is normally, but not always, the most important life-safety tactic. Extinguishment is both an operational priority and a tactical objective.

There are times when there are only a few easily rescued victims. When resources are limited, it may be best to rescue these victims before attempting to reach other possible victims. However, this is a very difficult decision. Victims who are in the greatest danger may not be visible, and reports from occupants are often incorrect.

When developing an incident action plan, it is best to categorize tactical objectives in terms of their relative importance as primary and secondary (and possibly tertiary) objectives. For example, extinguishing the fire would normally be a primary objective, as would search and rescue operations. Most property conservation measures would fall under secondary objectives. The idea is to provide an effective plan for deployment, with primary objective tasks and tactics being assigned prior to secondary objective tactics, when resources are limited.

Once available resources are at the scene, all priorities can be handled simultaneously. If the fire operation will continue for more than a few minutes, the IC should request enough staffing and equipment to handle the immediate life-safety, extinguishment, and property conservation activities, plus an allowance for a tactical reserve. Anticipate the unexpected.

Remember, the key to a successful fire-ground operation is keeping it simple. Incident action plans should be simply stated and concise. This chapter emphasizes the development of an incident action plan leading to offensive or defensive tactics. Five scenarios (a dwelling, an apartment building, an industrial property, a church, and a high-rise residence) are presented to provide examples of incident action plan development.

Determining Life-Safety Needs

Chapter 2, *Procedures, Preincident Planning, and Size-Up*, discusses the importance of life safety during preincident planning and size-up. The primary objective of structural firefighting is saving lives; therefore, life safety is the first consideration in developing an incident action plan. For this reason, the IC may expose fire fighters to greater risk when savable lives are threatened.

Evaluating Structural Conditions

Structural conditions bear heavily on the offensive/defensive decision. Even with sufficient resources, an interior attack should not be conducted in an unsafe building. Fire fighters should never enter a building that is in danger of imminent collapse, and they should maintain a collapse zone. However, rescue efforts may be justified in a building that is currently structurally sound but may ultimately fail as continuing fire damage undermines the structure. In this situation, conduct an offensive attack to assist occupants in exiting the building, with a reevaluation of the structure and fire conditions once the building is evacuated. If extinguishment is being accomplished during the initial offensive attack, offensive tactics will normally be continued. However, if the fire has not been contained or extinguished during the initial offensive attack, after the rescue effort has been accomplished, the expected benefit has changed from life safety to saving property. There is less reason to stay in a building of questionable stability, because the benefit component of the risk-versus-benefit analysis has changed from life safety to saving property. Proficient ICs are always weighing risk against expected benefits. Staying on the right side of this equation requires the development of an alternative incident action plan. If the fire in a building of questionable structural integrity is not contained when all life-safety tasks are completed, an alternative defensive plan (plan B) should be at hand and resources ready for implementation. It is a good idea to assign a planning section chief to develop a defensive plans as soon as a transition from offensive to defensive is anticipated. A major part of this alternative plan consists of moving fire fighters and apparatus out of the collapse zone, as well as deciding the appropriate position and type of fire streams for the defensive attack.

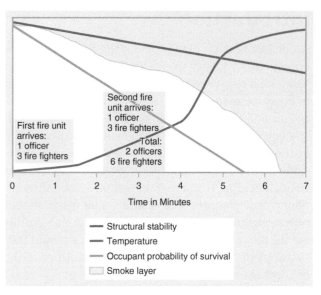

FIGURE 3-1 Resources versus possible life-safety benefit.
© Jones & Bartlett Learning.

Fallacy	Fact
The three tactical objectives are mutually exclusive and must be performed in order of priority: life safety, extinguishment, property conservation.	The three tactical objectives are interrelated; extinguishment is usually an important tactic in accomplishing the life-safety objective. When resources are sufficient, all three objectives can be accomplished simultaneously.

Estimating Resource Capability and Evaluating Resource Requirements

Comparing resource capability to incident requirements is part of the size-up process. The resources that are required to fight a fire offensively are different from those needed for a defensive battle. Generally, it takes more staffing to conduct an offensive operation. Lives and property are best saved by conducting an offensive attack. However, a lack of necessary resources could lead to a defensive decision, even when an offensive attack is clearly the better approach.

The IC must consider what will be needed to conduct an offensive attack early in the plan development process. In making the offensive-versus-defensive decision, the IC must apply sound risk management principles to ensure fire fighter safety. **FIGURE 3-1** charts time and resources versus occupant probability of survival in terms of structural stability, temperature, and smoke layer. At the beginning of the incident, there is a high probability of saving occupant lives and extinguishing the fire. As the temperature increases in an unvented structure, the smoke layer begins to fill the enclosure while the structure weakens, thus the probability of saving lives decreases, and extinguishment is more difficult. In other words, the probable benefit is lower and the risk is higher. In the early stages of the fire, no fire department personnel would be on the scene. As time passes, fire department resources become available, but the probable benefits of life safety and extinguishment decrease.

Each fire is different; thus, Figure 3-1 is a general depiction rather than an actual fire model. Underwriters Laboratories (UL) conducted tests comparing time to flashover in a living room containing modern furniture versus a living room containing "legacy" furnishings. For purposes of the study, "legacy" was defined as the fire load typically found in a residential occupancy in the mid-20th century. With all variables other than the fire load held constant, the room with modern contents flashed in 3 minutes 40 seconds compared to 29 minutes 30 seconds for the legacy room in the one-story residential structure (Kerber, 2010). Even this very credible research is not universally applicable to home fires. Many factors determine the rate of progression to an untenable temperature or immediately dangerous to life and health (IDLH) toxicity level. It is important to point out that if the occupant immediately calls dispatch to report the fire, units responding within the guidelines of NFPA 1710, *Standard for the Organization and Deployment of Fire Suppression Operations, Emergency Medical Operations, and Special*

Operations to the Public by Career Fire Departments, would usually arrive at or after flashover in a residential fire (NFPA 1710, 2020). Many factors affect fire dynamics and the progression to flashover, ventilation being a major factor, as pointed out in later research by UL (Kerber, 2013).

Developing an Offensive or a Defensive Incident Action Plan

The entire operation is governed by the offensive/defensive decision. A casual observer should be able to determine whether the tactics being applied are interior (offensive) or exterior (defensive). However, which tactics are being used may not always be obvious. During the UL residential vertical ventilation research, an exterior hose stream was operated from the exterior through an opening for a short period of time (usually 15 to 60 seconds) to soften the target prior to making entry for an offensive attack (Kerber, 2013). Many fire service leaders now advocate this tactic. After studying the research by UL and follow-up New York City Fire Department experiments, the potential value of this tactic is clear. However, there is concern that this tactic will be applied under circumstances that could be problematic. Under certain conditions, the time spent softening the target from the exterior could be spent making entry into the structure to apply water directly on the fire and potentially locate victims as the fire attack hose line advances into the structure.

Master streams are generally considered defensive tools. However, they can be used to support rescue efforts in large, complex structures with fire separations. Master streams are sometimes used to prevent fire from entering critical evacuation routes, such as a fire escape, or to cover exposures. When master streams are being used during an offensive operation, coordination through command is critical. An operation that begins as an offensive attack is sometimes transitioned to a defensive attack, but the actions taken during either attack must be coordinated as offensive or defensive; they must never be both. Chapter 8, *Offensive Operations,* and Chapter 9, *Defensive Operations,* discuss offensive and defensive attacks in detail. Whenever it is safe to do so, an offensive attack is preferred.

Formulating an Incident Action Plan

The IC sets the objectives, decides on the tactics necessary to achieve objectives, and assigns units to complete the tasks associated with each objective. The incident action plan is the focus of the entire operation. All tactics are directed toward completing the objectives, and each objective is directed toward accomplishing the overall incident action plan. The incident action plan should be simple and understandable, as shown in **FIGURE 3-2**.

Written incident action plans are required when an incident extends past a single operational period or if a unified command is established. Most structure fires do not require formal written incident action plans. However, critical aspects of an incident should be recorded to assist in the transfer of command and in documenting the incident in the written report and critique. This could be a tactical worksheet or other media. At a minimum, the IC should track on-scene resources, current assignments, known hazards, and benchmarks completed. Including a rough sketch of the fire scene and an organizational chart are highly recommended. Typically, an assistant to the IC maintains the written documents. As the incident grows, the planning section chief usually updates the information. When staffed, the planning section tracks progress and continuously develops the incident action plan and alternative plans. **FIGURE 3-3** shows a tactical worksheet.

The IC establishes the objectives, making strategic determinations for the incident based on the requirements of the jurisdiction. In the case of a unified command, the incident objectives must adequately reflect the policies and needs of all the jurisdictional agencies involved. The incident action plan should cover all tactical and support activities required during the operational period.

An incident action plan becomes more important as the incident grows in size. However, there should be an incident action plan for every structure fire, and units operating on the scene should be coordinated using this plan.

Deployment

Writing a list of tactical objectives is useless unless sufficient resources are assigned to accomplish the objectives. Experienced ICs know that after making assignments, they must follow up and request status reports. Sometimes an assignment is made but not achieved. Experienced company-level officers provide status reports, as appropriate, without being prompted. However, when reports are not received in a timely manner and the assigned company has had ample time to complete the assignment, it is incumbent upon the IC to request a status report. If status reports are requested too frequently, the status report communications can interfere with efficient operations.

FIGURE 3-2 Developing an incident action plan.
© Jones & Bartlett Learning.

NOTE

Ordering it done and getting it done are not always the same.

Reproduced from: © 2011, IAFC Safety, Health and Survival Section. All rights reserved.

The following five scenarios illustrate the process involved in developing an incident action plan. In these scenarios, assume that all companies are staffed with a minimum of four personnel (an officer, a driver, and two fire fighters; NFPA 1710, 2020). If your department operates fire companies with fewer than four responding members, two or more units should be combined under the supervision of a single officer. Likewise, more tasks may be accomplished if more than four members are assigned to a company. For example, some departments staff quint apparatus with six or more fire fighters, allowing the company to simultaneously obtain a source of water, advance an attack hose line, and perform ventilation activities.

Scenario 1: Single-Family Detached Dwelling

Fires in the home make up the vast majority of fires. Sometimes department standard operating procedures (SOPs) spell out company duties for the entire first-alarm assignment. It could be said that the IC is on autopilot when SOPs are followed, but going on autopilot can be dangerous. No fire should ever be considered routine. On the other hand, there is no need for unnecessary interference in an operation that is

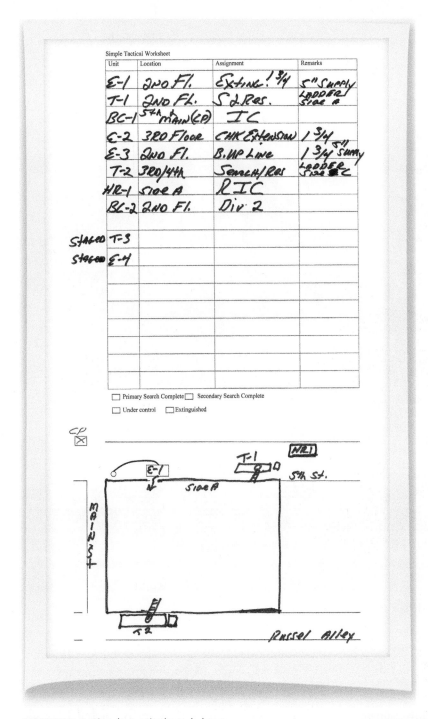

FIGURE 3-3 Simple tactical worksheet.
© Jones & Bartlett Learning.

going well. Most ICs feel more comfortable handling a small fire than a larger, more complicated fire, yet the process that is used to develop an incident action plan for a small residential fire is the same as that used for a large and complex fire. Even when the fire operation is going well, the IC should focus on completing the tactics necessary to meet strategic objectives, while simultaneously developing a plan B as a precaution should the situation change.

It is 3:30 PM on Saturday, July 14. It is 90°F (32°C) and very humid. The fire was reported by a neighbor who noticed smoke coming from the roof. This residential area has a strong water supply system. The first-arriving engine company is making entry, and the truck company has placed their aerial ladder on the roof. The status of the occupants is unknown. Conditions are as shown in **FIGURE 3-4**.

Risk-Versus-Benefit Analysis

Expected Benefits

Saving the lives of the occupants and saving property. There is a chance that the occupants were not at home at the time of the fire, but this cannot be known until verified by a primary and secondary search. The exact location of the fire is unknown at this time. There appears to be a good chance of extinguishment.

Expected Risk

Moderate. A life-safety benefit is the primary reason for conducting an offensive operation. There is a risk associated with any offensive operation, but the risk to fire fighters is usually less in a single-family dwelling than it would be in a larger, more complex occupancy. The fire pictured in Figure 3-4 appears to be limited. This could be a ventilation-limited fire. The risk is somewhat increased due to the unknown location and size of the fire. There is a chance that the fire originated in the first floor or basement and spread to the attic.

> **NOTE**
> Benefit as used here is the expected or potential benefit to occupants or owners. Rescuing occupants would be a life-safety benefit.

> **NOTE**
> Risk, as used in the risk-versus-benefit analysis, refers to the risk to fire fighters, not the risk to occupants.

FIGURE 3-4 Fire in a single-family detached dwelling.
Courtesy of Denny Baker, Cincinnati Fire Department.

Structural Stability

Primary failure not expected until near full involvement. This is a frame structure. Frame buildings are fairly unstable structures and contribute significant fuel to the fire. However, the combustible characteristics of the building provide stability clues. Frame structures will normally have considerable fire involvement before collapse, with the exception of truss roof and engineered floor construction. For this reason, if the fire is in a modern residential building, extreme caution should be used if the IC decides that it is necessary to vent the roof. Whenever possible, it is a good idea to provide a stable platform using an aerial, an elevated platform, or a roof ladder.

Offensive/Defensive Decision

Offensive. Given the possibility of saving lives and property benefits with a moderate risk to fire fighters, a decision is made to initiate an offensive attack.

Resource Needs

Initial staffing is needed for (1) finding the interior fire, (2) advancing a standard preconnected attack hose line while controlling ventilation by minimizing the door opening, and (3) simultaneously performing the primary search with coordinated ventilation. An initial rapid intervention crew (IRIC) should remain outside the building. In most operations, two or more fire fighters make entry with the hose line, with the IRIC remaining outside. In this case, one fire fighter and the pump operator are assigned as the IRIC. The IRIC fire fighter should assist with the hose deployment or other exterior tasks and be ready to join his or her company upon establishment of a formal rapid intervention crew (RIC). The IRIC concept usually applies to companies that arrive significantly ahead of other arriving companies and must go to work prior to additional companies arriving on the scene. It is recommended that a company dedicated to rapid intervention be dispatched on the initial alarm and designated as such. The truck company should begin the primary search and prepare to vent on orders from the IC. As the operation progresses, anticipate the need for an additional search and rescue crew, a backup hose line, a dedicated RIC, and tactical reserve. The size of the tactical reserve should take into account the need for rehabilitation due to hot and humid weather conditions.

Incident Action Plan

- Make entry; find and extinguish the fire.
- Establish an IRIC followed by a dedicated RIC.
- Search and rescue in basement, first floor, and attic (appears to be dormer rooms in attic area).
- Vent heat, smoke, and gases using horizontal and/or vertical venting.
- Advance a hose line to the, attic and check for extension on second floor and attic.

Deployment

- *First engine company.* Provide an initial status report, and verify command. Complete a 360-degree size-up of the house. Establish a water supply, force entry, and attack the fire on the first floor with a 1¾-in. (44-mm) hose line. Two members remain outside the building as the IRIC. The fire fighter assigned to the IRIC could control the entry door and assist feeding hose as the attack team advances.
- *First truck company.* Act as search and rescue teams on fire floor. Ventilate as needed, possibly using horizontal ventilation and/or opening the roof, depending on the location of the fire.
- *Battalion chief.* Transmit a status report, assume command, and establish a stationary command post as IC.
- *Second engine company.* Obtain a second source of water. Advance a 1¾-in. (44-mm) hose line into the building as a backup hose line to the first engine. If the first engine has the fire under control, the second engine can advance the hose line above the fire to check for fire extension. The first-in truck company should assist in checking for extension.
- *Second truck.* Advance to the floor above the fire for search and rescue and check for fire extension.
- *Third engine company.* Function as RIC. A two-person IRIC is the absolute minimum. While department SOPs vary regarding which company is assigned as the RIC, most departments assign at least four members (a four-person fire company) as the RIC.
- *Medic unit.* Establish a safe treatment area and stand by to transport injured fire fighters or civilians.
- *Tactical reserve of at least two companies.*

NFPA 1710, 5.2.4.1.1, establishes a minimum staffing of 16 fire fighters for a fire in a 2000 ft^2 (186 m^2) two-story, single-family structure with no basement and no exposures in immediate proximity (17 minimum if an aerial device is used; NFPA 1710, 2020). A NIST staffing study (NIST, 2010) used a building constructed to meet the single-family structure described in NFPA 1710. This study proved the need for the staffing levels enumerated in NFPA 1710.

It is obvious that this deployment covers only the most basic tasks. If the first-arriving engine company is successful in extinguishing the fire, the entire operation is favorably changed, and fewer fire fighters and companies will be needed. This is the reason for placing "extinguish the fire" as the highest strategic priority, thus accomplishing the life-safety objective by extinguishing the fire and ventilating the smoke, heat, and gases. A written incident action plan would not be required for this incident.

Scenario 2: Fourteen-Unit Apartment Building

Apartment buildings typically endanger more occupants than fires in single-family dwellings, and accounting for building occupants is nearly impossible. The increased life-safety problem is essentially found by multiplying the number of apartment units times the number of occupants typically found in a single-family house.

Many apartment buildings (including the one pictured here) have nonfirestopped attics that allow fire

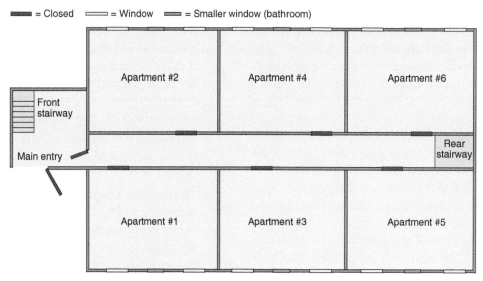

FIGURE 3-5 Plan view of apartment building first floor.
© Jones & Bartlett Learning.

to spread unimpeded throughout the attic. Shared basement areas used for tenant storage are the norm. The attic and basement areas can require large flows of water for extinguishment. Most apartments are made up of small individual compartments that are within the capabilities of preconnected attack hose lines carried by the most fire departments.

This fourteen-unit apartment building has two basement-level walk-out apartments on side Charlie (Apartments 1 and 2). The remainder of the basement is undivided with individual storage areas assigned to each tenant plus a laundry room. There are six apartments each, on the first and second floors, connected by a hallway running the length of the building. Stairways to the second floor are located on each end of the building **FIGURE 3-5**.

It is 7:30 AM on Sunday, May 6. Temperature is 55°F (13°C), and winds are light and variable. Dispatch reports the call was received from Apartment 12 on the second floor, reporting visible smoke.

Fire fighters on the first-arriving engine company encounter smoke issuing from the first-floor hallway as they advance a 1¾-in. (44-mm) hose line through the hallway entrance on side Alpha **FIGURE 3-6A**. Fire and heavy smoke can be seen issuing from side Delta **FIGURE 3-6B**. The first-arriving truck company is conducting a 360, checking all sides of the building's exterior for occupants and preparing for forcible entry and search of the fire floor.

Risk-Versus-Benefit Analysis
Expected Benefits
Saving lives of the occupants and saving property. Given the day of the week (Sunday) and time of day (7:30 AM) there is a good chance that occupants will be at home and many will be sleeping. There is no report of anyone escaping or notifying occupants of the fire. Smoke conditions in the hallways may result in occupants seeking safe refuge inside their apartments. With no evidence of occupants escaping the building, an assumption is made that all apartments are occupied. Information regarding fire extension is also unknown, but the presence of moderate smoke at the main entry indicates that smoke and possibly fire have migrated out of the apartment of origin to the hallway and possibly to other parts of the building.

Expected Risk
Moderate. This fire presents a greater risk to fire fighters than the fire in Scenario 1 due to what appears to be a more advanced fire in a more complex and larger building.

Structural Stability
No signs of imminent structural collapse are evident, although this building probably has a wood-truss roof structure. If fire extends into the attic, rapid extension will occur, endangering residents on the second floor and causing a potential roof collapse. A roof collapse presents a danger to fire fighters working on the second floor, but intervening wall structures will limit the collapse area.

Offensive/Defensive Decision
Offensive. An offensive attack is in progress, and this appears to be the correct decision.

A

B

FIGURE 3-6 Fire in an apartment building. **A.** Side Alpha. **B.** Side Delta.
Courtesy of Mike Carry and Jeff Neal, Cincinnati Fire Department.

Resource Needs

Three attack hose lines and possibly four are needed, with an initial and backup water supply. An RIC is needed, as are search and rescue crews on all levels of the building. Additional horizontal or positive-pressure ventilation may be required. A tactical reserve and medical units are also required.

Incident Action Plan

- Extinguish fire in apartment of origin.
- Extinguish fires in other apartments on the first floor.
- Extinguish fires on the second floor and attic.
- Search and rescue first floor.
- Search and rescue second floor.
- Search and rescue basement.
- Establish a dedicated rapid intervention crew (RIC).

Deployment

- *First battalion chief.* Transmit a status report, assume command, and establish a stationary command post.
- *First engine company.* Establish a water supply. Continue advancing fire hose line to apartment of origin and attack the fire on the first floor.
- *First truck company.* Conduct search and rescue operations on the first floor.
- *Second engine company.* Advance a 1¾-in. (44-mm) hose line into the first floor check for extension, extinguishing fires in other areas of the first floor. Obtain a second source of water.
- *Second truck company.* Conduct a search and rescue operation on the second floor.
- *Third engine company.* Advance a 1¾-in. (44-mm) hose to the second floor, checking for fire extension in second floor apartments and the attic.
- *Third truck company.* Complete 360 of building, placing ground ladders as needed. Function as the dedicated RIC.
- *Fourth engine company.* Search and rescue of basement and basement apartments while checking for fire extension.
- *Two medic units.* Establish safe treatment area and stand by to transport injured fire fighters or civilians.
- *Tactical reserve of at least three companies.* There is a good chance that the search and rescue effort will need to be augmented, especially if occupants are found who need assistance in evacuating the building.

Scenario 3: Supreme Meat Packing Company— Processing Plant Complex

The fire hazard challenge in large industrial complexes varies widely. Many industrial facilities have tremendous fire loads, dangerous processes, and hazardous materials inside large, complicated buildings. Locating specific areas within a large complex is nearly impossible without a preincident plan drawing and/or assistance from employees. A working fire deep inside the complex may not be visible from the street entrance to the facility. In most cases, arrangements are made during preincident planning to have a representative from the facility meet the fire department and direct them to the reported fire area. However, it is not good practice to rely entirely on plant personnel. At the BASF chemical plant fire, discussed in Chapter 1, *Organizing, Coordinating, and Commanding Emergency Incidents,* the plant safety supervisor (a retired fire fighter) was designated to meet the first-arriving fire company at a preselected area. On the day of the fire, the plant safety supervisor was seriously injured and found lying in the street in front of the plant. On other occasions, security personnel assigned to meet the fire department became involved in notifying occupants of the fire or using fire extinguishers. Preincident planning, including an overview of the property, is essential to successful operations in large industrial complexes.

The industrial complex used as an example of complex and formal preincident plans in Chapter 2, *Procedures, Preincident Planning, and Size-Up,* is the setting for this scenario **FIGURE 3-7**.

A fire is reported in Building H. It is Wednesday, May 5, at 10:00 AM. Weather conditions are cool, dry, and calm. According to the preincident plan narrative, a maximum of 254 employees could be at the plant during normal working hours. There is no employee accountability system.

The engine company (Engine 1) checks the preincident plan drawing, proceeds to Building H, forward laying a large-diameter hose line to establish a water supply. Engine 1 reports visible fire on the west side of the building and attempts to enter the building via the loading dock located between Buildings H and A with a charged 2½-in. (64-mm) hose line. They are unable to advance on the fire and decide to back out of the building. As they retreat, an attempt is made to determine if any employees remain in the building by

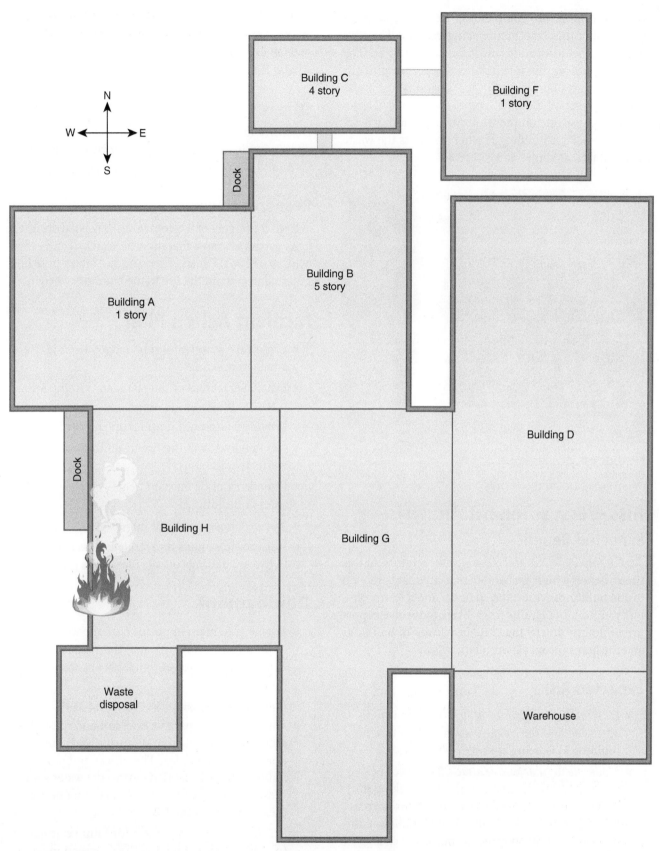

FIGURE 3-7 Supreme Meat Packing Company layout.
© Jones & Bartlett Learning.

verbally calling into the building and doing a quick check using their thermal imaging camera. They are unsuccessful in locating anyone in Building H. The Engine 1 Officer provides a status report recommending a defensive operation to the IC, who is just arriving at the scene. Engine 1 repositions their 2½-in. (64-mm) hose line on the west side, conducting a defensive attack on Building H where it meets the waste disposal unit as shown in **FIGURE 3-8**.

FIGURE 3-8 West side of Building H.
Courtesy of David J. Jones, Cincinanti, Ohio.

Risk-Versus-Benefit Analysis
Expected Benefit
Saving employee lives and reducing property loss. At this time, there is a high probability that occupants remain inside buildings within the complex and a low probability in Building H. The fire is likely to involve exposure buildings due to the close proximity of buildings and the access areas between buildings.

Expected Risk
- *Extreme for an offensive attack*. The fire is well advanced and the metal-truss roof structure on Building H is being attacked by the fire.
- *Moderate for a defensive attack*. The space between Buildings A and H is limited, making it difficult to attack the fire from positions beyond the collapse zone. Large-exposure buildings combined with confusing layouts make search and rescue and fire operations hazardous.

Structural Stability
Expect collapse of Building H and waste disposal unit. Replace first engine's attack hose line on the west side with an unstaffed master stream or an elevated master stream located farther away from the building. Evaluate all exposure buildings starting with Buildings G, A, and B.

Offensive/Defensive Decision
Defensive for Building H. Attempt entry and offensive operations in other buildings, prioritizing Buildings G, A, and B.

Resource Needs
A large contingent of fire companies is needed due to the large geographic area that must be searched and evacuated, as well as the many exposure and attack hose lines required to contain the fire to the building of origin.

Incident Action Plan
- Undertake search and rescue operations in the attached buildings.
- Evacuate all areas of the plant, accounting for as many employees as possible.
- Provide a large and uninterrupted water supply.
- Contain and limit the fire to Building H and the waste disposal unit.
- Provide exterior master streams on all accessible sides of Building H.
- Advance hose lines into Buildings G, A, and B.
- Advance hose lines into strategic positions between buildings on all sides of the plant.

Deployment
- *Engine 1*. Remain in position on the west side of Building H, staying out of the collapse zone. Apparatus operator to remain at apparatus pump panel.
- *Battalion chief*. Establish command as IC.
- *Engine 2*. The apparatus operator and fire fighter establish a second large-diameter water supply on north side of complex. The officer and other fire fighter establish the IRIC. After the water supply is established the four members of Engine 2 are then assigned as the RIC.
- *Truck 1*. Position apparatus for future elevated master stream, protecting the intersection between Buildings H and A. Crew enters Building A via the north side dock from the loading dock, conducting search and rescue operations. Evacuate occupants and evaluate fire spread into Building A.

- *Engine 3.* Advance 2½-in. (64-mm) hose line into Building A. Conduct search and rescue operations. Evaluate fire spread into Building A and protect south wall exposed to Building H.
- *Truck 2.* Set up elevated master stream outside the collapse zone on the west side of Building H
- *Battalion 2.* Assigned as west division supervisor (companies assigned to Buildings H, A, G, and B).
- *Department safety officer.* Assigned as incident safety officer.
- *Chief of department.* Assume command, reassigning Battalion Chief 1 as operations section.
- *Engine 1.* Deploy a master stream at the opening to Building A to protect wall exposed to Building H.
- *Engine 4.* Advance 2½-in. (64-mm) hose line into Building B to protect wall exposed to Buildings H and G; conduct search and rescue operation.
- *Heavy Rescue 2.* Assigned to Building G, conduct search and rescue operations. Evacuate occupants and evaluate fire spread into Building G.
- *Truck 3.* Evacuate Building A.
- *Engine 5.* Evacuate Building D, then move to and evacuate Building F.
- *Engine 6.* Evacuate Building C, then move to and evacuate Building F.
- *Battalion 3.* Assigned as evacuation group supervisor (companies assigned to Buildings A, J, D, E, F, and C).
- *Heavy Rescue 2.* Establish second RIC.
- *Engine 7 and Truck 4.* Establish large-diameter hose water supply; set up master streams on southeast corner and south side of Building H. Truck 4 Officer assigned as exterior attack group supervisor.
- *Medical branch.* Treat and transport occupants as necessary. Establish safe area for evacuees.
- *Engines 8 and 9.* Provide additional water supply on the south side of the complex via large-diameter hose. Assist in setting up master streams.
- *Four engine companies and two truck companies.* Staged as a tactical reserve. First company officer in staging assigned as staging manager.

Scenario 4: Church Fire

Older, gothic-style churches are characterized by high ceilings, with huge concealed spaces above the main body of the church. One of the largest church fires in history occurred in 2019 at Notre Dame Cathedral in Paris. Newer churches come in a variety of styles. Both old and new churches have a common characteristic: a large open area to accommodate the congregation and altar. This one factor means that total and catastrophic structural collapse of the roof is highly probable once a fire reaches the structure supporting the roof.

The large open area in the main body of the church requires an unusually high rate of flow because of the high ceiling. (Fire flow calculations are explained in Chapter 8, *Offensive Operations.*) It will be necessary to use 2½-in. (64-mm) or 3-in. (76-mm) handlines or portable master stream appliances with solid stream nozzles to obtain the necessary reach if an interior attack is attempted. An offensive attack on a well-involved church fire is a high-risk operation, especially if the roof structure is involved.

Many people congregate during services, especially on religious holidays. During these times, life safety is a key tactical consideration. There are subsequent long periods of time when these buildings are unoccupied, allowing a fire to gain considerable headway before it is discovered. Few older churches are protected by fire suppression systems or automatic alarms.

In this scenario, the twin bell towers in a gothic-style church are involved in fire **FIGURE 3-9**. **FIGURE 3-10** provides a layout of the entire church property.

FIGURE 3-9 Church fire.
Courtesy of David J. Jones, Cincinnati, Ohio.

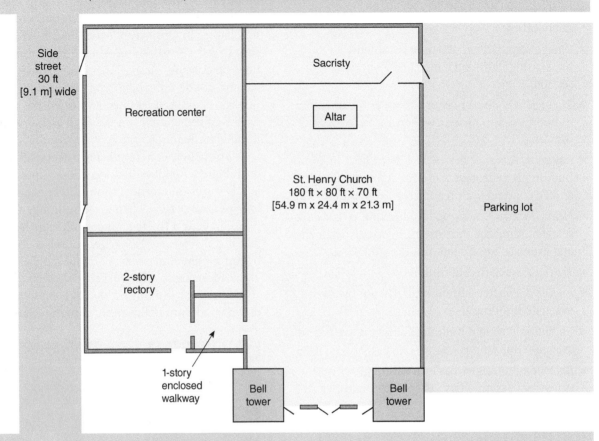

FIGURE 3-10 Plan view of church property.
© Jones & Bartlett Learning.

It is 9:00 PM Wednesday, October 5. Weather is cool; winds are light and variable following an earlier thunderstorm. A passerby reports fire in the bell towers. The first-arriving engine company (Engine 1) confirms a large volume of fire in the bell towers.

Risk-Versus-Benefit Analysis

Expected Benefits

Saving property. The church is not having services at this time, and the fire has gained considerable headway. There is a low probability that anyone is in the church. The pastor confirms that the church is locked for the night. This historic gothic church has many stained-glass windows and artwork that is considered priceless.

Expected Risk

Extreme. The high bell towers combined with a large volume of fire make collapse of the towers probable. Fire extension to the interior of the church is likely but unknown at this time. Electrical wires close to the front of the church present a hazard.

Structural Stability

Partial failure inevitable. As previously stated, the bell towers should be expected to fail and collapse. The height of the bell towers indicates a fairly large collapse zone that may be beyond the effective reach of exterior fire streams. If the fire has extended into the concealed space above the church interior, the church roof should also be expected to collapse.

Offensive/Defensive Decision

Defensive. Structural collapse potential is great; no one should enter the bell tower area. It may be possible to force entry to the rear of the church on side Delta to check for fire extension and make certain that the church is locked and everyone is out of the building. In checking for extension, evidence of fire in the roof area above the main body of the church may not be apparent. If fire is evident, it is usually too late to save the roof; therefore, it is preferable to make a defensive attack on the bell towers using streams placed beyond the collapse zone and away from electric wires in an attempt to prevent fire extension into other nearby buildings.

Resource Needs

A large water supply and multiple elevated master streams are needed.

Incident Action Plan

- Extinguish the fire in the bell tower.
- Maintain a wide fire perimeter.
- Check the interior of the church via the rear door if this can be safely accomplished.

Deployment

- *Engine 1.* Officer: take command, establishing a command post; remaining company members: supply water to Truck 1's elevated master stream.
- *Truck 1.* Force entry to the rear of the church; check for visible fire extension; make certain the church is unoccupied, then return to Alpha–Bravo corner with Crew 1.
- *Battalion Chief 1.* Assume command from Engine 1's officer upon returning to Engine 1.
- *Truck 2 and Engine 2.* Set up a second elevated master stream operating into the bell tower on the Alpha–Bravo corner of the church. Elevated master stream supplied by Engine 2.
- *Battalion Chief 2.* Assigned as Division Alpha–Bravo.
- *Truck 3 and Engine 3.* Set up a third elevated master stream operating into the bell tower on side Alpha.
- *Heavy Rescue Unit 1.* Advance a hose line into the one-story enclosed walkway between the church and rectory and check for extension along the wall separating the church and recreation center
- *ALS 1 and ALS 2.* Remain on standby at scene.
- *Truck 4 and Engine 4.* Set up a fourth elevated master stream operating into the bell tower on the Alpha–Delta corner of the church supplied by the third engine company.
- *Engine 1 apparatus operators.* Remain at pump panels.
- *Engine 2.* Assigned as RIC 1.
- *Engine 3.* Assigned as RIC 2.
- *Department safety officer.* Assigned as incident safety officer.
- *Third battalion chief.* Assigned as Division Alpha–Delta supervisor.
- *Two engines and two trucks.* Staged as a tactical reserve.

As previously stated, defensive operations require less staffing than interior operations. Once master streams have been set up, minimal personnel are needed to maintain the fire flow. However, they should remain on the scene as a tactical reserve or to complete extinguishment after the collapse danger has passed. Fire fighters on standby status should be assembled in a central area well away from the fire to prevent freelancing.

In the church fire example, the IC should direct the planning section to develop a plan that changes from a defensive to an offensive operation if the bell towers are extinguished before the fire enters the main church structure. However, there is a real possibility that the fire has already entered the concealed spaces above the church ceiling, which will require a larger defensive operation with little hope of saving the building. The stability of the bell tower structures is unknown, and they pose a threat to anyone near or in the church. A limited degree of risk is acceptable as long as there is property to be saved.

If a defensive operation is initiated for the entire church, there is a possibility that the only savable property will be external exposures. Therefore, the IC should reduce the risk to as near zero as possible. This may mean using unstaffed master streams or relocating staffed master streams farther from the church. Daytime operations tend to be safer than nighttime operations. In this case it may be best to remain in a defensive attack until after daybreak. Use the elapsed time to carefully plan a strategy that protects fire fighters, while assuring extinguishment and isolating the surrounding area.

Scenario 5: High-Rise Apartment Building

High-rise building fires often pose an extreme life hazard. Office buildings may contain thousands of people, and large residential buildings can house hundreds of families. The mere fact that the fire could be located several floors above grade level and beyond the reach of aerial apparatus increases the complexity and difficulty in rescuing occupants and deploying fire streams. In addition, many floors above the fire could be occupied, making it necessary to conduct search and rescue operations on many different levels. The actual fire conditions could be similar to those in a single-family detached dwelling; however, the challenges that are presented will be much more complex. Chapter 12, *High-Rise Buildings,* is dedicated to the special strategies and tactics needed to combat high-rise fires.

Fires in residential high-rise buildings will typically be limited to a small area in the building unless the fire occurs in a common area or the apartment door is left open or is damaged. Smoke, however, can spread far from the fire, endangering a large number of residents. Most apartments in residential high-rise buildings are located on a single floor. However, some high-end modern high-rise apartment buildings include apartments that span two floors with unprotected openings between floors. These are usually large open-layout apartments and generally require a higher rate of flow.

The high-rise involved is a 14-story, public housing apartment building of fire-resistive construction **FIGURE 3-11**. This building is not protected by a sprinkler system; however, standpipes are located in the two stairways located at the end of each floor. Standpipe intakes are located next to the main entrance at street level.

FIGURE 3-11 High-rise fire.
Courtesy of Bill Strite, Cincinnati, Ohio.

The first floor of this building is partially below grade and houses a laundry, activities rooms, and vending areas. The second floor contains an in-house security office, alarm room, two restaurants, business offices, medical facilities, and small retail shops. The building has 10 apartments per floor on Floors 3 to 14 (120 apartments) and primarily houses elderly occupants. The building has three stairways located at the end of each floor, and a stairway and two elevators located near the center on each floor.

In this scenario, the fire occurs in the living room area on the 13th floor of a 14-story public housing building. The fire is in Apartment 1408 on the floor named Floor 14, as shown in **FIGURE 3-12**. Note: Many high-rise buildings do not have a floor named as Floor 13. This is due to superstitions regarding the number 13 as being unlucky.

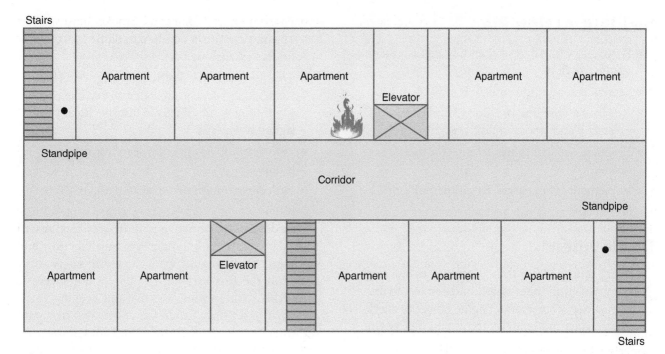

FIGURE 3-12 High-rise floorplan.
© Jones & Bartlett Learning.

The building is equipped with a standpipe system, but is not protected by an automatic sprinkler system. According to the preincident plan, elevators are equipped with a fire fighter emergency operation mode. Using an elevator with a fire in progress is a calculated risk. Elevator safety requires that fire fighters dismount the elevator at least two floors below the fire. The risk associated with using elevators may not be warranted for a fire on the sixth floor. Chapter 12, *High-Rise Buildings,* offers a more complete discussion of the use of elevators by fire fighters.

The fire originated in the living room. It is 7:00 AM on Saturday morning. The weather is mild and wind is calm.

Risk-Versus-Benefit Analysis
Expected Benefits
Saving the lives of the occupants and saving property. There is a good chance that most residents will be at home and asleep. Occupants on the floor above the fire and those on the fire floor are in imminent danger.

Expected Risk
Moderate to high. The risk in this building is somewhat greater than that in a smaller residential building, due to the number and location of occupants who are beyond the reach of aerial ladders and towers and the difficulty in ventilating the building or softening the target using an exterior stream. The layout of this building is fairly straightforward in comparison to some high-rise buildings. The fire-resistive construction and multiple stairways reduce the risk to some extent.

Structural Stability
Failure not expected. The structure is of fire-resistive construction; therefore, it is unlikely that a moderate fire would seriously threaten the stability of this building.

Offensive/Defensive Decision
Offensive. Given the high likelihood of many occupants being endangered on the fire floor and floor above, a decision is made to initiate an offensive attack.

Resource Needs
Resource needs are much greater than those in low-rise residential properties. It will take longer to reach the fire floor, as confirmed in research conducted by the NIST (NIST, 2013). The backup hose line is also more critical in this situation. The early estimation is for three companies to be assigned to the fire floor and two to the floor above. Property conservation, especially below the fire, is also important, but it is a secondary objective.

Incident Action Plan

- Force entry (if necessary) and extinguish the fire on the fire floor.
- Search, rescue, and ventilate Floors 14 and 15.
- Advance a hose line to Floor 15 and check for extension and smoke conditions.
- Evacuate occupants and arrange for shelter (as needed).
- Implement property conservation on Floor 12 and below.

Deployment

- *First engine and first truck.* Combine as one company. Establish water supply and connect to fire department connection. Engine officer to check alarm panel and establish command. The engine company officer and two fire fighters walk the stairs, connect to the standpipe outlet on Floor 12. The engine company advances into the Floor 14 hallway with a hose line to attack the fire. The truck officer and two fire fighters start the initial search and rescue operation on the Floor 14.
- *First battalion chief.* Take command and request a second alarm.
- *Second engine.* Proceed to Floor 11 via the elevator and connect to the standpipe and advance a backup hose line into Floor 14).
- *Third engine.* Advance to Floor 12 and stand by as the RIT.

The IC now orders all second-alarm companies' responding units to report to the staging area for further assignment upon arrival.

- *Second truck.* Assigned to search and rescue/evacuation of Floor 14.
- *Second battalion chief.* Assigned as Division 14 and coordinated the fire control and search efforts on Floor 14.
- *Fourth engine.* Connect to the standpipe on Floor 12 and advances a hoseline to Floor 14 to check for fire extension.
- *Third truck.* Truck 3 officer is assigned as the search and rescue group supervisor. The remaining company members are not specifically assigned to an area.
- *Heavy rescue.* Assigned to search and rescue group.
- *Fifth engine.* Assigned to search and rescue group.

In addition to the assigned units, a tactical reserve should be established to relieve exhausted crews and/or augment extinguishment and life-safety operations. If the first-arriving companies are successful in quickly extinguishing the fire and keeping smoke and toxic gases from extending to threaten occupants on other floors, some of the companies shown may not be needed. However, the IC should maintain a tactical reserve in the staging area to relieve fatigued members and to be prepared to handle unforeseen circumstances. When confronted with a working fire in a high-rise building, additional resources should be requested on arrival.

High-rise residences come in all sizes, including buildings that are much larger than the one in this example. As the size and complexity of the building increase, the danger to both fire fighters and occupants also increases. Numerous other factors, such as extreme temperatures, strong winds, blocked stairways, unsafe elevators, and locked passageways further complicate operations. The danger is substantially decreased if the building has a properly installed, working sprinkler system.

Many high-rise residential buildings house elderly residents. This special population can be partially debilitated, making evacuation more labor intensive. Furthermore, these elderly occupants are much more prone to injury and death from the products of combustion.

Wrap UP

CHAPTER SUMMARY

- An incident action plan outlines major tactical objectives and provides the central focus for operations.
- The incident action plan must be flexible, adapting to changing conditions.
- Primary strategic considerations are always life safety, extinguishment, and property conservation.
- Knowledge of the structure's components and the extent of fire involvement are essential in assessing structural conditions.
- Lives and property are best saved by conducting an offensive attack.
- A defensive attack should be used whenever the risk-versus-benefit analysis indicates the risks associated with an offensive operation outweigh the possible benefits (e.g., interior conditions are judged to present an extreme danger to fire fighters, and there is no reasonable chance to save lives).
- A written action plan is needed when the incident extends beyond a single operational period. Rarely will written incident action plans be required at structure fires. Written incident action plans are primarily used during large-scale wildland fires or during natural disasters. However, tactical worksheets aid the IC in tracking resources and in the transfer of command.
- The incident action plan must be followed up with status reports, as changing conditions may require changes to the plan.

KEY TERMS

immediately dangerous to life and health (IDLH) Exposure to airborne contaminants that are likely to cause death or immediate or delayed permanent adverse health effects or prevent escape from such an environment. Examples include smoke or other poisonous gases at sufficiently high concentrations.

incident action plan The objectives for the overall incident strategy, tactics, risk management, and member safety that are developed by the incident commander and updated throughout the incident.

risk-versus-benefit analysis The process of weighing predicted risks to fire fighters against potential benefits for owners/occupants and making decisions based on the outcome of that analysis.

unified command Command structure used when multiple agencies and/or jurisdictions respond to an emergency, in which the role of incident commander is shared by representatives of various responding jurisdictions and/or agencies.

SUGGESTED ACTIVITIES

Suggested Activities 1 to 5 are continuations of the five scenarios used earlier in this chapter. In each scenario the conditions on arrival will be the same as shown and explained in the five scenarios. The alarm card for each of the incidents remains constant for all five scenarios **TABLE 3-1**.

All engine, truck, and heavy rescue companies are staffed with an officer, driver, and two fire fighters. Medic units are staffed with two paramedics. The only chief officer with a chief's aide is the fire chief. The deployment and tactical questions also remain constant for each of the first five Suggested Activities. A full one-alarm response has been dispatched to each scenario. Assume the role of the first-arriving district chief (District Chief 1). Answer the

TABLE 3-1 Responding Fire Department Fire Companies, Fire Officers, and EMS Resources

	Engine Companies	Truck Companies All Equipped with 100-ft (30.5-m) Aerial Ladder	Other Units
1st Alarm	Engine 1 Engine 2 Engine 3	Truck 1 Truck 2	District Chief 1 Paramedic 1
2nd Alarm	Engine 4 Engine 5 Engine 6	Truck 3 Truck 4	District Chief 2 Operations assistant chief Safety officer Paramedic 2 Heavy Rescue 1
3rd Alarm	Engine 7 Engine 8 Engine 9	Truck 5	Fire chief Personnel assistant chief District Chief 3 Paramedic 3 Heavy Rescue 2
4th Alarm	Engine 10 Engine 11 Engine 12	Truck 6	Paramedic 4
5th Alarm	Engine 13 Engine 14 Engine 15	Truck 7	

© Jones & Bartlett Learning.

following questions for each of the first five Suggested Activities:

- Tactical Questions
  - List the potential risks to fire fighters.
  - List the potential benefits for occupants and nearby observers.
  - Based on the risk-versus-benefit analysis:
    - Would you begin with an offensive or defensive attack?
  - Explain why you chose an offensive or defensive attack.
  - Alternatively, would you employ an operation that is not exactly offensive or defensive, for example, a marginal attack (control fire until all savable occupants are in a safe location)?

- Structural Stability
  - Estimate the potential for structural failure:
    - Imminent collapse 100%
    - High possibility of collapse
    - 50/50 chance of collapse
    - Unlikely
    - Near zero probability
- Resource Needs
  - Would you return some of the companies on the first alarm?
    - If yes, what companies are needed?
  - Would you request additional assistance?
    - If yes, what additional fire or other units are needed (e.g., 2nd alarm, paramedics, heavy rescue)?
- Incident Action Plan
  - Describe the overall action plan, for example:
    - Attack the fire via the front door.
    - Perform search and rescue of floors 1, 2, and 3.
    - Perform roof ventilation.
    - Provide a RIC.
    - Establish rehabilitation area.
    - Etc.

1. Single-Family Detached Dwelling

   It is 3:30 AM on Sunday, January 14; the temperature is 10°F (−12°C).

   Dispatch received a call from a passerby stating that smoke could be seen at the front door porch.

   Refer to Figure 3-4.

2. Fourteen-Unit Apartment Building

   It is 7:00 AM on Sunday, July 4; the temperature is 80°F (27°C).

   Dispatch received a call reporting smoke from the second floor window on the Alpha side.

   Refer to Figures 3-5 and 3-6.

3. Meat Processing Plant

   It is 10:00 AM on Wednesday, May 5; weather conditions are cool, dry, and calm.

   Dispatch received a call from the plant stating there was a fire on the west side of Building H.

   Refer to Figures 3-7 and 3-8.

4. Church Fire

   It is 9:00 PM on Wednesday, October 5; weather is cool; winds are light and variable following an earlier thunderstorm.

   A passerby reports fire in the bell towers.
   Refer to Figures 3-9 and 3-10.

5. High-Rise Fire

   It is 4:00 AM on Monday, July 4.

   The fire is in Apartment 1408 on the 13th floor; it is reported by the in-house security guard.

   (This building does not have an actual 13th floor; this floor is named Floor 14.)
   Refer to Figures 3-11 and 3-12.

6. Critique a recent fire in your jurisdiction, and evaluate the operation using the following criteria:

   A. Identify a reasonably expected benefit, and compare the actual benefit to the risks taken.

   B. Was the structure sound, or was it in danger of collapse?

   C. Was the operation offensive or defensive?

   D. Was this the correct strategy? Explain why or why not.

   E. Were there sufficient resources to conduct a safe and effective offensive or defensive attack?

   F. List the objectives of the incident action plan in order of priority.

7. Assume the role of a battalion chief responding as the IC from a fire station with an engine and truck company. The fire department has the following:
   - 30 engine companies
   - 15 truck companies
   - 2 heavy rescue companies
   - 10 advanced life support (ALS) units
   - 5 on-duty battalion chiefs
   - Off-duty battalion chiefs, the chief of department, five assistant chiefs, and one safety chief are subject to recall.

   The water supply is from a city water system. Hydrants are closely spaced with high flows.

   The total first alarm response is three engine companies, two truck companies, a heavy rescue company, and you as the battalion chief. All companies are staffed with an officer, a driver, and two fire fighters.

   It is Sunday, August 24, at 3:00 am. Conditions on arrival at this ranch-style house are as shown in **FIGURE 3-13**.

   Apply the process used to evaluate the five scenarios in this chapter:
   - Analyze the risks versus the benefits.
   - Evaluate structural stability.
   - Determine the strategy (offensive or defensive).
   - Estimate resources needed.
   - Develop an incident action plan.

FIGURE 3-13 Ranch-style house fire.
Courtesy of Bill Strite, Cincinnati, Ohio.

REFERENCES

Kerber, Steve. 2010. *Impact of Ventilation on Fire Behavior in Legacy and Contemporary Residential Construction*. Northbrook, IL: Underwriters Laboratories Firefighter Safety Research Institute.

Kerber, Steve. 2013. *Study of the Effectiveness of Fire Service Vertical Ventilation and Suppression Tactics in Single Family Homes*. Northbrook, IL: Underwriters Laboratories Firefighter Safety Research Institute.

National Fire Protection Association. 2020. *NFPA 1710: Standard for the Organization and Deployment of Fire Suppression Operations, Emergency Medical Operations, and Special Operations to the Public by Career Fire Departments*. Quincy, MA: NFPA.

National Institute of Standards and Technology. 2010. *Report on Residential Fireground Field Experiments*. Washington, DC: NIST.

National Institute of Standards and Technology. 2013. *Report on High-Rise Fireground Field Experiments*. Washington, DC: NIST.

Courtesy of Bill Strite, Cincinnati, Ohio.

CHAPTER 4

Company Operations

LEARNING OBJECTIVES

- Compare the division of labor concept as used in industry to company-level deployment.
- Describe structural firefighting functional assignments.
- Enumerate engine and ladder company fire-ground functions and tasks.
- Analyze tactics for the first-arriving fire company in relation to life safety.
- Apply engine and ladder company tasks to coordinating and controlling company-level deployment.
- List situations where splitting companies may be acceptable.
- Evaluate the positive and negative aspects of preassigning tasks, tools, and fire-ground positions.
- Assess proper and improper ventilation methods with regard to achieving the operational priorities of life safety, extinguishment, and property conservation.
- Given the fire and victim locations, determine the best location for ventilation.
- Describe safe and efficient positioning of apparatus.
- Apply engine and ladder company tasks at a structure fire.
- Analyze company operations at a structure fire scenario.
- Develop an incident action plan using engine and ladder company functional assignments.

Courtesy of Bill Strite, Cincinnati, Ohio.

Case Study

The following table lists the responding fire companies, fire officers, and emergency medical services (EMS) resources available in a particular area.

Responding Fire Companies, Fire Officers, and EMS Resources			
	Engine Companies	**Truck Companies**	**Other**
Initial Response	Benton Engine 1 Benton Engine 2 Tomtown Engine 1	Benton Truck 2 Tomtown Truck 1	Benton Duty Chief Benton Fire Chief (responding from home) Benton EMT ambulance
2nd Alarm	Tomtown Engine 2 Tomtown Engine 3	Carolton Truck 1	Tomtown Fire Chief Tomtown Heavy Rescue
3rd Alarm	Carolton Engine 1 Carolton Engine 3 Bartlettville Engine 6	Bartlettville Truck 1	Carolton Fire Chief Bartlettville Medic Unit

© Jones & Bartlett Learning.

It is 7:30 AM on Sunday, August 1. A neighbor calls dispatch reporting a fire at 100 Fifth Street in Benton in a single-family residence. A one alarm is dispatched: Benton Engine 1 responds from Station 1 at 250 Third St. (approximately one-eighth mile from the fire location). There are only two fire stations in Benton; all fire companies are staffed with a lieutenant, driver, and two fire fighters. The Benton Duty Chief is located at Station 2. Station 2 is approximately 9 miles from Station 1. Travel time from Station 2 to the fire is approximately 15 minutes under favorable weather and traffic conditions. Fire conditions on side Charlie are as shown in the photo.

1. You are the company officer on Engine 1. Would you employ an offensive or defensive strategy?
2. List the tasks assigned to all members of your company, including yourself.
3. List and prioritize assignments to be carried out by responding units.
4. What tasks should be carried out by responding engine companies?
5. What task should be carried out by responding truck companies?
6. Would you request additional units? If so, explain why they are needed and assign specific tasks to each responding unit.

Introduction

The primary responsibility of all companies working at the scene of an emergency is to work within the overall incident action plan. Freelancing cannot, and should not, be tolerated. Standard operating procedures (SOPs), provide guidelines that allow first-in units to begin operations. Following these instructions provides the incident commander (IC) with additional time to evaluate tactical options while maintaining control during those first few critical minutes on the fire ground. Assignment of company-level operations should be included in the department's SOPs. This chapter focuses on identifying and coordinating engine and truck company tasks.

Companies and crews are supervised by a company officer, thus reducing the span of control by having one person report for the entire crew. In terms of both incident scene safety and operational efficiency, the importance of having a strong command system cannot be overstated.

The **division of labor principle** was the basis for the Industrial Revolution. The division of labor concept is not strictly adhered to in the fire service because fire fighters must be cross-trained to accomplish many different fire-ground tasks. The division of labor concept stresses the need for preassigning duties (functions) at an emergency to ensure that, before the incident occurs, everyone is familiar with what will be expected of them during an emergency.

Members of hazardous materials teams, paramedics, and heavy rescue squads are required to complete specialized training and certification requirements that ensure proficiency in their particular disciplines. However, all fire fighters should have a general understanding of these special duties, just as members of the special teams should understand the duties of line fire fighters. At a minimum, line fire fighters should be able to function as part of either an engine or a truck company crew and be trained in basic emergency medical and hazardous materials mitigation skills.

The incident action plan governs company-level operations. Engine company tasks during an offensive attack focus on the interior and involve "rescue as you go" tactics while applying water to extinguish the fire. During a defensive attack, the engine company sets up master streams and applies water to nearby buildings, deluges the main body of fire, or both.

Truck companies expend a great deal of effort ventilating, laddering, forcing entry, and conducting primary and secondary searches during an offensive attack. However, during defensive attacks, truck company duties are typically limited to setting up elevated master streams and support activities.

Simply put, engine company work involves applying water to the fire. Truck companies perform search and rescue activities, as well as assisting engine companies by gaining entry, laddering, and controlling the fire through coordinated ventilation.

Heavy rescue companies primarily function as a search and rescue company on the fire ground but can be assigned to a variety of other duties, such as deploying an additional attack.

Engine Company Tasks

An engine company is the basic firefighting unit. On close examination, engine company operations can be dissected into the following functions:

- Performing rescue operations
- Establishing a water supply
- Advancing and operating hose lines (extinguishment)

In most cases, an engine company is first to arrive at the scene. The first-in officer initiates the attack and provides a means of moving water from a water supply source to the fire.

Life safety—not only the safety of the victims, but also that of the fire fighters—is the first priority of everyone on the fire ground. Engine companies usually accomplish life-safety objectives by placing attack hose lines in position to protect victims and rescuers and by providing safe evacuation routes.

Engine companies advancing hose through the main entrance of the structure sometimes encounter occupants attempting to exit the building or in some cases occupants who are unconscious. It may also be possible to conduct a cursory search and rescue as hose lines are being advanced into the fire area. Occupants may only be in need of direction and guidance, rather than actual physical assistance. In these cases, the engine crew will do whatever is necessary to rescue occupants.

When the location of victims is unknown, the determination by an engine company officer to position hose lines or to search for occupants is difficult. This tactical decision is based on the following concerns:

- Immediate danger to the occupants
- Available staffing and resources
- Time before additional resources arrive
- Extent of fire involvement
- Equipment available to perform the rescue

As an example, the fire shown in **FIGURE 4-1** is in a three-story apartment building with a reported basement fire. It is 9:00 AM on a Sunday. The first-due engine company (Engine 1) is located 2 miles from the

FIGURE 4-1 Apartment building fire with visible occupants.
© Jones & Bartlett Learning.

fire scene. The next closest fire station houses an engine and truck company and is located 5 miles from the incident. All companies are staffed with an officer, driver, and two fire fighters. Upon arrival, Engine 1 discovers that the basement fire has extended out of the basement and into the stairway. They also observe occupants on apartment balconies awaiting rescue. Fire walls divide the building into three separate fire areas. Each apartment has rated fire doors leading to the apartment and sliding glass doors leading to a patio. The truck company is not yet on the scene.

Priorities at this scene may include the following:

- Rescue occupants on balconies using a ground ladder.
- Identify which occupants would be the highest priority and develop your strategy based on saving those in greatest danger. Remember, some occupants may be inside their apartments and not visible.
- Conduct an interior search and evacuation.
- Extinguish the fire or confine it to the stairway and basement.

Of these options, extinguishing the fire has the greatest potential to save lives. The occupants on the balconies are not in imminent danger. Occupants on the first floor of the fire building are in greatest danger, but the fire door will provide a measure of safety. If necessary, they can escape or be easily rescued via windows that are at grade level. Occupants on the second and third floor of the fire building will be safe for a time provided they take advantage of the fire door leading to their apartment from the hallway. Occupants in the areas of the buildings beyond the fire wall have the protection of the fire wall and fire door, as well as the patio door. Fire and smoke will have to penetrate all three barriers before these occupants are in imminent danger.

The basement fire may present a challenge, but containing the fire in the stairway should be well within the capabilities of a standard preconnected hose line.

If there is a potential rescue that can be made, life safety should always be the top priority. The incident action plan should direct all resources (either directly or indirectly to a successful evacuation of all endangered occupants. In the apartment building example, the engine company should save lives by extinguishment or fire containment—an indirect means of saving lives.

Truck Company Tasks

The terms "ladder company" and "truck company" are synonymous. It is not necessary to have an aerial ladder or elevated platform at every fire, or even in every fire department. However, it is necessary to have truck company *functions* assigned to a specific group of fire fighters on the fire ground. General truck company duties should be preassigned. Fire operations are more often unsuccessful because of poor or nonexistent truck company work than due to a lack of water.

The following tasks are normally assigned to truck companies:

- Conduct search (primary and secondary)
- Rescue trapped victims
- Ventilate
- Force entry
- Ladder the building
- Check for fire extension
- Access concealed spaces

As this list indicates, truck companies are responsible for a wide variety of tasks when an offensive attack is in progress; therefore, proper staffing is crucial for companies charged with carrying out truck company duties. Other than providing ladder pipe and other elevated master streams, truck company activities are limited during defensive attacks. In low-staffing situations, the truck company may need to assist the engine company in initial placement of hose lines during an offensive attack. However, the tendency is for the truck company to continue handling hose lines, even when adequate engine company personnel are available. This is not the most effective use of truck personnel.

A Note about Staffing

Many departments find it difficult to provide adequate fire company staffing. A standard engine or truck company should be staffed with a minimum of four fire fighters to operate effectively. Some departments assign a driver to an elevated platform or aerial apparatus with no additional staffing. This apparatus provides a means of accessing upper levels of a building, but should not be considered a truck company unless additional staffing is provided. Additional staffing is often provided by members responding in ambulances, private automobiles, and other means.

Coordinating Company Operations

For engine companies, rescue is typically an indirect activity. Getting hose lines between the victims and the fire makes rescue or self-evacuation possible. Extinguishment or reducing heat production improves safety for fire fighters and occupants. For truck companies, rescue may be a more direct activity. Truck companies will generally ventilate to control the spread of heat, smoke, and fire while conducting the primary search and removing any victims. It is imperative that ventilation be coordinated with the engine company attacking the fire. The engine company should be in position and ready to attack the fire prior to venting. Venting will generally increase fire intensity and often changes the fire and smoke flow path. Truck companies will also ladder the building to establish alternative evacuation routes.

FIGURE 4-2 illustrates how an engine company and a truck company coordinate rescue and ventilation.

FIGURE 4-2 Engine and truck company coordinated rescue.
© Jones & Bartlett Learning.

It is imperative that engine and truck company rescue operations be coordinated. Conducting a search and rescue operation without the protection of a hose line controlling the fire is extremely hazardous. For example, the truck crew operating on the third floor in Figure 4-2 would be in an extremely hazardous location if the engine company crew were not attacking the fire on the floor below. Likewise, the outside venting at the second floor would tend to increase fire intensity if they vented the window prior to the fire attack team getting water on the fire. It is critical that all crews coordinate activities and keep the IC informed of their location and status.

The IC must coordinate all activities on the fire ground. Control and coordination are vital if the engine company is unable to control the fire or a decision to change from an offensive to a defensive attack is made. Truck companies will in many cases be working in areas above the fire while the engine company controls the fire. If the fire is not controlled, truck companies working inside the structure need to be notified, and, if at all possible, their retreat should be protected by the engine company. When companies are operating above ground level, ladders should be placed to provide alternate egress paths. A responding truck company or a proactive rapid intervention team can accomplish ladder placement.

Safety and control dictate that operating units work as groups (companies or crews). The first-in truck company at an offensive operation is the exception to that rule. Given the variety of tasks and the fact that many of these tasks can be performed by two-member teams, it is permissible for truck companies to split into separate crews. At a working structure fire, two members may go to the roof to ventilate, while a second two-member crew may conduct the primary search. In addition, once the first task is completed, these "truckies" may reposition to perform other ladder company tasks. For example, fire fighters who are assigned to the roof may also perform property conservation on the floor below the fire. Once these initial operations are accomplished, truck company members should be reunited as a company or full crew under the direction of the truck company officer. It is not freelancing when a company officer splits a crew into separate teams. Each crew should have a team leader who can communicate with the company officer. In some departments, the crew leader will be a higher-ranking company member (such as a sergeant); in others, the crew leader will be a fire apparatus operator or senior fire fighter.

Other exceptions to the general rule of maintaining company unity under the direct control of the company officer include the first-arriving company splitting into an inside crew and an outside crew in compliance with two-in/two-out rules (NFPA 1500, 2018; Occupational Safety and Health Administration, 2004). When a company is split into two or more crews, the NIMS organizational chart should account for each crew. There are also times when the apparatus operator should remain at the apparatus to operate pumps or an aerial device.

In Figure 4-2, the engine company is divided into inside and outside crews. The two members of the inside crew are advancing and operating the hose line to attack the fire. The outside crew is the **initial rapid intervention crew (IRIC)** and is staffed by the pump operator at the apparatus and the second fire fighter establishing the water supply from the hydrant.

All companies should remain together as a unit whenever possible. Maintaining company unity facilitates accountability because the officer can immediately verify the safety of the entire crew.

As has been pointed out, general truck company duties should be preassigned, usually to fire fighters arriving on aerial elevated platform or quint apparatus. This brings about a point of discussion: preassignment of specific tasks, tools, and positions. Many departments establish SOPs that require truck company members to carry specific tools. For example, the officer carries a utility bar (e.g., Halligan, Hux bar), another member an axe, and still another member a salvage cover or rope rescue equipment. Are these good SOPs?

In a high-rise office building, is the same equipment needed for a fire on the first floor during working hours as is needed for a fire on the top floor when the building is unoccupied and secured? Forcible entry tools will be needed in many cases, but will the same tools be needed at every fire?

Most jurisdictions have a wide variety of occupancies, from single-family dwellings to high-rise structures, and carrying the same tool into every situation, or even the same tool for different fires within the same structure, may not prove to be a productive policy. The best that can be said for the practice of preassigning tools and equipment is that establishing default SOPs for tools has merit. If there is no reason to bring a different tool (e.g., nothing-showing situations), bring the assigned tool. Preassignment of tools should be based on experience, and it should be limited to one or two members, rather than the entire crew having preassigned tools. Company officers must be prepared to change tool assignments according to circumstances.

The same reasoning can be applied to preassigned tasks such as search and rescue, ventilation, and forcible entry. Ventilation may not be required or may have negative consequences under certain conditions. In many cases, the fire self-vents. When ventilation is warranted, its priority changes depending on fire conditions, building characteristics, location of the fire, location of the occupants, and available ventilation options. Primary search may require the entire truck crew or could be accomplished by the engine company as it advances.

Preassigning truck company positions, unless done in a general way, with company officers having the latitude to use discretion, may result in inefficient use of resources. Assigning fire fighters to the roof when the fire is several stories below the roof simply places them in a useless—and often dangerous—position when they could be accomplishing critical truck company activities. Few fire departments have sufficient staffing to place fire fighters in unnecessary positions when a working fire is in progress.

Fire officers and fire fighters should be highly trained, rational human beings. SOPs that limit their ability to make reasonable decisions are generally counterproductive. Fire fighters who are assigned to truck companies should bring tools into the structure, but the tools should be those needed for the specific operation. The roof assignment may be extremely important in many fires but useless in others. Forcible entry may be the single most important fire-ground task when victims are trapped in secured areas or when the fire attack crew cannot make entry. Other times, windows and doors may be open, making forcible entry an unnecessary task. Given the fact that timing ventilation and forcible entry can make the difference between success and failure, the preferred approach is to have "truckies" attending to necessary truck work while engine crews are securing a water supply and advancing hose lines.

Quint and Quad Companies

Some departments staff multifunction apparatus that are equipped to perform both engine and truck company operations. These are usually referred to as **quint** or **quad** companies. A quint is equipped to provide the following five functions: water, pump, hose, ground ladders, and an aerial device **FIGURE 4-3**. A quad includes all of these functions except the aerial

FIGURE 4-3 Quint apparatus.
Courtesy of Captain Nick Morgan, St. Louis Fire Department.

device. Some departments provide sufficient staffing on this single apparatus to achieve both engine and truck company objectives, while other departments do not provide the staffing necessary to perform both engine and truck company tasks. For example, a quint staffed with an officer, driver, and four fire fighters could establish a two-person IRIC. The officer and two fire fighters can advance an attack hose line to the fire while the other two fire fighters initiate a primary search of the fire area. The driver will operate the apparatus pumps or ladder. An alternative would be to utilize the additional two fire fighters as an IRIC if the additional responding companies are a considerable distance away. The IRIC fire fighters could place ladders and eliminate barriers to fire fighter exit until the arrival of a formal rapid intervention crew (RIC). Having a minimum of six fire fighters assigned to a quint company allows the versatility of using the pumps and ladder in many different ways. With only four fire fighters assigned to a quint, the department, via SOPs or the company officer, must make a decision to be either a truck company or an engine company. It is essential that SOPs describe actions the quint company is to take when it is staffed with fewer than six fire fighters and is the first apparatus on the scene. In most cases, this first-arriving quint company will function as an engine company if staffing does not permit functioning as a fully staffed quint company. However, there are occasions when the aerial device can be used to rescue visible occupants. Even if the aerial device is not being used on arrival, the quint apparatus should be positioned for future use of the aerial device. Fire fighters assigned to quint apparatus must remain aware that this apparatus is likely to be used as an engine company, truck company, or both.

Ventilation

Ventilation is one of the IC's most important tactical considerations, and it requires close coordination between fire fighters who are venting and attack crews. Ventilation should not occur until the interior crews are ready to apply water to the fire. Proper ventilation can have a positive effect on all three fire-ground priorities (life safety, extinguishment, and property conservation). However, ventilation is a double-edged sword. Improper ventilation can adversely affect all three priorities. **TABLE 4-1** lists the effects of both proper and improper ventilation. **FIGURE 4-4** shows proper vent locations.

It must be remembered that any form of venting, including opening an exterior door, will increase the oxygen supply and change the flow path. In many cases, the fire is oxygen deficient, and the increase in oxygen created by the ventilation opening can result in a dramatic increase in fire growth. This is yet another reason to have a charged hose line staffed and in place whenever interior operations are being conducted.

TABLE 4-1 The Effects of Proper and Improper Ventilation		
	Proper Ventilation	**Improper Ventilation**
Life safety	■ Pulls fire away from trapped occupants or their means of egress	■ Provides air to a ventilation-limited fire, causing rapid increase in temperature ■ Draws fire toward victims or extends fire through their exit path
Extinguishment	■ Limits fire spread by channeling fire toward nearby openings and allows fire fighters to safely attack the fire	■ Spreads fire into previously undamaged areas ■ Causes a backdraft or lets the fire gain headway while hose lines are being advanced
Property conservation	■ Limits smoke, heat, and water damage by allowing an interior attack and removing the products of combustion, which are often the leading cause of damage	■ Causes excessive and unnecessary damage ■ Does not remove the damaging products of combustion, but rather spreads them throughout the building

© Jones & Bartlett Learning.

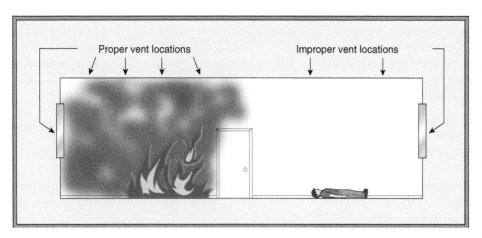

FIGURE 4-4 Vent locations.
© Jones & Bartlett Learning.

Apparatus Positioning

Apparatus positioning and the fire company's assignment are directly related. The positioning of apparatus must be consistent with the company's objective. If aerial ladders or towers are needed, the first-arriving truck company will position for aerial or tower placement. All companies should make a conscious effort to keep the fire zone as accessible and safe as possible.

The company officer (or IC) should always think about positioning. If the first-in engine company will be used to lay attack hose lines, position the apparatus on the same side of the street as the fire and as close to the curb as possible, with the side discharge gates aligned with the desired entry point, provided there is a low probability of building collapse. In most residential fires, the first-due engine should pull past the front of the structure, providing a three-sided view and leaving the front of the structure for ladder truck access. Positioning of this first-arriving apparatus should be addressed in SOPs and, among other things, depends on the layout of the hose bed.

Department SOPs generally address actions to be taken by the second-in engine company. SOPs for this company may allow more discretion on the part of the company officer. Some departments specify that the first-in and second-in engine companies secure separate sources of water. Other departments permit an **attack pumper** configuration in which the first engine company goes directly to the front of the building and the second engine company provides the water supply. It must be remembered that working from the limited water supply carried on a pumper increases the chance that the company attacking the fire will exhaust their water supply prior to extinguishment, and a continued operation is based on an assumption that the second-due engine company will arrive at the scene and supply water in a timely manner. Each department must decide what works best for its jurisdiction.

For purposes of discussion, when the second engine company is supplying the attack pumper from a hydrant, it should select the suction intake that will allow the apparatus operator to get close to the curb while keeping the intersection open.

If the first-in truck company will be using its aerial device, the truck must be positioned to obtain a safe operating angle and reach the desired location for roof access, rescue, ventilation, or other tasks. Whenever possible, the aerial should be placed in a position to access two sides of the building. The first-due truck should always be positioned to utilize the aerial for upper level rescue or roof access. Getting tools, ladders, and other equipment to the building is a labor intensive job if the truck is parked too far from the entry point. In positioning the truck, it is also important to remember that on some apparatus, ladders feed off the rear of the truck. When this is the case, be certain that access to the rear of the truck is not blocked.

FIGURE 4-5 shows an example of apparatus positioning for an attack pumper operation with the aerial ladder being used to gain access to the roof.

Some common errors when positioning apparatus include the following:

- Placing aerials under wires where they cannot be safely raised
- Placing aerials and platforms in unsafe and unstable positions
- Not allowing enough room to extend the outriggers
- Apparatus/staff cars blocking access to the fire area or front of the building
- Pumpers placed in positions where preconnected attack hose lines are difficult to lay
- Apparatus unnecessarily blocking streets or fire hydrants

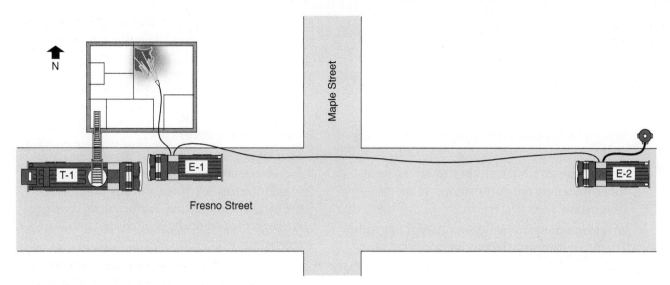

FIGURE 4-5 Coordinated attack pumper operation.
© Jones & Bartlett Learning.

- Unnecessarily blocking streets with large-diameter hose
- First-arriving companies not securing essential attack positions
- Positioning apparatus in dead ends or other locations where they cannot be quickly repositioned
- Companies assigned to staging responding through the fire area, causing unnecessary congestion (Staging operations must be defined in department SOPs.)

Other vehicles should be positioned well outside the operational area.

Wrap UP

CHAPTER SUMMARY

- At the company level, the span of control is maintained by having the company officer manage the assigned crew.
- The division of labor concept stresses the need for preassigning duties (functions) at an emergency scene.
- An engine company is the basic firefighting unit. Engine company operations include conducting rescue, establishing a water supply, and operating hose lines.
- General truck company duties include search and rescue, ventilation, forcible entry, laddering, and reconnaissance.
- Truck company activities are varied and numerous during offensive attacks but are limited during defensive attacks.
- Safety and control dictate that operating units work as companies or crews.
- Maintaining company unity improves safety and accountability; however, it is permissible to split the first-arriving company to abide by the two-in/two-out rule. Furthermore, the first-arriving truck company is often split into two crews.
- Communication between company officers or crew leaders and the IC must be maintained.
- Preassigned tools, tasks, and operations should be adjusted to specific circumstances; strict adherence to tool and task SOPs can sometimes be counterproductive.
- Offensive firefighting involves a coordinated effort of hose placement and ventilation.

KEY TERMS

attack pumper The first-arriving engine company that goes directly to the fire building without securing a water supply.

division of labor principle The principle that an incident or task should be broken down into smaller, more manageable tasks and personnel assigned to complete those tasks (developing job skills in a concentrated area to allow for more productivity).

initial rapid intervention crew (IRIC) A standby team of at least two fire fighters located outside the hazard area available to provide assistance to fire fighters operating within the hazard area until a formal RIT can be established.

quad A multifunction apparatus that is equipped to provide for the following four functions: water, pumps, hose, and ground ladders. In other words, this is a quint company minus the aerial ladder.

quint A multifunction apparatus that is equipped to perform both engine and truck company operations. It is equipped to provide the following five functions: water, pumps, hose, ground ladders, and an aerial ladder.

SUGGESTED ACTIVITIES

1. Using first-alarm resources from your department or a department with which you are familiar, assign tasks to the first-arriving and subsequent-arriving companies for the scenario shown in **FIGURE 4-6** and described as follows:
   - Time, day, and date: 7:00 AM, Sunday July 7
   - Weather: partly cloudy and warm
   - Structure: two-story apartment building
   - Conditions on arrival: as shown in Figure 4-6
   - Area water supply: good; hydrants spaced every 500 ft (152 m)

FIGURE 4-6 Apartment building with third-floor balcony on fire.
Courtesy of Bill Strite, Cincinnati, Ohio.

2. Using first-alarm resources from your department or a department with which you are familiar, assign tasks to the first-arriving and subsequent-arriving companies for the scenario shown in **FIGURE 4-7** and described as follows:

   - Time, day, and date: 2:00 PM, Tuesday January 7
   - Weather: clear and cold (20°F [–7°C])
   - Structure: one-story print shop attached to a four-story automobile parts warehouse
   - First alarm response: three engine companies (including Engine 23), 1 truck company, and a district chief
   - Conditions on arrival: as shown in Figure 4-7
   - Area water supply: good; hydrants spaced every 500 ft (152 m)

   You are the captain of Engine 23 with a driver and two fire fighters assigned to the company.

   A. Provide an initial report to dispatch.

   B. What is your initial action as Engine 23 Captain?

   C. Would you request assistance beyond the initial alarm? Explain your answer and, if you request assistance, indicate how much assistance would you request on arrival.

FIGURE 4-7 Print shop attached to automobile parts warehouse.
Courtesy of Captain Nick Morgan, St. Louis Fire Department.

3. Using the resources from your department or a department with which you are familiar, assign tasks to the companies that normally respond on the first alarm.

   - Time, day, and date: 4:00 PM, Saturday June 23
   - Weather: partly cloudy and warm (75°F [24°C])
   - Structure: one-story with basement, ordinary construction
   - Occupancy: Auto-body shop
   - Exposures:
     - Side Alpha (view as shown is from Side Alpha): parking lot)
     - Side Bravo: medical supply distributor
     - Side Charlie: parking lot

- Side Delta: auto parts manufacturer (doors lead to auto parts manufacturer from auto body shop.
- The first-alarm response: three engine companies (including Engine 23), one truck company, and a district chief
- Conditions on arrival as shown in **FIGURE 4-8**
- Area water supply: good; hydrants spaced every 500 ft (152 m)

FIGURE 4-8 Auto body shop.
Courtesy of Bill Strite, Cincinnati, Ohio.

REFERENCES

National Fire Protection Association. 2018. *NFPA 1500: Standard for Fire Department Occupational Safety and Health Program.* Quincy, MA: NFPA.

Occupational Safety and Health Administration (OSHA). 2004. "Section 1910.134: Respiratory protection." *Code of Federal Regulations*, title 29. Published February 17, 2004. https://www.ecfr.gov/cgi-bin/retrieveECFR?gp=&SID=bcb5f28d754bd42e5ca662c9a8e8bd1c&mc=true&n=pt29.5.1910&r=PART&ty=HTML#se29.5.1910_1134.

Courtesy of David J. Jones, Cincinanti, Ohio.

CHAPTER 5

Fire Fighter Safety

LEARNING OBJECTIVES

- Explain how applying a risk-versus-benefit analysis is the key to safe and effective fire-ground operations.
- Describe how employing effective fire-ground strategy and tactics affects safety on the fire ground.
- Identify and analyze the major causes involved in on-duty fire fighter fatalities related to health, wellness, fitness, and vehicle operations.
- Evaluate the effects of proper and improper ventilation during offensive fire attacks.
- Define the term *occupational injury* as used by the National Fire Protection Association (NFPA).
- Analyze the trend in the number of fire fighter on-duty deaths and injuries over a 37-year period.
- Define frequency and severity as they relate to fire fighter injuries.
- Compare and contrast fire trends and fire fighter on-duty deaths.
- Describe the relative risk to fire fighters when combating fires in different occupancy types.
- Describe how investigating and analyzing "near miss" accidents can reduce the number and severity of fire fighter injuries and fatalities.
- Discuss risk management principles applied to the fire ground.
- List and compare the provisions in NFPA 1500, *Standard on Fire Department Occupational Safety, Health, and Wellness Program*; Chapter 8, *Offensive Operations,* and the IAFC Rules of Engagement.

- Use a probability analysis to assess the occupied status of a building based on time, occupancy, and other factors.
- List the variables affecting the time to flashover.
- Estimate collapse time based on burn time, fire intensity, content load, and construction type.
- List three important questions that must be answered prior to conducting roof operations at a structure fire.
- Describe the characteristics and possible components of a green building.
- Examine the difference between a managed retreat and an evacuation due to an imminent hazard.
- Evaluate the difference between modern lightweight and older structural components.
- Use experimental data and case studies to analyze the risk-versus-benefit assessment related to basement fires.
- Discuss and contrast prefire and fire conditions that contribute to structural collapse.
- Examine the hazards presented by suspended ceilings.
- Compare construction methods in terms of structural stability, fire extension, and fuel contribution.
- Develop zones and perimeters around a structure fire.
- Define and explain the five time segments from ignition to effective action.
- Evaluate the survivability, structural stability, and flashover from ignition to effective action.
- Compute the staffing necessary to achieve the tasks enumerated in NFPA 1710, *Standard for the Organization and Deployment of Fire Suppression Operations, Emergency Medical Operations, and Special Operations to the Public by Career Fire.*
- Define and compare flashover and backdraft.
- Explain the relationship between the National Incident Management System (NIMS) and a fire fighter accountability system.
- List situations when a personal accountability report should be initiated.
- Explain the importance of alternative egress for fire fighters conducting an offensive attack.
- Define and describe the role of a rapid intervention crew (RIC).
- Determine the number of personnel to be assigned to the RIC based on the size and complexity of the building and incident.
- Using the LUNAR acronym construct an emergency message for a fire fighter needing assistance.
- Explain measures that can be taken to improve the chances of survival when fire fighters are lost and out of air in a large building, including supplying air to a fire fighter who has exhausted his or her air supply.
- List tools that should be available to a RIC.
- Compare the advantages and disadvantages of a mobile RIC versus a stationary RIC.
- Evaluate hazards to fire fighters during overhaul operations.
- List factors the incident commander should consider when formulating an incident action plan to be used during overhaul.
- Describe hot and cold weather rehabilitation.

Courtesy of the Bryan Fire Department.

Case Study

At 8:13 PM, on February 15, 2013, two career fire officers were killed and two career fire fighters were injured combating an advanced fire in a 7400-ft^2 (687-m^2) assembly hall. The fire was observed burning at the roof level on arrival. The parking lot was empty and there was no indication that the building was occupied. Two members of the first-arriving fire crew advanced a small-diameter hose line into the building in an unsuccessful attempt to extinguish the fire. No attempt was made to remove ceiling tiles or use a thermal imaging camera to determine if the fire visible at the roof was in the overhead false ceiling space.

Eventually, the attack crew was forced to retreat due to a low air supply. The fire fighter was able to follow the fire hose to safety, but for unknown reasons, the fire officer was unable to follow the fire hose out of the building. The trapped fire officer transmitted a call for help. A two-person rapid intervention crew (RIC) entered the building, following the hose line, and found the fire officer, who was responsive. The RIC and fire officer being rescued were caught in a flashover as they made their way toward the exit; the fire officer did not survive. The RIC officer later succumbed to his injuries while undergoing treatment at a burn center.

1. How do occupancy and time of day affect the risk-versus-benefit analysis?

2. Explain the importance of a risk-versus-benefit analysis with regard to fire fighter safety.

3. Evaluate the risk to fire fighters entering the building using a 0 to 10 scale. A risk of 0 indicates there is no risk to fire fighters entering the building, whereas a grade of 10 would mean an almost certain risk of death for fire fighters entering the building.

4. Evaluate the potential life hazard to occupants using a 0 to 10 scale.

5. Using a 0 to 10 scale, in the given scenario, what is the probability that occupants are inside the building at this time?

6. What are the positive and negative effects of ventilation during a structure fire?

7. What influence does compartment size have on fire fighter safety?

8. How can fire companies determine whether fire has entered a concealed space?

Data from: NIOSH Fire Fighter Fatality Investigation Summary [F2013-04], May 20, 2014, "Two Career Lieutenants Killed and Two Career Fire Fighters Injured Following a Flashover at an Assembly Hall Fire—Texas," https://www.cdc.gov/niosh/fire/pdfs/face201304.pdf.

Introduction

Fire departments are dedicated to saving lives and property from the perils of fire. Saving lives is the highest tactical priority at the incident scene. The fire fighter's life is valued as being as important as any other life, and the fire fighter is the most valued resource of any fire department. Too often, fire fighters lose their lives in the process of saving the lives of others. Even more tragic are incidents where fire fighters lose their lives when there are no lives or property to be saved.

The guiding objective of this text is preparing the fire officer to take command at structure fires by fully using available resources in a safe *and* effective way. Being safe *or* effective is relatively easy. The first-arriving company officer or chief officer assuming the role of incident commander (IC) only concerned with fire fighter safety would apply defensive streams from a distance beyond the collapse zone at all structure fires. In contrast, the fire officer only interested in being effective might routinely apply an aggressive offensive interior attack regardless of circumstances. Being safe *and* effective requires fire officers to carefully analyze the risk versus the benefit. Fire fighter safety is closely related to the risk-versus-benefit analysis discussed in the section, *Developing an Incident Action Plan*, in Chapter 3. Probably the most important safety factor at the scene of a structure fire is a well-organized operation based on a solid risk-versus-benefit analysis utilizing effective tactics.

In examining fire-ground incidents resulting in on-duty injuries and deaths, it can be seen that improper firefighting tactics were a contributing cause at many of these incidents. Extinguishing the fire improves safety for both occupants and fire fighters. In examining several cases, it was found that inadequate fire flows due to poor water supplies, hose kinks, or improper pump operation allowed the fire to gain in intensity resulting in injuries to fire fighters exposed to extreme heat and smoke conditions. In addition to extinguishing the fire, it is necessary to protect fire fighters by placing hose lines in close proximity to all areas where fire fighters are working within a structure. Proper ventilation can improve operational safety, but in many cases ventilation results in increased fire intensity unless hose lines are charged and in position to knock down the fire. Poor ventilation combined with a lack of protective hose lines is a deadly combination. A properly coordinated fire attack with proper ventilation greatly reduces the risk of fire fighter injury.

It is imperative that proper tactics be applied at structure fires to protect fire fighters. Proper training, combined with solid fire-ground procedures are an absolute necessity. Too often, a failure to complete basic fire-ground tasks, such as efficient hose line evolutions or pump operations, results in fire fighter injuries or deaths.

This chapter describes how NFPA 1500, *Standard on Fire Department Occupational Safety, Health, and Wellness Program,* and other standards are applied to assessing fire fighter safety at the incident scene (NFPA 1500, 2018). It also relates fire fighter safety to prefire and fire conditions to include burn time, fire intensity, and structural stability, in addition to the correlation between response time and progression to flashover. This chapter also discusses some of the factors relating to fire fighter safety, using statistics, as well as a series of case studies and established practices.

Incident Safety Officer

The incident safety officer is a critical component of ensuring fire fighter safety at an incident. The incident safety officer should survey the entire incident scene and has the authority to immediately stop unsafe operations. Duties of the incident safety officer include:

- Report to the command post and confirm the "incident safety officer" assignment (e.g., "Chief Jones reporting, do you want me to take safety?")
- Determine the strategy (offensive, defensive, transitional) and tactics being employed at the incident scene.
- Evaluate the operation by systematically visiting every area where fire fighters are deployed, beginning with the areas presenting the greatest potential hazard.
- Share safety concerns with the officer responsible for that specific area. If necessary, stop the operation and immediately notify the IC and other officers in the chain of command.
- After completing a tour of all areas where fire fighters are deployed, return to the command post and provide the IC with a status report.
- Continue touring the fire ground throughout the fire, suggesting changes in strategy and tactics as required.

At the scene of a major and widespread fire, the incident safety officer is often the only person with a complete view of the fire scene. On rare occasions it is necessary to appoint subordinate safety officers to adequately cover a major and/or widespread incident.

Many times, fire fighters are injured during overhaul, therefore the incident safety officer must continue to monitor the situation and recommend relief forces and rehabilitation as needed. The incident safety officer should also be included in developing a post-incident analysis.

Detailed information on the incident safety officer role can be found in *Fire Department Safety Officer*, by the National Fire Protection Association (NFPA) and David W. Dodson.

Fire Fighter Injuries and Fatalities

Identifying and analyzing how fire fighters are killed and injured is critical to reducing fire fighter injuries and deaths. The NFPA has compiled and analyzed fire fighter on-duty fatality statistics since 1977. The annual report of on-duty fire fighter fatalities for the previous calendar year is published each year in the July/August edition of the *NFPA Journal*. The annual report of on-duty fire fighter injury is published in the November/December issue of the *NFPA Journal*. The charts and graphs used here are taken from data provided by the NFPA as well as other sources.

A necessary part of any statistical study involves defining what is meant by various terms and describing and analyzing the results. Various organizations define fire fighter injuries and deaths differently than the NFPA and use other methodologies to gather data, thus yielding different results.

NFPA 1500, *Standard on Fire Department Occupational Safety, Health, and Wellness Program* defines the term occupational injury as an injury sustained during the duties, responsibilities, and functions of a fire department member (NFPA 1500, 2018). If the injury proves fatal, or if any illness that was incurred as a result of actions while on duty proves fatal, then the death is counted as an on-duty fatality. On-duty fatalities are associated with specific on-duty activities and are reported as of the date of injury, or

> **NOTE**
> The fire resulting in the largest loss of fire fighters was the World Trade Center fire in 2001.

> **NOTE**
> An FDNY study found that nearly 9000 fire fighters who were exposed to 9/11 World Trade Center dust may be at greater risk for cancer than those who were not exposed, and the city's World Trade Center Registry found small increases in the rates of prostate cancer, thyroid cancer, blood cancer, and multiple myeloma among nearly 34,000 rescue and recovery workers, compared to New York State residents.
>
> *Source*: Never Forget Project. http://neverforgetproject.com/statistics. Accessed January 4, 2020.

onset. One of the real values in statistical analysis is defining trends. Statistical data occurring over a long period are more meaningful in determining trends than research limited to a few years. For this reason, this text uses NFPA statistics exclusively to evaluate fire fighter injuries and deaths.

In the case of fire fighter injury and fatality reports, the goal is to determine whether conditions related to on-duty injuries or health issues are improving or worsening. For this reason, large loss of life incidents, such as the September 11, 2001, World Trade Center incident, are not included in the printed data in the NFPA report. However, it is also important to evaluate large loss of life incidents, to evaluate methods of reducing catastrophic incidents.

TABLE 5-1 lists the deadliest incidents resulting in eight or more fire fighter fatalities.

TABLE 5-1 Incidents Resulting in Eight or More Fire Fighter Fatalities

Date	Fatalities	Occupancy	Location
9/11/2001	340*	World Trade Center	New York City, NY
4/16–4/17/1947	27	Ship explosion	Texas City, TX
12/22/1910	21	Meat packing plant	Chicago, IL
2/17/1882	19	Fireworks plant	Chester, PA
7/19/1956	19	Refinery	Dumas, TX

Date	Fatalities	Occupancy	Location
3/17/1890	13	Store	Indianapolis, IN
1/15/1895	13	Warehouse	Butte, MT
12/21/1910	13	Leather manufacturer	Philadelphia, PA
3/10/1941	13	Theater fire	Brockton, MA
7/10/1893	12	Cold storage warehouse (Columbian Exposition)	Chicago, IL
9/5/1896	12	Theater	Benton Harbor, MI
10/17/1966	12	Multiple mercantile	New York City, NY
7/5/1973	12	Railroad tankers	Kingman, AZ
10/19/1857	10	Stores and warehouses	Chicago, IL
10/28/1954	10	Chemical plant	Philadelphia, PA
7/23/1984	10	Oil refinery	Romeoville, IL
6/19/1867	9	Theater	Philadelphia, PA
11/9/1872	9	The Great Boston Fire	Boston, MA
4/9/1894	9	Theater	Milwaukee, WI
2/3/1939	9	Restaurant	Syracuse, NY
1/17/1950	9	Construction site	Colorado Springs, CO
1/28/1961	9	Cold storage warehouse	Chicago, IL
6/17/1972	9	Building under renovation	Boston, MA
6/18/2007	9	Furniture store/warehouse	Charleston, SC
4/17/2013	9	Fertilizer plant	West, TX
7/9/1850	8	Wharf fire/conflagration	Philadelphia, PA
10/26/1913	8	Rubber products manufacturer	Milwaukee, WI
4/18/1924	8	Leather product storage	Chicago, IL
8/1/1932	8	Hotel	New York City, NY
7/9/1943	8	Mixed commercial	Chicago, IL
8/17/1975	8	Oil refinery	Philadelphia, PA

Wildland fires have been excluded from this chart, for the purpose of focusing on structural fires.
*On 9/11/2001, 343 employees of the FDNY were killed: 340 fire fighters, two paramedics, and a chaplain. For the purposes of its annual fire fighter fatality report, NFPA only lists the count of fire fighters.
© Jones & Bartlett Learning.

In examining the charts in **FIGURE 5-1**, it is clear that the number of on-duty fire fighter deaths is trending downward over this 42-year period.

It is essential that fire departments closely examine injuries, fatalities, and near miss accidents that occur in their department and elsewhere, and then

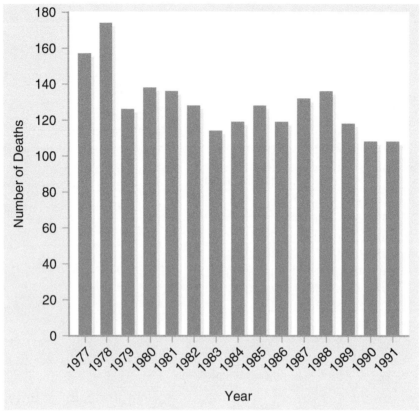

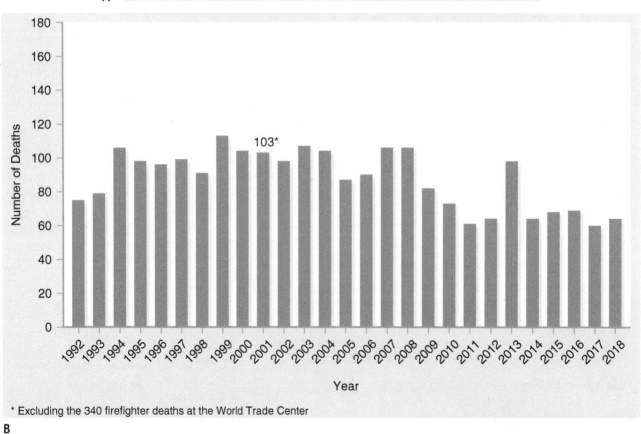

FIGURE 5-1 A. On-duty fire fighter deaths, 1977–1991. **B.** On-duty fire fighter deaths, 1992–2018.
Data from NFPA, Firefighter Fatalities in the United States – 2018, June 2019.

take action to prevent future injuries. Several organizations publish fire investigative reports that analyze individual fires that resulted in fire fighter fatalities, including the NFPA, National Institute for Occupational Safety and Health (NIOSH), U.S. Fire Administration (USFA), and others. These reports provide information, which, if properly used and analyzed, can prevent future on-duty deaths and reduce injury **frequency** and **severity**.

It is highly recommended that all fire departments appoint an individual as department safety officer. In large departments, there may be more than one person designated as department safety officer. Department safety officers should analyze working fires not only in their jurisdiction, but also include working fires where members of their department respond. A case study is useless if it is not shared, therefore training based on case studies should be provided to all members of the department. In addition, case studies from other credible organizations should be analyzed during training sessions.

Many measures have been taken to improve fire fighter safety since 1977 when the NFPA began tracking fire fighter line-of-duty deaths. Personal protective equipment and breathing apparatus used by fire fighters in the present day are vastly improved compared to protective gear used in 1977 when the NFPA began gathering data on fire fighter injuries and fatalities.

In 1987 the first edition of NFPA 1500, *Standard on Fire Department Occupational Safety and Health Program*, was adopted with subsequent revisions every 5 years. Following the provisions outlined in NFPA 1500 can substantially reduce injury frequency and severity (NFPA 1500, 2018).

NFPA 1500 provides the requirements for safety measures to be taken at the incident scene as well as protective clothing, apparatus, fire stations, medical qualifications and others. These measures represent the absolute minimum. All fire fighters and officers must be familiar with this standard to be effective in protecting their most valued resource: the fire fighter. Several important fire-ground safety issues are enumerated in the NFPA 1500 standard (NFPA 1500, 2018):

- Risk management principles must be applied to fire-ground operations.
- The IC is responsible for overall safety at the incident scene.
- An incident management system must be used at all emergency scenes.
- The IC maintains command and control of all operating forces within a common strategy based on situation analysis. The situation analysis must be ongoing with changes in strategy commensurate with the changing situation.
- An adequate number of personnel must be available to implement the strategy.
- Preestablished standard operating procedures (SOPs) must be implemented.
- An accountability system must be used at all incidents.
- Rapid intervention crews (RICs) must be provided at all stages of the incident, beginning with the first-arriving unit using a two-in/two-out process.
- Inexperienced members must be directly supervised by more experienced members.
- Minimum basic and continued fire training must be provided in accordance with SOPs.
- Medical treatment and rehabilitation must be available as needed.
- Full personal protective clothing must be worn.
- Self-contained breathing apparatus (SCBA) must be worn in hazardous atmospheres, atmospheres that are suspected of being hazardous, or atmospheres that may rapidly become hazardous.
- Personal alert safety system (PASS) devices must be worn and activated before entering a hazardous area.
- Postincident analysis is necessary for significant incidents or those causing serious injury or death.
- Prevention and mitigation of fire fighter exposure to fire ground toxic contaminants

NFPA 1500 contains a detailed description of minimum requirements (NFPA 1500, 2018). Standards issued by the Occupational Safety and Health Administration (OSHA) also address measures to be taken to protect fire fighters (OSHA 29 CFR 1910, 2019).

There is no doubt that attention to safety and the attitude toward safe operations has improved greatly within the fire service. Unfortunately, even with these improvements the fire service is still experiencing a large number of on-duty deaths and injuries. Most troubling is the fact that the fire service continues to experience a large number of fire fighter deaths on the fire ground **FIGURE 5-2**.

Considering all that has been done to improve fire fighter safety, the number of fire fighter fatalities is a very disturbing statistic. Could this fatality rate be due to a lack of fire experience and training? Fire fighters are now expected to perform many additional emergency tasks as compared to the 1970s. Each of these tasks necessitates extra training. As a result, many departments find it difficult to provide adequate structural firefighting training and education to make up for the lack of fire-ground experience. Nothing

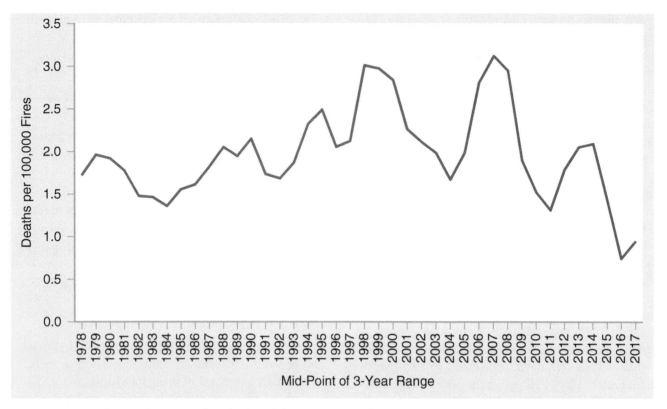

FIGURE 5-2 Death rates for noncardiac deaths while operating inside structure fires—rolling 3-year average.
Source: National Fire Protection Association, Applied Research, Quincy MA, November 2019.

completely takes the place of experience, but task- and tactic-oriented fire training and education can improve safety during fire-ground operations.

The fire ground is becoming more hazardous. Lightweight construction combined with heavy fire loads (including large quantities of plastics) in larger buildings significantly increase the risk to fire fighters. Engineered lightweight roof and floor systems, present in residential and larger commercial buildings, are discussed later in this chapter. Lightweight construction methods create a substantial hazard to fire fighters.

Modern personal protective equipment (PPE) provides a higher level of protection, thus permitting fire fighters to make a closer approach to the fire. Advancing farther inside the building and closer to a more intense fire places fire fighters at greater risk. The everyday home fire provides experience that can be used on larger fires, but different strategies and tactics are required to meet the challenges posed when a large, complex building is on fire. Occupancy plays a distinct role in fire fighter safety.

Never consider any fire routine. Most fire fighter fatalities occur in home (dwellings and apartments) fires because most fires occur in these occupancies. In 2017, there were 379,000 residential structure fires, accounting for 76 percent of all structure fires, an increase of 7500 fires from 2016. Of these fires, 262,500 occurred in one- and two-family homes, accounting for 53 percent of all structure fires. Another 95,000 fires occurred in apartments, 19 percent of the structure fire total. The total number of home fires for 2017 was 357,000. There were also 120,000 nonresidential structure fires in 2017, an increase of 15 percent from 2016. However, once a fire occurs, the risk to fire fighters is much greater in nonresidential occupancy types, as shown in **FIGURE 5-3**.

The probability of fatality is much greater in a manufacturing occupancy than in a residence. Comparing the on-duty death rate of 22.5 deaths per 100,000 fires in manufacturing occupancies to the 2.6 deaths per 100,000 fires in residential occupancies indicates that the chances of a fire fighter fatality is 8.7 times greater when combating a manufacturing occupancy fire.

When examining statistical information, it is important to consider the impact of a single fire where many lives are lost. For example, if the 343 fire fighter line-of-duty deaths at the World Trade Center were added to the stores/offices category, the bar representing these occupancies would be off the chart. Vacant structures were once listed separately and represented the highest risk to fire fighters. The NFPA now statistically categorizes vacant structures according to their former use.

It is important to avoid a single-family residential mindset. Escape routes are closer and easier to find in most single-family residential structures as compared

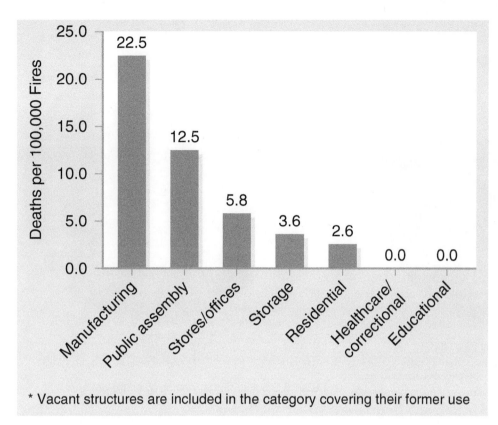

FIGURE 5-3 Fire fighter fire-ground fatalities by occupancy type, 2013–2017. Note: "Vacant" is no longer treated as a separate category; vacant buildings are including in the occupancy category of their intended or former use (e.g., a vacant warehouse is included in the storage category).
Source: NFPA, Firefighter Fatalities in the United States – 2018, June 2019.

to larger buildings where more time is needed to escape and a higher probability of disorientation exists. Depleting the SCBA air supply or an SCBA failure often results in survivable smoke inhalation injury in a one- or two-family dwelling fire. The loss of air supply is more likely to be fatal in a larger, more complex building. Even a defensive posture may not be sufficient to protect fire fighters when a building contains an extra-hazard fire load, as was the case in the West, Texas, ammonium nitrate storage facility.

FIGURE 5-4 and several figures that follow show the number of deaths by type of duty. Given the lower number of fires and increased activity in nonfire emergencies such as EMS, the fire ground remains a very dangerous workplace. In analyzing fire fighter on-duty deaths, it is important to consider the nature and cause of the injury leading to fire fighter fatalities, as shown in **FIGURE 5-5** and **FIGURE 5-6**. **FIGURE 5-7** shows that cardiac events and asphyxiation are the two leading causes of on-duty fire fighter fatalities at structure fires. Sudden cardiac deaths are responsible for almost one third of fire fighter fatalities at structure fires. Many of these deaths are preventable through proper medical evaluation and wellness programs.

Deaths by asphyxiation require further study to determine common denominators. Much has been written about air management. Part of the problem is depleting the SCBA air supply while working on the interior at a structure fire. Under the strenuous work conditions experienced on the interior at a working structure fire, an air cylinder rated at 30 minutes usually provides an air supply of approximately 20 minutes. Low air alarms should sound with 33% of the air remaining or approximately 10 minutes of escape time. A notice of 10 minutes is probably sufficient in a typical dwelling, provided the fire fighter immediately retreats when the alarm sounds. When a fire fighter is disoriented or working in a larger building, 10 minutes may not provide sufficient escape time. The accountability officer can track "on-air" time, notifying fire fighters of approximate time remaining to allow more escape time. Technology now available from breathing apparatus manufacturers shows promise. Several SCBA manufacturers offer air management tools for the fire ground. These computer-based systems use radio telemetry to monitor the fire fighter's air consumption rate and amount of available air remaining in the air cylinder. The system will also notify the monitor when the fire fighter's low air alarm is activated or even when the member is motionless and the PASS is sounding. The fire fighter assigned to monitor SCBA air can advise command when companies are low on air and need to be rotated out of the fire zone.

The advent of 45-minute and 60-minute open-circuit SCBAs provides additional time and increases the escape time. Unfortunately, higher-pressure and larger-volume cylinders weigh more and sometimes have a higher profile, which increases the workload and decreases mobility.

Evaluating fire fighter deaths in regard to situations and causes is essential to a continued decrease in the number of fire fighter fatalities. Moreover, the company officer and IC can also affect non–life-threatening injuries that often lead to pain, suffering, and disability.

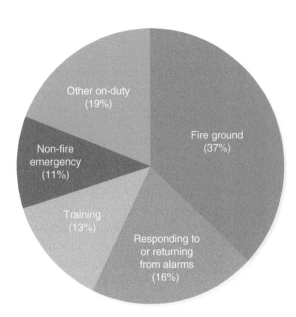

FIGURE 5-4 Fire fighter deaths by type of duty, 2009 to 2018.
Source: National Fire Protection Association, Applied Research, Quincy MA, November 2019.

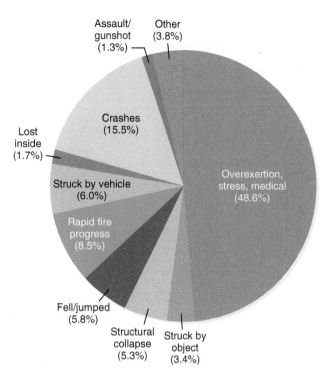

FIGURE 5-5 Fire fighter deaths by cause of injury, 2009 to 2018.
Source: National Fire Protection Association, Applied Research, Quincy MA, November 2019.

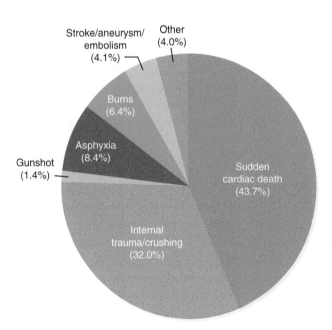

FIGURE 5-6 Fire fighter deaths by nature of injury, 2009 to 2018.
Source: National Fire Protection Association, Applied Research, Quincy MA, November 2019.

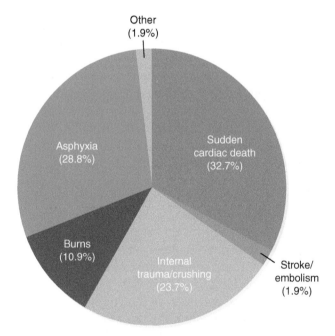

FIGURE 5-7 Fire fighter deaths by nature of injury—structure fires only, 2009 to 2018.
Source: National Fire Protection Association, Applied Research, Quincy MA, November 2019.

Incident Summary

West Fertilizer Explosion

On April 17, 2013, ten emergency first responders (ranging in age from 26 to 52 and all male) were killed when a burning fertilizer plant, containing an estimated 40 to 60 tons of ammonium nitrate, exploded just outside the city limits **FIGURE IS5-1**. The explosion occurred less than 20 minutes after the emergency responders arrived on scene. The victims included five volunteer fire fighters from the city's volunteer fire department, and four volunteer fire fighters from three neighboring volunteer fire departments who were attending an emergency medical services (EMS) class in the city. One off-duty career fire captain and two civilians who responded to offer assistance to the volunteer fire department were also killed by the explosion. The victims were among a number of first responders engaged in fire suppression and support activities and were in close proximity to the burning structure when the explosion occurred. Five other volunteer fire fighters with the city's fire department were injured. The two civilians were providing nonsuppression support to the fire department when they were killed by the blast. Three civilians living nearby also died as the result of the blast.

The U.S. Chemical Safety Board (CSB) estimated that total damages could exceed $230 million. There are differing opinions regarding the cause of the fire; in its preliminary findings, the CSB stated, "the explosion resulted from an intense fire that led to the detonation of the ammonium nitrate. The Environmental Protection Agency (EPA) estimates that 13,000 facilities similar to West Fertilizer pose threats to communities throughout the United States."

This fatal fire and explosion point out the importance of preincident planning. It is interesting to note that the largest loss of life in an ammonium nitrate explosion also occurred in Texas. On April 16, 1947, in Texas City, ammonium nitrate was being transferred from a chemical plant onto a ship when a fire erupted on board. A tremendous explosion killed hundreds, including 27 members of the Texas City Fire Department.

A **B**

FIGURE IS5-1 Aerial view of the site before the explosion (**A**) and after (**B**).
Courtesy of the Bureau of Alcohol, Tobacco, Firearms, and Explosives

Data from: NIOSH Fire Fighter Fatality Investigation Summary [F2013-11], November 12, 2014, "9 Volunteer Fire Fighters and 1 Off-Duty Career Fire Captain Killed by an Ammonium Nitrate Explosion at a Fertilizer Plant Fire—Texas," https://www.cdc.gov/niosh/fire/pdfs/face201311.pdf; and Durso, Jr., Fred, 2014,"In West's Wake," *NFPA Journal* March/April 2014. https://www.nfpa.org/News-and-Research/Publications-and-Media/NFPA-Journal/2014/March-April-2014/Features/NFPA-400.

Most minor injuries and close calls (near misses) go unreported. Many of these near misses could have resulted in serious injuries or death in slightly different circumstances. An example of a near miss would be a fire fighter falling through a collapsed floor and being rescued without injury or failing to report a minor injury. These near misses are good indicators of common hazards and merit examination. With this in mind, the Department of Homeland Security, the International Association of Fire Chiefs (IAFC), and the International Association of Fire Fighters (IAFF) have developed websites designed to share this vital information. Many fire departments and local unions also collect and provide near miss information.

Fire administrators must be committed to fire fighter safety and provide the necessary procedures, training, and equipment. ICs have overall safety responsibility during an incident and must make the crucial offensive/defensive attack decision at structure fires, then monitor, organize, coordinate, and provide safety measures described throughout this text. An incident safety officer allows the IC to better concentrate on tactics while the incident safety officer evaluates on-scene safety. Appointing an incident safety officer is essential at large-scale fires. Company officers have a responsibility to properly supervise members of their company. During the early stages of an operation, a company officer is often the IC.

Fire fighters must take personal responsibility for their safety and the safety of members working with them by following procedures, maintaining firefighting skills, and properly using the equipment provided. Sustaining a high level of physical fitness is important not only to the individual member but to the entire fire-ground operation.

Risk Management Applied to the Fire Ground

Probably the most important element of the incident safety program is applying risk management to fire-ground operations. NFPA 1500 specifically addresses risk management principles to be applied at the incident scene (NFPA 1500, 2018):

8.4.2.1 The concept of risk management shall be utilized on the basis of the following principles:

(1) Activities that present a significant risk to the safety of members shall be limited to situations where there is a potential to save endangered lives.

(2) Activities that are routinely employed to protect property shall be recognized as inherent risks to the safety of members, and actions shall be taken to reduce or avoid these risks.

(3) No risk to the safety of members shall be acceptable when there is no possibility to save lives or property.

(4) In situations where the risk to fire department members is excessive, activities shall be limited to defensive operations.

This text continually stresses the importance of developing an incident action plan based on a risk-versus-benefit analysis, and further, it describes how to apply NFPA 1500. This important standard must be applied to every situation. The single most important ability the fire officer must acquire is being skilled at recognizing the point at which the risk to fire fighters' lives outweighs the possible benefits of saving lives and property. Too many fire fighters are injured or killed while undertaking body recoveries or wetting down the ruins. The decision to employ a defensive approach based on the risks outweighing the possible benefits can be difficult. One of the clearest illustrations of this difficulty can be found in the Worcester, Massachusetts, warehouse fire, where the IC made a decision to abandon interior operations, knowing that six missing fire fighters who remained in the building would almost certainly perish.

The first statement in NFPA 1500-8.4.2.1(1) reads, "Activities that present a significant risk to the safety of members shall be limited to situations where there is a potential to save endangered lives." (NFPA 1500, 2018). This statement can be somewhat ambiguous and has been the subject of much discussion. The controversy heightens with the provision of an exception to the two-in/two-out rule in the face of an imminent life-threatening situation. The degree of risk associated with entering a burning building lacking the safety provided by the two-in/two-out provision is obviously higher than the risk taken when two fire fighters are outside the building ready to come to the assistance of the interior crew using the buddy system. Therefore, the perceived benefit would have to be more absolute to justify the risk of entry without an initial rapid intervention crew (IRIC).

It is relatively easy to visualize conditions presenting an imminent life-threatening situation. Or is it? An argument could be made for urgency in a late-night residential fire with an assumption that occupants are at home sleeping and possibly unaware of the fire, or already in serious trouble. This process could be continued with assumptions that the possibility of a threat to life always exists until a primary search verifies that no one is endangered. A warehouse or office building could also be occupied at night. Even vacant buildings

Incident Summary

Worcester, Massachusetts, Warehouse Fire

In Worcester, Massachusetts, on December 3, 1999, six career fire fighters died in a vacant, six-floor, maze-like, cold storage and warehouse building while searching for two homeless people who accidentally started the fire and escaped prior to the arrival of the fire department **FIGURE IS5-2**. Approximately 30 minutes after the first alarm was struck, two fire fighters searching for victims and checking for extension sounded an emergency message. A personal accountability report confirmed two missing fire fighters. The search-and-rescue operation was expanded to find the two missing fire fighters and the homeless people, who were thought to be in the building. Four additional fire fighters became disoriented during this part of the operation. The fire was believed to have had 30 to 90 minutes of preburn time. After more than an hour on the scene, a company on the interior notified command of structural problems and an arson investigator on the exterior reported the fire was venting from the roof. A defensive attack was ordered at approximately 1 hour and 45 minutes after the initial alarm. The six missing fire fighters perished.

FIGURE IS5-2 The combination of a massive fire load and the maze of rooms in this Worcester, Massachusetts, warehouse contributed to the death of six fire fighters.
© Paul Connors/AP Images.

Data from: NIOSH Fire Fighter Fatality Investigation Summary [99 F-47]. September 27, 2000, "Six Career Fire Fighters Killed in Cold-Storage Warehouse Building Fire—Massachusetts," https://www.cdc.gov/niosh/fire/pdfs/face9947.pdf. Accessed November 25, 2019.

---

can be occupied by the homeless or others. When a fire breaks out in a vacant structure, there is a possibility that the person who caused the fire is still in the building.

The concept of probability is important to the risk management process. The first-arriving officer must consider the possibility of people being in the building not as a yes-or-no proposition, but rather as a degree of probability. So, how does the fire officer evaluate the probability that savable people are in the building?

The probability that a nursing home is occupied is very high at any time of the day or night, as is the potential for an occupied dwelling at 2:00 AM. Finding someone in an office complex at 2:00 AM on a Saturday night or Christmas Eve would be less likely, but still possible. Thus, occupancy type, time of day, day of week, and time of year are all factors in determining the probability of the structure being occupied. The probability of the building being occupied is essential to analyzing risk versus benefit.

The NFPA 1500 committee undoubtedly expects the fire officer to exercise judgment in regard to the issue of an imminent life-threatening situation.

NOTE

Sometimes it is necessary to enter vacant structures to determine the life-safety hazard or extinguish a minor fire. Fire officers must be extremely careful when confronted with a well-involved fire in a vacant structure.

Fallacy	Fact
Vacant buildings are always unoccupied. Always apply a defensive attack.	While the probability that a vacant building is occupied is low, there is a chance that someone will be in the building. The only way to know for sure is to conduct a primary search.

Fire Intensity

Time and occupancy factors are not the only considerations. Fire intensity and building construction are equally important in determining the probability of saving lives and the risk to fire fighters. The hazards to fire fighters entering a given area increase as the fire progresses toward flashover. Throughout the process, the building is getting weaker, to the point of complete failure. No one can determine exactly when flashover will occur or when the building will fail. However, there are useful approximations. In post-flashover fires, the chance of occupant survival is minimal within the flashover compartment.

In the past, the time from ignition to flashover was given as 10 minutes. The actual time can vary significantly, depending on a number of variables, including the following:

- Compartment size
- Available air supply (ventilation)
- Ignition source
- Fuel type
- Fuel geometry
- Geometry of the enclosure
- Distance between fuel packages
- Location of the fuel

Studies by Underwriters Laboratories (UL) comparing modern furnishings to legacy (mid-20th century) furnishings indicate that flashover is likely to occur prior to the arrival of the first-due fire company in residential occupancies (Kerber, 2010). Other studies by UL and the National Institute of Standards and Technology (NIST) indicate that the fire may become ventilation limited prior to flashover. In this case, fire fighters can expect a rapid acceleration when air is supplied via doors, windows, or other openings.

Flashover (discussed in Chapter 2, *Procedures, Pre-incident Planning, and Size-Up*) is the transition from a fire that is growing, to a fire where all of the exposed surfaces have ignited, via ignition of one type of fuel to another. Because the potential harm from a flashover is so great, fire fighters must recognize conditions that might indicate the possibility of a flashover and take steps to prevent it whenever possible.

Several computer programs can provide rough estimates of when flashover will be reached in a compartment. However, given the wide range of variables that must be entered into these computer programs, it is difficult to provide a reliable time frame that the fire officer could use on the fire ground. Larger-volume compartments generally take longer to flashover. A backdraft (sometimes called a smoke explosion) occurs when sufficient oxygen (air) is introduced into a vapor-fuel–enriched atmosphere.

Ventilation can radically affect the time to flashover. A fire in an oxygen-starved room could go into the decay phase and not flash over; adding air to the room could result in a backdraft or rapid fire acceleration. Most fire fighters think of ventilation in terms of purposeful venting. Many times, self-venting occurs when the fire breaches containment, such as when windows fracture. Unintentional venting occurs when fire fighters open doors or windows to gain access.

Wind can have an adverse effect by providing an abundance of oxygen to rapidly increase fire intensity. Most studies of wind-driven fires have been conducted in high-rise buildings or simulated high-rise scenarios. However, wind can have an adverse impact on operations in a building of any size, and the lessons learned from these studies also apply to other situations. Wind blowing toward a vent opening tends to reverse the normal flow path due to an increase in pressure at the opening. If a ventilation outlet is available on another side of the building, the fire and smoke will naturally flow toward that outlet; as an example, wind blowing into the rear of the building through an open window when fire fighters open the front door to make entry. Higher wind speeds increase fire and smoke movement within the structure. Fire fighters in this ventilation stream (flow path) are in great peril. This potential risk is one of many reasons why the company making initial entry should control the door and have a charged hose line in place before making entry.

Fuel Load

The building's fuel (fire) load consists of fuels provided by the contents and combustible building materials. Most of the construction materials used in wood-frame buildings will burn, thus creating a large fuel load. Combustible building materials are limited in fire-resistive and noncombustible construction. The primary fuel load for most structure fires is made up of the combustible contents. The combustible contents in commercial occupancies vary widely, even among occupancies of the same type. For example, the fuel load in a warehouse storing plastic products would be high compared to the fuel load in a warehouse containing metal parts in cardboard cartons. Sprinkler calculations take fuel load into account. Although some residential properties may contain excessive contents, most often average fuel loads can be anticipated. The residential fuel load has increased significantly due to people having more possessions and the amount of plastic and high-fuel-capacity contents.

UL conducted a series of 15 full-scale experiments in residential properties to examine the effect of the increased fuel load, as well as the geometry and room size, in modern residential buildings (Kerber, 2010). They found that the fuel load in modern buildings is much greater than in legacy residential buildings of the mid-20th century. This increase and more volatile fuel load dramatically reduce the time to flashover. In UL experiments, the time to flashover in a living room furnished with modern furniture occurred in 3.5 minutes, compared to 29.5 minutes for the same room furnished with legacy furnishings. This study and later studies also examined the effects of ventilation, including tests to determine when windows and doors would fail due to fire exposure. Armed with this knowledge, it is incumbent on fire departments to adjust their tactics. This is explained in detail in Chapter 8, *Offensive Operations*.

Incident Summary

Mixed-Use Industrial Fire, New York

A fire in a mixed-use industrial building burned for several days and drew a massive response from fire departments and other emergency response personnel before operations were concluded and the building was turned back to its owners.

The first crews were dispatched to the scene following a call from the complex shortly before 7:30 AM. On arrival, crews were escorted to the fire location, a large building that covered an area of 1 million square feet. Employees waiting outside the building directed crews to the location of the fire, and the IC observed a fully involved fire after opening a door to investigate. At that point, fire broke out on the exterior of the building **FIGURE IS5-3**.

Crews established a water supply and positioned an aerial platform and deck gun for fire attack.

Incident command had already called for a second alarm before arrival and issued a request for mutual aid from nearby communities. Due to the intensity of the fire, crews remained outside the structure and mounted a defensive operation. News reports indicated that sections of the roof and walls collapsed and residents reported hearing explosions. Over 100 fire fighters were reported to be involved in extinguishment efforts.

The fire originated in a storage area of a plastic recycling business. Employees thought that they had successfully extinguished the fire with fire extinguishers, but it flared up and spread quickly through the storage area. The presence of cars, boats, and recreational vehicles in an adjoining vehicle storage area contributed significantly to the intensity of the fire.

FIGURE IS5-3 Smoke rising from the fire at the mixed-use building in New York.

Campbell, Richard, November 1, 2018, "Firewatch: Fire Incidents from Across the Country," *NFPA Journal* January/February 2018. https://www.nfpa.org/News-and-Research/Publications-and-media/NFPA-Journal/2018/November-December-2018/News-and-Analysis/Firewatch. Accessed November 25, 2019.

(continues)

Incident Summary (Continued)

Due to a large smoke plume and concerns about airborne contaminants, nearby residents were advised to shelter in place. Hazardous materials teams and environmental authorities set up several air particulate monitors and sampling units, which were adjusted with changing wind conditions, and an absorbent boom was deployed along a nearby creek to capture runoff from firefighting operations. The state health department distributed fact sheets with health-related information to neighborhood residents.

As the fire continued to burn for a second day, heavy smoke and a change in wind direction led to the evacuation of one neighborhood, with the evacuation order remaining in effect for just over 24 hours.

Heavy equipment was brought to the site to remove sections of the building and provide fire fighters with direct access for fire suppression. Master stream and aerial operations continued over a 3-day period before the fire was determined to be under control, but extinguishment efforts continued for several more days as the equipment opened up hot spots.

Initially, a broken light bulb was cited as the cause of the fire, but news reports 10 months after the fire indicated that investigators classified the cause of the fire as undetermined.

The building, which previously housed a steel mill, was a single-story structure that stood 50 ft (15.2 m) in height and was built in a triangular shape. The structure was constructed with steel walls and roof framing, a concrete floor, and a roof deck of steel and wood. The building was not equipped with sprinkler protection.

One fire fighter suffered a broken ankle while laying hose outside the building. No other civilian or responder injuries were reported.

The fire caused an estimated $1 million in damage to the building, valued at $2.5 million. No estimates were available for damage to building contents.

Campbell, Richard, November 1, 2018, "Firewatch: Fire Incidents from Across the Country," *NFPA Journal* January/February 2018. https://www.nfpa.org/News-and-Research/Publications-and-media/NFPA-Journal/2018/November-December-2018/News-and-Analysis/Firewatch. Accessed November 25, 2019.

Building Design Loads

Loads imposed on a building are divided into live, dead, seismic, wind, snow, and ice loads. Building loads affect structural stability. Unusually high building loads can result in premature collapse. Loads placed on lightweight roof and lightweight floor structural assemblies are particularly dangerous under fire conditions. A fire fighter walking on a roof creates a live load that is concentrated on a fairly small area the size of their boot. Using a roof ladder distributes this live load over a much greater surface area, reducing the chance of penetrating a weakened section of roof. Working from an aerial ladder or platform to perform ventilation or other roof tasks is much safer. Likewise, placing a ladder on a stairway or floor that could be structurally compromised distributes live loads created by fire fighters over a larger area.

Structural Stability

Structural failure can occur at any time. Always consider structural stability when sizing up the fire scene. Fire intensity, burn time, content loads, construction methods and materials all affect structural stability. A general rule called the "20-minute rule" was based on all of these factors except the content load. The **20-minute rule** states that when a heavy volume of fire is burning out of control on two or more floors for 20 minutes or longer, structural collapse should be anticipated. This rule is based on Type III (ordinary construction). A frame structure may be completely consumed in less than 20 minutes, and unprotected steel can fail in less than 20 minutes when subjected to a heavy volume of fire. Fire entering a concealed truss space causes failure in a relatively short time, as has happened in several fires that killed fire fighters. Heavy-timber or fire-resistive structures subjected to the same volume of fire would be expected to withstand fire attack longer than a frame structure.

Masonry construction is generally thought to be stable under fire conditions, but generalizations are often in error. For example, masonry buildings with exterior stars are generally more prone to collapse **FIGURE 5-8**. The visible star on the exterior is an indication that the building wall is being supported by a metal rod attached to interior structural components. These building are known to be subject to premature collapse.

There is a subjective element in the 20-minute rule—heavy volume of fire. What is perceived as a heavy volume of fire by one fire officer is a moderate volume of fire to another. There is also reason to question the "two or more floors" part of the rule. A

FIGURE 5-8 Brick building with metal star support. The stars appear between the second and third story windows and above the first escape, near the roof.
Courtesy of Ben Klaene.

heavy volume of fire on a single floor could also result in collapse. Is 20 minutes an exact time? Absolutely not! Likewise, the length of burn time at heavy volume is subjective, as there is no way of knowing the exact time of ignition or the intensity of burn before arrival. However, time, intensity, and fire volume factors must be considered in determining whether it is safe to enter or be near a structure.

An IC confronted with a well-involved fire should "start the clock" when making a decision to conduct operations in the offensive mode. If, after a given period of time, the fire is still not being brought under control, the incident action plan should be reviewed and the operation possibly changed to a defensive attack.

Given the amount of information that an IC must assimilate, it is easy to lose track of how much time has passed on the fire ground. NFPA 1500 requires dispatch centers to notify command every 10 minutes until the fire is knocked down, the incident becomes static, or the IC cancels the notification (NFPA 1500, 2018). This notification helps the IC track elapsed time. It is critical that burn time takes into account the time the fire was burning prior to notification and while units were responding, not just the "time on scene."

Transitioning from an offensive to a defensive strategy is the ultimate test of command. This transition phase is a dangerous time on the fire ground. When changing from an offensive to a defensive operation, a managed retreat is best. Engine companies operating hose lines should provide protection while other companies, especially companies above the fire, evacuate the building. Planning and safety officers can be of great assistance to the IC in making the decision to go defensive and assist in managing a retreat. This is one reason this text recommends staffing these positions

as soon as possible. However, determining the need to retreat and managing the process is ultimately the IC's responsibility.

Whenever signs of imminent collapse are observed, the offensive operation must be abandoned immediately. All units must be notified and immediately move to the exterior. Usually, an announcement is given over the radio followed by a preplanned signal such as several long blasts on an air horn (10 three-second blasts, for example). New SCBA technology can notify each fire fighter of an evacuation order and then transmit an acknowledgement back to the IC. Forces operating on the outside must be moved back a safe distance from the building, usually a distance at least 1½ times the height of the building. In buildings with adequate fire walls and separate supporting structure, it may be possible to evacuate to, or move operations to, an adjoining compartment (e.g., a shopping center with fire walls separating stores).

Construction Methods and Materials

A wide variety of construction materials and methods have been employed in buildings over the years. These range from reinforced concrete to lightweight wood trusses. Because of the different materials, the behavior of buildings under fire attack will vary significantly. Broad assumptions can be made regarding structural stability based on construction methods and materials, such as a heavy-timber building will survive longer than a lightweight wood-truss structure. However, there are unknown factors, such as deterioration or repairs, that affect the reliability of assumptions based on construction type.

Fire spread inside of the building can occur through a variety of horizontal or vertical openings. Some buildings are constructed with large vertical shafts for utility lines or other reasons. At the Las Vegas MGM Grand Fire in 1980, fire spread vertically through seismic joints and stairways. In the NFPA report on this incident, a space 5/8 in. (1.6 cm) wide in one of the stairways that had not been properly sealed off was cited as one of the conditions that contributed to the spread of smoke and fire into the stairway (NFPA, 1982).

What the fire officer should learn from these incidents and others is that a building's performance under fire conditions can be unpredictable, and exterior appearances may not indicate interior conditions. Building renovations can dramatically affect how the structure performs under fire conditions. For example, openings made in floors to run utilities are not always properly sealed.

No building is completely immune to structural failure, but some buildings will withstand a large and intense fire without a catastrophic (total) collapse. Other structures are known to experience early collapse under intense fire conditions. When deciding on an offensive or defensive strategy or when placing companies for rescue or fire attack, the fire officer must take structural stability into account.

Modern construction methods conserve materials by using lighter-weight structural members that provide the same load-bearing capabilities as earlier construction methods that used massive structural members. The primary way this is accomplished is through various types of truss construction. Other forms of lightweight construction include lightweight metal C joists and engineered wooden I-beams. Lightweight structural systems take the place of large wood beams or steel I-beams. Lightweight construction methods are structurally sound under normal conditions. In fact, the added strength allows builders to place roof trusses 24 in. (61 cm) on center, whereas earlier methods required 16 in. (41 cm) on center supports.

However, a major problem arises because these lightweight structural members are adversely affected by fire much sooner than the more massive building materials. Compounding the problem is the fact that once the truss loses its triangular configuration, it loses its load-bearing capacity. Lightweight construction systems frequently fail under fire conditions, with little warning to the fire fighter. A well-involved fire in a truss space may not be detected until it breaks out of containment. Several fire reports mention that fire fighters working on the interior stated that there were few signs of a serious fire prior to collapse. NIOSH issued an alert identifying problems and solutions to the challenges presented by modern lightweight construction (NIOSH, 2005). Unprotected metal trusses used in noncombustible construction fail quickly when exposed to high heat FIGURE 5-9.

Lightweight wood-truss assemblies burn through sooner than structural assemblies that are built of more massive components. Metal-web wood trusses combine the problems of early failure and premature burn-through experienced in wood and metal trusses. Truss roofs above large open areas are particularly dangerous. A truss roof failure over a compartmented area may be partially suspended by walls separating the rooms below. A fire entering an undivided attic area can extend unimpeded throughout the entire attic. Truss roof collapse generally results in a large area of damage. More fire-resistive buildings can suffer partial interior collapses that are not often noticed by the IC or companies working in other areas of the building.

Structural connections can play a critical role in a building under fire attack. A large I-beam supported by a single bolt is only as strong as the single bolt. Tilt-Slab construction relies on the roof structure to connect the massive concrete slab walls and keep them vertical. Failure of the bar-joist roof truss will result in the failure of the slab concrete walls presenting a serious fire fighter safety issue. Wood-truss construction typically uses gusset plates in place of nails FIGURE 5-10A. Nails form a stronger connection under fire conditions, owing to the depth of their penetration into the wood. Gusset plates usually penetrate a fraction of an inch and form a large surface area to collect heat. As soon as the gusset plate teeth lose their strength or the fire burns through the wood to the depth of the gusset plate teeth, the plate releases from the wood, and the wood truss loses its stability FIGURE 5-10B.

Engineered wooden I-beams are now frequently used, based on the same logic as a steel I-beam. A 2 in. × 4 in. (approximately 50-mm × 100-mm) board is placed on the top and bottom chords with a piece

FIGURE 5-9 Fire attacking metal truss construction.
Courtesy of David J. Jones, Cincinnati, Ohio.

under normal conditions, but they lack reliability and stability when exposed to fire and heat and may even collapse under their own weight if deteriorated due to prolonged water damage.

Many older buildings have been modernized with suspended ceilings or by adding walls or paneling and installing a facade. Often these renovations are constructed using lightweight engineered materials. Be aware of renovations in your response district, because presumably "legacy" era buildings may very well contain modern building components. These renovations provide additional concealed spaces for the fire and present additional dangers to fire fighters. For example, a fire in Charleston, South Carolina, that killed nine fire fighters spread throughout a 5 ft (1.5 m) high concealed space between a suspended ceiling and the roof, and then suddenly broke out of containment, killing fire fighters who were working below.

Understanding the dynamics associated with structural collapse is critically important to the fire fighter. The IC and fire fighters working inside at a structure fire must pay close attention to signs of impending structural failure **FIGURE 5-11**. Many decisions to conduct the attack in the defensive rather than offensive mode are based on structural stability. Fire fighters and safety officers must pay close attention to signs of impending collapse. Once an engineered structural component is directly exposed to fire or high heat, little time remains to escape.

Building collapse is a real and present danger. An offensive attack is often abandoned in favor of a less hazardous defensive attack. However, defensive attacks can also be hazardous if fire fighters are in or near the collapse zone. Establishing a collapse zone at least 1½ times the height of the fire building is essential. Other hazards, such as electric service wiring, must also be considered.

A

B

FIGURE 5-10 A. Gusset plates. **B.** Fire-damaged gusset plates.
A. © David Huntley/ShutterStock, Inc. B. Courtesy of Thomas Lakamp, Cincinnati, Ohio.

of plywood or oriented strand board (OSB) between, thus increasing the load-bearing strength. Light-gauge steel C-channels and other lightweight structures are also used to support floors and roofs. These methods are satisfactory and even superior in some cases

FIGURE 5-11 Defensive fire attack with handheld hose line beyond the collapse zone.
Courtesy of David J. Jones, Cincinnati, Ohio.

Incident Summary

Incidents Involving Bowstring Truss Construction

On July 1, 1988, fire fighters responded to a fire occurring at a Ford dealership on River Street in Hackensack, New Jersey **FIGURE IS5-4**. The 1948 building contained an attic that stored automobile parts and janitorial supplies. The roof was 60-ton bowstring truss construction. The building contained portable fire extinguishers, but no sprinklers.

The initial call from employees of Hackensack Ford indicated that an exhaust system hose in the attic had fallen, burning to the floor, and had been extinguished with a portable extinguisher. However, heavy smoke was discovered in the attic several minutes later. Two engines and one truck company (a total of 10 fire fighters) initially responded.

Initially, the attic could not be accessed by fire fighters. After 35 minutes, the roof collapsed. Five fire fighters died: three died when the roof collapsed, and two others died as a result of being trapped inside and subsequently running out of air.

In 2009, a California fire fighter was seriously injured, resulting in multiple fractures, while operating a hose line under the overhang of a bowstring truss roof when the roof collapsed and struck him. Fire fighters encountered heavy fire conditions and started an offensive attack. The strategy was changed to defensive upon the arrival of the first chief officer; however, the strategy change was not fully communicated throughout the scene.

In 2010, two Chicago fire fighters were killed and 19 fire fighters were injured in the collapse of a bowstring truss roof in an abandoned commercial laundry. The fire fighters were engaged in suppression activities at a fire involving rubbish when the roof collapsed 16 minutes after arrival on the scene. The fire had been reported as under control shortly before the collapse. The building was under orders from the city to either repair or demolish the structure.

Modified from: "Fire Investigation Report: Fire Fighter Fatalities: Hackensack Ford," Bureau of Fire Safety, State of New Jersey. https://nj.gov/dca/divisions/dfs/reports/hackensack.pdf, Accessed November 25, 2019; Adely, Hannan, June 30, 2013, "The Hackensack Tragedy That Changed How Fires Are Fought," https://www.northjersey.com/story/news/bergen/hackensack/2017/08/31/archive-hackensack-tragedy-changed-how-fires-fought/616651001/, Accessed November 25, 2019; NIOSH Fire Fighter Fatality Investigation Summary [F2009-21], December 1, 2010, "Career Fire Fighter Seriously Injured from Collapse of Bowstring Truss Roof—California," https://www.cdc.gov/niosh/fire/reports/face200921.html, Accessed November 25, 2019; and NIOSH Fire Fighter Fatality Investigation Summary [F2010-38], July 6, 2011, "Two Career Fire Fighters Die and 19 Injured In Roof Collapse During Rubbish Fire at an Abandoned Commercial Structure—Illinois," https://www.cdc.gov/niosh/fire/reports/face201038.html, Accessed November 25, 2019.

FIGURE IS5-4 This deadly fire at a Ford dealership killed five fire fighters when the building collapsed.
© Al Paglione/The Record.

Roof Operations

Many departments automatically assign members to the roof at every fire. This can be a dangerous practice. A conscious decision should be made regarding roof safety. Lightweight roofs supported by trusses, C-channels, or other engineered construction methods are particularly dangerous. When a fire has self-vented, the roof is likely to be unsafe, and the need for roof operations is questionable. The same kind of risk-versus-benefit analysis that is made before deciding on interior operations should be made before placing fire fighters on the roof. The fire officer should ask four important questions before assigning fire fighters to roof operations:

- Are roof operations necessary?
- What do I hope to gain (benefit) by opening the roof? In many cases, roof ventilation is not necessary and sometimes has a negative impact on operations.
- How long has the fire been impinging on the roof components?
- What is the safest method of operating on this roof (e.g., aerial device, roof ladder)?

In addition, limit the number of fire fighters working on the roof, and use an aerial ladder or other work platform. If two fire fighters can perform necessary roof operations, there is no reason to place more than two people on the roof. Some departments prohibit roof operations on truss roofs or at specific occupancies as part of their standard operating procedures (SOPs).

Truss roofs and/or concealed spaces have played a major role in fire fighter fatalities at structure fires as indicated in a report based on data from several agencies. **TABLE 5-2** lists a number of fires where truss roofs and/or concealed spaces were contributing causes according to fire reports analyzing fire fighter on-duty deaths.

TABLE 5-2 Fire Fighter On-Duty Deaths with Roof Collapse and/or Concealed Spaces as Contributing Factors

Location	Occupancy	Year	Number of Fire Fighter Fatalities
New Jersey*	Automobile dealership	1986	5
Tennessee***	Church	1992	2
Indiana	Hotel/office	1992	2
Connecticut*	Carpet store	1996	1
Virginia*	Auto parts store	1996	2
Georgia**	Church	1998	1
California**	Commercial	1998	1
Mississippi**	Strip mall	1998	2
Texas*	Church	1999	3
Texas**	Restaurant	2000	2
Arkansas**	Church	2000	4
South Carolina**	Residential	2001	1
Wisconsin**	Commercial	2001	1
Oregon**	Auto parts store	2002	3
Tennessee**	Retail store	2003	2

(continues)

TABLE 5-2 Fire Fighter On-Duty Deaths with Roof Collapse and/or Concealed Spaces as Contributing Factors (Continued)

Location	Occupancy	Year	Number of Fire Fighter Fatalities
Texas**	Vacant residential	2005	1
South Carolina**	Furniture store	2007	9
Michigan**	Vacant residential	2008	1
Illinois**	Vacant commercial	2010	2
Indiana**	Church	2011	1
California **	Single family dwelling	2011	1
Illinois**	Vacant commercial	2011	2
Tennessee**	Restaurant	2012	1
Pennsylvania**	Commercial	2012	1
Wisconsin**	Theater	2012	1
Georgia**	Residential	2013	1
Texas**	Restaurant	2013	4
Indiana**	Commercial	2014	1

* NFPA Investigative Report available (www.nfpa.org)
** NIOSH Fire Fighter Fatality Investigation available (www2a.cdc.gov/NIOSH-firefighter-face/)
*** USFA Technical Report Series report available (https://apps.usfa.fema.gov/publications/display.cfm)
NFPA Investigative Report available (www.nfpa.org); NIOSH FireFighter Fatality Investigation available (http://www2a.cdc.gov/NIOSH-firefighter-face/); USFA Technical Report Series report available (https://apps.usfa.fema.gov/publications/display.cfm)

Incident Summary

Texas Apartment Building Fire

In a 2011 apartment building fire in Texas, a fire lieutenant was fatally injured when he fell through a fire-weakened roof into the attic while performing vertical ventilation **FIGURE IS5-5**. The fire originated on the first floor of this two-story building and was rapidly contained. However, the fire had extended via the exterior walls into the attic, damaging the truss roof structure.

Data from: NIOSH Fire Fighter Fatality Investigation Summary [F2011-20], June 27, 2012, "Career Lieutenant Dies After Being Trapped in the Attic After Falling Through a Roof While Conducting Ventilation—Texas," https://www.cdc.gov/niosh/fire/reports/face201120.html, Accessed November 25, 2019.

FIGURE IS5-5 Side Alpha of the gable roof where the incident occurred.

NIOSH Firefighter Fatality Investigation 2011-20. Career Lieutenant Dies After Being Trapped in the Attic After Falling Through a Roof While Conducting Ventilation – Texas. https://www.cdc.gov/niosh/fire/reports/face201120.html. Accessed November 25, 2019.

> **NOTE**
> The practice of always assigning fire fighters to roof operations often places fire fighters in unnecessary peril. Never conducting roof operations denies the IC a valuable tactic.

Green Construction

Green construction is becoming more prevalent. Green roofs are being installed on new and existing buildings. The green roof consists of layers of materials supporting soil used to grow vegetation on a roof, which also provides roof insulation. This roof design not only makes roof ventilation more difficult or impossible, it adds substantial weight to the roof. Of particular concern is the additional roof load added to an existing roof that was not designed to carry the weight.

Solar panels have been used for many years but have become increasingly common, with many photovoltaic (PV) solar arrays installed on roofs. In California, PV panels will be required in new home construction beginning in 2020 (Chappel, 2018).

Solar panels not only add to the roof load, they create a potential ignition source, make roof ventilation more difficult, and present an electrical shock hazard that is difficult to deenergize **FIGURE 5-12**. Disconnecting the electrical service to a building does not deenergize the direct current roof system, which connects to an inverter that converts direct current from the solar panel to alternating current used in the building. Standards pertaining to installation and maintenance of PV panels are now being promulgated to reduce the fire hazard (Fire Protection Research Foundation, 2014). However, many existing solar panel installations do not meet these evolving standards. It is imperative that fire fighters familiarize themselves with the layout and operation of solar panels.

Other green energy materials such as alternative insulation materials, lightweight construction methods, and wind turbines also present special problems.

Floor Construction

The use of trusses and other lightweight support systems were once limited to roofs but are now commonplace in floor construction, creating some of the same premature collapse problems experienced in lightweight roofs. Fire fighters must realize that lightweight building materials generally accelerate structural failure; however, older solid-beam construction materials are not immune to collapse.

Changes that have occurred in the density and dimensions of newer lumber have compounded the problem of structural failure in wood construction. Today's lumber is less dense and smaller in size. UL compared older 2×8s to newer 2×10s and found that the older 2×8s had a cross-sectional dimension of 1.75 in. by 7.56 in. (4.4 cm by 18.9 cm; cross-sectional area of 13.23 in.^{2}). The new 2×10s had a cross-sectional dimension of 1.5 in. by 9.125 in. (3.8 cm by 22.8 cm; cross-sectional area of 13.6875 in.^{2}). The new 2×10s had a cross-sectional area slightly larger than that of the old 2×8s. The density of the new lumber was 32.5 lb/ft^3 compared to 36.9 lb/ft^3 for the older lumber. UL also conducted furnace tests on older and newer dimensional lumber and found that older 2×8s did not collapse until after 18 minutes (Kerber, 2012).

FIGURE 5-12 Solar panels are a potential ignition source and present additional hazards.
NFPA Journal, January/February 2014.

UL Basement Fire Study Abstract: Understanding and Fighting Basement Fires

Many fire fighters have been injured or killed while trying to extinguish a basement fire or a fire on a level below them. Prior research has shown basement fires present a high risk to fire fighters. This risk stems from unexpected floor collapse and high heat. Prior research also indicated that the methods that fire fighters have traditionally used to determine the structural integrity of the floor offer little value with lightweight construction. Past experiments in small basements have indicated that the most effective method of fighting a basement fire may be from the exterior of the building.

This study went beyond earlier research by increasing the size of the basement and incorporating three different ventilation and access conditions to the basement. Those access conditions include the following:

1. No exterior access to the basement
2. Limited exterior access to the basement
3. Exterior access to the basement

The results of the experiments show the importance of identifying a basement fire, controlling ventilation, and flowing an effective hose stream into the basement from a position of advantage, as soon as possible.

These experiments highlighted the importance of identifying a basement fire during size-up and subsequently choosing the appropriate tactics that coordinate ventilation with suppression. In all experiments, the basement fires were ventilation limited. Additional ventilation without suppression was shown to increase the hazard to any occupants trapped in the structure.

Various nozzles and appliances were used to flow water into the basement. Water streams applied through the floor, through a small window remote from the seat of the fire, and through a basement-level access door controlled the fire and reduced the hazard throughout the structure.

Effective water application into the basement cooled the fire gases to prevent flashover, slowed the destruction of the structure, and reduced the hazard from fire. This action made entry conditions into a basement with active burning possible for a fully protected fire fighter. Effective water application also supported search operations and reduced the threat from heat and toxic gases for any trapped occupants. Occupants isolated from the fire environment by a closed door or other means were provided additional protection when compared with conditions in rooms open to the fire environment.

Source: Madrzykowski, Daniel, and Craig Weinschenk, August 3, 2018, "Understanding and Fighting Basement Fires," UL Firefighter Safety Research Institute, https://ulfirefightersafety.org/docs/Understanding_and_Fighting_Basement_Fires.pdf, Accessed January 27, 2020.

Basement Fires

Basement fires have always presented a difficult and dangerous problem, especially basements that are completely underground. Various tactics have been used to fight basement fires. Probably the most often used, and the most difficult and dangerous, is advancing a hose line down the interior stairway from the first floor to the basement. Opening the door at the top of the stairway leading to a basement fire creates a vertical vent opening **FIGURE 5-13**. The fire and products of combustion will flow toward, into, and out of this chimney-like vent. If a basement fire has gained considerable headway, it may be impossible and most certainly ill advised, to use the stairway/chimney.

If the basement is partially above ground, other tactics may be available such as operating hose streams through a basement window to control and limit the fire by softening the target. This tactic may allow entry into the basement via the stairway for final extinguishment. Pulsing and indirect attacks are tactics that could be used to knock down the main body of fire in a basement that is partially above grade. Chapter 8, *Offensive Operations*, explains these tactics.

Some basements are accessible at ground level. Usually, houses built on hillsides where the hill slopes downhill from front to back have the main entrance at grade level on side Alpha, but may also have a walk-out basement at grade level on side Charlie. Garages at basement level or separate basement entrances can also provide direct access to the basement. A 360 evaluation of the structure by the first-arriving company is imperative to identify these basement entrances. Availability of a grade-level basement entrance allows a direct fire attack, but venting opportunities may be limited. When walk-out basement doors are available, they should be the first choice for direct fire attack in a basement.

Softening the target by applying water through a window or from the exterior through a grade-level basement entrance may provide the cooling effect needed to allow fire fighters to enter the basement for final extinguishment. Venting a basement fire through openings on the first floor is likely to spread the fire to areas near the vent opening unless water is applied into the basement prior to making vent openings.

Fire fighters are at significant risk of injury or death when fighting fires in basements or floors below

CHAPTER 5 Fire Fighter Safety 161

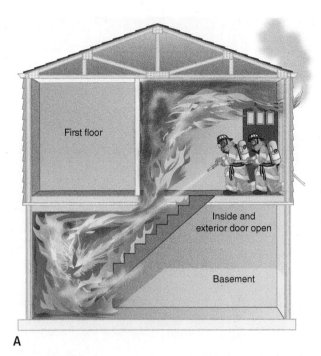
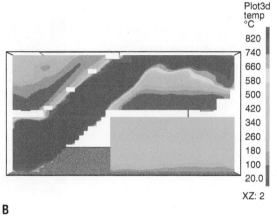

FIGURE 5-13 A. Basement fire venting via stairway and open door leading to exterior. **B.** NIST fire model showing flow from basement to first floor at a Washington, DC, townhouse fire.
A. © Jones & Bartlett Learning. **B.** Courtesy of NIST.

Incident Summary

Ohio Basement Fire

On April 4, 2008, a captain and fire fighter lost their lives in a basement fire in a two-story, 2050 ft^2 (190 m^2) single-family residential building in Ohio **FIGURE IS5-6**. Companies were dispatched to an automatic alarm at 6:11 AM. The first-arriving engine company reported on scene at 6:23 AM. Response was delayed because the building was located in a wooded area 450 ft (137 m) from the roadway, making it difficult to locate. Upon arrival, moderate smoke was visible. Building occupants pointing the way to the basement stairs reported that the fire was in the basement and everyone was out of the house.

FIGURE IS5-6 Site of the 2008 Ohio basement fire.
NIOSH Fire Fighter Fatality Investigation Summary [F2008-09], July 29, 2009, "A Career Captain and a Part-time Fire Fighter Die in a Residential Floor Collapse—Ohio, https://www.cdc.gov/niosh/fire/pdfs/face200809.pdf, Accessed November 25, 2019.

The company captain and a fire fighter donned their SCBA and entered the residence through the open front door with a 1¾-in. (44-mm) hose line. After making entry, the captain ordered the hose line charged and, with the assistance of the fire fighter, proceeded down the basement stairway. At the bottom of the stairs, they encountered heat and could see a glow to their left, but they needed additional hose to reach the fire area. Another fire fighter assisted in pulling additional hose at the top of the stairway and through the front door. Upon returning to the top of the stairway, this fire fighter witnessed the captain calling a mayday, but the captain was unable to communicate with units on the exterior. Smoke was now getting black and heavy and was pushing out the front door, whereupon the IC ordered an evacuation of the building. An engine company and the established rapid intervention crew (RIC) accessed the basement through the grade-level walk-out door in the rear of the building. The engine company was able to extinguish the fire, and the RIC found both victims under a collapsed floor area. The victims apparently fell through the floor from the first floor after backing out of the basement.

Data from: NIOSH Fire Fighter Fatality Investigation Summary [F2008-09], July 29, 2009, "A Career Captain and a Part-time Fire Fighter Die in a Residential Floor Collapse—Ohio, https://www.cdc.gov/niosh/fire/pdfs/face200809.pdf, Accessed November 25, 2019.

Incident Summary

Delaware Arson Floor Collapse

On September 24, 2016, a lieutenant and two fire fighters died due to a floor collapse in a row house during a structure fire **FIGURE IS5-7**.

At 2:56 AM, two engines, a squad, a ladder truck, and two battalion chiefs were dispatched to a report of a residential structure fire with persons trapped. The ladder company arrived on scene at 3:01 AM and reported heavy fire showing from the rear of the structure. The engines arrived simultaneously and both laid supply hose lines from different directions.

Battalion 2 arrived on scene and assumed command. Crews from both engines and the ladder went to the front door and entered the first floor at approximately 3:07 AM. At approximately 3:09 AM, the ladder officer, a fire fighter from Engine 1, and a fire fighter from Engine 2 fell into the basement. At 3:10 AM, a mayday was transmitted for a floor collapse and fire fighters in the basement.

Fire fighters were able to get the fire fighter from Engine 1 out of the basement at 3:18 AM. Crews were able to get the fire fighter from Engine 2 to the base of an attic ladder placed in the basement. Crews in the basement were also searching for the ladder officer, plus fire fighters were exiting and entering the basement. The injured fire fighter from Engine 2 moved away from the attic ladder and crews were unable to locate the fire fighter.

Two fire fighters from the squad entered the basement from side Charlie and found the lieutenant from the ladder near the Alpha–Delta corner. The two fire fighters from the squad pulled the ladder 2 officer toward the doorway on side Charlie. As the squad members and the ladder lieutenant moved to within 4 to 6 ft (1.2–1.8 m) of the doorway, a second collapse

FIGURE IS5-7 Rescue efforts at the Delaware row house where line-of-duty deaths occurred due to floor collapse.
NIOSH Firefighter Fatality Investigation F2016-18. Arson Fire Kills Three Fire Fighters and Injures Four Fire Fighters Following a Floor Collapse in a Row House—Delaware. https://www.cdc.gov/niosh/fire/pdfs/face201618.pdf. Accessed November 26, 2019.

occurred, trapping one squad fire fighter and the ladder lieutenant under the debris. The other squad fire fighter was pushed toward the doorway and escaped.

The squad fire fighter was removed from the structure at 3:29 AM and pronounced deceased at a trauma center due to asphyxiation and thermal burns. The ladder officer was located in the debris pile and transported to a trauma center where he was pronounced deceased due to asphyxiation and thermal burns. The fire fighter from Engine 2 was located and removed from the structure at 3:48 AM and transported by air ambulance to a trauma center and then a medical burn center. The Engine 2 fire fighter succumbed to the injuries 2 months later due to complications from thermal burns.

NIOSH Firefighter Fatality Investigation F2016-18. Arson Fire Kills Three Fire Fighters and Injures Four Fire Fighters Following a Floor Collapse in a Row House—Delaware. https://www.cdc.gov/niosh/fire/pdfs/face201618.pdf. Accessed November 26, 2019.

---

grade level. The increased risk is due to limited entry and egress, working above the fire, weakened floor structures in the fire's flow path, unknown fire load, ventilation issues, utility panels, hanging wires, utility meters, and appliances. These risks can lead to fire fighter entrapment from floor collapse, burns, and asphyxiation. Fire departments should conduct a complete 360 size-up to locate possible exterior attack positions and reassess fire conditions prior to conducting interior operations.

Performing a complete walk-around can be a challenge when buildings are located on steep hillsides. However, every effort should be made to do a complete survey, especially when dealing with a suspected basement fire. Several of the basement fires listed in **TABLE 5-3** had basement access at the rear.

TABLE 5-3 Fire Fighter On-Duty Deaths with Floor Collapse as a Contributing Factor*

Location	Occupancy	Year	Number of Fire Fighter Fatalities	Fell Into Basement Due to Floor Collapse
Kentucky	Residential	1997	1	Yes
Pennsylvania	Residential	1997	2	No
Ohio	Residential	1998	2	No
District of Columbia	Residential	1999	2	No
Alabama	Residential	2000	1	Yes
Ohio	Residential	2001	1	Yes
Illinois	Residential	2001	2	No
North Carolina	Residential	2002	1	Yes
New York	Residential	2002	2	Yes
Massachusetts	Residential	2003	1	No
Pennsylvania	Residential	2004	1	Yes
New York	Residential	2005	1	No
Maryland	Residential	2006	1	No
New York	Mercantile	2006	2	Yes
Wisconsin	Residential	2006	1	Yes
Indiana	Residential	2006	1	Yes
Tennessee	Residential	2007	1	Yes
Ohio	Residential	2008	2	Yes
Illinois	Residential	2008	1	Yes
Pennsylvania	Residential	2008	1	No
New York	Mercantile/Residential	2009	2	Yes
New York	Mercantile/Residential	2009	2	Yes
New York	Residential	2013	1	Yes
Maryland	Residential (vacant)	2014	1	Yes
South Dakota	Residential	2015	1	Yes
Ohio	Residential	2015	1	Yes
Delaware	Row House	2016	3	Yes

*Based on the 2012 UL report and other more recent NIOSH reports. (NIOSH Fire Fighter Fatality Investigation Summary [F2016-18]. November 9, 2018. "Arson Fire Kills Three Fire Fighters and Injures Four Fire Fighters Following a Floor Collapse in a Row House—Delaware." https://www.cdc.gov/niosh/fire/pdfs/face201618.pdf. Accessed November 26, 2019.)

Unfortunately, in many of these cases, companies and the IC were unaware of the grade-level access to the basement.

An earlier study by UL in 2012 examines fire fighter safety when combating basement fires (Kerber, 2012). This comprehensive report addresses collapse time for various floor support systems commonly used in residential construction as well as the effect of floor load, fuel load, and venting on collapse time. As expected, lightweight support systems failed much earlier than the dimensional lumber. An earlier fire research project sponsored by the NFPA reached the same conclusion (Grundahl, 1992).

In the UL study, collapse times were compared to estimated response times. It was determined that collapse of all the floor systems tested would have occurred either before fire units arrived or while they were at the scene. Does this mean that fire fighters should never enter a building with a known basement fire? No! Instead, it should be understood that under the conditions of the experiment with no fire attack, collapse should be expected within the estimated response times used during the research. Many basement fires are minor fires involving laundry appliances or small quantities of combustible materials. Basement fires are often confined to clothes dryers or heating ventilation and air conditioning (HVAC) equipment.

Every basement fire is different. Basement fires range from small fires involving limited combustibles or equipment typically found in a residential basement to a raging inferno involving the entire basement area of a large industrial building with an extra-hazard fuel load. Obviously, a small fire could extend to become a larger fire, but a small, contained fire should not affect the structural stability of the floor above or require special tactics. A larger fire might be controllable, but special tactics such as softening the target prior to making entry should be employed.

Large commercial buildings should undergo preincident planning, allowing the fire department to make specific decisions in advance of a fire. For example, a decision could be made not to enter the basement in a large industrial building if there are indications of a working fire. This plan would most likely be based on fuel content, size of the basement, and the floor support system.

Venting in the absence of effective suppression had a negative effect on structural stability during the UL test fires. Fires generally consume the available oxygen then go into decay when the fire is ventilation limited. On its face, this would be an argument for staying out of the building because fire operations sometimes result in unintentional venting when gaining access to the fire area. The problems with a "no-entry" policy are (1) occupants remaining in the fire area would be at great risk due to the lack of oxygen and toxic gases, and (2) all basement fires would result in total destruction of the building. Enclosures with high-heat conditions and a lack of oxygen are ripe for a rapid intensification, flashover, and backdraft. After a careful risk-versus-benefit analysis, the fire officer should determine whether interior operations are reasonable. If a decision is made to conduct an offensive attack, fire fighters should enter the building cautiously with a charged hose line, expecting a rapid increase in fire intensity when opening doors and windows that vent the basement.

According to the UL study, standard on-scene measures used by fire fighters to test a floor for structural stability (sounding the floor, thermal imaging, and floor sag) provide some indication of an impending collapse, but are *very* late indicators that are not entirely reliable. If any of these procedures reveal a weak flooring system, prompt and immediate withdrawal from the area should be compulsory. In several of the NIOSH investigations listed in Table 5-3, a "spongy" floor or roof was noted prior to collapse. These are *very* late indicators, and thermal images are not reliable indicators of impending floor collapse.

The worst-case scenario when fighting a basement fire is falling through the first floor into a well-involved fire in the basement below. A close second would be opening a door leading from the first floor to the basement and being subjected to a superheated flow path. Fire reports and experiments confirm two basic concepts of fire dynamics. First, heat and heated gases will travel upward, then horizontally, following the path of least resistance to reach equilibrium. Second, a ventilation-limited fire will rapidly increase in intensity when air is introduced. When there is a working basement fire, opening the front door and the door at the top of the stairs leading to the basement will allow fire, smoke, heat, and toxic gases to travel upward via the vertical opening (stairway), then horizontally out the front door.

The standard practice of checking the door for heat and being ready to stop venting by controlling the door is especially important when opening doors leading to basements.

After two Washington, DC, fire fighters lost their lives at a basement fire in a town house, NIST conducted research using fire models to determine heat conditions before and after opening windows and doors (Madrzykowski and Vettori, 2000). The model (shown in Figure 5-13B) indicated that there was a distinct flow path from the basement up the stairway and out the front door. Temperatures on the basement stairway were as high as 1508°F (820°C). On the first floor, ceiling temperatures were in the same range as in the stairway. However, floor temperatures were 68°F (20°C).

Floor collapse is a major concern when combating a basement fire. Just as with roof operations, there is a danger that fire fighters working on the floor above will fall through the floor into an inferno below, or that when conducting a direct attack in the basement, a floor collapse from above could kill, injure, or entrap fire fighters below.

Based on the findings of the UL study and the number of on-duty fire fighter fatalities, some in the fire service recommend that fire fighters never enter the first floor of a building with a working fire in the basement. This is a safe approach, but is it effective? Obviously when conditions indicate an impending collapse, staying out of the building is the only option. However, there are also good reasons to conduct operations inside the building on the first floor or basement. Several of the on-duty death reports from NIOSH listed in Table 5-3 revealed the following:

- There were savable occupants or reports of occupants in the fire building upon arrival.
- Fire fighters did not know the fire was in the basement. In some cases, the initial size-up included smoke or fire visible at the roof when the point of origin was in the basement.
- Visual clues of fire were absent or indicated a minor fire.
- There was unknown direct access to the basement at grade level at the rear or side of the building.
- Floors were supported by dimensional wooden beams, or the construction method for the floor was not mentioned. There is no doubt that the percentage of lightweight flooring systems will increase; however, some contractors continue to use dimensional lumber to support floors.

Knowing common construction methods used in the department's response area is important to safe and effective fire operations. This is discussed further in Chapter 2, *Procedures, Preincident Planning, and Size-Up*. In some jurisdictions, building codes require that when the supporting members of the floor separating the basement from the first floor are of lightweight construction, it must be protected with a fire-rated basement ceiling assembly. Fire resistance ratings are most often accomplished by installing gypsum wallboard on the basement ceiling. Clearly, this requirement provides additional time for occupant escape or rescue, as well as extending the time available to conduct an offensive attack. The primary reason for using lightweight construction materials is to reduce construction costs. Applying gypsum wallboard to the ceiling increases the construction cost. Therefore, many contractors use dimensional lumber to avoid this additional expense. As a result, areas with fire-rated basement ceiling requirements will have fewer lightweight floor assemblies between the basement and first floor. It is essential that fire fighters make every effort to know construction methods in their response area. Houses within subdivisions will usually have similar construction methods and materials throughout.

Finishing a basement is a common way to expand the living space within a residential property. Gypsum wallboard is sometimes applied to the ceiling in the finished basement area regardless of code requirements; however, finished living areas are more likely to be occupied by residents and often contain a substantial fire load.

Prefire Conditions and Fire Conditions

Several factors must be taken into consideration in evaluating the collapse potential of a building that is under fire attack. These factors fall into two general categories: building prefire conditions and current fire conditions.

Prefire Conditions

Prefire conditions include the type of construction, as was mentioned previously. In addition, the following prefire conditions can contribute to structural collapse in a building that is heavily involved in fire:

- Weight. Live and dead loads placed on floors, walls, and roofs can include the following:
  - Air-conditioning units
  - Green roofs
  - Solar panels
  - Tanks containing liquids
  - Large signs and marquees
  - False fronts (facades)
  - Cantilever appendages
  - Heavy machinery
- The type, location, and arrangement of combustible/flammable fuel loads inside the building. Modern home furnishings create a much larger fuel load than legacy furnishings (Kerber, 2010).
- Damage to the building's structural support system from previous fires, weather, or partial collapse.
- Building renovations, especially those that have had adjoining structures removed.
- Buildings or areas in poor repair, including vacant structures that may have deteriorated over time, creating a hazard.

- Buildings with long spans, such as churches and warehouses. These buildings should be carefully evaluated for collapse potential.
- Roofs and floors supported by trusses, wooden I-beams, C-channels, and other lightweight construction.

All of these factors can be determined in advance through preincident planning.

Fire Conditions Leading to Structural Collapse

Building conditions leading to structural collapse are sometimes difficult to read, and structural failures have occurred without warning **FIGURE 5-14**. However, the proficient fire officer must recognize signs of imminent danger and maintain the span of control necessary to react quickly when fire fighters are endangered.

As was mentioned previously, time and fire intensity are major factors. Signs of structural collapse include, but are not limited to, the following:

- Bulging, cracked, or unsupported walls.
- Walls leaking water or smoke.
- Falling bricks.
- Floors holding large volumes of water or stock soaked with water, thereby increasing the weight bearing on the floor. An indicator would be large quantities of water entering the building but little water draining out of the building or materials that are obviously absorbing water.
- Movement in floors or the roof. If it feels as though it is unsafe to walk on, it probably is.
- Any other signs of structural movement, including unusual noises.
- Vertical structural members (e.g., columns, walls) that are out of plumb.

Fire Extension

Some buildings limit fire spread and contain the products of combustion better than others. As the fire enters concealed spaces, it can extend to remote locations. This **extension** can result in a sudden increase in heat intensity when the fire breaks out of containment, self-vents, or is vented by fire fighters. A sudden increase in temperature can occur at a location remote from the original fire and result in fire breaking out at multiple locations. Such extension is especially problematic when resources are limited and there is a need to attack the fire from several positions. Fire fighters performing search and rescue operations without a charged hose line are at particularly high risk, as are units working above the fire floor. The immediate availability of a charged hose line is more than a means of extinguishment; it is also an important safety factor. If there is any chance that the fire has spread into, or is likely to advance into, an area where fire fighters are working, a charged and staffed hose line is an imperative.

> **NOTE**
>
> Arsonists or terrorists may purposely set traps, shut off sprinkler systems, and use flammable liquids or explosives, resulting in unusually rapid fire spread and building collapse. Fire departments should track arson behaviors, and police agencies should communicate credible terrorist threats to the fire department.

Fire extension can cut off the primary means of egress for fire fighters. Fires that enter and are confined to concealed ceiling areas are particularly prone to getting behind the fire fighter **FIGURE 5-15**. Fire fighters must check concealed spaces in floors, ceilings, and walls to determine whether the fire is in these areas.

Thermal imaging cameras (TICs) can be used to find hot spots in concealed spaces, but these areas should be opened up if there is any doubt regarding fire extension. There is a high likelihood of a ventilation-limited fire in

FIGURE 5-14 Fire conditions leading to structural collapse.
© Jones & Bartlett Learning.

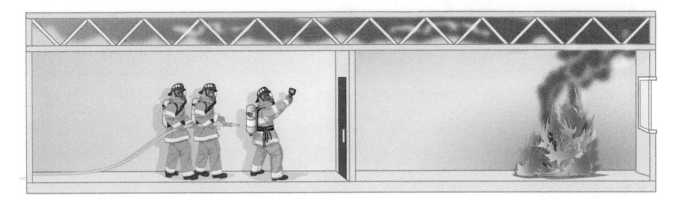

FIGURE 5-15 Fire entering truss space in an enclosed room.
© Jones & Bartlett Learning.

a concealed space. The fire may be in a decay phase, and as a result, it may not be producing high temperatures. Insulation, plaster, gypsum board, finish flooring materials, or other building materials may also result in low temperature readings on the surfaces exposed to the view of the TIC. It is usually best to start by opening a small inspection hole to examine a concealed space, thus limiting oxygen to a possibly oxygen-starved fire.

Coordinated venting will direct the fire out of the building and protect fire fighters; improper venting pulls fire toward fire fighters or into areas behind fire fighters. Increasing the air supply to a ventilation-limited fire will increase fire growth and intensity; therefore, a charged hose line must be in place to limit fire spread prior to opening concealed spaces.

Suspended ceilings are used to hide open construction methods such as trusses, thus creating a large, undivided, concealed space. Most suspended ceilings are supported by wires attached to lightweight metal gridwork that is known to prematurely collapse under fire conditions. The area between the suspended ceiling and actual ceiling or roof provides a convenient location for lighting fixtures, ventilation equipment, wiring, and other utilities. Heavy light fixtures, fans, and equipment are often located in this area, creating a serious falling-object hazard if the gridwork supporting the ceiling fails. Coaxial television cables, communication system wires, conduit, and the supporting gridwork also present an entanglement hazard.

The truss floor, like the truss roof, creates a large, concealed space and is less fire resistive than heavier solid-beam construction **FIGURE 5-16**.

FIGURE 5-16 Truss floor assembly.
Courtesy of Captain David Jackson, Saginaw Township Fire Department.

Incident Summary

Ohio Residential Fire Flashover

One fire fighter died and two other fire fighters were injured in a flashover at a residential fire as they entered the front door to attack the fire with an uncharged 1¾-in. (44-mm) hose line. The preconnected hose line was seven 50-ft (15-m) sections in length, which is beyond the maximum length recommended by hose and nozzle manufacturers for 1¾-in. (44-mm) hose. The pump operator charged the hose line, but kinks in the hose prevented water from getting to the nozzle.

Data from: NIOSH Fire Fighter Fatality Investigation Summary [F2003-12], January 7, 2005, "Career Fire Fighter Dies and Two Career Fire Fighters Injured in a Flashover During a House Fire—Ohio, https://www.cdc.gov/niosh/fire/reports/face200312.html. Accessed November 26, 2019.

TABLE 5-4 Comparing Construction Types

Construction	Type	Structural Stability	Fire Extension Probability	Fuel Contribution
Fire resistive	I	Outstanding	Low	Low
Noncombustible	II	Poor	Average	Low
Ordinary	III	Average	Average	Average
Heavy timber	IV	Excellent	Low	Average
Frame	V	Poor	Average to high	High

© Jones & Bartlett Learning.

TABLE 5-4 compares building construction methods in terms of structural stability, fire extension, and fuel contribution. It is important to remember that the five building categories listed do not represent all possible construction methods. For example, noncombustible buildings are further subdivided into protected and nonprotected. Variations occur within all types of building categories. Buildings being renovated or remodeled may contain structures different from those prescribed by the original construction method. Additions and alterations sometimes result in what can best be described as a hybrid building (a combination of construction types).

The noncombustible building is sometimes mistaken for fire-resistive due to the nonburning characteristics of the structure. The differences are immense. The noncombustible building will normally be masonry or metal on the exterior with lightweight metal trusses as a roof structure. These buildings often have large open areas with long structural spans. Once the fire enters the truss space, expect imminent roof collapse. Unfortunately, fire fighters fail to pay attention to this fact and continue to lose their lives in these buildings. Modern big-box retail stores, such as Walmart, Sam's Club, and Lowe's are generally of noncombustible construction with large open areas, however most of these buildings are protected by an automatic sprinkler system. A properly installed and operating sprinkler system will usually control fires in these stores. However, it is important to remember that many code variances or "trade-offs" may have been allowed because the building was constructed with sprinkler protection. If the sprinkler system is out of service or ineffective, these buildings should be considered non–code-compliant structures. Building management will sometimes arrange materials in a way that shields combustibles or results in a fire load beyond the designed capacity of the sprinkler system. If the sprinkler system is not controlling the fire in a building of this type, consider the hazards involved in entering this large-span truss space with a heavy fire load.

A complete discussion of structural collapse is beyond the scope of this text. Several textbooks discuss the different aspects of building design and how it can affect fire-ground operations. *Brannigan's Building Construction for the Fire Service* by Glenn Corbett and Francis Brannigan is considered a companion text to this one (Corbett and Brannigan, 2021). Fire officers should be familiar with the information contained in the Corbett–Brannigan text and how this information applies to their respective jurisdictions.

Hazard Control Zone

Establishing a collapse zone is critical when structural stability is the reason for a defensive attack. Construction features combined with fire factors indicate the most probable type of structural failure. Given that the IC is always working with incomplete and imperfect information, it is impossible to accurately predict the type of collapse and resultant collapse zone. The only safe collapse zone is one that is equal to the height of the building plus an allowance for scattering debris. A good general rule when determining collapse zone distances for most buildings is to establish an area 1½ times the height of the fire building. This sometimes presents a dilemma if the safe zone is beyond the street width, meaning that effective defensive positions are within the collapse zone. When a defensive stand is a reasonable alternative, positions at the corners of buildings are normally safer than those on the flat side of a wall. Consider using unstaffed ground monitors to reduce the risk of placing personnel in exposed

Incident Summary

Charleston, South Carolina, Furniture Store Fire

On June 18, 2007, at 7:07 PM, the Charleston, South Carolina, Fire Department was dispatched to a fire in an enclosed loading dock behind a furniture store. The structure was a non–sprinkler-protected, one-story commercial furniture showroom and warehouse facility totaling over 51,500 ft² (4785 m²) that incorporated mixed construction types.

The store/warehouse facility had been renovated and expanded a number of times. The original structure was constructed in the 1960s as a 17,500-ft² (1626-m²) grocery store with concrete block walls and lightweight metal bar joists (metal roof trusses) supporting the roof. After being converted to a retail furniture store, the original structure was expanded by adding a 6970-ft² (648-m²) addition on the right (Delta side) and a 7020-ft² (652-m²) addition to the left (Bravo side). Both additions were attached to the original exterior with large interior doors between the buildings. A 15,600-ft² (1449-m²) warehouse was then added to the rear of the main showroom **FIGURE IS5-9A**.

The main showroom measured 9 ft (2.7 m) from the floor to a suspended drop ceiling and approximately 14 ft (4.3 m) to the roof, creating almost 5 ft (1.5 m) of concealed space above the suspended ceiling. Using a TIC or removing suspended ceiling tiles during initial entry into the furniture store might have revealed the presence of fire or heated products of combustion in this concealed area earlier in the fire, resulting in the use of different tactics.

The rear warehouse measured 29 ft (8.8 m) from the floor to the roof with rows of metal storage shelving that contained a variety of modern furniture items.

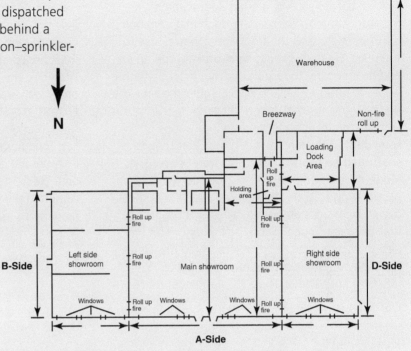

FIGURE IS5-9A Illustration of the floor plan of the furniture showroom and warehouse facility that was the site of the 2007 Charleston, South Carolina, fire.
Courtesy of NIOSH.

FIGURE IS5-9B One-story commercial furniture showroom and warehouse facility that was the site of the 2007 Charleston, South Carolina, fire.
Source: National Institute for Occupational Safety and Health (NIOSH). Nine Career Fire Fighters Die in Rapid Fire Progression at Commercial Furniture Showroom, South Carolina. https://www.cdc.gov/niosh/fire/reports/face200718.html. Accessed April 3, 2020.

The fire department's initial response was three fire companies with three fire fighters per company and two chief officers (well below the response

(continues)

Incident Summary (Continued)

levels recommended in NFPA 1710) at the time of the fire (NFPA 1710, 2020). The first chief officer arrived at 7:09 PM and verified a fire behind the furniture store on the loading dock that connected a furniture store to the front and a warehouse to the rear. This first-arriving chief officer ordered an engine company to respond to that location with a 1½-in. (38-mm) hose line to combat the fire. The second-arriving chief officer checked the furniture store at the front and found no indications of fire spread into the store at that time. Later reports confirmed fires in the store and warehouse. The modern furniture provided a large fuel load. At approximately 7:26 PM (19 minutes after the original dispatch), a store employee called 911 reporting that he was trapped inside the building. This employee was rescued. At this time, three hose lines were inside the main showroom: an initial 1½-in. (38-mm) hose line, a 1-in. (25-mm) booster hose line, and a 2½-in. (65-mm) hose line. All three hose lines were pulled off a single pumper that was supplied by another pumper through a single 2½-in. (65-mm) supply line approximately 1850 ft (564 m) long.

Six fire companies working inside the furniture store became disoriented when heavy smoke obscured vision as the fire rapidly intensified **FIGURE IS5-9B**. Several calls for help were received from interior crews along with at least one mayday. Fire conditions were so severe that rescue attempts by other fire companies were unsuccessful. The nine fire fighters remaining inside the building were caught in the rapid fire progression and were killed.

NIOSH Fire Fighter Fatality Investigation Summary [F2007-18]. February 11, 2009, "Nine Career Fire Fighters Die in Rapid Fire Progression at Commercial Furniture Showroom—South Carolina," https://www.cdc.gov/niosh/fire/reports/face200718.html, Accessed November 26, 2019.

---

positions. A risk-versus-benefit analysis is essential. The crucial question that any IC must ask is:

What could I potentially save in relation to the risk being taken?

Obviously, no building is worth a fire fighter's life; therefore, imminent risk to a fire fighter's life to save a building is unacceptable. Also, remember that nothing should be risked to save what is already lost.

When total collapse is imminent, the collapse zone represents a **no-entry zone** that no one is permitted to enter, regardless of the level of protective clothing. No-entry zones can also exist within buildings, especially when roof or floor structures are suspect (see case study). In addition, no-entry zones would include other areas containing imminent hazards such as falling glass, areas containing atmospheres within or near the flammable range, and any other area that the IC or safety officer deems too hazardous to enter.

Collapse and no-entry zones are not the only safety considerations regarding access. The concept of limiting access to the fire scene is defined in a variety of ways. It seems appropriate to extend the lessons learned from hazardous materials responses regarding zones, as similar zones are possible at structure fires. In this text, working areas are called hazard control zones, whereas a wide area beyond the working zones is known as the **fire perimeter**. The fire perimeter is usually staffed by police, who keep unauthorized people away from the scene, as in an isolation area at a hazardous materials incident. Incident conditions must be considered when determining the dimensions of the fire perimeter. Two blocks in all directions beyond the building on fire is a good general rule for the fire perimeter.

Within the **hazard control zone**, there could be several subdivisions. In most cases there will be a **cold zone** where personal protective clothing is not required (similar to the cold zone at a hazardous materials incident). The command post, as well as other staff and command functions, would be based in this safe area. The cold zone would also include rehabilitation and medical treatment areas.

The **hot zone** would be an operating area, considered safe only for individuals wearing appropriate levels of personal protective clothing (much like the hot zone at a hazardous materials incident). The IC and safety officer have a responsibility to establish and enforce the hot zone. Everyone has a responsibility to abide by their decision.

It is not always necessary to establish a **warm zone** during a structure fire. A warm zone is established when different levels of protective clothing are needed for various areas. It is an intermediate zone (between the hot and cold zones). For example, fire fighters working close to a structure (warm zone) may need to be in full protective clothing, but not

Incident Summary

Indiana Roof Collapse

On August 5, 2014, a 40-year-old male volunteer assistant fire chief died after being trapped under a roof collapse while fighting a fire in a commercial storage building **FIGURE IS5-10**.

The county dispatch center transmitted Box 9101 for county Fire Station 91 at 2059 hours to a septic tank cleaning business for a confirmed commercial structure fire. The fire chief of Fire Station 91 (Chief 9101) communicated to the county dispatch center that the response was incorrect. A fire station from another county was first due at this address. Note: The boundary for both fire stations runs through the center of this property. Chief 9101 also relayed to the county dispatch center that Fire Station 91 would continue their response. Chief 9101 was the first unit on the scene at 2105 hours in a vehicle designated as Battalion 9 and assumed command.

The fire was in a pole barn–style building with metal siding and a roof with wood-truss supports and a pan ceiling (a metal ceiling that blocks the truss, creating a cockloft). Heavy fire was showing through the roof on side Bravo and side Charlie of the structure when the first-due company arrived. After a brief conversation with the assistant fire chief (victim), the IC decided to open the doors on the north end (side Alpha) of the building to set an unmanned ground monitor to keep the contents of the building cool. Access was made through both a doorway and overhead door on the north side. Smoke conditions were light with good visibility.

The assistant fire chief was assigned to side Alpha. A defensive fire attack was initiated. The assistant fire chief was one of three fire fighters who had entered side Alpha of the structure to stretch a 2½-in. (65-mm) hose line to protect equipment and acetylene cylinders. The crew was operating approximately 50 ft (15.2 m) inside the structure and

FIGURE IS5-10 The 2014 roof collapse that occurred in Indiana.
Source: National Institute for Occupational Safety and Health (NIOSH). Death in the line of duty . . . https://www.cdc.gov/niosh/fire/pdfs/face201418.pdf. Accessed May 14, 2020.

then decided to change the 2½-in. (65-mm) nozzle to a portable ground monitor (deck gun). During the changeover, one fire fighter left the interior to go outside and charge the hose line. The fire was already in the overhead truss system above the assistant fire chief and the fire fighter, and the fire was likely concealed by the ceiling.

As the third fire fighter got to the overhead door, a loud crash occurred. The truss system failed and the ceiling and roof assembly collapsed on the assistant fire chief and fire fighter. The assistant fire chief was killed by the collapsing truss system. The fire fighter, who suffered a broken leg, was able to crawl under some equipment before being rescued by a RIC from Squad 18.

Source: NIOSH Fire Fighter Fatality Investigation Summary [F2014-18], July 18, 2018, "Volunteer Assistant Chief Killed and One Fire Fighter Injured by Roof Collapse in a Commercial Storage Building—Indiana," https://www.cdc.gov/niosh/fire/reports/face201418.html, Last reviewed August 7, 2018. Accessed December 11, 2019.

---

breathing air from their SCBA. Members entering the (hot zone) would be required to don their SCBA face pieces. When the fire is in a larger building, donning the SCBA face piece may be delayed until the fire fighter reaches an area that is or could be an immediately dangerous to life and health (IDLH) atmosphere. Again, this intermediate area is referred to as a warm zone. For example, the hot zone for a fire in a multistory building would normally include the fire floor, floors above the fire, and one or two floors below the fire. Lower floors could be the warm zone.

FIGURE 5-17 illustrates hazard control zones, with the building's interior being considered the hot zone. Fire fighters entering the building would be expected to be in full turnout gear, breathing from their SCBA. The area on all sides of the exterior would be the cold zone, and the fire perimeter would be located beyond

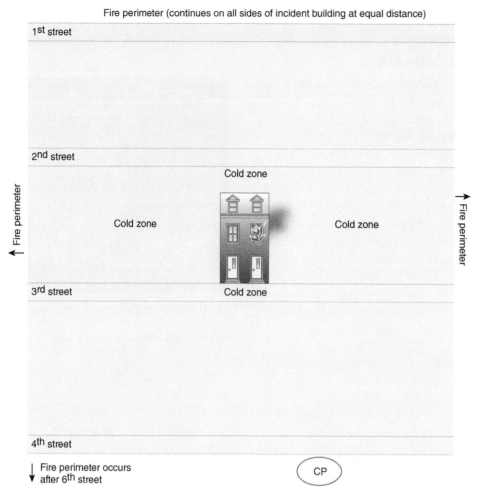

FIGURE 5-17 Hazard control zones—offensive attack.
© Jones & Bartlett Learning.

the cold zone. The fire perimeter would be maintained by police tasked with keeping unauthorized people away from the scene. The actual distances for the zones would vary according to the hazards at the incident.

If the fire is not contained and an exterior (defensive) attack becomes necessary, the hot zone is moved far enough away from the structure to place fire fighters outside the collapse zone and in fresh air. The collapse zone then becomes a no-entry zone, as shown in **FIGURE 5-18**.

Electrical Hazards on the Fire Ground

Electrical hazards are common at the fire scene and other emergencies. Wires and electrical equipment are often damaged at a fire or other emergency scenes, creating an electrical shock hazard. Electrical equipment is often found inside and outside at a structure fire.

Shock Hazard Warning Precautions

The following material is from David W. Dodson's text, *Fire Department Incident Safety Officer* (Dodson, 2016). Any fire fighter who discovers or suspects an electrical shock hazard (e.g., wires down, damaged electrical equipment) should take immediate action to warn others. The following steps should be covered by department SOPs or standard operating guidelines, and all members should be trained to perform them:

1. Immediately warn others using an urgent or priority radio message.
2. Act as guard or sentry to warn others until the IC can assign resources.
3. Use a box light, lantern, or other scene light to illuminate the hazard at night.
4. Flag the area as a no-entry zone using red-and-white diagonal-striped barrier tape

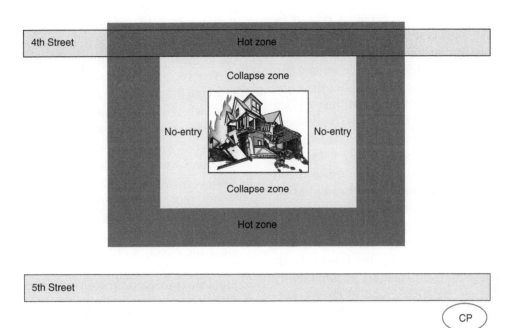

FIGURE 5-18 Hazard control zones—defensive attack.
© Jones & Bartlett Learning.

(10 ft [3 m] minimum clear zone for ground hazards, greater when wet).
5. For overhead power pole fires or hanging wires, clear a zone equal to the distance between two poles.
6. Ensure that the local power company has been notified to respond. (pg. 107)

Failure to adequately warn other fire fighters of a known electrical hazard has been a contributing factor in fire fighter electrocutions. Following these steps can prevent this from happening again.

Other steps that fire fighters can take to protect themselves from electrical hazards are as follows:

- Treat all downed power lines as if they are energized.
- Keep a safe distance from downed power line—at least 10 ft (3 m).
- Wear the appropriate personal protective clothing and footwear.
- Do not attempt to move or cut downed wires.
- Avoid walking or standing in pooled water when downed power lines are present.
- When fighting fires near downed power lines, do not use solid-stream nozzles. Never apply water directly to electrical equipment that is burning or arcing. Because water is a good conductor of electricity the current may flow back through the hose stream to the nozzle and cause serious injury. Using a high-pressure fog nozzle to break up the water stream can help prevent electrical current from flowing back to the nozzle. (pg. 107)

NOTE

Beware, some buildings, including homes, are equipped with backup power equipment. Auxiliary power is often provided via a natural gas, LP gas, or gasoline/diesel-powered generator on the premises. In most cases, it is best to wait until the equipment is deenergized prior to entering areas with downed electrical equipment!

Relationship of Time, Fire Intensity, and Structural Stability

As mentioned earlier in this chapter under *Fuel Load*, the UL study titled *"Impact of Ventilation on Fire Behavior in Legacy and Contemporary Residential Construction"* proves that modern synthetic fuels found in residential occupancies today burn faster and progress to flashover conditions much sooner than fuels found in mid-20th-century homes, defined as "legacy" furnishings (Kerber, 2010). The UL flashover experiments were first conducted in two side-by-side rooms that were each 12 ft by 12 ft by 8 ft (3.7 m by 3.7 m by 2.4 m) high. A second experiment was conducted in rooms 14 ft^2 (1.3 m^2) larger than in the first experiment. In both cases, the fire compartment fuel load consisted of typical living room furnishings. One room was furnished with legacy furnishings and the other with modern furniture. For the most part, the structural firefighting tactics that are in common use today are based on materials found in homes 50 or more years ago. These furnishings no longer exist in the typical 21st-century home. An opening 8 ft (2.4 m) wide by 7 ft (2.1 m) high remained open during the experiments to allow better viewing of fire behavior. This opening provided more combustion air than would be present from average residential openings, where air is supplied by way of smaller window and/or door openings. As a result of these experiments UL concluded the following (Kerber, 2010):

> The modern and the legacy rooms demonstrated very different fire behavior. It was very clear that the natural materials in the legacy room burned slower and produced less energy and smoke than the fast burning synthetic furnished modern room. The times to flashover show that the flaming fire in a room with modern furnishings leaves significantly less time for occupants to escape the fire. It also demonstrates to the fire service that in most cases the fire has either transitioned to flashover prior to their arrival or became ventilation limited and is waiting for a ventilation opening to increase burning rate.

This is an important research finding. Today, fires in residential occupancies are indeed different than in the past. When the fire is well vented, there is little chance to save occupants remaining in the compartment of origin within a residential occupancy. With the 7 ft (2.1 m) by 8 ft (2.4 m) opening, the modern room experienced flashover in 3 minutes and 20 seconds in the original test room (3 minutes and 30 seconds in the slightly larger room), compared to a flashover time of 29 minutes and 30 seconds in the first burn in the legacy-furnished room. During the second burn, the legacy-furnished room did not reach flashover conditions. There is a high probability that these fires in a less-ventilated room would go into decay as the fire consumed available oxygen.

From a risk-versus-benefit analysis, the probability of saving lives in the fire compartment is greatly diminished when fire fighters arrive after flashover. However, occupants may be savable in areas separated by barriers such as closed interior doors or located remote from the fire. Larger rooms with different furnishings (fuel load) may take longer to reach flashover conditions, allowing more time to conduct rescue activities on the interior.

Fire fighters arriving at a structure fire need to consider the effect of providing additional air by opening doors, windows, and roofs, which could lead to a rapid increase in combustion in a ventilation-limited fire compartment. Introducing air in such situations creates flow paths that endanger not only the remaining occupants but also fire fighters.

NOTE

The fuel load in today's homes burns much faster and contributes more fuel than in homes of the past. This factor, combined with lightweight construction, places fire fighters and occupants at significantly greater risk.

FIGURE 5-19A shows a relationship between response time, structural stability, and survivability of occupants in a well-ventilated legacy living room. **FIGURE 5-19B** shows a well-ventilated modern living room.

FIGURE 5-19C shows a ventilation-limited fire in a modern living room. The relationships shown in these three charts are generally based on measurements taken during the UL research. However, the heat, toxicity, lack of oxygen, and response time are all highly dependent on a number of factors. During research, constants such as room size and ventilation are controlled to compare variables such as fire conditions produced by legacy furnishings versus modern furnishings. In actual fires there are few or no known constants, and each variable has an effect on other variables. For example, an uncontrolled fire in a room similar to the one used by UL during this

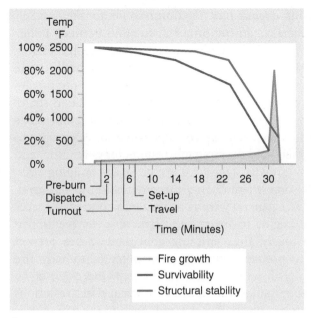

A

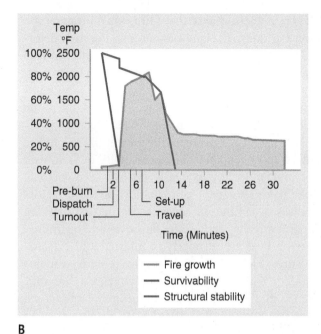

B

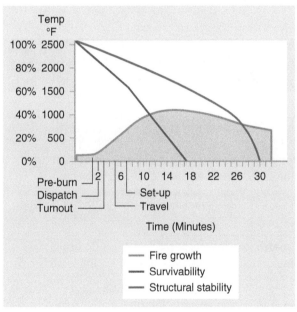

C

FIGURE 5-19 A. Flashover, survivability, and collapse versus response time—well-ventilated fire (legacy living room). **B.** Flashover, survivability, and collapse versus response time—well-ventilated fire (modern living room). **C.** Flashover, survivability, and collapse versus response time—ventilation-limited fire (modern living room).
© Jones & Bartlett Learning.

research would probably be attached to other rooms, quite often rooms without doors or open doors, which would affect ventilation, temperature, and thus rate of fire growth.

Structural stability and survivability were not tested during the UL experiments. However, there is little chance of occupant survival under flashover conditions. Figure 5-19B shows the relationship between survivability and heat conditions in a well-ventilated room with modern furnishings. Therefore, the survivability part of the graph in Figure 5-19B would be considered accurate under the given conditions. In this well-ventilated fire involving modern furnishings, flashover occurs prior to the arrival of the fire department and there is little chance of rescuing the occupants who are in the fire compartment. However, a less-ventilated room may go into the **decay phase** due to a lack of oxygen or fuel (as illustrated in Figure 5-19C).

The building structure is minimally affected during the **ignition phase** and early **growth phase** of the fire. Structural stability is compromised as the fire continues to burn through the growth phase to the **fully developed phase**.

Structural stability is highly dependent on other factors such as construction methods and materials (e.g., wood truss versus an older dimensional timber supporting system). Another factor is construction materials that affect fire extension into walls, ceilings, and roofs.

The legacy fire in Figure 5-19A progresses slowly, and the probability that occupants will be able to safely escape is high. In addition, fire fighters have time to rescue those who fail to escape. In the less-ventilated fire, which goes into the decay phase due to a lack of oxygen (air), there may well be savable occupants, although the effects of toxic gases from the burning synthetics and lack of oxygen will significantly reduce their chance of survival. The same building with a lighter fuel load (e.g., a vacant building) could burn out due to a lack of fuel.

There is also a chance that a fire would progress at a slower rate due to the ignition source and/or early detection. For example, a fire ignited by a cigarette will take longer to develop into a flaming fire than the UL experimental fires using a flaming ignition source. This additional time could result in alerting occupants due to the smell of smoke or activation of a smoke detector.

The longer the fire burns, the greater the risk to fire fighters and occupants. The probability of flashover, structural failure, low oxygen levels, high levels of toxic smoke/fire gases, or a combination of these factors increases over time. In some cases, the first-arriving fire company may arrive after flashover and be limited to saving lives and property in internal and external exposures. However, response time remains a major life-safety and extinguishment factor, as a fire may progress at a slower rate due to a number of factors. As the fire continues to burn, the risk also increases in other parts of the fire building or adjoining buildings.

Structural collapse could occur before flashover, particularly in a large building where localized heating of the area above the fire could result in structural failure before the entire area reaches its flashover temperature. In Figure 5-19A, the line marked "structural stability" starts at the top left corner of the chart indicating 100% structural strength. Downward deflection of the structural collapse line represents a weakening of the structure to the point of collapse where the line angles steeply downward. Structural stability will be different for every fire. Thus, this line represents a best guess. Newer structures with lightweight engineered-frame construction or in which fire enters a truss space are likely to collapse much sooner than well-maintained buildings made with dimensional lumber.

The line marked "survivability" is an estimation of the probability of occupant survival. As smoke and toxic gases reach the floor level, the probability of occupant survival is very low. Occupant survival is improbable in a post-flashover compartment, and even fully protected fire fighters will quickly succumb under flashover conditions. The survivability line will always reach the bottom of the chart at or before flashover, as temperatures necessary for flashover are lethal. In ventilation-limited fires, heat could result in fatal injuries, but a lack of oxygen and toxic gases usually plays a larger role than in fires that rapidly progress to flashover. Likewise, the probability of occupant survival is low after a structural collapse due to fire. A different fuel configuration, a different ventilation profile, or a different type or size of structure could radically change the time value and curve configuration. The principle to keep in mind is that fire growth and time are critical factors.

Brannigan's Building Construction for the Fire Service makes an important distinction between building fires and structural fires (Corbett and Brannigan, 2021). Fires involving the contents of the building are categorized as building fires, whereas fires involving actual structural members are considered structure fires. Although this differentiation is not made in this text, but the concept is useful. According to Corbett and Brannigan's terminology, ignition and early growth stage fires are generally building fires, but somewhere during the progression to flashover, the fire generally finds its way into concealed spaces and begins to involve the structure. As mentioned previously, fires entering concealed spaces present many problems, including the potential to harm fire fighters by cutting off their egress. **TABLE 5-5** lists the time variables from ignition to when effective actions are taken on the fire ground. Each of the variables is subject to change depending on local resources and fire conditions.

TABLE 5-5 Time Components: Ignition to Effective Operations

	Definition	Time Objective
Preburn Time	From ignition until the fire is reported to the public safety answering point (PSAP)	Unknown variable No time objective
Dispatch Time	From public call until fire units are notified	64 seconds to 106 seconds (NFPA 1221, 2019)
Turnout Time	From when fire units are notified until apparatus leaves the station	80 seconds (NFPA 1710, 2020)
Travel Time	From when first engine leaves the station until they arrive at the scene	240 seconds; this is for first-arriving engine (NFPA 1710, 2020)
Setup Time	From when fire units arrive at the scene until fire units take effective action	2 minutes (NFPA 1720, 2020)

© Jones & Bartlett Learning.

Time: Ignition to Effective Actions

The time from ignition to effective actions is critical. The goal is to arrive prior to flashover and intervene, thus interrupting the fire's progression to flashover. Depending on conditions, progression to flashover in small enclosures will be rather fast and may well occur prior to the arrival of the fire department. In this case, the objective is to contain the fire to the original flashover compartment and areas already involved in fire. In analyzing the time from ignition to effective action, it is necessary to consider the five components shown in Table 5-5. These times can vary depending on alarm systems, travel distance, proficiency of the attack team, and other factors.

Detection/Transmission

The first time segment shown in Figure 5-19 and Table 5-5 is preburn time, which is the time from ignition until the fire is reported to the dispatch center. Unless a supervised detection/alarm system is in the fire area, this time component is dependent upon a person discovering the fire and then calling in the alarm. This time element will vary greatly. A fire in an occupied area where everyone is awake will probably result in less detection/alarm time than one in an unoccupied building where the fire is not noticed until it shows on the exterior, which may be during the fully developed phase. When detection relies on a person detecting then transmitting an alarm, the detection/transmission time can be very long. There have been many cases in which a fire burned undetected until it reached sufficient intensity to break through a window or roof. Likewise, fire investigations of large, uncontrolled fires often note a delayed alarm, resulting from occupants investigating the source of the fire or trying to extinguish the fire before notifying the fire department. If the building is protected by a properly installed and operating supervised detection/alarm system, this time can be estimated based on experience. Otherwise this time component is an unknown.

Dispatch Time

The second time segment is the dispatch time. This includes the time for the dispatcher to take the call, select the units for the assignment, and then dispatch the companies (process the call).

NFPA 1221, *Standard for the Installation, Maintenance, and Use of Emergency Services Communications Systems*, requires that the dispatcher answer a call from the public within the following time limits (NFPA 1221, 2019):

- Time to answer
  - Less than or equal to 15 seconds: 90%
  - Less than or equal to 20 seconds: 95%
- Event processing
  - Less than or equal to 60 seconds: 90%

Turnout Time

The third segment is turnout time. This is the time from the receipt of the alarm by the fire department until the apparatus crosses the front door sill of the fire station or otherwise begins response toward the scene. Turnout time can differ greatly between fully staffed stations and on-call stations.

NFPA 1710 defines these parameters as follows (NFPA 1710, 2020).

> **3.3.64.8** Turnout Time. The time interval that begins when the emergency response facilities (ERFs) and emergency response units (ERUs) notification process begins either by an audible alarm or visual annunciation or both and ends at the beginning point of the travel time.
>
> **4.1** Fire Department Organizational Statement
>
> **4.1.2.1 (2)** 80 seconds turnout time for fire and special operations response and 60 seconds turnout time for EMS response.

Travel Time

Travel time is defined in NFPA 1710 4.1.2.1 (3)-(6) as "the time interval that begins when a unit is en route to the emergency incident and ends when a unit arrives at the scene" (NFPA 1710, 2020).

> **4.1.2.1 (3)** 240 seconds or less travel time for the arrival of the first engine company at a fire suppression incident.
>
> **(4)** 360 seconds or less travel time for the arrival of the second company with a minimum of 4 personnel at a fire suppression incident.
>
> **(5)** for other than high-rise 480 seconds or less travel time for the deployment of an initial full alarm assignment at a fire suppression incident.
>
> **(6)** for high-rise, 610 seconds or less travel time for the deployment of the initial full alarm assignment at a fire suppression incident.

In addition, the NFPA 1710 department organizational statement includes time requirements for emergency medical incidents (NFPA 1710, 2020).

Travel time depends on distance, road conditions, terrain, traffic, and other factors. Average response times can be established by using computer models. If computer models are not available, reasonably accurate estimations can be determined by using empirical methods, such as the International Organization for Standardization (ISO) formula, which states that $1.7 \times \text{distance} + 0.65 = \text{travel time}$, or the RAND Corporation's average speed of 35 miles per hour (56 km/hr). Administering NFPA 1710 requires record-keeping for turnout and travel time. These historical data can also be used to calculate response times within the response area.

Setup Time

Setup time is the time necessary to position the apparatus, advance the first hose line into an attack position, and apply water. Staffing levels and training radically affect this part of the response time. The two-in/two-out rule is included in NFPA 1500 and has been adopted by the Occupational Safety and Health Administration (NFPA 1500, 2018; OSHA, 2019). This rule, in effect, changes the setup time. Now four people must be on the scene, with two of them positioned outside the hazard area.

It is highly recommended that each fire department time its initial operations during training using department equipment and SOPs. In many cases, a significant amount of time can be saved through practicing, providing more accessible equipment, and revising SOPs. NFPA 1410, *Standard on Training for Emergency Scene Operations*, provides guidance on evaluating and timing a variety of fire-ground operations (NFPA 1410, 2020).

These five time segments are valid for the initial response only. Preburn time does not figure into on-scene calls for assistance, and the dispatch time is generally less when fire units are calling for help, as the dispatcher does not have to ask questions of the caller. Turnout and response times can also be reduced for subsequent calls for assistance by placing units on alert status, moving them into vacated stations, or placing them in staging. Setup time will change depending on the task assignment. These times are shown in Figure 5-19.

Adequate Number of Personnel

Setup time is related to staffing. If the initial company is staffed with less than four personnel and an imminent life-threatening situation does not exist, the attack should be delayed until additional personnel arrive. As discussed previously, NFPA 1500 stipulates a minimum of four fire fighters as an initial crew at a working structure fire. The standard allows fewer than four in imminently life-threatening circumstances. NFPA 1500, 8.8.2.10 reads as follows (NFPA 1500, 2018):

> **8.8.2.10** Initial attack operations shall be organized to ensure that, if on arrival at the emergency scene, initial attack personnel find an imminent life-threatening situation where immediate action could prevent the loss of life or serious injury, such action shall be permitted with less than four personnel.

From a commonsense point of view, a minimum crew size of four is logical, three being a compromise in situations of imminent danger. This is based on one person at the pump panel maintaining a reliable but limited water supply. The pump operator also acts as the outside safety backup during initial operations with a single crew on the scene. The buddy system is required whenever fire fighters enter a burning building; therefore, two fire fighters would be required to make entry for fire suppression and/or rescue. Suppression is usually the best first action, and it takes at least two people to move and operate a standard attack hose line. When practical, one of the two fire fighters making entry (usually the company officer) would conduct a complete walk-around of the structure while the other fire fighter making entry would begin stretching the initial attack hose line toward the point of entry. When department SOPs direct the first-arriving engine company to secure a source of water, the fourth person would be the hydrant or water supply person. In most cases, the fire fighter assigned to the hydrant is the second person outside using a two-in/two-out rule conforming with IRIC. This person should also be able to control the entry door to limit air intake as the interior crew enters the building.

There are situations in which one or two fire fighters could start operations by setting up the pump, advancing a hose line toward the building, or securing a water supply. This would be acceptable under the NFPA 1500 safety standard, provided that entry was delayed until a team of at least two could enter the building with two remaining outside (NFPA 1500, 2018).

There is also a possibility of a first crew of one or two fire fighters raising ladders to rescue people from the exterior. This would meet the standard, as the fire fighters would not be making entry into the structure. However, this is a low-probability scenario. OSHA regulations require a two-in/two-out attack configuration for the initial attack when fire fighters enter a fire area but with an imminent life-threatening situation exception. The determination of imminent life-threatening situation is often somewhat subjective, thus inviting second guessing.

Assembling a minimum four-person team is highly recommended when conducting an offensive attack. Defensive attacks may require fewer people, but they are generally not effective in saving occupants and often result in a total loss of the building of origin.

The initial attack is designed to immediately save lives by extinguishing the fire or making a quick rescue. Many additional tasks are required to save lives and protect property. Some of these include the following (in no particular order):

- Additional attack hose lines to meet flow requirements
- Attack hose line above the fire
- Attack hose line to concealed spaces
- Backup for the initial attack hose line
- Exposure protection
- Forced entry
- Laddering the building
- Opening up of concealed spaces
- Salvage or property conservation
- Search and rescue of the area around the fire
- Search and rescue of the area immediately above the fire
- Search and rescue of other areas
- Ventilation
- Utility control

Not all of these tasks will be needed at every working structure fire, and not all need to be performed by a separate team of two fire fighters. The exact number of fire fighters needed to perform these tasks will vary depending on conditions as well as the size and complexity of the building. It is obvious that many fire fighters will be needed to attack the fire, ventilate, perform the primary search, provide backup for inside crews, force entry, perform salvage, and, where necessary, protect exposure buildings.

In addition to all the tasks being performed to rescue occupants, control the fire, and reduce property damage, a fire company should also be assigned as the rapid intervention team (RIT). NFPA 1710 indicates the following (NFPA 1710, 2020):

5.2.4.1.1 (8) At a minimum, an initial rapid intervention crew (IRIC) assembled from the initial attack crew and, as the initial alarm response arrives, a full and sustained rapid intervention crew (RIC) established (4).

5.4.5 The fire department shall have the capacity to institute a RIC, consisting of personnel trained and equipped as specified in NFPA 1407, during all special operations incidents that would subject members to immediate danger or injury in the event of equipment failure or other sudden events, as required by NFPA 1500.

NFPA 1710 establishes minimum staffing levels in terms of tasks to be accomplished and the number of personnel needed to accomplish specific tasks by occupancy. For example, NFPA 1710 states the following staffing requirement for a single-family dwelling (NFPA 1710, 2020):

5.2.4.1.1 The initial full alarm assignment to a structure fire in a typical 2000 ft^2 (186 m^2) two-story single-family dwelling without basement and with no exposures shall provide for the following:

(1) Establishment of incident command outside of the hazard area for the overall coordination and direction of the initial full alarm assignment with a minimum of one member dedicated to this task. (1)

(2) Establishment of an uninterrupted water supply of a minimum of 400 gpm (1520 L/min) for 30 minutes, with supply line(s) maintained by an operator. (1)

(3) Establishment of an effective water flow application rate of 300 gpm (1140 L/min) from two handlines, each of which has a minimum flow rate of 100 gpm (380 L/min) with each handline operated by a minimum of two members to effectively and safely maintain the line. (4)

(4) Provision of one support member for each attack, backup or exposure line deployed to provide hydrant hookup and to assist in laying of hose lines, utility control, and forcible entry. (2)

(5) Provision of at least one victim search-and-rescue team with each such team consisting of a minimum of two members. (2)

(6) Provision of at least one team, consisting of a minimum of two members, to raise ground ladders and perform ventilation. (2)

(7) If an aerial device(s) is used in operations, one member to function as an aerial operator to maintain primary control of the aerial device at all times. (1)

(8) At a minimum, an initial rapid intervention crew (IRIC) assembled from the initial attack crew and, as the initial alarm response arrival, a full and sustained rapid intervention crew (RIC) established. (4)

(9) Total elective response force should be a minimum of 16 (17 if an aerial device is used.)

Note that the NFPA 1710 requirements listed earlier are for a somewhat typical one-family house. NFPA 1710 also addresses the following (NFPA 1710, 2020):

- Open air strip shopping centers
- High-rise initial full alarm assignment capabilities
- Fire alarm notification assignment
- Additional alarm assignments
- EMS functions

Many skeptics doubted the need for 16 or 17 fire fighters and a minimum 4-person crew at a typical residential fire with no exposures. In response, the NIST built the 2000 ft^2 (186 m^2) two-story structure described in NFPA 1710 and conducted over 60 full-scale experiments in the structure to determine the effectiveness and safety of various crew sizes and different staggered times (interval between arriving crews; NIST, 2010). Technical fire service experts identified 22 separate tasks that should be conducted at this single-family residential fire. Each task and the overall time to completion were timed during the experiments. This scientific study proved the need for the staffing requirements identified in NFPA 1710.

The data for two-, three-, and four-person crews with a short interval between arriving companies showed a dramatic difference in the time to complete the 22 identified tasks. The two-person crew took 22 minutes and 16 seconds to complete the tasks, the three-person crew required 20 minutes and 30 seconds, and the four-person crew completed all 22 tasks in 15 minutes and 44 seconds.

NIST also conducted a study in a high-rise building that is discussed in Chapter 12, *High-Rise Buildings*.

Tactical Reserve

Planning ahead to obtain required resources is crucial. The number of fire fighters needed to safely and effectively combat a working structure fire can be significant. Tactical efficiency can reduce the number of people necessary to fight the fire.

For example, immediate application of water coordinated with ventilation will assist the primary search team, allowing them to cover a much larger area with fewer people. This same tactic also permits the attack team to extinguish the fire and move to *another task more rapidly*.

After the IC staffs all of the attack and support positions, there is a need to establish a tactical reserve. The size of the reserve force depends on the stage of the incident and the number of units working, as well as the type of incident. It can be accurately said that if all units are assigned, you need additional personnel.

Correlation between Elapsed Time and Progression to Flashover

The main point to be made in studying the flashover/survival/collapse charts is that time is a critical factor (see Figure 5-19). Over time, the opportunity to rescue occupants decreases and the chance of structural collapse increases. From a fire fighter safety viewpoint, the worst scenario is one where the fire department arrives near the end of the buildup to flashover with occupant lives at risk in a large, undivided area. Once flashover occurs, the chance of survival is near zero in the flashover compartment. The risk to fire fighters has increased because of the fire intensity and the potential for significant structural damage. This is a situation in which risk management is critical to fire fighter safety and survival.

In general, it can be said that the smaller the compartment, the sooner the progression to flashover or backdraft. Most of the fuel load in a structure fire will be in the form of solid materials. Ordinary solid materials must be preheated to the point at which vapor fuel is released, through a process called pyrolysis.

The vapor fuel will ignite if there is an available and sufficient ignition source. However, as the fire progresses, it preheats solid fuels that release unignited pyrolytic vapors. If the fire area is enclosed, pyrolytic vapors and "products of combustion" such as smoke and carbon monoxide will accumulate. If there is sufficient oxygen available when the contained fuels reach ignition temperature, a flashover occurs. If there is a lack of oxygen, the fire could go into the decay phase. However, if air is introduced into this superheated and fuel-rich atmosphere, flashover or a backdraft explosion could occur. Fire fighters must use extreme caution when opening doors and windows when the fire is ventilation limited; the fire will almost surely gain intensity and possibly result in flashover or backdraft.

Small, enclosed rooms quickly consume available fuel and/or oxygen, creating potential flashover or backdraft conditions. In a small room, the fuel packages will tend to be close to each other, allowing heat radiation to raise their temperature toward ignition sooner. The volume of a smaller room allows the overall temperature to rise more rapidly than in a larger room, decreasing the likely time to flashover or backdraft.

Observing smoke and temperature conditions is critical when entering small, enclosed areas. Sometimes these areas simply burn out before entry. Initially, flashover and backdraft tend to be the most critical problems when encountering a working fire in a small-volume compartment **FIGURE 15-20**. The chance of burning all available fuel or using up all the available oxygen is much less in larger areas.

Flashover and backdraft will usually occur much later, if at all, in large fire compartments. However, the concealed space created between suspended ceilings and truss roofs may be a relatively small-volume area owing to the height of the area. In fact, many backdrafts occur in concealed spaces. One such incident on December 17, 1960, at the Parkmoor Bowling Alley in Louisville, Kentucky, resulted in the deaths of three fire fighters. Two fire fighters were fatally injured in an incident investigated by NIOSH due to a possible

FIGURE 5-20 Post-flashover fire in a single apartment.
Courtesy of Mike Carey and Jeff Neal, Cincinnati Fire Department.

Incident Summary

Tennessee Partial Roof Collapse

On June 15, 2003, a 39-year-old male career lieutenant (Victim #1) and a 39-year-old male career fire fighter (Victim #2) died while trying to exit a commercial structure following a partial collapse of the roof, which was supported by lightweight metal trusses (bar joists) **FIGURE IS5-11**. The victims were part of the initial entry crew searching for the fire and possible entrapment of the store manager.

Both victims were in the back of the store operating a handline on the fire that was rolling overhead above a suspended ceiling. A truck company was pulling ceiling tiles searching for fire extension when a possible backdraft occurred in the void space above the ceiling tiles. Victim #1 called for everyone to back out due to the intense heat. At this point, the roof system at the rear of the structure began to fail, sending debris down on top of the fire fighters.

Victim #1 and Victim #2 became separated from the other fire fighters and were unable to escape. Crews were able to remove Victim #2 within minutes and transported him to a local hospital where he succumbed to his injuries the following day. Soon after Victim #2 was removed, the rear of the

FIGURE IS5-11 A photo of the partial roof collapse, indicating Side Bravo.
Source: NIOSH.

building collapsed preventing further rescue efforts until the fire was brought under control. Victim #1 was recovered approximately 1½ hours later.

Source: NIOSH Fire Fighter Fatality Investigation Summary [F2003-18], July 26, 2004, "Partial Roof Collapse in Commercial Structure Fire Claims the Lives of Two Career Fire Fighters—Tennessee," https://www.cdc.gov/niosh/fire/reports/face200318.html, Accessed December 11, 2019.

backdraft that contributed to a roof collapse and entrapment as ceiling tiles were removed from below in a 6670 ft² (620 m²) retail store area (NIOSH, 2004). This was also a problem in the previously discussed Charleston, South Carolina, Furniture Store Fire.

Most buildings have more than one enclosed area or compartment. It may be possible to conduct search and rescue operations in other compartments after the compartment of origin has flashed over. For example, suppose a fire occurs in the apartment building shown in Figure 5-20. The apartment of origin sustains flashover, but surrounding apartments are not yet involved. Survival chances for occupants remaining in the apartment of origin are near zero, but occupants in other areas near the fire and/or behind closed doors within the apartment of origin could be rescued. In this multistory building, there is a critical need to quickly get water on the fire and perform post-flashover rescue operations on the fire floor and on the floor above the fire.

Fire-Ground Operations

Communications

Communications are the lifeblood of any command organization. Without communications, the situation is at best perplexing, and at worst, chaotic. NFPA 1500 requires each fire fighter working inside the structure, whether in the hot zone or the warm zone, to be provided with a portable radio (NFPA 1500, 2018). Providing radios to every fire fighter has been a documented lifesaver on several occasions. Radio discipline can prove to be a significant challenge, but it is imperative when everyone on the scene is assigned a radio. If everyone is talking simultaneously, it will be impossible to communicate assignments or emergency messages. Proper use of the radio in exchanging important information in a clear, calm manner is essential to good scene management and fire fighter safety.

Modern radios have the capacity to operate on a wide variety of channels. This is important to interoperability and can improve on-scene communications. However, it is important to keep the communications network simple so all units can be promptly notified of an emergency condition or critical need.

While operating in a hot or warm zone, all personnel shall be equipped with a portable radio provided by the fire department.

Frequent progress reports are essential to the IC, who should have a good overall understanding of the operation in progress. Interior crews and crews working in areas not visible to the IC are the eyes and ears of the IC. Progress reports should also provide everyone on the fire ground with the location as well as information on other aspects of the fire that relate to their own particular operations.

Fire fighters in need of assistance transmit a "mayday." A special radio alert tone indicating a mayday is an advisable enhancement. It is important that SOPs spell out the response to a mayday. The tendency is for all crews to stop what they are doing to go to the assistance of fire fighters who are declaring the mayday. If everyone responds to the mayday call, however, there will be too many people in the area of the mayday. In most cases, the RIC will be assigned to rescue the fire fighters who are in danger. However, there are some commonsense exceptions, such as when another fire crew is in the immediate area of the problem.

Critical functions such as fire suppression will not be accomplished if all personnel focus on the fire fighter rescue operation. Controlling or extinguishing the fire provides the time needed for the rescue operation. Fire officers must understand the importance of remaining within the command system, and the IC must remain in total control at all times. Specific assignments should be made during mayday rescue operations so that all tasks can be coordinated.

Departments should establish an emergency evacuation signal. One method is to use apparatus air horns to signal the retreat, such as ten 3-second blasts. Radio channels in use at the scene should also transmit an emergency evacuation message. Again, SOPs should be in place describing actions to be taken during an emergency evacuation. Just as the hose line controlling the fire buys critical time to accomplish a rescue, it can also provide additional time to evacuate and protect fire fighters above the fire. In rare cases is a "drop everything and run" evacuation warranted. The drop everything and run tactic is more often used during defensive operations. For offensive operations, an organized retreat is generally a better alternative.

Command and Control

It is important to point out that the National Incident Management System (NIMS) is also a safety system. For a detailed discussion of command and control see Chapter 1, *Organizing, Coordinating, and Commanding Emergency Incidents*. Taking measures to have everyone working toward a common goal in an organized fashion is the basis of a safe and effective operation. Conversely, freelancing leads to injuries and fatalities. NIMS is the foundation for accountability and rapid intervention procedures.

Accountability

An **accountability system** must be established on the fire ground for two purposes. The first is to ensure that everyone entering the area has a specific assignment. This is done to eliminate freelancing on the fire ground, which has led to several fire fighter fatalities.

The second purpose of an accountability system is to track all personnel at the scene and to identify the location of any missing personnel if a catastrophic event should occur, such as a building collapse. It also serves to identify whether any personnel or crews are overdue so that search and rescue operations can be initiated.

A good organizational structure accounts for all personnel operating at the scene and assigns responsibility for each crew with a reasonable span of control. Crew unity is an essential part of this process. Crew members should not be separated within the structure. The company officer or crew leader should be able to account for all members of their assigned crew working within the hazard area. Splitting crews inside the hazard area is inviting disaster. However, during offensive operations, it is acceptable to have an apparatus operator or other crew members working outside the hazard area while the officer and remaining crew members are inside the structure.

Many department SOPs allow the first-arriving truck company to split into a ventilation crew and a search and rescue crew. This should be a temporary arrangement, with someone supervising each crew. Each of the split crews must have a radio to enable them to provide progress reports, receive assignments, and transmit a mayday if needed. Even then, the crews should reunite as a company as soon as possible.

In addition to the accountability provided by NIMS, a separate system is needed to track and account for everyone at the scene **FIGURE 5-21**. Several systems are available for use, and most are effective if they are properly used.

Department accountability procedures must address (1) the initial accountability process used prior to the arrival of an assigned accountability officer and (2) how the formal accountability officer should include units already on scene. Personnel assigned as accountability officers should receive appropriate training, including simulations in which the system is implemented. The IC may assign units that are en route to the scene or to a staging area. These units will not report to the command post. Provisions must be made to include all units at the scene in the accountability system.

If a situation deteriorates rapidly, the IC should call for a personnel accountability report (PAR). SOPs generally call for PARs in the following situations:

FIGURE 5-21 Accountability officer at the command post with the incident commander.
Courtesy of Bill Strite, Cincinnati, Ohio.

- Whenever the IC or safety officer thinks it is necessary
- When the IC changes from an offensive to a defensive attack
- When sudden changes occur, such as backdraft, flashover, or collapse
- When the fire is extinguished

Routine requests for a PAR at a designated time or whenever there is a change in strategy reinforces the concept of personnel accountability and ensures that officers are aware of the location of the personnel assigned to them.

Because accountability procedures become more important as the incident increases in size and complexity, it is very likely that mutual aid resources will be at the scene. Therefore, a regional approach to an accountability system is the only logical choice.

Assigning an accountability officer is critical. It may be necessary to add a fire company to the initial alarm to ensure an accountability officer is available. The remaining company members could provide additional command assistance (accountability, SCBA monitoring/air consumption) as well as adding personnel to the RIT.

The safety officer cannot effectively be the accountability officer. The safety officer position is mobile. The accountability officer should be at a fixed location. The safety officer should ensure that accountability is established as well as enter into the accountability process by keeping the accountability officer informed of company locations. Likewise, the accountability officer is a valued informational resource for the IC, safety officer, and RIC.

Safety Officer

There is no exact time or situation that requires the assignment of a safety officer at a structure fire. Each department should develop SOPs outlining when this staff position must be established. In general, the safety officer should be separately staffed whenever the IC can no longer effectively monitor safety at the scene. This situation usually happens fairly early, as the IC should be located at a fixed command post. The safety officer monitors all areas where fire fighters are working. The IC should be stationary, whereas the safety officer is mobile.

Alternative Egress

Most large buildings have multiple egress routes, as required by building and fire codes. When this is the case, the IC should send crews into each stairway to manage the stairs. In high-rise fire operations, lobby control manages the stairs and elevators. (See Chapter 12, *High-Rise Buildings,* for a complete explanation of lobby control.)

This procedure will help to ensure that alternative routes that can be used during an emergency are identified and under the control of the fire department. The lobby control function can also be useful in multistory buildings that are not technically classified as high-rise structures.

The interior stairs are the preferred means of access and egress, but they are sometimes untenable or inadequate. When available, fire escapes provide additional means of escape and can also provide fire fighters with access to, and escape routes from, upper floors of the building. Fire escapes would be a second choice as a means of egress/access, because using fire escapes requires fewer fire fighters than positioning fire department ground ladders, and *if properly maintained*, fire escapes are safer. The building shown in **FIGURE 5-22** has two fire escapes and three stairways, providing alternative egress from every location within the building. However, if the use of a fire escape is included in the preincident plan, the fire escape

A

B

C

FIGURE 5-22 Apartment building with multiple means of egress. **A.** Side Alpha. **B.** Side Bravo. **C.** Side Charlie.
Courtesy of Ben Klaene.

should be thoroughly inspected on a regular basis and its capacity should be noted. A proper inspection often requires more than a visual inspection, because the support structure may be embedded in the building or otherwise hidden.

A common practice in the past was to assign the first-due truck company to "claim the building" by placing a ladder somewhere (anywhere) on the

FIGURE 5-23 Apartment building with multiple means of egress—side Charlie.
Courtesy of Ben Klaene.

building. This practice had a fundamental purpose of providing an alternative means of egress for occupants and fire fighters. Unfortunately, the purpose was often forgotten and the practice was eliminated as staffing on truck companies was reduced. Fortunately, this practice is making a comeback, with many departments now "laddering the building" to provide an alternative means of escape for fire fighters working inside a burning building. A proactive RIT company can place ladders providing alternative egress for fire fighters operating above ground level.

Effectively laddering the building is an important task that should be accomplished early in the operation. Placing a ladder to the one-story roof area shown in **FIGURE 5-23** would provide an alternative egress from several apartments on the second floor of this building.

SOPs should address laddering the building, possibly calling for ladders to be placed at predictable locations, such as the center window at the rear of the building or at the end of hallways. If ladders cannot be placed at a prescribed location or if SOPs do not address where ladders are to be placed, it is important that the exterior crews placing ladders communicate the location of the ladders to the crews working on the interior.

Just as codes require two separate and distinct means of egress, there should be provision for a second way out for fire fighters. The alternative egress should be remote from the first. Access to the roof is generally by ladder, but there should also be a second means of egress from the roof, usually a second ladder placed remote from the first. Some departments assign laddering the building to the RIC. The pros and cons of this and other RIC assignments will be discussed in the next section.

Rapid Intervention Crews

NFPA 1710 defines IRIC and RIC as follows (NFPA 1710, 2020):

> **3.3.53** Rapid Intervention Crew (RIC). A dedicated crew of at least 1 officer and three members, positioned outside the IDLH, trained and equipped as specified in NFPA 1407, who are assigned for rapid deployment to rescue lost or trapped members.
>
> **3.3.53.1** Initial Rapid Intervention Crew (IRIC). Two members of the initial attack crew, positioned outside the IDLH, trained and equipped as specified in NFPA 1407, who are assigned for rapid deployment (i.e., two in/two out) to rescue lost or trapped members.

In the initial stages of an operation, with only one crew operating at the scene, the pump operator and one additional fire fighter are generally assigned as the IRIC. This outside crew must know the position of the fire fighters working inside the structure. In most cases this crew will be the driver and hydrant fire fighter arriving on the first engine company. The IRIC is at a distinct disadvantage compared to a dedicated RIC team in performing an actual rescue. RIC activities are extremely staffing- and labor-intensive.

A two-person RIC crew will be ineffective and quickly exhausted as they search for fire fighters, provide air and attempt rescues. As additional first-alarm fire companies arrive, a minimum of four fire fighters with full turnouts should be assigned as the dedicated RIC for the incident.

Studies conducted by the Phoenix Fire Department in the aftermath of the Southwest Supermarket Fire determined that rescue of a downed fire fighter will require at least 12 fire fighters.

> **NOTE**
>
> For additional information and statistics related to maydays, review the Project Mayday reports at http://projectmayday.net/. This website contains in-depth mayday report information from a wide variety of paid and volunteer fire departments. It is a must-read for all fire fighters.

The RIC should assemble necessary tools and equipment as prescribed in department SOPs. This tool list will be based on department operations, tools available, types of buildings protected, and other factors. The RIC must have an additional SCBA cylinder with connections to supply air as stated in NFPA 1500 (NFPA 1500, 2018):

> **8.8.5** At incidents where any SCBA being used is equipped with a RIC universal air connection (UAC), the RIC shall have the specialized rescue equipment necessary to complete the RIC UAC connection to a supplied air source.

Tools commonly carried as part of the RIC cache include:

- Air cylinder for transfill
- Rescue ropes, search ropes, guideline ropes
- Thermal imaging camera (TIC)
- Patient carrier, webbing, or harness
- Collapsible ladder for above- and below-grade rescues
- Forcible entry tools
- Wire cutters and other hand tools
- Lighting equipment

A TIC is an indispensable tool when the exact location of the fire fighter(s) needing assistance is unknown. The RIC officer should determine if the situation calls for tools that may not be listed as part of standard RIC equipment. For example, a building equipped with hardened steel security bars might require a special saw or cutting torch. The construction type, occupancy, fire location, and other factors will also indicate the type of rescue that may be needed as well as the tools necessary to perform the rescue. Additional tools and equipment needed because of building features could be known in advance and thus identified in the preincident plan.

A RIC that is assigned non-RIC tasks is at a severe disadvantage. The dedicated RIC can determine the best access locations, assemble tools and equipment, and generally preplan potential rescue operations according to conditions. They are also immediately available for assignment with full breathing apparatus. The dedicated RIC monitors conditions and provides assistance to the safety officer.

The availability status of the RIC poses a dilemma. A RIC stationed at a central location near the tool cache would be able to immediately respond with tools and equipment needed to rescue fire fighters calling for assistance. However, there is a benefit in having the RIC do a complete walk-around to determine the location of alternative means of ingress and egress and other building features (i.e., a basement and basement access) that should be considered if a mayday occurs. While doing a walk-around, the RIC can also improve egress by forcing doors (being careful that the means of doing so does not create an unwanted vent opening), placing ladders, and removing obstacles. A mobile RIC can survey the building and provide valuable information to the IC and incident safety officer. Mobile and stationary RICs both have advantages and disadvantages. Some departments opt for a combination of the two by assigning some RIC members to reconnaissance and improving egress while others remain at the equipment cache location. When writing RIC procedures, it is important to consider the advantages and disadvantages of each type, based on RIC staffing and other factors in the response area. If a preincident plan is available for the building, the RIC should have access to the preincident plan.

The United States Fire Administration (USFA) conducted a study of fire department RIC operations (Williams and Stambaugh, 2003). Eighty-three (mostly career) departments responded to the study. The number of members assigned to the RIC (when a RIC is established) and RIC operations vary greatly. The fire service is still learning the best procedures to use when rescuing our own. Even the name RIC is not standardized. Crews assigned to rescue fire fighters are sometimes titled RIT (rapid intervention team), RAT (rapid assistance team), FAST (fire fighter assist and search team), and go team, among other names.

Serious questions have been raised regarding the need for additional staffing for the RIC. Most departments now use the two-out team as an IRIC but assign a full fire company as a RIC as the operation continues. Experience indicates that a two-person IRIC is inadequate, except in situations where fire fighters are easily located and able to escape with minimal assistance. A four-person RIC may be inadequate in a large building or where fire fighters need to be extricated.

When notified that a fire fighter needs assistance, the RIC must locate, extricate, remove/assist, and provide medical attention for fire fighters trapped inside a building. The sooner the RIC reaches trapped fire fighters and is able to get them to a place of safety, the better the chance of survival. In a large structure, this could require establishing multiple RICs at various locations around the building.

Some departments categorize RIC operations as a "rapid removal" or an "extended operation." Rapid removal would normally consist of finding and assisting fire fighters to safety. In many cases, simply assisting a disoriented fire fighter to a safe area is all that is needed. Other times a fire fighter may be unconscious or debilitated, but located near a safe area or near the exterior. In these cases, it may be best to quickly remove the fire fighter rather than trying to establish an air supply and performing other special rescue techniques. Extended operations may require two or more RICs. The first RIC would make every effort to locate fire fighters needing assistance while a

Incident Summary

Phoenix Supermarket Fire

In Phoenix, Arizona, a fire on the exterior of a large supermarket extended into a concealed space between the ceiling and roof in the storage area **FIGURE IS5-12**. A crew of fire fighters entering the store encountered worsening smoke conditions as they advanced. The low air supply warning sounded on two SCBAs, causing the crew to retreat. As they retreated, three members lost contact with the hose line that they had advanced into the store, and they became disoriented. One member was able to reestablish contact with the hose and find the way out. The other two fire fighters remained disoriented. One of the disoriented fire fighters heard another crew nearby and followed the sound to their location and ultimately to safety. The third member was found, but several crews were required to remove him, thus increasing his time in the hazard area. He was transported to the hospital, where he was pronounced dead.

FIGURE IS5-12 Debris fire fighters encountered while trying to find and rescue a missing fire fighter. Fire fighters needing rescue in large commercial structures may require more staffing and time to find and remove.
Photograph by Robert Duval, © National Fire Protection Association, Inc. used with permission.

Data from: Duval, Robert F., and Stephen N. Foley, 2002, "Supermarket Fire, Phoenix Arizona, March 14, 2001, 1 Fire Fighter Fatality." NFPA Fire Investigations. Quincy, MA: NFPA.

second team stands ready to extricate the fire fighters once they are found.

Fire departments may or may not require RICs to have a hose line as part of their standard procedure. Extending a hose line to the area where fire fighters need assistance can substantially delay search and extraction operations. However, it may be necessary to protect trapped fire fighters or to provide a path to the exit by extinguishing the fire. Some departments assign a truck or rescue company as the RIC and an engine company as the engine-RIC or RIC-assist. The engine-RIC provides hose line protection. Other departments assign more members to the RIC to provide sufficient personnel to find and extract the fire fighter(s) requiring rescue and to protect all involved fire fighters. If the RIC deploys a hose line or another unit is assigned to protect RIC operations, it is best to provide a separate water supply for the RIC hose line, as hose line failure may be the reason for the call for assistance.

In a high-rise operation on an upper floor, the RIC and RIC equipment should be located one or more floors below the fire floor so RIC members do not expend their air supply while on standby, but are near areas where fire fighters are working. The RIC should not be located in an area where they need to activate their SCBA while on standby.

Many times, training under simulated fire conditions will point out deficiencies in procedures that can be changed to improve RIC operations. During training sessions, it is essential to develop a variety of scenarios in different buildings and use full protective clothing as you would while performing a rescue. RIC procedures should be practical under fire conditions, and they should stress a commonsense approach. It is much easier to walk through various scenarios, practical rescue techniques, and self-rescue techniques during training than under the stress of an actual mayday.

Hands-on RIC training is essential. All members of the department must be trained in rapid intervention techniques. Often a fire fighter will become incapacitated due to a medical emergency or an injury and their best chance of survival is immediate rescue by the companies surrounding them at the time of the incident. Training on downed fire fighter drags, movement, and air transfill may negate the need for deploying the RIC and provide the best chance of survival for the downed fire fighter. A Columbus, Ohio, fire lieutenant suffered a medical emergency in the basement of a single-family dwelling. The lieutenant's crew immediately moved him up the narrow basement steps and out onto the porch where they immediately began resuscitation efforts including cardiopulmonary resuscitation (CPR). The lieutenant enjoyed a full recovery due to the diligence of his crew members and their dedication to training on rapid intervention skills.

Company Safety Responsibilities

It is important to point out that the RIC is not a substitute for safe and effective operations. Formulating an incident action plan based on a risk-versus-benefit analysis, combined with good tactics and company-level attention to safety will reduce the need for emergency rescues. Tactics such as extinguishing the fire and venting the building can greatly reduce the possibility of fire fighters being trapped or disoriented.

Safe interior operations include the following:

- Maintaining crew integrity during interior operations with at least two members working together and in contact with one another
- Controlling and coordinating ventilation
- Providing hose line protection in areas where fire fighters are working
- Providing a means of communication to the exterior and among units working in an area
- Maintaining contact with the hose when operating a hose line
- Maintaining contact with the wall or rope when operating without a hose line
- Placing a fire fighter, signal light, or audible signal at the area leading to the room where fire fighters are working

Despite these measures, the unexpected can occur. Fire fighters must be taught self-survival techniques, including how to operate the emergency features on SCBA and PASS.

Declaring a Mayday

When a fire fighter or group of fire fighters suspect they are in trouble and might need assistance, they should immediately transmit a mayday call reporting their situation. When requesting assistance, fire fighters should give their exact location, if possible. If they do not know their exact location, they should give the best estimate (e.g., We are located on the second floor on side Charlie."). Fire fighters should state the problem, such as "out of air," "disoriented," "trapped in collapse," "entangled," "fell through floor," "fell through roof," etc. They should also state the number of fire fighters needing assistance and company number. Transmitting a mayday with a location report should be practiced during training sessions until it becomes routine.

Even after requesting assistance, it is sometimes possible to continue self-rescue efforts or take measures that could extend survival time, such as finding a source of outside air at a window or under a door. If fire fighters requesting assistance are merely disoriented and not in imminent danger, it may be best for them to activate their PASS and conserve air while awaiting assistance. In this case, isolation from the fire area (e.g., closing the door to an area of safe refuge) can extend survival time.

> **NOTE**
> Firefighters transmitting a mayday should include the following information (the LUNAR mnemonic):
> L - Location
> U - Unit
> N - Name
> A - Assignment/Air
> R – Resources

The PASS device should be activated immediately after a mayday is declared. It is possible to increase the efficiency of the PASS by directing it away from the fire fighter's body. Trapped members can sometimes indicate their location by tapping on doors, directing a light around the room, or otherwise signaling their location. Breaking or opening a window leading to the exterior can help fire fighters regain their orientation and provide a better description of their location when calling for assistance. However, the window will become a vent opening and may draw fire and smoke toward the fire fighters unless the disoriented fire fighters have isolated their location by closing doors and other openings.

A team of fire fighters in Cincinnati depleted their air supply deep inside a smoke-filled warehouse/manufacturing building. They were able to extend their survival time by breathing air from a crack at the bottom of a metal overhead door that was locked on the exterior. They used a spanner wrench to bang on the door to indicate their location. They were eventually rescued and survived without serious injury.

Opposing Fire Streams

Hose streams are important safety tools, but they can also do much harm when the flow path is changed, placing fire fighters or victims in danger. The UL study comparing the impact of ventilation on fire behavior in legacy versus modern furnishings found that "while the fog stream 'pushed' steam along the flow path, there was no fire 'pushed'" (Kerber, 2010). Earlier editions of this text included a description of hose lines, especially fog streams pushing fire. Hose streams do affect the flow path. The later UL study on vertical ventilation further examined the "pushing" effect of fire streams, resulting in the following statement (Kerber, 2013):

> You cannot push fire with water. The previous [Underwriters Laboratories] study (Kerber S. 2010) discussed the concept of pushing fire in the data analysis. Since the release of that study there has been a lot of discussion about this and stories from well-respected fire service members where this did happen or was perceived to happen. The specific fires that were being recalled by the firefighters were discussed in detail. In many of these conversations, the firefighters were in the structure and in the flow path opposite the hose line. In most cases where firefighters experienced fire moving over their heads, fire attack crews were advancing on the inside and not applying water from the outside into a fully developed fire. All of the current experiments described in this report were designed to examine the operations and the impact of the initial arriving fire service units so we did not, do not, and will not suggest that firefighters should be in the position where they are in a flow path opposite the hose line. However, there are times when this may happen, so the experience of these firefighters should not be discounted. During our discussions, four events could have been witnessed which may have had the appearance of pushing fire:
> (1) A flow path is changed with ventilation and not water application.
> (2) A flow path is changed with water application.
> (3) Turnout gear becomes saturated with energy and passes through to the firefighter.
> (4) One room is extinguished, which allows air to entrain into another room, causing the second room to ignite or increase in burning.

Whenever possible all interior hose lines should attack from the same access point. If this is not possible or practical, communications between units attacking the fire is essential.

Master streams, especially fog streams, improperly operated on the exterior can change the flow path by disrupting the vent process. Offensive and defensive attacks should not be conducted at the same time in the same building. The IC assigns units so that everyone is working either offensively or in a defensive mode.

> **NOTE**
>
> The limited application of water from the exterior directed to the interior fire area to quickly reduce fire intensity in preparation for an offensive attack is not considered a defensive tactic.

Personal Protective Clothing

In 1977 when NFPA first published annual Fire Fighter Line of Duty Death reports, the fire ground was much different. Many of the safety measures now considered routine were nonexistent or only implemented as the IC thought necessary. Personal protective clothing was also much different in the 1970s.

Summer gear used by many departments included the following:

- Aluminum helmet that absorbed some impact as it was crushed on the fire fighter's head
- Filter mask and a seldom-used chemical rebreather mask
- Denim jacket (with or without insulation) in warm weather

Cold weather attire was a bit more protective and included the following:

- Lining under helmet
- Lined rubber coat
- Heavier gloves
- Rubber hip boots (lined night pants were optional and often purchased by the fire fighter)
- Lined boots (optional)

Today, fire fighters are completely encapsulated in PPE. Modern helmets provide shock absorption and can be specified to also include communications capability. Turnout coats and pants are engineered and constructed to quantify the measure of external thermal protection provided and internal heat released. Boots have evolved from sloppy-fitting rubber boots to leather boots designed to fit like an athletic shoe.

Figure 5-1 at the beginning of this chapter indicates that there has been a dramatic reduction in the number of fire fighter fatalities over the past 42 years. Personal

FIGURE 5-24 Fire fighter in full turnout gear.
Courtesy of Bill Strite, Cincinnati, Ohio.

protective clothing undoubtedly played an important part in this reduction in the number of line-of-duty deaths. The total number of deaths prior to 1977 is unknown. Looking back 50 years at fire fighter protective clothing, it is obvious that protective clothing played an important part in reducing fire fighter injuries and fatalities **FIGURE 5-24**.

Over the years there have been pronounced improvements in the level of protection provided by personal protective clothing **TABLE 5-6**. NFPA and other safety standards now address minimum standards for PPE used on the fire ground.

Even though personal protective clothing is vastly improved, there are still far too many inhalation, burn, and cutting injuries on the fire ground. Often, these injuries occur because fire fighters are not properly using PPE. However, there is also an attitude of invincibility due to the increased protective attributes of modern protective clothing.

Many departments now use longer-duration SCBAs that permit the fire fighter to stay in the IDLH atmosphere longer, which increases the time available to escape; however, this equipment also adds to the fatigue factor. Positive pressure within the face piece eliminates many of the inhalation problems of the past, but can also result in a depletion of the air

TABLE 5-6 Comparison of Fire Fighter Personal Protective Equipment in 1970 versus 2020		
Protected Body Part(s)	**1970**	**2020**
Respiratory	Filter mask or chemical rebreather	Self-contained breathing apparatus (SCBA)
Head	Aluminum helmet* with head band	Synthetic helmet with thermal resistance and shock absorbing insert
Ears and neck	Helmet flap that partially covered ears Coat collar that partially covered neck	Fire-resistive filter hood that covers ears and neck and absorbs products of combustion
Hands	Summer: lightweight gloves with no thermal lining Winter: heavier-weight gloves with thermal lining (to warm hands)	Fire-resistive gloves with thermal lining and moisture barrier
Upper body	Summer: denim jacket Winter: rubber coat with thin liner	Fire-resistive coat with thermal lining and moisture barrier
Lower body and legs	Summer: polyester pants Winter: water-resistant pants with liner	Heavy-duty fire-resistive pants with thermal lining and moisture barrier
Feet	Summer: work boots, shoes, rubber boots Winter: rubber boots, some with light lining	Leather firefighting boots with athletic design

*Metal helmet provided some impact protection as it was crushed on fire fighter's head.
Courtesy of Ben Klaene.

supply due to leakage. Likewise, the turnout ensemble has higher thermal ratings and is more cut resistant than in the past. Helmets and boots are improved, and the use of PASS and protective hoods is the norm.

These elements of protective clothing all have limitations. For example, the minimum thermal protective performance (TPP) value is 35 for protective coats and trousers. Time in a heated environment gradually erodes the thermal protection depending on the temperature. At near flashover temperatures, a TPP of 35 only provides protection against second-degree burns for approximately 17.5 seconds. NFPA 1971, *Standard on Protective Ensembles for Structural Fire Fighting and Proximity Fire Fighting*, provides specific test methods and requirements for each element of the protective clothing ensemble (NFPA 1971, 2018). Many departments simply develop PPE specifications referencing the minimum levels of protection in the NFPA 1971 standard. A better approach is to specify greater protective levels if available when flexibility is not a negatively affected.

New studies have determined that fire fighters are continually exposed to carcinogenic material trapped in PPE. Many departments have implemented SOPs that require fire fighters to be decontaminated after structure fires and to place PPE in plastic bags prior to reentering the fire apparatus. A second set of PPE for each fire fighter is recommended to allow for proper cleaning of the contaminated gear. Fire fighters should also wipe down their face, neck, and groin with body wipes as soon as possible to remove contaminants from these susceptible areas. Some departments have also implemented a hood exchange policy where the safety officer or IC carries spare hoods in their vehicle and swaps them with the fire fighters after every working fire.

While dirty fire gear was once considered a badge of honor, it is now known to be a silent killer that attacks the fire fighter. An initial decontamination should be established on the fire scene to remove the large fire debris on the PPE. All components of the fire protective ensemble should be then thoroughly cleaned following manufacturer's specifications after every fire using the appropriate washing and drying procedures—including the helmet **FIGURE 5-25**!

The IC should establish the level of protective clothing necessary to enter hot and warm hazard control zones. The safety officer should be consulted on this issue.

FIGURE 5-25 It is important that all components of the fire protective ensemble be thoroughly cleaned.
Courtesy of David Mullis.

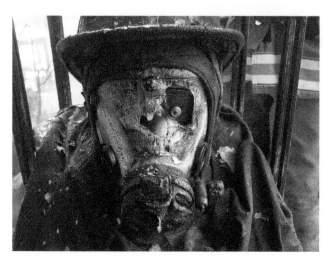

FIGURE 5-26 Face piece lens damaged by high heat.
Courtesy of NIST.

NFPA 1500 addresses the use of SCBA in stating the following (NFPA 1500, 2018):

7.10.7 When engaged in any operation where members could encounter atmospheres immediately dangerous to life or health (IDLH) or potentially IDLH or where the atmosphere is undefined or hazardous (including overhaul), the fire department shall provide and require all members to use SCBA that has been certified as being compliant with NFPA 1981.

7.10.8 Members using SCBA shall not compromise the protective integrity of the SCBA for any reason when operating in IDLH, potentially IDLH, or undefined or hazardous atmospheres (including overhaul) by removing the facepiece or disconnecting any portion of the SCBA that would allow the ambient atmosphere to be breathed.

Studies indicate that the SCBA face piece lens may fail under high-heat conditions. After reports of several face piece lens failures, NIST conducted tests subjecting the SCBA lens to high levels of heat that could be expected under high-heat conditions (572°F [300°C]) **FIGURE 5-26** (Mensch et al., 2011). One example of conditions leading to high heat cited in the report was opening the door to a room and causing a rapid increase in temperature as fire extends through the doorway. It is recommended that SCBA face piece lenses be checked for thermal damage on a regular basis.

Thermal ratings vary for different components of the protective clothing ensemble. Fire fighters must understand the limitations of all parts of their PPE. When faced with an extremely hot interior fire, fire fighters can reduce the heat level by applying water prior to entering the fire compartment. In some cases, water is applied from the exterior prior to making entry.

The protective clothing performance ratings assume that the PPE is dry, properly maintained, and replaced when worn or damaged. Normal department maintenance is necessary, but each individual fire fighter should check their equipment after every use. It is also recommended that protective clothing, including the SCBA face piece lens, be checked when members are in rehabilitation.

RIC and self-rescue methods sometimes involve making openings in interior or exterior walls. Typically, these are small openings such as the 14.5 in. (36.8 cm) between wall studs. Moving through this small space requires fire fighters to remove the SCBA backpack while leaving the face piece intact. This procedure does not necessarily result in the fire fighter breathing the ambient atmosphere, and at times, it may be necessary. However, consider making a larger opening, which would permit a faster exit that would not potentially compromise any part of the protective clothing ensemble.

Overhaul

A tendency to "dress down" during the overhaul phase of an operation can lead to unnecessary injuries and long-term detrimental health effects. Even after airing out the structure, removing the SCBA face piece is a dangerous practice. Residual burning materials produce toxic gases, and concentrations of fire gases will be present within the building for a considerable time after fire control.

A 2010 UL report provides further evidence of the toxic inhalation hazards faced by fire fighters when removing the SCBA during overhaul operations (Fabian, 2010). Air quality monitoring was conducted at actual fire scenes during active firefighting and for 30 minutes following fire suppression. This UL research proves that fire fighters are exposed to toxic and potentially carcinogenic off-gases that are known to result in both acute and chronic adverse health effects. The burning contents found in structure fires today include more synthetic materials that produce an abundance of toxins. Some of the toxins emitted by common household furnishings and structural components during and after extinguishment are as follows:

- Arsenic
- Benzene
- Carbon monoxide
- Formaldehyde
- Phenols
- Hydrochloric acid
- Hydrogen cyanide
- Formic acid

The study found that in addition to gaseous/vaporized toxins, the air and surfaces within a postfire compartment contain smoke particles and condensed liquids that when inhaled pose a long-term health threat. Many of the smoke particles collected by UL during overhaul were too small to be visible with the naked eye, leading UL researchers to a conclusion that "clean" air was not really that clean. Even if open areas within the building were free of smoke and toxins, research conducted by NIOSH found that dead air spaces within a building contained toxic by-products of combustion long after the building was aired out (UL, 2010).

Many fire fighters make a decision to remove their SCBA based on the observable presence or absence of smoke. Other departments take a more scientific approach, using portable gas-measuring instruments (usually a carbon monoxide meter) to determine the safety of the atmosphere. This approach, although more scientific, is no more reliable unless the only toxic material present is the one measurable by the available instrument. Even scientific studies, such as the 2010 UL study, often miss products that are not known to be present. Some in the fire service recommend that a lighter-weight respirator be used during overhaul. This was the subject of laboratory research conducted in 2007 that concluded that the respirators used during the testing do not fully protect personnel during overhaul exposure (Anthony et al, 2007).

After a working structure fire and subsequent operations to gain access to concealed spaces, there are often many other hazards and dangers (see the earlier box in the Fire Fighter Injuries and Fatalities section, regarding the FDNY study regarding exposure to 9/11 World Trade Center dust). Therefore, removing any part of the turnout ensemble inside the fire building is an unsafe practice.

The time between suppression and final **overhaul** is an ideal time to rotate personnel through rehabilitation while the building airs out. This recommended rehabilitation between active suppression and final overhaul is not meant to preclude the need for protective clothing, including the SCBA, or earlier rehabilitation when gaining control of the fire takes longer.

The IC should use the time between active firefighting and overhaul to develop an overhaul plan (a nonemergency incident action plan). Safety considerations include, but are not limited to the following:

- Structural damage and structural stability
- Smoke and airborne contaminants
- Cutting hazards (broken glass, jagged metal, protruding nails)
- Holes in floors
- Damaged stairways
- Utility hazards (gas, oil, electric)
- Overhead hazards (damaged ceiling, loose materials)
- Visibility (provide lighting or wait until there is adequate natural light)

When the pre-overhaul inspection reveals imminent hazards or risks that cannot be determined, overhaul should be delayed. On some occasions, overhaul is conducted using heavy equipment. The building shown in FIGURE 5-27 was heavily damaged by fire,

FIGURE 5-27 Delayed overhaul.
Courtesy of Ben Klaene.

and the structural stability could not be determined. Master streams remained in place for several days and the area was secured. Eventually, heavy equipment was used to raze the building.

When the pre-overhaul inspection finds no serious imminent hazards, the identified hazards should be discussed with crews assigned to that area. If possible, mark the hazardous conditions using barrier tape, ladders, or other means. See Chapter 10, *Property Conservation,* for more information on overhaul. Crews reentering the building should be on-air and in full protective clothing.

Rehabilitation

NFPA 1584, *Standard on the Rehabilitation Process for Members During Emergency Operations and Training Exercises,* provides procedures for hot and cold weather rehabilitation (NFPA 1584, 2015).

Rehabilitation can be divided into three phases:

1. Preincident hydration and preparation
2. Incident rehabilitation
3. Postincident recovery

Thirst is a sign of mild dehydration; members should not wait until they are thirsty to ingest liquids. While awaiting calls, members should prehydrate by drinking 6 to 8 ounces of water every 6 hours plus liquids that are ingested with meals. This practice is especially important during hot and humid weather conditions. However, water or sports drinks should not be ingested in extreme quantities. Members should avoid activities that cause extreme fatigue or exhaustion while waiting for a response.

On-scene rehabilitation can be formal or informal depending on weather conditions, length of time on scene, and activity level. Working structure fires tend to require high energy levels; thus, they cause fatigue at a rapid rate. Informal rehabilitation usually takes place at the company apparatus. An apparatus used for rehabilitation should be in the cold zone so members can "dress down" while resting and rehydrating. Resting members should not be placed in areas where they will be breathing exhaust fumes. Formal rehabilitation involves setting up an area for rehabilitation. This area should provide shade and mechanical cooling equipment during hot weather, and it should be a warm location during cold weather. The formal rehabilitation area must have water or sports drinks available. If members are on the scene for an extended time, healthy food should also be provided.

Rehabilitation efforts shall include the following:

- Relief from climatic conditions
- Rest and recovery
- Active and/or passive cooling or warming as needed for the incident type and climate conditions
- Rehydration (fluid replacement)
- Calorie and electrolyte replacement, as appropriate for longer-duration events
- Medical monitoring
- Member accountability
- Release

NFPA 1584 stipulates that rehabilitation should be provided after depletion of a second 30-minute or 45-minute SCBA air cylinder, depletion of a single 60-minute SCBA air cylinder, or after 40 minutes of strenuous work without SCBA (NFPA 1584, 2015). The standard does allow these times to be adjusted depending on work or environmental conditions.

During temperature extremes, rehabilitation becomes a more critical issue, and longer rest periods may be needed. Hot weather rehabilitation could be outside using water mist and/or fans in a shaded area. Research regarding cooling techniques recommends active cooling techniques such as forearm immersion (Selkirk, McLellan, and Wong, 2004). **FIGURE 5-28** shows the effect of humidity during warm weather

Temperature (°F)

Relative Humidity (%)	80	82	84	86	88	90	92	94	96	98	100	102	104	106	108	110
40	80	81	83	85	88	91	94	97	101	105	109	114	119	124	130	136
45	80	82	84	87	89	93	96	100	104	109	114	119	124	130	137	
50	81	83	85	88	91	95	99	103	108	113	118	124	131	137		
55	81	84	86	89	93	97	101	106	112	117	124	130	137			
60	82	84	88	91	95	100	105	110	116	123	129	137				
65	82	85	89	93	98	103	108	114	121	128	136					
70	83	86	90	95	100	105	112	119	126	134						
75	84	88	92	97	103	109	116	124	132							
80	84	89	94	100	106	113	121	129								
85	85	90	96	102	110	117	126	135								
90	86	91	98	105	113	122	131									
95	86	93	100	108	117	127										
100	87	95	103	112	121	132										

Likelihood of Heat Disorders with Prolonged Exposure or Strenuous Activity
Caution ■ Extreme Caution ■ Danger ■ Extreme Danger

FIGURE 5-28 National Weather Service (NWS) heat index chart.
Courtesy of National Weather Service/NOAA.

and the probability of heat stroke at elevated temperatures combined with high humidity.

Cold weather requires that, as a minimum, blankets or other coverings be provided to fire fighters in rehabilitation. Wind adds to the hazard during cold weather as shown in **FIGURE 5-29**. A better option for hot or cold weather rehabilitation is an air-conditioned or heated area. This area can be a building, bus, or other climate-controlled facility.

An additional EMS unit, preferably an advanced life support (ALS) unit, should be assigned to rehabilitation when a formal rehabilitation is implemented to evaluate fire fighters entering and exiting rehabilitation **FIGURE 5-30**. This unit's responsibility is to determine when members can returning to action, to treat injuries, and transport personnel when necessary. Rehabilitation protocols should be spelled out in department SOPs.

Before fire fighters are released from the rehabilitation area to return to active firefighting duties, their vital signs must be checked and be at safe levels. Vital signs to be checked should, as a minimum, include pulse, respiration, and temperature. Consult the department physician when establishing reentry guidelines.

Maintaining company unity and the accountability system is important as members are cycled through rehabilitation. If a company member cannot be returned to duty, the accountability tracking system should be revised accordingly.

FIGURE 5-29 National Weather Service wind chill chart.
Courtesy of National Weather Service/NOAA.

FIGURE 5-30 Rehabilitation area.
Courtesy of Bill Strite, Cincinnati, Ohio.

Wrap UP

CHAPTER SUMMARY

- Fire fighter on-duty deaths have decreased dramatically in recent years.
- Fire fighters and fire officers should be familiar with minimum safety measures described in NFPA 1500.
- NFPA 1500 requires a risk-versus-benefit analysis at all structure fires.
- Time from ignition to flashover can vary significantly, depending on a number of variables related to the compartment size, ventilation, and fuel load.
- The fuel load in modern residential occupancies tends to be much greater and faster burning than fuel loads in the past.
- Construction materials and methods affect structural stability under fire conditions.
- Green buildings present unique problems under fire conditions.
- Fire fighters must understand the dynamics associated with structural collapse and recognize the signs of impending structural failure.
- Fire entering concealed spaces can extend to remote locations and break out with a sudden increase in intensity.
- Fire extension can cut off the primary means of egress for fire fighters.
- The area between a suspended ceiling and the roof/ceiling is a common concealed space that allows fire to extend unnoticed.
- A safe collapse zone is equal to the height of the building, plus an allowance for debris scatter usually calculated as a distance equal to 1½ times the height of the fire building.
- Accountability for all people entering the fire zone is essential.
- Structural stability decreases as time and fire intensity increase.
- Smaller compartments progress to flashover faster than larger compartments.
- Proper staffing requires that an adequate number of fire fighters be available to complete all necessary fire-ground tasks and to form a tactical reserve.
- Response to a mayday by the rapid intervention crew (RIC) must be orderly and organized.
- An emergency evacuation signal is crucial to warn of imminent collapse.
- Accountability should not be assigned to a safety officer; the safety officer must be mobile, while accountability is a stationary position.
- The safety officer monitors all areas where fire fighters are working.
- Safety is everyone's responsibility at all times at the fire scene, whether or not someone has been specifically assigned the role of safety officer.
- Fire fighters in doubt regarding the need to transmit a mayday should immediately declare a mayday.
- Removing any part of the turnout ensemble inside the fire building, even during overhaul, is an unsafe practice.
- On-scene rehabilitation is especially important in temperature extremes, inclement weather, or extended incidents.

KEY TERMS

20-minute rule A general rule used for estimating the length of time until structural collapse occurs. It states that when a heavy volume of fire is burning out of control on two or more floors for 20 minutes or longer in a building of ordinary construction, structural collapse should be anticipated.

accountability system A system established on the fire ground to ensure that everyone entering the area has a specific assignment, to track all personnel at the scene, and to identify the location of any missing personnel if a catastrophic event should occur.

cold zone An area where personal protective clothing is not required; the command post, rehabilitation, and medical treatment should be located in this zone.

collapse zone The area endangered by a potential building collapse; generally considered to be an area 1½ times the height of the involved building.

decay phase The phase of fire development in which the fire has consumed either the available fuel or oxygen to a point that the fire begins to diminish in intensity.

extension Fire that moves into areas not originally involved, including walls, ceilings, and attic spaces; also, the movement of fire into uninvolved areas of a structure.

fire perimeter A wide area beyond the hazard control zones, usually staffed by police to keep unauthorized people away from the scene.

frequency As related to fire fighter injuries, a measure of how often an injury occurs; for example, sprains and strains are the most frequent fire fighter injuries.

fully developed phase The phase of fire development at which the fire is free-burning and consuming much of the fuel.

growth phase The phase of fire development at which the fire is spreading beyond the point of origin and beginning to involve other fuels in the immediate fire area.

hazard control zone The area in which emergency responders are working; it can be subdivided into no-entry, hot, warm, and cold zones.

hot zone An operating area considered safe only for individuals wearing appropriate levels of personal protective clothing; established by the IC and safety officer.

ignition phase The phase of fire development at which the fire is limited to the immediate point of origin.

no-entry zone An area that is unsafe regardless of the level of personal protective equipment and that must be cleared of all personnel, including emergency response personnel.

occupational injury An injury sustained during the duties, responsibilities, and functions of a fire department member. (NFPA 1500)

overhaul Examination of all areas of the building and contents involved in a fire to ensure that the fire is completely extinguished.

prefire conditions Factors that can contribute to a collapse in a building that is heavily involved in fire; these factors include construction type, weight, fuel load, damage, renovations, deterioration, support systems, and related factors such as lightweight truss ceilings and floors.

pyrolysis The chemical decomposition of a compound into one or more other substances by heat alone; the process of heating solid materials until combustible vapors are emitted.

rehabilitation The process of providing rest, rehydration, nourishment, and medical evaluation to members involved in extended incident scene operations and/or extreme weather conditions.

severity The extent of an injury's consequence, usually categorized as death, permanent disability, temporary disability, and minor.

warm zone An intermediate area between the hot and cold zones where personal protective equipment is required, but at a lower level than in the hot zone.

SUGGESTED ACTIVITIES

1. Residential fire

It is Sunday, June 23 at 6:30 AM. Skies are clear with 10 mi/h (16 km/h) winds from the west (note: the left side of the photo is west).

A basement fire is reported in a residence at 4 Collin Drive **FIGURE 5-31**. You are the lieutenant on the first-arriving engine company (Engine 10) with a four-person crew: yourself, a driver, and two fire fighters. Your response time is approximately 4 minutes.

A district chief, engine, and truck company are also responding from the Headquarters Station 1 (District 1, Engine 14, and Truck 14). Their typical response time to this location is 14 minutes. Mutual aid companies and medic units are available, but expect a long response time (20 to 30+ minutes) after receiving the request for assistance.

Upon arrival, smoke is visible at the east side first-floor windows. Lighter smoke is coming

FIGURE 5-31 Residential basement fire.
Courtesy of Ben Klaene.

out of the top of the garage doors. The smoke plumes are rising on an angle on the right side of the house.

A woman approaches you reporting, *"My husband and I were sleeping in in our bedroom on the first floor when the smoke alarm woke us up. There was lots of smoke so I called you guys."*

You ask, *"Is your husband still in the house?"*

She replies, *"Yes, my husband went to get the kids out of their bedrooms on the second floor and I haven't seen him since. Before he went upstairs, he opened the basement door; smoke poured out of the basement. The smoke was so heavy on the first floor that I could hardly breath or find my way around. He yelled at me and said 'Get the heck out of the house, right now! I will get the kids.' I think he went upstairs to get our five kids and I haven't seen him since."*

As the officer of Engine 10:
- Provide a status report to dispatch.
- List the actions would you take immediately. Justify your actions
- The district chief arrives with Engine 14 and Truck 14. Assume the role of district chief.
- Explain your strategy and tactics, considering the location and report from Engine 10.

2. Residential fire

Several possible tactics used to combat a working basement fire are discussed in this chapter. Evaluate each of the possibilities listed here using the scenario in Suggested Activity 1.

State the advantages and disadvantages of each of the flowing tactics:
- Advancing a hose line through the front door to the second floor in an attempt to rescue the five children.
- Advancing a hose line from the first floor to the basement via an interior stairway leading to the basement.
- Advancing a hose line from the exterior into the basement via the garage door or other opening leading directly to the basement.
- Applying water from an exterior above-grade level using an opening such as a basement window.
- Using a piercing or cellar nozzle to apply water from the first floor into the basement.

3. Use the most recent NFPA fire fighter fatality report to determine the type of duty as well as the nature and cause of injury most likely to result in an on-duty fire fighter fatality.

4. Use the most recent NFPA fire fighter injury report to determine the most common cause of injury. Using a scale of 1 to 10, where 1 is most severe, estimate the probable severity for various injury types (e.g., heart attack on the fire ground = 10).

5. Use a fire report from NFPA, NIOSH, USFA, or other source to evaluate the risk to fire fighters conducting an offensive attack. Classify the incident using the four risk/benefit categories from NFPA 1500, 8.4.2, in the section *Risk Management Applied to the Fire Ground*.

6. Assume the role of safety officer for the fire shown in **FIGURE 5-32**. It is 4:00 PM. Weather conditions are 95°F (35°C) with high humidity and calm winds. The operation began as an offensive attack at 2:00 PM. An emergency evacuation was ordered at 2:45 PM. A PAR was conducted with all fire personnel accounted for at that time. The collapse shown in Figure 5-32 has just occurred. List the concerns you have and recommend precautions to the IC.

7. Use the following scenarios to analyze and discuss the probability of the building described being occupied at the time of the alarm. Express the probability of occupancy as a percentage, where 0% means there is no chance anyone is inside the building and 100% indicates that the building is definitely occupied by at least one person. Explain your answer.

   A. Single-family residential building at 2:00 AM, Sunday, July 10, clear and mild weather conditions

   B. Abandoned five-story building in the center of the city at 2:00 AM, Sunday, December 10, very cold and windy weather conditions

   C. High-rise hotel in Michigan vacation resort area at 2:00 PM, Saturday, August 1, sunny and warm weather conditions

   D. Nursing home at 1:00 PM, Wednesday, July 13, sunny and warm weather conditions

   E. High school, 1:00 PM, Monday, October 2, mild and rainy weather conditions

   F. High school, 1:00 PM, Monday, July 11, hot and rainy weather conditions

8. Use the scenarios in Question 5 to compute the staffing needed to safely and effectively conduct an offensive attack.

9. You are assigned to manage rehabilitation at a multialarm fire. The County Fire Chief's Association has worked with numerous area communities and businesses to acquire an 8 ft (2.4 m) × 40 ft (12 m) rehabilitation van with interior and exterior lighting as well as a heating, ventilation, and air conditioning system. There is a door in the front and rear and a partition dividing the rear 20 ft (6 m) area from the front 20 ft (6 m) area. It is 95°F (35°C) with a relative humidity of 85%. What is the National Weather Service heat index rating temperature?

   The rehabilitation van has just arrived with two paramedics staffing the unit. They are awaiting your instructions. Describe the preferred rehabilitation area, e.g. at least 10 ft (3 m) from the fire building, flat, shaded area, on a side street. List services that should be provided in a formal rehabilitation shelter (e.g., monitor vital signs). List food and liquids that should be provided inside the rehabilitation vehicle.

FIGURE 5-32 Partial collapse of a large four-story commercial building.
Courtesy of Captain Nick Morgan, St. Louis Fire Department.

REFERENCES

Abbott, Don, and Bev Abbott. 2015. "CERT (Command Emergency Response Training)." Don Abbott's Project Mayday website. http://projectmayday.net/. Accessed December 13, 2019.

Anthony, T. Renee, Philip Joggerst, Leonard James, Jefferey L. Burgess, Stephen S. Leonard, and Elizabeth S. Shogren. 2007. "Method Development Study for APR Cartridge Evaluation in Fire Overhaul Exposures." *The Annals of Occupational Hygiene*. 51 (8): 703–16.

Chappel, Bill. December 6, 2018. "California Gives Final OK to Require Solar Panels on New Houses." NPR website. https://www.npr.org/2018/12/06/674075032/california-gives-final-ok-to-requiring-solar-panels-on-new-houses. Accessed November 25, 2019.

Corbett, Glenn P., and Francis. L. Brannigan. 2021. *Brannigan's Building Construction for the Fire Service* (6th ed.). Burlington, MA: Jones & Bartlett Learning.

Dodson, David W. 2016. *Fire Department Incident Safety Officer*. Burlington, MA: Jones & Bartlett Learning.

Fabian, Thomas, Underwriters Laboratories, University of Cincinnati, and Chicago Fire Department. April 1, 2010. *Firefighter Exposure to Smoke Particulates*. Northbrook, IL: Underwriters Laboratories.

Fire Protection Research Foundation. *Property Insurance Research Group Forum on PV Panel Fire Risk*. June 8, 2014. Quincy, MA: FPRF.

Grundahl, Kirk. 1992. *National Engineered Light Weight Construction Fire Research Technical Report*. Quincy, MA: National Fire Protection Research Foundation;

Kerber, Steve. 2010. *Impact of Ventilation on Fire Behavior in Legacy and Contemporary Residential Construction*. Northbrook, IL: Underwriters Laboratories Firefighter Safety Research Institute.

Kerber, Steve. 2012. *Fire Service Collapse Hazard Floor Furnace Experiments*. Northbrook, IL: Underwriters Laboratories Firefighter Safety Research Institute.

Kerber, Steve. 2012. *Improving Fire Safety by Understanding the Fire Performance of Engineered Floor Systems and Providing the Fire Service with Information and Tactical Decision Making*. Northbrook, IL: Underwriters Laboratories Firefighter Safety Research Institute.

Kerber, Steve. 2013. *Study of the Effectiveness of Fire Service Vertical Ventilation and Suppression Tactics in Single Family Homes*. Northbrook, IL: Underwriters Laboratories Firefighter Safety Research Institute.

Madrzykowski, Daniel, and Robert L. Vettori. April 2000. *Simulation of the Dynamics of the Fire at 3146 Cherry Road NE, Washington DC*. May 30, 1999 [NISTIR 6510]. Gaithersburg, MD: National Institute of Standards and Technology.

Mensch, Amy E., George G. Cajaty, Barbosa Braga, and Nelson P. Bryner. 2011. *Fire Exposures of Fire Fighter Self-Contained Breathing Apparatus Facepiece Lenses*. Gaithersburg, MD: National Institute of Standards and Technology.

National Fire Protection Association. 1982. Fire at the MGM Grand. *Fire Journal* January: 19–37. https://www.nfpa.org/-/media/Files/News-and-Research/Resources/Fire-Investigations/FIR_1980_11_21_hotel3.ashx?la=en. Accessed December 16, 2019.

National Fire Protection Association. 2015. *NFPA 1584: Standard on the Rehabilitation Process for Members During Emergency Operations and Training Exercises*. Quincy, MA: NFPA.

National Fire Protection Association. 2018. *NFPA 1500: Standard on Fire Department Occupational Safety, Health, and Wellness Program*. Quincy, MA: NFPA.

National Fire Protection Association. 2018. *NFPA 1971: Standard on Protective Ensembles for Structural Fire Fighting and Proximity Fire Fighting*. Quincy, MA: NFPA.

National Fire Protection Association. 2019. *NFPA 1221: Standard for the Installation, Maintenance, and Use of Emergency Services Communications Systems*. Quincy, MA: NFPA.

National Fire Protection Association. 2020. *NFPA 1410: Standard on Training for Emergency Scene Operations*. Quincy, MA: NFPA.

National Fire Protection Association. 2020. *NFPA 1710: Standard for the Organization and Deployment of Fire Suppression Operations, Emergency Medical Operations, and Special Operations to the Public by Career Fire Departments*. Quincy, MA: NFPA.

National Fire Protection Association. 2020. *NFPA 1720: Standard for the Organization and Deployment of Fire Suppression Operations, Emergency Medical Operations, and Special Operations to the Public by Volunteer Fire Departments*. Quincy, MA: NFPA.

National Institute for Occupational Safety and Health. July 26, 2004. Fire Fighter Fatality Investigation Summary [F2003-18]. Partial Roof Collapse in Commercial Structure Fire Claims the Lives of Two Career Fire Fighters—Tennessee. https://www.cdc.gov/niosh/fire/pdfs/face200318.pdf. Accessed December 16, 2019.

National Institute for Occupational Safety and Health. 2005. *NIOSH Alert Preventing Injury and Deaths of Fire Fighters Due to Truss System Failures*. Atlanta, GA: NIOSH.

National Institute of Standards and Technology. 2010. *Technical Note 1661, Report on Residential Fireground Field Experiments*. Gaithersburg, MD: NIST.

Occupational Safety and Health Administration (OSHA). 2019. "Section 1910: Occupational Safety and Health Standards." *Code of Federal Regulations*, title 29. https://www.ecfr.gov/cgi-bin/text-idx?SID=9d828d4d1a251a74520045c33956cd4c&mc=true&tpl=/ecfrbrowse/Title29/29cfr1910_main_02.tpl. Accessed December 16, 2019.

Selkirk, Glen A., Tom M. McLellan, and Joanna Wong. 2004. "Active Versus Passive Cooling During Work in Warm Environments While Wearing Firefighting Protective Clothing." *Journal of Occupational and Environmental Hygiene* 1 (8): 521–31.

Underwriters Laboratories Safety Research Institute. 4-1-10. https://ulfirefightersafety.org/docs/EMW-2007-FP-02093.p.

Williams, James, and Hollis Stambaugh. March 2003. *Special Report: Rapid Intervention Teams and How to Avoid Needing Them* [USFA-TR-123]. Emmitsburg, MD: U.S. Fire Administration.

CHAPTER 6

Life Safety

LEARNING OBJECTIVES

- Explain the relationship between life safety and extinguishment.
- Determine the applicability of vent, entry, isolate, search, (VEIS) and exterior rescue tactics.
- Discuss the positive and negative aspects of ventilation in regard to life safety.
- Evaluate ventilation options as they relate to fire location and given a scenario with several vent options, select the best option.
- List and evaluate rescue options.
- List rescue priorities in terms of occupant proximity to fire.
- Define a mass-casualty incident.
- Explain the emergency medical services (EMS) function at a large structure fire with multiple casualties.
- Describe conditions that impact life-safety staffing requirements.
- Evaluate tactics at a fire scenario where a large number of occupants are in need of rescue.
- Compare and contrast the positive and negative effects of entering an enclosed fire area.
- Use a scenario to select and describe proper ventilation techniques.
- Use a scenario to describe and apply rescue options.
- Use a scenario to evaluate priorities as they relate to occupant proximity to the fire.
- Use a scenario to estimate staffing requirements at a fire in a structure occupied by a large number of people.

Courtesy of David J. Jones, Cincinanti, Ohio.

Case Study

On November 2, 2010 at 11:07 PM, a one-alarm fire response consisting of two engine companies, three truck companies (one truck company assigned as rapid intervention team), one heavy rescue company, and two district chiefs responded to a reported structure fire.

The first unit on the scene reported a "working fire" in a two-story house with heavy smoke and fire visible on the first floor. Two occupants (grandfather and granddaughter) were reported trapped on the second floor. The heavy volume of fire and smoke on the first floor prevented using the stairway to gain access to the second floor. The engine companies attacked the first-floor fire while truck companies used ground ladders to gain access to the second floor. Truck companies encountered heavy smoke conditions, but were able to locate the two occupants and rescued them via ground ladders.

Both victims were resuscitated at the scene and then transported to the hospital; both survived.

1. Would you consider the response to this fire as adequate, excessive, or inadequate?

2. When responding to a working fire with reported trapped occupants, what options are available to the first-arriving engine company reaching the scene prior to arrival of other fire units?

3. In this scenario, what other rescue options were available, other than using ground ladders, to remove the second-floor occupants?

4. After rescuing the two known occupants at this fire, how would you determine that all occupants have made it to safety?

5. When confronted with a working fire and occupants above the fire (not necessarily this fire), what are some of the options available to the incident commander (IC)?

6. Explain necessary tactical assignments if this fire had occurred on the first floor of a 10-story, 80-unit apartment building.

7. What factors affect the probability of extinguishment by the first-arriving engine company?

8. Under what conditions would it be acceptable to abandon a fire attack with people trapped?

Introduction

In this chapter, the most important of all fire-ground functions—life safety—is discussed. Life safety presents the ultimate challenge to the incident commander (IC). Decisions of whether to control the fire, remove the victims, or both might seem obvious. However, during the heat of battle, the IC is faced with numerous high-priority challenges. Deciding how best to assign initial limited resources at the incident scene is difficult. Initial strategic decisions at the incident scene must be made quickly on the basis of the situation and the resources available to the IC. Understanding the interrelationship of various life-safety tactics discussed in this chapter is essential for an IC. Rescuing occupants is not the only way to save lives at a structure fire. Fire suppression and ventilation are also used to accomplish this most important mission.

Although fire extinguishment is generally considered the second priority, it is an essential part of most rescue operations. Fire fighter safety is a critical component of the life-safety priority. The protection provided by fire streams, combined with proper ventilation, is an important part of protecting fire fighters as well as occupants.

NOTE

In the vast majority of cases, the best life-safety tactic is extinguishing or controlling the fire.

Evaluating the Probability of Extinguishment

Determining the probability of extinguishment is a major factor in life-safety decisions. If the fire can be quickly extinguished, the top life-safety priority is providing the required rate of flow. Only on rare occasions is it best to rescue visible occupants rather than controlling or extinguishing the fire.

The case study at the beginning of this chapter involved a coordinated operation. A district chief, engine company, and truck company all responded from a nearby fire station. The district chief established command while the engine company attacked the fire on the first floor and the truck company rescued the two occupants on the second floor via a ladder.

Upon arrival at the fire scene, the fire officer must take several factors into account. A common tactical error is prioritizing victims according to visibility. Occupants who are still inside the building may be in grave danger, and it may be best to rescue people who are at windows via the interior stairs *after* the fire is under control or extinguished.

In some cases, it is possible to implement a defend-in-place tactic if the fire can be quickly controlled. Electing to leave people in a burning building is a calculated risk that the fire officer or IC must be prepared to make on the basis of fire conditions, available resources, and the extent of danger to victims. Sheltering in place is a temporary rescue tactic addressing the immediate rescue problem. Shelter areas must be assessed for tenability, including the presence of toxic gases.

A fire may require a long lead time to suppress or even be impossible to contain when large-volume flows are needed. This chapter's Case Study provides an example of how difficult the fire control/rescue decision can be. Rate-of-flow calculations can assist the fire officer in making this critical decision. These calculations are explained in Chapter 8, *Offensive Operations*. Large interior flow requirements from handheld hose lines translate into larger staffing needs, which, in turn, may require more time to assemble the required resources.

If the fire cannot be extinguished immediately, building features such as fire doors and fire walls can aid in extending the time available for evacuation by providing a barrier between victims and the fire. Hose lines can also be used to protect egress routes. It is critical to ensure that sufficient personnel and resources are available to provide the required water flow. Opening a fire door to apply water at less than the required rate of flow could have severe consequences. The opening that is made to make entry will allow the fire and smoke to extend into other noninvolved areas at an accelerated rate.

It is essential to advance a charged hose line into position to extinguish the fire; this is usually accomplished by the first-arriving engine company crew. The crew advancing the hose line will assist victims they come upon but cannot be expected to perform a complete primary search of the area surrounding the fire. Their primary responsibility is extinguishment.

The floor or floors above the fire are critical areas that must also be searched. The number of crews

Fallacy	Fact
Occupants who are visible at windows are in the greatest danger and are always the first rescue priority.	People who are visible at windows might or might not be in grave danger. Occupants who are unable to reach a window are likely to be in more danger.

needed for search and rescue depends on the number of floors to be searched, the size of the building, fire intensity, smoke conditions, and occupant status.

A single crew can search and evacuate an entire floor in a single-family detached dwelling where the occupants are able to assist themselves. If occupants are missing or unable to make their own way to safety, it may well take an entire crew to rescue one individual plus an emergency medical services (EMS) crew to provide treatment and transportation once the victim is removed from the building.

A high-rise building may have many floors above and below the fire that need to be searched. Floor areas in these buildings are generally large. More crews will be needed when confronted with large areas and/or when several floors are to be searched. There is sometimes a tendency to assign multiple companies to search a floor before getting at least one company in each exposed area above the fire to determine evacuation, fire, and smoke conditions. In most situations it is best to get at least one company into all areas on the fire floor and above as soon as possible. Once all critical areas have been assigned, additional units should then be assigned to each area as needed to complete search and rescue activities.

Only in the most extreme circumstances should the first-in engine crew do anything other than advance an attack hose line. Generally, more lives will be saved by controlling the fire than by performing ladder rescues or knocking on doors to evacuate. Often, the only hope for occupants in the immediate fire area is fire control, especially if the fire is between them and their way out. What does all of this mean? The first-in engine crew *should not* be included in the staffing needed for rescue and evacuation. If multiple hose lines are needed to confine the fire, add the staffing needed to operate additional fire lines to the total staffing requirement.

Rescue versus Fire Attack

In rare circumstances, the best tactical choice for the first-arriving unit may be to perform interior or exterior rescues prior to attacking the fire. This would be the exception rather than the normal procedure. An immediate fire attack is not an option when an apparatus without pumping capabilities is first on the scene (e.g., a heavy rescue or ladder truck without pumps). If occupants requiring rescue are visible, placing a ladder or other exterior rescue method may be the best first action.

One technique that is sometimes used when victims are known to be inside the building, but not visible from the exterior, is the Vent Enter Isolate Search (VEIS) tactic. Indications for a VEIS include the following:

- Strong evidence that a savable victim remains inside the building
- Known location of victim
- Victim location directly accessible through exterior opening such as a window
- Victim unable to be safely rescued via interior stairway

The vent part of VEIS involves opening the window or other exterior access point. Interior entry is then made and the space is isolated by closing the door to the room or space in order to avoid drawing the fire into the search area. The space is then searched for the occupant. The fire fighter then exits in the same manner that was used to gain access to the space. Operating without hose line protection is precarious, especially on the fire floor or above. The risk to fire fighters and occupants is greatly increased when a vent opening is made to enter the search area.

If the fire is in the room being entered, both the victim and fire fighters performing the rescue are in greater danger, and opening the window will greatly intensify a ventilation-controlled fire. Closing the door is now the accepted means of controlling the flow path.

If the fire is located in an area beyond the room where entry is made, it may be possible to isolate the area by closing doors leading to the area involved in fire, while leaving the access window open to vent the room being searched. Whenever possible a fire fighter should remain at the point of entry to assist in victim removal. This fire fighter can control the access point and aid interior fire fighters in escaping.

In the majority of cases, a standard fire attack coordinated with ventilation and rescue is the best strategy when an engine company or quint is first on the scene, or due to arrive within a few minutes of the first-arriving nonpumping unit. Only in special situations, such as when other first alarm units will be delayed and only one or two occupants remain in the building at a known location, would a VEIS operation be warranted. Additionally, in circumstances where extinguishment is known to be beyond the capabilities of initial responders, a VEIS operation may be the best choice. If occupants are known to be inside the building but are not visible at windows, balconies, etc., initiating interior rescues without the protection of a hose line is more hazardous than when a quick entry and search are possible. The deeper within the interior of the building and the closer to the fire, the more dangerous the VEIS operation becomes.

Assessing the Ventilation Profile

Ventilation has a positive component and a negative component. If fire fighters are working in an enclosed area when a window breaks, any fire not isolated from the vent opening will gain in intensity, but the products of combustion will be vented out of the area. For this reason, ventilation must be coordinated with the placement of attack hose lines to control the fire.

In most cases, ventilation makes the job of finding the fire and victims much easier. However, under certain conditions, improper venting could produce a backdraft, and the fire fighter or occupants could be seriously injured or killed by the resultant fire and explosion. Furthermore, a vent opening made between fire fighters or victims and their path to egress could prove fatal. Any opening in, or directly connected to the fire area, will provide ventilation (e.g., opening a door to gain entry to the interior of the building).

Multiple openings often result in air flowing into one opening and smoke and fire exiting the other vent. For example, if the wind is blowing against the rear of a building when high heat fractures rear windows, the fire may be driven back into the building. Fire fighters opening the front door would then complete a flow path causing fire, heat, and smoke to exit the same door they are entering.

Ventilation is done to relieve the products of combustion, allowing fire fighters to advance on the fire. When venting to support extinguishment, it is important to coordinate hose line placement with ventilation. Close coordination of venting means that the hose line is ready to quickly overcome the increase in combustion that will likely occur.

Venting can be an effective life-safety procedure. When venting for life-safety purposes, the principle is to draw the fire, heat, smoke, and toxic gases away from victims, stairs, and other egress routes.

A common life-safety venting tactic involves opening a scuttle at the top of a stairway. This can be an effective means of venting the stairs because the smoke will generally rise, due to the natural stack effect. However, there is a negative component to this method of vertical venting: If a door to the fire area is opened, the vent opening will draw the fire and smoke from the fire compartment into the stairs. (See Chapter 2, *Procedures, Preincident Planning, and Size-Up.*) Vent openings made in the wrong location may draw fire, heat, and toxic gases into areas containing fire fighters and occupants. **FIGURE 6-1** is a plan view of a one-story warehouse building with overhead doors identified as A, B, C, D, and E. Roof vents are labeled 1, 2, 3, and 4. Given the location of the fire, the best possible vertical vent would be the roof vent directly over the fire (vent 1). However, the roof could be weakened in this area or the fire intensity below could cause injury when the roof is opened. Vent 2 would be an acceptable alternative, although it would spread the fire into unburned areas. Cutting open the roof in the area near the fire would also be acceptable, but cutting through the roof of this commercial building would cause additional damage and could be significantly more difficult than opening a roof in a residential building. Opening vents 3 or 4 would be unacceptable as they would pull the fire toward fire fighters operating the hose line. Ventilation through door C would be best if horizontal venting were chosen, with the understanding that opening this door might be hazardous due to the sudden release of fire through the door. Door D would be the next best horizontal opening. Doors A and B could be used, but opening either one would not be an ideal choice as venting through these doors would spread the fire. Opening door E could have disastrous effects as it would pull the fire toward fire fighters advancing on the fire and through a large portion of the warehouse. When venting, wind direction and velocity, as well as the location of the fire should be considered.

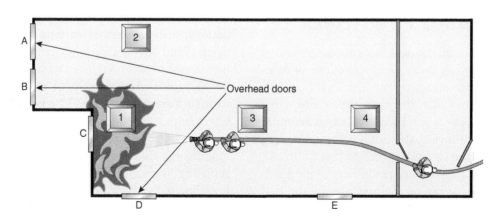

FIGURE 6-1 Plan view of a one-story warehouse.
© Jones & Bartlett Learning.

Strategically placed hose lines can offset the negative effects of improper venting, but drawing the fire toward occupants and fire crews is *never* a good tactic. At times, particularly when a large rate of flow is required, the hose line will not be able to stop the forward progress of an improperly vented fire.

When venting to save lives, it may be necessary to vent before having a hose line ready. However, this is a risky tactic that should be used only as a last resort. Building features can retard fire growth and prevent extension into occupied areas if they remain closed. For this reason, it is usually best to avoid opening doors and windows in the fire area before charged hose lines are in position.

Fire fighters may break windows as they conduct their primary search. This venting tactic is often necessary; however, when possible, it is better to open the windows rather than breaking them. Reversible venting is the preferred method, because once a window or door is shattered the vent opening cannot be easily closed. If the windows or door are opened so that they can be tightly closed again, negative, unintended consequences caused by the vent opening can be reversed by closing the window or door.

In making openings for ventilation, it is important to consider the fact that the fire and heated gases will naturally move upward, horizontally, and then downward. Flow paths will be from high pressure to low pressure and from hot to cold.

> **NOTE**
> Regardless of the intended results, the fire is bound by the rules of physics and chemistry.

Positive-pressure ventilation has become popular and is a useful tool. The key to successful positive-pressure venting is to control outlet openings. If too many doors and windows are opened, positive-pressure venting will prove ineffective. As with all other ventilation methods, there is a negative side to positive-pressure venting: The fire can be directed toward victims, toward their escape routes, or into unburned areas.

Analyzing the Available Rescue Options

Occupants should be rescued by the safest, shortest, easiest, and most direct route. Victims who are in danger above or below grade level should be evacuated by using the interior stairways whenever possible.

A **defend-in-place concept** is now being used in occupancies such as high-rise buildings and healthcare facilities in which occupants are moved away from the fire area but remain in the structure. Many large buildings are constructed in a manner that would allow a defend-in-place strategy. Some buildings, particularly high-rise structures, evacuate occupants sequentially to better manage travel in stairways. For example, with a fire on the 15th floor of a 30-story building, occupants on the 14th, 15th, 16th, and 17th floors would be directed to evacuate the building or go to a lower floor within the building. All other occupants would remain in place. If the fire was not immediately controlled, other occupants would be directed to evacuate the building.

A defend-in-place, or sequential evacuation, requires the cooperation of building occupants. Many occupants of the World Trade Center Fire on September 11, 2001, survived by acting contrary to the evacuation plan. People are aware of this incident and may be more likely to disobey evacuation instructions.

The decision of whether or not to leave people inside a burning building may be particularly perplexing on floors above the fire that are not yet threatened, especially if stairwell capacity is needed for evacuation of more endangered occupants or for moving fire control forces and equipment to the fire area.

When deciding on an evacuation, the following questions arise:

- Would the occupants be safer remaining in their present locations, which have not been invaded by the products of combustion?
- Would it be better to evacuate them through smoke-filled corridors and stairways?

In several fatal fires victims were found in the corridors and stairways, but people who remained in their rooms were not injured. If a defend-in-place strategy is employed, the spaces must be monitored for products of combustion that cannot be seen. Recent experience has found levels of hydrogen cyanide (HCN) and carbon monoxide (CO) well above acceptable levels in locations where occupants were sheltered in place despite no signs of products of combustion.

> **NOTE**
> Making a decision to defend in place, when the fire is not yet under control, is a calculated risk taken by the IC with the lives of occupants.

Next, we will discuss some of the most common means of building evacuation from upper floors. They are listed in preferential order; for example, fire

Incident Summary

North York, Ontario

In a fatal high-rise fire in North York, Ontario, six victims were found in the stairways in the upper stories of a 29-story building **FIGURE IS6-1**. The fire occurred on the fifth floor.

Data from: Residential High Rise, North York, Ontario, NFPA Fire Investigation Report. Quincy, MA: NFPA.

FIGURE IS6-1 The door to the stairway where victims were found in the North York, Ontario, Fire.
© 1995, National Research Council Canada.

escapes are generally preferred over ground ladders. The overriding factors in determining the best means of evacuation are based on safety and efficiency. Large numbers of occupants can be directed down the interior stairs with little or no assistance from fire fighters. All of the other listed methods are more difficult, require more staffing, and are more hazardous. However, when fire or smoke enter the interior stairway, alternative egress may be required. Rescue via an aerial device is usually safer and requires fewer fire fighters than rescue using ground ladders.

Interior Stairs

Interior stairs are preferred over all other means of egress from upper floors when they are tenable. The safest and easiest way to move occupants is by using the stairs, which are intended for that purpose. Any other method is considered a less desirable alternative, and the use of alternative methods should be justified by special circumstances.

Fire Escapes

Fire escapes are poor substitutes for interior stairs, and their structural integrity is sometimes questionable, especially when large numbers of people are attempting to use them. However, when properly maintained, they are the second-best option. If the fire escape is structurally sound, it is preferred over all methods other than the interior stairway.

In many cases, occupants choose to use the fire escape prior to the arrival of the fire department. Once occupants are on the fire escape, fire fighters must facilitate their movement to safety. Often this involves assisting or directing people who are impeding movement on the fire escape.

Fire or smoke pushing out of window or door openings leading to the fire escape will slow or stop movement **FIGURE 6-2**. If fire fighters are inside the building, they may be able to close windows that are emitting smoke or close interior doors to keep smoke and fire from entering rooms with openings to the fire escape. On rare occasions, it may be necessary to apply water from the exterior to knock down the fire threatening the fire escape. In this case, use a transitional

FIGURE 6-2 Smoke impeding escape via fire escape.
Courtesy of David J. Jones, Cincinnati, Ohio.

attack (explained in Chapter 8, *Offensive Operations*) to minimize endangering occupants who are still inside the building. If fire that is impinging on the fire escape cannot be contained, it may be necessary to remove occupants from the fire escape via a ladder or assist them in reaching the interior stairway.

Drop ladders are typically difficult to climb or descend and sometimes will not release due to rust or paint. If the drop ladder or counterbalanced stair is in the up position, fire crews should attempt to pull it into the down position, being cognizant that these ladders may drop suddenly. Placing ground ladders in position next to the fire escape at the second-floor level provides an alternative or second means of escape from the fire escape. It is best to place the ground ladder on the side opposite the fire escape ladder.

Aerial Ladders and Elevated Platforms

Following interior stairs and fire escapes, the next rescue preference is aerial ladders or elevated platforms. As noted previously, aerial ladders and platforms can be used in conjunction with evacuations via the fire escape. There are two schools of thought when using aerial devices for rescue. Aerials can move larger numbers of people from a single area in a shorter period of time. However, elevated platforms provide a more secure, less stressful rescue for the occupant, and articulated platforms can reach areas that an aerial would have difficulty reaching. Fire departments should survey hazard areas before purchasing apparatus, matching apparatus to identified needs.

Ground Ladders

Ground ladders follow aerials and elevated platforms on the rescue preference list. Ground ladders are generally less stable, do not have the reach of most aerial ladders and require more personnel to place in position. However, in certain circumstances, such as when buildings are set back from the street or when two or three fire fighters can quickly access the second or third floor, ground ladders are preferred over aerial devices. As noted previously, ground ladders can provide an alternative pathway to the ground level from a fire escape.

When fire and/or smoke are between the victim and the interior stairs, aerial ladders, elevated platforms, and ground ladders are the best alternative when a quick knockdown is not possible or when extinguishment tactics further endanger occupants. Again, remember that if the stairways and corridors leading to the stairways are relatively clear of smoke, use the stairs.

When the fire threat seems real to the occupants, they will sometimes go to a window or other visible position to await rescue. They may or may not have checked the stairs before deciding to wait at the window. Do not attempt a ladder rescue just because people are at the window. Reassure them that you are there to help, but determine whether they must be moved and whether stairways are available. Whenever possible, it is best to send fire fighters into the building to check the stairways and evaluate the need for ground ladder rescues. Fire fighters on the interior are in the best position to determine the need for evacuation, the best rescue tactics, as well as evaluating which occupants are in greatest danger.

Raising and positioning ground ladders may require several fire fighters. A ground ladder rescue will involve at least one fire fighter assisting the occupant and probably another trying to assist the person onto the ladder with a third fire fighter at the foot of the ladder.

> **NOTE**
>
> Rescues and evacuations via interior stairs, fire escapes, aerial apparatus, and ground ladders are considered acceptable, and even the preferred means of egress under certain circumstances. Elevators, scaling ladders, ropes, helicopters, and air bags are all questionable rescue methods that are potentially hazardous.

Elevator Rescues

An interesting question arises regarding the use of elevators for rescue. Fire safety professionals warn against occupants using the elevators as a means of escape. However, they could be safely used when under the fire department's control. Under certain conditions, the use of elevators for evacuation is justified in buildings that are subdivided with suitable fire-resistive construction. If an elevator is remote and separated from the fire area with an auxiliary power supply, using it for rescue purposes, especially to evacuate immobile occupants, may be the best tactic. The use of elevators in the immediate fire area is hazardous to everyone, including fire fighters.

Rope Rescues

The use of ropes for rescue purposes can be justified only in extreme cases when occupants are beyond the reach of aerials and elevated platforms. Before resorting to this last-option rescue technique, every effort should be made to perform an interior rescue. Rope rescues are extremely slow and dangerous and require specialized equipment and expertise. There is an added hazard to the rescuer and the victim because of the inherent danger of technical rescue operations. This rescue method is a tactic of near last resort.

Helicopter Rescues

Before deciding on rooftop rescues, the IC should ask whether the occupants who have reached the roof are safe. Many times it is possible for occupants to wait out the fire on the roof. Most roofs provide a difficult operating platform for helicopters, owing to obstructions that include rooftop facades, antennas, satellite dishes, and ventilation equipment **FIGURE 6-3**. These obstructions make it impossible for the helicopter to land on the roof. Even if the roof is clear of obstructions, it may not be a good landing platform because of the hot air currents caused by an intense fire.

Because of the limited number of people who can be safely transported in a helicopter, the choice of who is going to be rescued and who will stay behind will also need to be made. Victims may think they are in immediate danger and believe that they cannot survive long enough to wait for another trip. They may attempt to climb into the helicopter, overloading it beyond its safe operating capacity. The helicopter has been used to advantage in some rescue situations, but its usefulness has been overstated. At many fires, dramatic helicopter rescues were completely unnecessary.

Some departments use helicopters to place fire fighters on the roof. These fire fighters then assist and reassure occupants on the roof or descend down the stairway to perform rescues. They could also use the standpipe for extinguishment. Helicopters can provide quick and less fatiguing access to upper floors in a high-rise building. However, fire fighters typically use the path of ingress as their egress. A helicopter that drops fire fighters on the roof may not be able to return later and remove them from a position above the fire floor. If your department anticipates the use of helicopters, continuous training of both fire personnel and helicopter crews is absolutely essential. Most private helicopter owners will resist getting involved in rescue work because they lack rescue equipment, training, and insurance coverage when performing rescues.

Classifying Evacuation Status

During the early stages of an operation the IC is forced to make staffing estimates based on incomplete or even inaccurate information. As companies are deployed, crews will give status reports, request assistance, or possibly release fire fighters for other assignments. Information becomes more complete and more reliable as fire crews provide reconnaissance by entering endangered areas.

Fires with a significant life-safety challenge reinforce the need to build a command structure right from the start of the incident, so that a systematic search-and-rescue operation can be made without duplication of effort. When searching a large building with many rooms, a marking system is highly recommended. Many departments use plastic door hangers or other marking procedures to indicate rooms that have been searched. Even a simple chalk-marking system can be effective. For example, some use chalk or other medium to mark an "X" on the door leading to an enclosed area to indicate the area has been searched. A half "X" (/) denotes that the search is underway in that compartment. An "X" on a door leading to a floor or opposite the elevator indicates the entire floor has been searched **FIGURE 6-4**. Whatever method is chosen, fire fighters must know the system, use it regularly, and have the necessary materials immediately available to carry out the marking system.

FIGURE 6-3 Rooftop obstructions.
Courtesy of Ben Klaene.

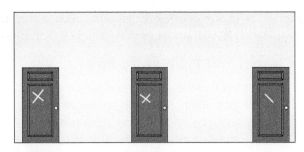

FIGURE 6-4 Door marking system.
© Jones & Bartlett Learning.

Adopting a marking system as an standard operating procedure (SOP) is highly recommended.

Occupants who escape on their own may need direction, emergency medical treatment, a place of safe refuge, or other assistance. The care of these victims is usually a nonemergency function that can be delayed if necessary, until personnel have verified that no one is in immediate danger.

Other victims may not be aware of the fire or of egress routes. They may need direction to safety, or fire fighters may need to provide a safe egress route for them. These victims will then be able to escape with little assistance. It may not be necessary to escort them all the way to the outside or, for that matter, to a place of safe refuge. When there are a limited number of emergency personnel available, victims should be provided direction and permitted to evacuate on their own whenever possible, allowing available fire fighters to perform the more crucial task of search and rescue.

Sometimes it is possible to designate specific stairways as evacuation routes and other stairways for fire operations. This is a good tactic because fire fighters advancing hose lines through the stairs to the fire area will allow smoke to enter the stairs through the partially open door and impede egress because of the hose laying on the stairs. Six occupants of the Cook County Administration Building in Chicago perished while attempting to exit via the stairway being used by fire fighters to attack the fire. The Cook County Administration Building Fire is discussed further in Chapter 12, *High-Rise Buildings*.

A practical problem occurs in attempting to channel the flow of occupants into a designated stairway. In most cases, fire forces will not be in total control of the evacuation until they reach each floor to direct escaping occupants to the stairway of choice. There may be confusion regarding which stairway is to be used unless doors leading to stairways are plainly marked, e.g. Stairway A. Simply designating the stairway is not enough; specific instructions must be given to both fire fighters and occupants. Some large buildings have voice annunciation systems or public address systems that permit the IC to relay messages to evacuees.

Incident Summary

St. Louis High-Rise

A fire occurred in a high-rise building with elderly occupants in St. Louis, Missouri. An attempt was made to attack the fire from both stairwells, but because of the number of elderly people coming down one of the stairwells, this was not possible. Whenever the door to the fire floor was opened, the smoke and heat entered the stairway, endangering the occupants. Once the flow of residents was stopped, it was possible to enter the floor from the second stairway to fight the fire and rescue an injured fire fighter. However, this operation was delayed because of the danger in which the occupants were placed whenever the door was opened.

Data from: Edward R. Comeau, personal correspondence following the investigation.

Immobile or unconscious victims will place additional requirements on the rescuers. In most cases these victims must be physically removed to the outside or to a place of safe refuge. This effort usually requires a minimum of two fire fighters (often more) to rescue one victim.

A commonly held fallacy is that the occupants will be able to provide rescuers with reliable information concerning occupant status and that the IC will be able to account for the building's occupants.

Fallacy	Fact
It is possible to account for all the occupants of the building without actually conducting a thorough and complete primary search.	The only way to be sure the building is completely evacuated is to conduct a primary search throughout the entire structure.

The *only* reliable way to verify evacuation status and fire conditions is for fire fighters to enter the structure and systematically check every room. Occupancies such as primary schools and healthcare facilities conduct regular fire drills, and staff should have control over students and patients. But even in these occupancies the staff may not account for visitors in the building. Some industrial settings have evacuation plans that include accountability. However, the mobility of employees within the facility may make these plans less effective than would be possible in a school or healthcare facility where the location of

Incident Summary

Bremerton, Washington, Apartment Complex Fire

An early morning fire in a large apartment complex in Bremerton, Washington, is an example of the difficulty in determining how many victims remain inside the burning structure. The fire broke out at approximately 6:00 AM, and a number of people had already left for work. The apartment complex was four stories high, with balconies on the outside. A number of people had to be rescued from these balconies by ladders because their primary escape route had been blocked by the fire. During the initial stages it was feared that possibly 20 people were unaccounted for and still trapped in their apartments. During the course of the day a more complete accounting was made, and it was ultimately determined that there were four people missing who had died in the fire.

Data from: Comeau, Edward R. *Apartment Fire, Bremerton, Washington*, NFPA Fire Investigation Report. Quincy, MA: NFPA. 1999. https://www.nfpa.org/-/media/Files/News-and-Research/Resources/Fire-Investigations/Bremerton.ashx?la=en. Accessed February 7, 2020.

Incident Summary

Chapel Hill Fraternity Fire

The fire chief in Chapel Hill, North Carolina, faced problems accounting for occupants when a fraternity building caught fire. The fire occurred on a Sunday morning that was also graduation day. A number of the residents had already moved out for the semester, but no one was sure who or how many or whether any visitors were in the building. It was originally feared that as many as 25 people were inside the building. Ultimately, it was determined that five people had been trapped and killed by the fire.

Data from: Isner, Michael. Fraternity Fire, Chapel Hill, North Carolina, NFPA Fire Investigation Report. Quincy, MA: NFPA.

---

occupants is known. In the typical apartment building some of the occupants may not be home when the fire occurs, and others may have escaped and then left the scene.

Whenever possible, the IC should drive around the fire building to see all sides before setting up a command post. Fire crews should view the exterior, looking for occupants in need of assistance. However, this outside reconnaissance should not be done at the expense of the fire attack or interior search. Reports from occupants may be unreliable, but they should not be ignored. In many cases, occupants will be able to specifically direct rescuers to the location of a victim.

Flashover is the critical landmark; before flashover, rescue is possible, after flashover, rescue is highly improbable within the flashover compartment. Recognizing the fire stage is important to the life-safety effort. In the early stages of a fire (ignition and early growth phases), occupants tend to rescue themselves once they are aware of the fire. Later, as the fire becomes larger and produces greater quantities of smoke and toxic gases, occupants may be conscious but unable to exit until the fire is extinguished.

The most difficult victim to rescue is one who has been felled by the smoke, heat, or gases in an unknown location.

Fire conditions are related to many factors, such as the time that elapsed before the fire was detected, the time it takes to notify the fire department, and the response time. If a fire is extending through a structure, the threat to occupants is directly related to the following:

- Awareness of the fire
- Ability to escape—includes occupant physical condition
- Fire and smoke conditions in escape route
- Construction of building
- Provisions made in advance to facilitate egress

Once a fire is in progress, the IC has little control over these factors.

Critical time (the time available until the structure becomes untenable) will vary depending on numerous building factors, as well as the building's fuel load. No exact figures can be given, but experience indicates that fires spread rapidly upward in balloon-frame construction or, for that matter, in any construction that allows vertical fire spread to occur unimpeded. The time from ignition to full involvement in a balloon-frame structure will be short in comparison to a fire in a fire-resistive building. The fire-resistive building allows fire forces or occupants to confine the fire by using built-in features such as fire doors. An enclosed fire-resistive area can, however, quickly become untenable because the products of combustion, including heat and smoke, are not relieved as they would be in a less compartmentalized structure.

The Underwriters Laboratory (UL) study of modern fuel loads in homes indicates that fire progression to flashover will occur much more rapidly than in the past (Kerber, 2010). If the IC determines that time does not permit the rescue of all the occupants, victim triage should be established to rescue as many victims as possible. The importance of fire control to life safety cannot be overstated. Buying time with fire control tactics is a critical strategy that must be implemented whenever possible. If the fire is extinguished, or at least in the knockdown stage, the fire fighters and victims within the building have a much better chance to evacuate safely.

Estimating the Number of People Needing Assistance

Once the preincident plan, visual, and reconnaissance information have been considered, the IC should be able to evaluate life-safety requirements in terms of staffing and equipment.

Fire fighters assigned to search and rescue tasks should be equipped with forcible entry tools. Heavy metal doors with locks will extend the time and effort required to complete the primary search. In forcibly entering an interior area, it is important to consider the value of the opening being forced. A fire door is designed to resist the fire. It can contain a fire for a significant period of time or keep fire from extending to an exposed area. If the fire extends to threaten the hallway while the search is being conducted, the door can isolate fire fighters and victims from the fire for a period of time. Whenever possible, doors should be closed after the search is complete to limit fire spread and damage.

The fire attack crew should have hose lines in place and charged before forcing the door to the room or area of origin. In buildings with strong compartmentalization, conditions beyond the door are unknown, and extreme caution must be exercised when making entry. If the door is hot to the touch and evacuation is not complete, it may be best for the crews to stand ready at the door and allow the door to contain the fire while other crews complete the evacuation.

By making return trips to the fire area, fire fighters can rescue more than one person even in situations where carry-and-drag rescues are needed. However, this is demanding work and will quickly exhaust the available rescuers. In sizing up the personnel requirements, estimate the physical condition of the occupants, the number of occupants on the fire floor, and the travel distance to safety, as well as fire growth. Be proactive in requesting additional companies when faced with a potential large-scale rescue effort.

After estimating the numbers of victims needing to be rescued, the method of rescue and victim condition must be considered. Immobile or unconscious victims will require at least one rescuer for each victim or, more typically, two rescuers per victim.

Few, if any, fire departments could successfully evacuate a high-rise building with 1000 or more occupants in need of assistance. Furthermore, if all of the occupants of a high-rise building are placed in the exit facilities at the same time, more people might be placed in danger. Trade-offs must be made. An evacuation must be prioritized by removing the most endangered occupants first. Meanwhile, fire fighters can plan ahead to move potential victims in less immediate danger before they are felled by the products of combustion, allowing rescuers to conserve strength by providing direction or assisting victims to a safe area while they are still mobile.

Incident Summary

Bremerton, Washington, Apartment Complex Fire—Continued

In the apartment fire in Bremerton, Washington, that was summarized earlier, the first-arriving officer was faced with a large number of victims who were on their balconies, needing to be rescued. Their only means of egress was blocked by the fire. One person went so far as to lower a rope and slide down four stories, severely burning his hands.

The officer knew that he could not possibly rescue all of the people who were in danger by removing them from the fire. Therefore, he elected to advance a hose line into the courtyard to fight the fire. Thus, the fire spread was slowed until other units arrived on the scene and were able to begin rescuing the occupants using fire department ladders.

Data from: Comeau, Edward R. *Apartment Fire, Bremerton, Washington*, NFPA Fire Investigation Report. Quincy, MA: NFPA. 1999. https://www.nfpa.org/-/media/Files/News-and-Research/Resources/Fire-Investigations/Bremerton.ashx?la=en. Accessed February 7, 2020.

Surveying Floor Layout and Size

The size of the area to be searched and evacuated, as well as the location of stairways, halls, and fire escapes, has much to do with organizing the search effort. Often, the primary and secondary searches are done by assigning floors to each search team. Large buildings may be further subdivided by wing or building side.

Many buildings have mazelike designs, like the Beverly Hills Supper Club, shown in **FIGURE 6-5**. This is particularly true of buildings with multiple additions or office buildings that use cubicles for work spaces. In some cases, building additions result in confusing egress paths.

Fire fighters working in multistory buildings should, whenever possible, survey an uninvolved floor before ascending to the fire floor or floors above. The layout may be different from floor to floor, but some building features tend to remain constant, such as stairs, elevators, and standpipes. A few precautions are in order. The general layout of the space can be different on each floor. One floor may be wide open, while the next is subdivided into small offices or suites.

Another problem is consistency in floor numbering. Buildings that are constructed on a grade may have unusual floor numbers. Sometimes what appears to be the first floor of a building from one side is actually identified as a higher or lower floor. For example, the first floor main entrance to Good Samaritan Hospital in Cincinnati is the sixth floor. The terrain slopes away from the front of this hospital, and a section that was added to the building has a grade-level entrance six floors below the main entrance. To avoid the confusion caused by having different floor numbers at the same level, the entire hospital floor numbering system was changed to be consistent with the new addition. This same situation can exist in single-family detached dwellings that are built on a grade. In high-rise buildings there may be more than one set of windows on the first floor. The mezzanine may not be counted as a floor, and these buildings sometimes do not have a 13th floor. Each of these factors adds to the confusion. The grade-level sixth floor at Good Samaritan hospital or the absence of a 13th floor can be determined in advance through preincident planning.

In smaller buildings it is often possible to get a good idea of the interior layout from the exterior of the building. Large windows indicate areas that are most likely common areas (living or family rooms), whereas bedrooms tend to have smaller windows. Some buildings have windows at stairway landings that can be seen between floor levels, thus indicating the location of the stairway. Buildings of similar construction in an area will typically have similar floor layouts. For instance, the locations of stairs to the second floor of a single family detached dwelling tend to be near the main entry.

Even a small building that has been renovated or one that is at various grade levels on different sides can be difficult to search. These buildings can be deadly to fire fighters who go inside to conduct offensive operations.

Larger buildings, particularly large, open buildings, present an additional hazard to fire fighters

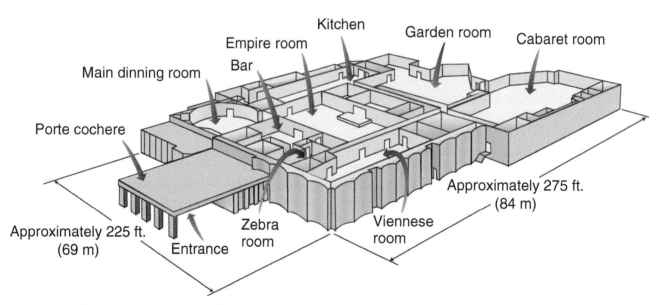

FIGURE 6-5 Floor plan of the Beverly Hills Supper Club in Southgate, Kentucky, where 165 people died in a 1977 fire.
© Jones & Bartlett Learning.

Incident Summary

Fires at Buildings on Sloping Grades

At two fires—one in Seattle, Washington, and another in Pittsburgh, Pennsylvania—a total of seven fire fighters were killed in buildings that were built on sloping grades. Because of the confusing topography, it was not clear to everyone on the fire ground how many floors the buildings had or the exact location of the crews operating inside of the buildings in relation to the fire.

A similar situation occurred in a California fire in a residence that appeared to be two-stories at the front, but had a basement and sub-basement that were not visible from the front of the building **FIGURE IS6-6**. The fire originated in the basement. Two fire fighters attacking the fire at the top of the basement stairway perished.

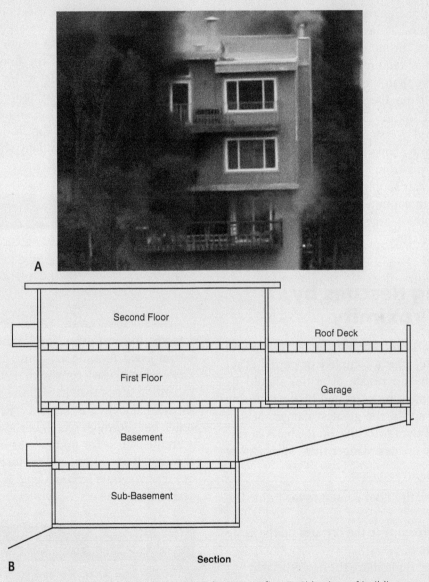

FIGURE IS6-6 A. Side Charlie showing basement fire. **B.** Side view of building showing basement and sub-basement.
Reproduced from: Fire Fighter Fatality Investigation Reports, NIOSH 2011-13.

Data From: Comeau, Edward R. *Warehouse Fire, Seattle, Washington*, NFPA Fire Investigation Report. Quincy, MA: NFPA. 1995. https://www.nfpa.org/-/media/Files/News-and-Research/Resources/Fire-Investigations/fiseattle.ashx?la=en. Accessed February 7, 2020; Isner, Michael, *Residential Fire, Pittsburgh, Pennsylvania*, NFPA Investigation Report. Quincy, MA: NFPA. 1995. https://www.nfpa.org/-/media/Files/News-and-Research/Resources/Fire-Investigations/fipittsburgh2.ashx?la=en. Accessed February 7, 2020; and NIOSH 2011-13. A career lieutenant and firefighter/paramedic die in a hillside residential house fire – California. Centers for Disease Control and Prevention, National Institute for Occupational Safety and Health, Fatality Assessment and Control Evaluation (FACE) Report F2011-13. https://www.cdc.gov/niosh/fire/pdfs/face201113.pdf. March 1, 2012. Accessed February 7, 2020.

and occupants when smoke conditions obscure vision. Most fire fighters have experience in homes and apartment buildings where most fires occur. The same tactics and procedures used in small, compartmented buildings could be fatal when used in a large, open structure with low visibility. Searching a small room using a right- or left-hand search can be safe and effective. A larger open area would require the use of a rope or other guideline and different search techniques.

A thermal imaging camera (TIC) can be used to quickly scan a small room prior to entry. Larger rooms may require the fire fighters conducting the search to advance with the TIC as they check around obstacles and advance throughout the compartment. A very large room may require substantial escape time, which may be more than the time available when the low-pressure alarm on the self-contained breathing apparatus (SCBA) sounds. The person in charge of the search may need to assign someone as timekeeper. New technology also allows the accountability officer to monitor air supply from a remote location. As one crew depletes their air supply a second crew will need to move forward to continue the search. Large compartments can produce large, overwhelming fires that can overrun a hose crew.

Prioritizing Rescues by Location/Proximity

After determining the total number of occupants and developing a strategy for a complete or partial evacuation, the search and rescue priorities should be established. This must be evaluated by determining which occupants are in the greatest danger. Search and rescue should then prioritize on the basis of rescuing those in the greatest danger first. The priority list is as follows:

1. People on the fire floor nearest to the immediate fire area
2. People in proximity to the fire area on the same level as the fire
3. People on the floor above the fire, especially immediately over the fire area
4. People on the top floor (unless fire conditions result in smoke stratification; see Chapter 12, *High-Rise Buildings*)
5. People on the floors between the floor above the fire and the top floor
6. People on the floors below the fire
7. People in nearby buildings
8. People outside (in the collapse or falling glass zones)

Note: Number eight in the list could be placed anywhere in the priority list, depending on the structural integrity of the building, falling glass, etc. Rescue priorities are shown in **FIGURE 6-6**.

Those in the greatest danger should be rescued first. An exception to this priority list would be when it is not possible to save everyone. In these situations, the IC should opt to save the largest number of people possible. This will require committing resources to areas where the largest number of occupants can be rescued. Another possible exception exists if one or more of the floors are known to be unoccupied, for example, a fire at midnight in the office building shown in Figure 6-6 where the only occupants are security personnel on the first floor and a nightclub on the top floor. Even in this case, the first-due engine company should be assigned to attack the second-floor fire.

> **NOTE**
>
> Most occupants either will escape on their own or, if given direction, will be able to evacuate without assistance.

The key to successful search operations is to be *systematic*. In most cases, the IC should follow the priorities listed above in assigning primary search areas. Crews should use a consistent method of marking and recording areas that have been searched so that all threatened areas are checked. The primary search is a quick but thorough search of the area. The secondary search ensures that no one was missed the first time through. If conditions and resources permit, a secondary search should be conducted as soon as the primary search is complete.

> **NOTE**
>
> The key to a successful search is to be systematic.

To ensure complete coverage during the secondary search, good practice dictates using a different crew. For example, if the first truck crew conducted the primary search of the fire floor and the second truck crew checked the floor above during the primary

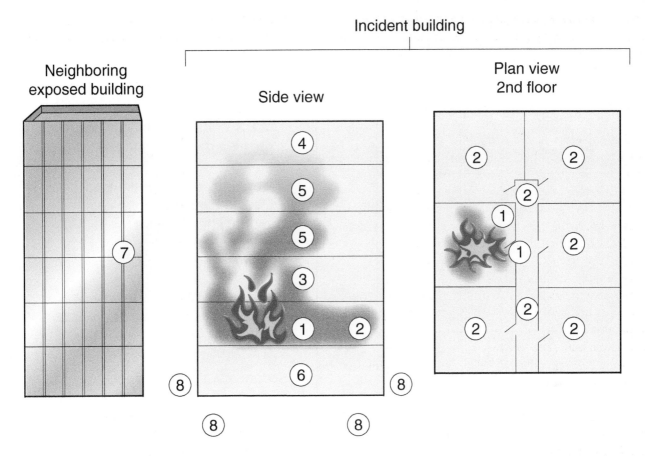

FIGURE 6-6 Rescue priorities.
© Jones & Bartlett Learning.

search, the secondary search assignments should be reversed (second truck crew assigned to the fire floor and the first truck crew to the floor above). Some fires require multiple search assignments, making this reassignment more complex. The officer in charge of the search and rescue group would manage the reassignment from primary to secondary search. **FIGURE 6-7** provides an example reassignment for the secondary search where five companies are assigned to primary searches in different areas of the fire building and exposure building.

The time and effort expended conducting the primary search dictates whether these same companies should be reassigned to secondary search or whether additional companies should be assigned to perform the secondary search. When necessary, additional secondary search companies can be summoned to the scene, but having a tactical reserve available in staging would be preferred. It may be possible to rotate companies in staging with companies who conducted the primary search.

The person managing the search and rescue effort should assign specific areas of responsibility, and then track these assignments and progress reports on a status board. When a large evacuation effort is required, the units involved in search and rescue will rely on communications to ensure that all areas are searched and to avoid duplication of effort. If at all possible, the search group should be assigned a separate radio channel or use telephones specifically designated for this purpose. It may also be possible to use the existing telephone system within the building as long as it is not affected by the fire.

Evaluating the Medical Status of Victims

The IC must have wide discretion in calling medical assistance to the scene. Medical assistance can be requested to treat and transport injured fire fighters or occupants. The relative hazard, number of potential victims, and type of incident will dictate the need for EMS at the scene. A general rule for calling assistance applies here: If you think you need help, you do.

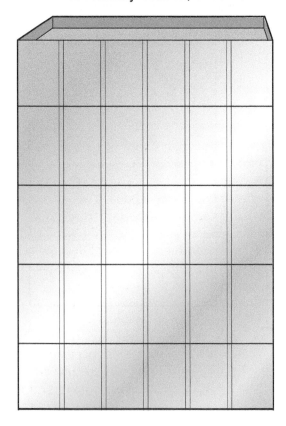

FIGURE 6-7 Primary and secondary search assignments.
© Jones & Bartlett Learning.

It is much better to have medical assistance standing by at the scene than to have injured fire fighters and occupants waiting for treatment while EMS is responding.

For most fire incidents, medical units should be set up at one or more designated locations within the cold zone or beyond. Transport vehicles should be parked with exit in mind and located near medical personnel ready to receive the injured. A fire in a multistory building may warrant moving medical personnel and equipment to a safe area inside the building. It may be a good idea to set up the medical triage/treatment area near the rehabilitation area. However, sufficient room should be available to physically separate the two areas.

Evaluating Victims in Mass-Casualty Incidents

Mass-casualty situations are usually the result of specific incidents, such as transportation accidents, hazardous materials releases, or natural disasters. Fires in buildings with large numbers of occupants have the potential to become mass-casualty incidents. Providing a system that includes triage, treatment, and transportation is a proven method of managing the medical component when large numbers of victims are encountered regardless of the type of incident.

Mass casualty is defined in terms of department and community capabilities. A mass-casualty incident occurs when one or more of the following situations exist:

- The number of victims and the nature of their injuries make the normal level of stabilization and care unattainable.
- The number of immediately available trained personnel and transportation vehicles are insufficient.
- Hospital capabilities are insufficient to handle all of the victims requiring care.

Triage, Prioritizing, and Transport

Triage is the first medical priority in managing a medical disaster. The first-arriving EMS personnel should not leave the scene until they are relieved of triage responsibilities. Emergency medical teams are accustomed to treating one or more individuals, followed by immediate transport. Just categorizing victims without treatment and transportation runs contrary to their normal role and must be overcome by rigid enforcement of triage procedures.

When staffing permits, it is best to physically separate the dead from living patients. The walking wounded should be directed to another location for treatment.

Once triage teams have prioritized the victims on the basis of their injuries, treatment teams will follow. Treatment teams first treat those needing immediate care (priority 1) and arrange for their transportation. After this, priority 2 and 3 patients can be treated. If sufficient personnel are on the scene, multiple teams can be formed to treat as many patients as possible. When available, paramedics or doctors should be assigned to treat the most seriously injured; EMTs and first responders can be assigned to take care of the remaining patients.

> **NOTE**
>
> One system of prioritizing patients uses the categories of 1: life-threatening priority, 2: serious, but not life-threatening priority, 3: walking wounded, and 4: deceased. Color-coded tags are also used to categorize victims.

Providing transportation for victims is an important function of the medical transportation officer. A drive-through arrangement for ambulances will keep traffic lanes open **FIGURE 6-8**.

At times, it is beneficial to use helicopters to transport patients. When using helicopters, the landing zones must be in safe areas, far enough away from treatment and triage areas that they will not interfere with those activities. Helicopters that are used to transport victims should not be flown over the fire area or command post. The noise and effect of the downdraft from the helicopter rotors can be extremely disruptive and the downdraft can create a hazardous condition.

Staging and categorizing transport capabilities of medical units are essential in managing transportation needs. Incoming EMS transportation vehicles should be directed to an established staging area during mass-casualty operations and requested as needed. It may be possible to transport several priority 2 or 3 victims in the same vehicle. Priority 1 victims often require advanced life support (ALS) personnel on board the transport vehicle. Minor care (priority 2 or 3) victims are usually staged near the site for later treatment. Depending on the nature of the injuries, it may be possible to transport priority 2 and 3 patients in buses or other non-ambulance vehicles.

Communication with hospital facilities is necessary to determine how many patients and what types of injuries can be treated at each facility. Hospitals determine their limits on the basis of personnel, staff expertise, and space. Hospitals must be prepared to exceed their normal patient capacity; however, they must also be trained to recognize when they have reached their emergency capacity.

Patients should be directed to hospitals and trauma care centers according to hospital capabilities. The medical planning officer (similar to the incident planning officer, but assigned to the medical branch or medical group) or scribe should document where and when each patient is transported. The number and severity of injuries may require that only priority 1 patients be transported. Field hospitals can be used to treat priority 2 and 3 patients. Even with transportation available, it is generally not a good idea to flood hospitals with the walking wounded.

During normal operations, EMS personnel can treat individual fire fighters and occupants. When the number of potential victims exceeds normal capacity for individual treatment, the IC is well advised to staff a medical branch or medical group to coordinate medical activities. When the number and/or condition of victims approach mass-casualty status, separate staffing of a medical branch is essential. **FIGURE 6-9** is a flowchart for a mass-casualty incident showing victims needing various levels of care.

The mass-casualty flowchart should be read from the top down. The objective to provide the best possible medical assistance within the limitations of available resources.

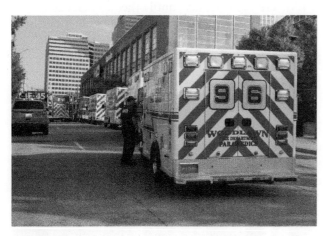

FIGURE 6-8 Ambulances staged on a city street.
Courtesy of David J. Jones, Cincinnati, Ohio.

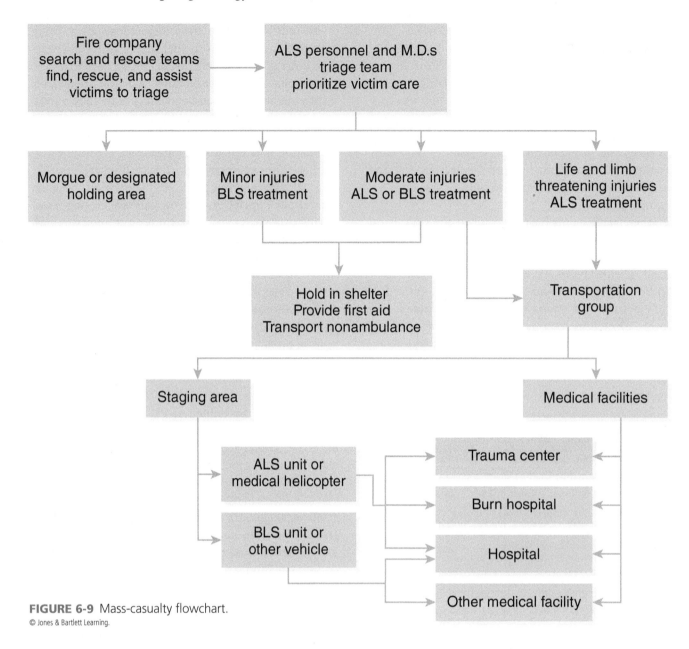

FIGURE 6-9 Mass-casualty flowchart.
© Jones & Bartlett Learning.

Evaluating the Need for Shelter

During weather extremes there is a need to rehabilitate fire fighters and provide shelter for occupants. Occupants of nursing homes, hospitals, and other places where special needs exist will require shelter even during normal weather conditions.

In many cases, evacuees will find their own shelter by leaving the scene to go home or to the home of a friend. When shelter is required, a nearby building, well outside the fire zone, should be used. A mass transit bus or school bus can also be utilized for shelter from the elements and placed strategically at the scene. Many charitable agencies such as the Red Cross generally have personnel available to assist in setting up and managing shelters. However, their response times may be fairly long, making it necessary to assign fire, police, or EMS crews to this task until other agencies arrive.

Estimating Life-Safety Staffing Requirements

The IC must have sufficient staffing to extinguish or at least control the fire, conduct search and rescue operations, treat and transport the injured, and remove victims to a place of safe refuge, while preventing re-entry. The number of fire fighters needed depends on several factors:

- Number of victims
- Rescue methods used

- Condition of victims
- Fire conditions
- Smoke conditions
- Victim mobility
- Weather
- Accessibility (need to force entry)

Additional staffing will be needed in the following situations:

- Victims are close to the immediate fire area.
- Victims have existing or fire-related physical impairments.
- The fire occurs during weather extremes, especially freezing weather.
- Evacuation routes other than the interior stairs must be used.
- It is necessary to force entry to rooms or hallways.

Evacuation status is determined by the following:

- Preincident planning information
- Alarm information
- Occupant information
- Visual observation
- Reconnaissance

Each fire department must develop SOPs based on local conditions and resources. Throughout most of the United States, crews, units, or companies are assembled according to apparatus type and function. Many different tasks are required on the fire ground; therefore, a division of labor approach is common practice. Fire department duties normally include the following:

- Water supply and application (engine company duties)
- Ventilation, entry, search and rescue, rapid intervention, and property conservation (ladder company duties)
- Search, rescue, and rapid intervention (rescue company duties)
- Triage, treatment, transportation, and rehabilitation (emergency medical duties)
- Planning, organizing, coordinating, and establishing command (IC duties)

This division of labor concept is important. Without preassignment of general duties, actions are delayed and/or duplicated. In a rescue scenario with occupants showing at windows, there would be a tendency to assist the visible occupants at the expense of addressing more critical tasks. In other cases, fire fighters might neglect ventilation, entry, and search in favor of water application.

Many references are available that complement the information presented here or describe actual tasks necessary to fulfill life-safety objectives. Suggested activities at the end of this chapter are designed to extend the learning experience by providing an opportunity to apply the material presented.

Summary

The most important fire-ground activity is saving lives. To accomplish this, the IC must evaluate a number of factors on arrival to ensure that rescues and evacuation are performed in the safest, most effective method possible. Sound risk management principles must be applied throughout the incident as conditions change to ensure that fire fighter safety is addressed while every reasonable effort is made to rescue those who are in danger.

The following is a sample of the information that must be considered and the numerous decisions the IC must make:

- Consider department SOPs and preincident plan information.
- Consider size-up factors.
- Determine the number and location of victims.
- Determine the number of personnel that are needed to perform rescues.
- Determine the resources needed to deliver the required rate of flow to extinguish the fire.

Much of the information needed to develop an incident action plan should be included in the department's SOPs and preincident plan for the property. Other decisions must be made at the scene, depending on information obtained through the IC's size-up. However, all of this information is interrelated and will require rapid decision making to ensure the safety of both victims and fire fighters.

Wrap UP

CHAPTER SUMMARY

- Life safety is the first incident priority.
- Suppressing or extinguishing the fire is often the best way to protect lives.
- The fire floor and floor(s) above the fire are critical areas, and searching them must be a high priority.
- Ventilation must be coordinated with the placement of attack hose lines to control the fire.
- Improper venting could produce a backdraft; vents opened between fire fighters or victims and their path to egress could be fatal.
- Ventilation strategy must consider the fact that the fire and heated gases will naturally move first upward, then laterally, and finally downward.
- Whenever possible, occupants should be rescued by the shortest, easiest, and most direct route using the interior stairways for above- or below-grade rescues.
- The Vent Enter Isolate Search (VEIS) technique is sometimes used when victims are known to be inside the building, but not visible from the exterior.
- A defend-in-place strategy may be a better option when occupants are located far from the fire zone or are not able to evacuate rapidly.
- Interior stairs are the best means of egress from above grade, but fire escapes, aerial ladders, elevated platforms, ground ladders, and other less-desirable alternatives may be used as circumstances require.
- Ground ladders are less desirable than aerial ladders and elevated platforms but can provide quicker access in some situations.
- Elevators should not be used for rescue except when they are verified as safe to use or when they are remote and separated from the fire area with an auxiliary power supply.
- Rope rescues are among the least desirable rescue options, as this tactic is often slow and dangerous.
- When considering helicopter rescues, remember that they are often unnecessary; landing zones may be obstructed and air currents above a burning structure may be hazardous to helicopter operations.
- The *only* reliable way to verify evacuation status is for fire fighters to enter the structure and systematically check every room.
- Before flashover, rescue is possible; after flashover, rescue is highly improbable within the flashover compartment.
- Fire fighters working in multistory buildings should survey an uninvolved floor before ascending to the fire floor or floors above, as some building features will be constant from floor to floor.
- Establish search and rescue priority by deciding who is in the greatest danger from the fire.
- The key to successful search operations is to be systematic.
- A system to expedite triage, treatment, and transportation is essential in mass-casualty situations.
- During weather extremes, shelter is required for rehabilitating fire fighters and evacuated occupants.
- Agencies such as the Red Cross generally assist in setting up and managing shelters, but fire fighters sometimes must begin the process of providing shelter until they arrive.
- The staff needed for life safety depends on the number of victims, available rescue options, condition of victims, and fire and smoke conditions.

KEY TERMS

critical time The time available until the structure becomes untenable.

defend-in-place A tactic utilized during a structure fire when it is very difficult to remove occupants from the building. Occupants are either protected at their present location or moved to a safe location within the building.

SUGGESTED ACTIVITIES

1. It is 8:00 AM on Sunday, October 15. The weather is overcast, and winds are light and variable.

   Assume the role of the officer of the first-arriving engine company (Engine 1) responding from a single company station. Station 2, located 4 miles from the incident is second due. Station 2 houses District Chief 2, Engine 2, Truck 2, and EMS 2. Also responding are Engine 3 (6 miles), Engine 4 (8 miles), and Truck 4 (8 miles). All companies are staffed by an officer, driver, and two fire fighters. The district chief does not have an aide and the EMS unit is staffed with two paramedics.

   On arrival, a heavy volume of fire can be seen blowing out the side door on side Bravo and smoke is pushing out the front door (side Alpha) as well as on side Bravo **FIGURE 6-10**. A fire hydrant is located immediately in front of the house. The Engine 1 driver is able to connect to the hydrant and establish a water supply at the front of the house (side Alpha) without assistance. The two fire fighters advance a 1¾-in. (44-mm) hose line to side Alpha. You enter the center door on side Alpha with the two fire fighters and quickly extinguish the fire that is blowing out the door on side Bravo.

   The objective of this activity is to apply and evaluate rescue priorities and tactics.

   A. Was attacking the fire the best option?
   B. Explain why or why not.
   C. List alternative actions available upon arrival.
   D. What is the probability that occupants remain in the building?

   Assume the role of IC as District Chief 2. Conditions on your arrival are as seen in Figure 6-10.

   E. Would you call for assistance beyond the initial alarm?
   F. Implement an incident action plan, and assign fire companies to carry out your plan.
   G. Develop an Incident Command System chart, including all units at the scene.
   H. Where would you locate the rapid intervention team (RIT) unit?
   I. Where would you locate the EMS unit?

2. A tenant using a charcoal grill on the first-floor patio applied additional charcoal lighter fluid to hot coals resulting in a large fireball that ignited the balcony above and spread to the third-floor balcony. Fire has entered the first-floor apartment where the patio fire originated **FIGURE 6-11**. It is July Fourth at 3:00 PM. It is hot and humid with calm winds.

   A. Where would you place the first hose line?
   B. Prioritize life-safety activities by area (e.g. third priority: third floor above apartment of origin.
   C. Where would you assign the first search and rescue crew?
   D. What tactics would you use to rescue occupants from various areas of the building?
   E. How would you rate the degree of danger for the first, second, and third floor apartments in the immediate fire area, as well as the apartments on either side of these

FIGURE 6-10 Conditions at the scene upon arrival.
Courtesy of David J. Jones, Cincinanti, Ohio.

FIGURE 6-11 Balcony fire that originated on the ground-level patio.
Courtesy of Bill Strite, Cincinnati, Ohio.

apartments on each level? Use a 0-to-10 scale, where 0 means the fire poses no danger and 10 indicates immediate life-safety tactics are required.

3. It is 2:00 AM. An alarm is transmitted for 1000 Rialto Blvd FIGURE 6-12. Weather conditions are fair, calm, and cold.

Upon arrival at the three-story building with a penthouse, 10 occupants have escaped and are located on the opposite side of the street; they appear to be suffering from smoke inhalation. These occupants report that there are still people in the building. Although there is no preincident plan for this building, your knowledge of the building gained during previous responses indicates that the building is primarily occupied by elderly people, many of whom are only partially mobile. TABLE 6-1 lists responding fire department fire companies, fire officers, and emergency medical resources.

All engine, truck, and heavy rescue companies are staffed with an officer, driver, and two fire fighters. Medic units are staffed with two paramedics.

FIGURE 6-12 Fire in a three-story building with penthouse.
Courtesy of David J. Jones, Cincinanti, Ohio.

TABLE 6-1 Responding Fire Department Fire Companies, Fire Officers, and Emergency Medical Resources

Alarm	Engine Companies	Truck Companies All Equipped with 100-ft (30.5 m) Aerial Ladder	Other Units
1	Engine 1 Engine 2 Engine 3	Truck 1	District Chief 1 Paramedic 1
2	Engine 4 Engine 5 Engine 6	Truck 2	District Chief 2 Operations assistant chief Safety officer Paramedic 2 Heavy Rescue 1
3	Engine 7 Engine 8 Engine 9	Truck 3	Fire chief District Chief 3 Paramedic 2 Heavy Rescue 1 Paramedic 2

© Jones & Bartlett Learning.

As the officer of the first-arriving engine company (Engine 1), you and a fire fighter advance a 1¾-in. (44-mm) hose line into the first floor via the front door. The pump operator and hydrant fire fighter remain outside as the two-out initial rapid intervention team (IRIC) team, and they secure a water supply.

Smoke conditions are moderate in the first-floor hallway, which runs the entire length of the building with apartments off both sides. As you advance the 1¾-in. (44-mm) hose line down the first-floor hallway toward the interior stairs, you encounter several elderly occupants who are self evacuating. You use the thermal imaging camera (TIC) to check the doors leading to apartments as you advance down the first-floor hallway. The doors leading to apartments are just above ambient temperature.

As Engine 1 Officer:

A. Provide a status report to dispatch.

B. List your available options prior to the arrival of District Chief 1 and other first-alarm companies.

C. Choose and implement an action plan.

As officer of Truck 1, arriving simultaneously with Engine 1:

D. Consider Engine 1's actions, and then explain your options as the first-arriving truck.

As District Chief 1:

E. Assign (reassign if necessary) all first-alarm companies.

F. Would you call for additional alarms? If yes, how many alarms?

G. Develop an incident action plan.

H. Develop an Incident Command System chart.

4. As chief of department you are dispatched on the second alarm to a structure fire **FIGURE 6-13**. Upon arrival a woman meets you at the front of the building reporting that her husband is still inside the house on the second floor.

Engines 1 and 2 have a water supply and have deployed fire lines to the rear of the building and made entry to the basement at the rear of the building (Figure 6-13). Engine 3 has entered the first-floor side Delta door with a fire line. Truck companies 1 and 2 have taken positions

FIGURE 6-13 Wood-frame house with two stories in front and three stories in rear. Smoke showing on side Delta (visible side) with flame showing at rear of building.
Courtesy of Bill Strite, Cincinnati, Ohio.

at the front of the house. Truck 1's aerial ladder is at the Alpha-Delta corner of the building extended to the roof. Truck 2 is setting up their aerial on side Alpha. Engines 4, 5, and 6 as well as Truck 3 are responding on the second alarm.

A. Provide a status report and assume command.

B. Develop an incident action plan.

C. Assign and/or reassign on-scene and responding companies.

D. Develop an Incident Command System chart.

5. Assume the role of a truck company officer. You arrive on the scene of the fire shown in **FIGURE 6-14**. A woman meets you at the curb reporting that her husband is still inside the house in the second floor bedroom at the rear of the house.

The truck is a straight stick aerial and is not equipped with hose or pumps. The engine company is out of service on an EMS response. The next closest engine company is expected to arrive in 5 minutes.

A. Provide an initial report and assume command.

B. Is this a Vent Entry Isolate Search (VEIS) situation?

FIGURE 6-14 A two-story building with attic, with fire visible on second floor at front of building.
Courtesy of David J. Jones, Cincinnati, Ohio.

FIGURE 6-15 Male resident (victim) hanging out third floor window with fire fighters at ground level. Aerial ladder raised to near victim.
Courtesy of Bill Strite, Cincinnati, Ohio.

C. Explain how you would rescue the husband or what other actions you would take prior to the arrival of the first engine company.

6. It is 9:00 AM on Sunday, January 21. Weather is cold with a slight wind. A call is received from a resident of the 24-unit Salvane Apartment complex reporting a smell of smoke in the building. The first-due engine and truck company arrive simultaneously. Smoke can be seen at the roofline and at the first-floor windows on side Charlie **FIGURE 6-15**. The engine company is advancing a hose line into the first floor on side Alpha.

The one-alarm response to this apartment complex is three engine companies, two truck companies, and a district chief. You are the officer of the first-arriving truck company and you order your aerial truck operator to raise the aerial to rescue the third floor occupant at the side Charlie window.

The trapped occupant is screaming, "*I have to get out of here right now.*"

You reassure him that help is on the way and point to the aerial ladder. You ask him if there is anyone else in his apartment. He replies that his wife is still inside.

As truck company officer you order one fire fighter to assist in rescuing the third floor occupant and decide to begin a search of the building with the remaining fire fighter.

A. Where would you begin your search and rescue operation, and how would you get to that location?

B. Describe your continued search and rescue pattern.

The district chief arrives, assumes the role of IC and requests a status report.

C. Provide a status report including a request for assistance in completing the search and rescue.

D. How much assistance would you request in conducting the search and rescue operation?

Now assume the role of the arriving Engine Company Officer:

E. Where would you begin your attack on the fire, and why?

F. As IC would you immediately request additional fire and EMS units? If yes, what would you request in addition to the companies already on scene and responding?

G. Describe your deployment, RIT, staging, EMS, and other resources.

REFERENCE

Kerber, Steve. 2010. *Impact of Ventilation on Fire Behavior in Legacy and Contemporary Residential Construction*. Northbrook, IL: Underwriters Laboratories Firefighter Safety Research Institute.

CHAPTER 7

Fire Protection Systems

LEARNING OBJECTIVES

- List information related to fire protection systems that should be included in a preincident plan for a protected building.
- List the two leading causes for sprinkler systems failing to operate.
- Compare residential sprinkler systems to commercial sprinkler systems.
- Explain why it is important to "let the system do its job" when conducting operations in a building protected by an automatic fire suppression system.
- List the various kinds of sprinkler systems and where these different types of systems are most likely to be installed.
- Recognize the differences between wet pipe, dry pipe, and deluge sprinkler systems.
- Compare the reliability of wet pipe sprinkler systems to dry pipe and deluge sprinkler systems.
- Explain the differences in loss when a fire occurs in a nonsprinklered system compared to a building with a wet pipe sprinkler system.
- Describe the differences in operating and method of extinguishment between a wet pipe sprinkler system and a fixed water spray system.
- List and describe the operation of different types of sprinkler systems as well as the advantages and disadvantages of each type of sprinkler system.
- Compare and contrast operations at a sprinkler-protected building with and without obvious signs of a fire or system operation.

- Describe fire department operations at a building protected by a deluge system.
- Identify, classify, and describe different types of standpipe systems.
- Describe fire department operations at a building equipped with a standpipe system.
- Explain discharge pressure differences in standpipe systems and how these differences affect operations.
- Describe the advantages and disadvantages of solid-bore and automatic nozzles when operating from a standpipe.
- Compute the pump discharge pressure needed to supply a fire line in a high-rise building equipped with a standpipe.
- Develop a list of standard standpipe equipment.
- List and describe fire protection systems other than sprinkler or standpipe systems.
- Explain fire department operations at facilities protected by a foam, carbon dioxide, dry chemical, and clean agent systems.
- Describe the hazards to occupants and fire fighters in enclosed areas where a carbon dioxide system has discharged.
- Define the term *interlock* and provide an example of an interlock on a carbon dioxide system.
- Explain why dry chemical kitchen hood systems are used in restaurant cooking areas.
- As it relates to company responses, explain the possible problems with habitual false alarm system activations.
- Develop a preincident plan for buildings protected by sprinkler, standpipe, foam, or nonwater-based systems.
- Evaluate operations at a fire in a building protected by a fire protection system.
- Examine the fire hazard potential in a nonsprinkler-protected building, then develop a position paper describing the increased safety obtained if a sprinkler system were installed.

Audrey McAvoy/AP Images.

Case Study

On July 14, 2017, three people lost their lives and at least 12 more were injured at the nonsprinklered Marco Polo apartments in Honolulu, Hawaii. Four of the injured including a fire fighter were transported to the hospital. Hundreds of people evacuated the building as smoke spread to other locations within the building.

The building was divided into seven sections. Sprinkler systems were not required in 1971 when the building was built. There are roughly 300 high-rise buildings on Oahu that were built before 1975 and are not required to have sprinkler systems, according to a survey conducted by the city Fire Department.

1. Write a position paper defending the passage of a code requirement that all high-rise buildings be protected by sprinkler systems.

2. List other code requirements that would better protect occupants and fire fighters in a high-rise building fire.

3. Assume the role of the first-arriving engine company officer with a four-person crew. The truck company from your station should arrive within minutes. Other units will start arriving in 5 minutes or more. Describe your actions and provide instructions to the truck company from your station. Assume the role of the highest-ranking officer on the scene (incident commander).

4. Develop an incident action plan. List the number of fire companies (assume each fire company is staffed with an officer, driver, and two fire fighters). How many emergency medical units, chief officers, and other emergency responders would be needed to control the fire, perform search and rescue, and assist residents in reaching safety?

5. Develop an Incident Command System (ICS) chart accounting for all on-scene emergency responders.

6. How would the operation and outcome differ had the building been fully protected by a code compliant, fully operational sprinkler system?

Introduction

When properly designed, maintained, and operating fire protection systems are present in a building, the offensive attack takes on an entirely different character. By properly using an installed fire protection system, the incident commander's (IC's) job can be made much easier, and the risk to fire fighters and civilians is significantly reduced. This chapter covers operations at properties protected by various types of fire protection systems.

Fire protection systems can greatly simplify firefighting efforts. When automatic fire protection systems are present, a considerable part of the offensive attack strategy involves properly supporting or using the system. If the building or area that is on fire is equipped with an automatic fire suppression system, the primary tactic is to support the system and let it do its job.

It is much safer to let the system handle the fire than to expose fire fighters to positions deep inside the building. Efforts should be directed toward maintaining the system in a fully operational status, while laying backup hose lines for final extinguishment. Preincident planning is essential for any building that is protected by an automatic fire suppression system, with the possible exception of a building containing a range hood system as the only fire protection system.

During preincident planning, fire fighters need to familiarize themselves with the general layout of the building. Emergency contact telephone numbers of owners and building managers or professional alarm service representative should be available to responders. The alarm room or alarm panel lockbox is an ideal place to keep this information.

A facility's lockbox is also a reliable and practical way to provide information about the system in addition to the contacts. However, many departments have onboard computer systems that can be used to relay premise history and other information. The people included on the emergency contact lists can be called for assistance in gaining access to all parts of the building, restoring the system to service, and having employees assist in salvage operations. The location and operation of various water supply components should be known in advance, including the following:

- Main control valves
- Divisional control valves
- Fire pump (electric, diesel, other)
- Fire department connections (FDCs)
- Water supply (gravity tank, pump, public water system)
- Hydrant water supply

System limitations and peculiarities should be addressed during preincident planning.

Buildings equipped with fire protection systems, particularly sprinkler systems, are often permitted to reduce fire protection measures in other areas, such as increasing the fire load or reducing the exit capacity. If the system is out of service, these buildings no longer meet minimum code requirements and present a greater danger to occupants and fire fighters. Buildings under construction may delay installing fire protection systems, whereas a building being demolished may have had the fire protection systems removed, disabled, or shut off. The fire department should receive notification when a system is out of service and revise tactics accordingly.

Fallacy	Fact
Manually applying water is the best way to control a fire.	In a building equipped with a properly operating automatic fire suppression system, the best tactic is to preincident plan and let the system do its job.

Sprinkler Systems

Residential sprinkler system installations, as described in NFPA 13R, *Standard for the Installation of Sprinkler Systems in Low-Rise Residential Occupancies*, and NFPA 13D, *Standard for the Installation of Sprinkler Systems in One- and Two-Family Dwellings and Manufactured Homes*, are still fairly rare (NFPA 13R, 2019; NFPA 13D, 2019). Home sprinkler systems have a proven track record in saving lives and property. The fire death rate in homes protected by a wet pipe sprinkler system is 83% lower than in nonsprinkler-protected homes. The fire damage in homes with wet pipe sprinkler systems is 69% lower than in nonsprinkler-protected homes (Hall, 2012). The primary purpose of these life-safety systems is to allow additional escape time. The hardware requirements for life-safety systems are different from those of a sprinkler system in a place of assembly, industrial, or office occupancy. For example, fire pumps are not generally included in the NFPA 13R and 13D systems. The tactics described next apply to all buildings that are protected by sprinkler systems; however, some modifications are necessary because of the differences in the sprinkler standards applied to older and home systems.

Commercial-type automatic sprinkler systems meeting the requirements of NFPA 13, *Standard for the Installation of Sprinkler Systems* (NFPA 13, 2019) have an exceptional record in controlling fires. Large losses of life and property are practically nonexistent in buildings equipped with a properly designed, maintained, and operating sprinkler system **FIGURE 7-1**. When large-loss fires do occur in these properties, some degree of human error is generally involved **FIGURE 7-2** and **FIGURE 7-3**.

NFPA 13E, *Recommended Practices for Fire Department Operations in Properties Protected by Sprinkler and Standpipe Systems*, lists some frequent causes of sprinkler system failure that fire fighters should be prepared to resolve, stating the following (NFPA 13E, 2020):

4.1.1 Fire department personnel should be knowledgeable of and prepared to deal with the following three principal causes of unsatisfactory sprinkler performance:

(1) A closed valve in the water supply line

(2) The delivery of an inadequate water supply to the sprinkler system

(3) Occupancy changes that render the installed system unsuitable

Although rare, large losses do occur in sprinkler-protected buildings. There are times when an explosion or extremely fast-moving fire overwhelms a

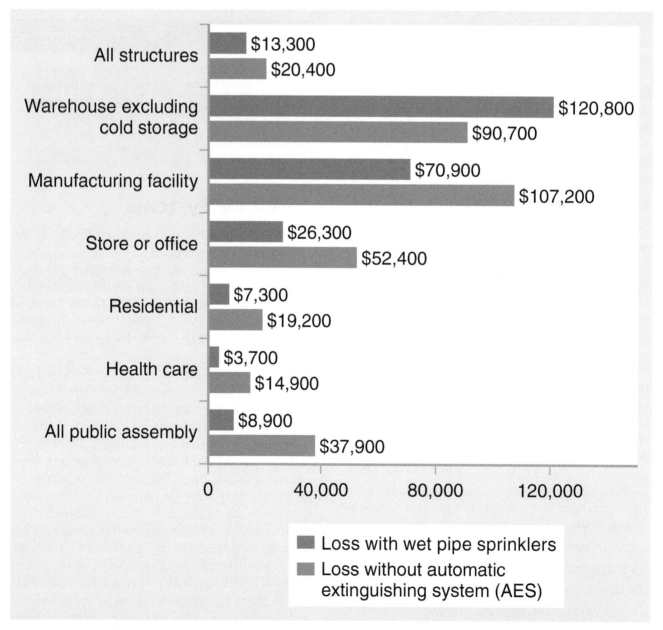

FIGURE 7-1 Amount of damage per fire with wet pipe sprinkler system versus without automatic extinguishing equipment, 2010–2014.

Reproduced from: Ahrens, Marty, July 2017, U.S. *Experience With Sprinklers: NFPA Research*, Quincy, MA: National Fire Protection Association; Table 4, p. 19.

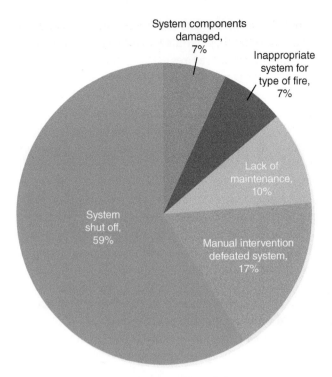

FIGURE 7-2 Reasons that sprinklers fail to operate, 2010–2014.
Reproduced from: Ahrens, Marty, July 2017, U.S. *Experience With Sprinklers: NFPA Research*. Quincy, MA: National Fire Protection Association; Figure 11, p. 6.

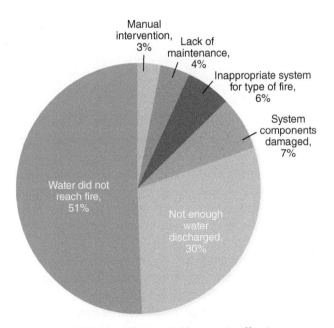

FIGURE 7-3 Reasons that sprinklers are ineffective, 2010–2014.
Reproduced from: Ahrens, Marty, July 2017, U.S. *Experience With Sprinklers: NFPA Research*. Quincy, MA: National Fire Protection Association; Figure 12, p. 6.

sprinkler system. Sometimes improper design, such as a building feature constructed below the sprinkler, may shield the fire from sprinkler water. A recent renovation in a high-rise office building prevented the sprinkler system from extinguishing the fire because a drop ceiling was installed below the sprinkler system. The fire consumed the maintenance office but was extinguished before it could extend into the office space.

Losses also occur when the sprinkler system is shut down or otherwise out of service. When the system or part of the system is out of service, the system should be tagged as specified in NFPA 25, *Standard for the Inspection, Testing, and Maintenance of Water-Based Fire Protection Systems*, which states the following (NFPA 25, 2020):

> **15.3.1** A tag shall be used to indicate that a system, or part thereof, has been removed from service.
>
> **15.3.2** The tag shall be posted at each fire department connection and system control valve, and other locations required by the authority having jurisdiction indicating which system, or part thereof, has been removed from service.

Proficient fire officers are familiar with the general operation of fire protection systems. Department standard operating procedures (SOPs) should address working in structures that are protected by fire protection systems.

The system that is most often encountered is the wet pipe sprinkler system, which is highly effective and the most reliable fire suppression system. The wet pipe sprinkler system is connected to a reliable water supply such as a water main or tank (or both). The water is distributed throughout the protected structure and applied to the fire through sprinkler heads. Individual sprinkler heads are self-contained detection/application devices, which account for their reliability. Valves control water distribution to sprinkler piping, and fire pumps may be needed to provide the necessary water pressure and volume to the system.

A **dry pipe sprinkler system** is used in areas that might be subject to freezing temperatures. The difference between wet and dry pipe sprinkler systems is that the piping in a dry system is filled with air that is under pressure instead of water. When a sprinkler head opens because of a fire, the air bleeds out of the system. This reduction in air pressure causes the main valve to open, flooding the system with water that ultimately discharges from the open sprinkler head. Because of this design, water may take longer to reach the fire than in a wet pipe system. Sometimes a wet pipe system is protected using antifreeze to avoid the delayed operation and installation of additional equipment.

A **preaction sprinkler system** is also filled with air that may or may not be under pressure. The one difference is that a sensing device, such as a smoke or heat detector, opens a valve, flooding the piping with water. If a sprinkler head has also fused, then water will come out of the sprinkler onto the fire. In some systems the air in the sprinkler piping will be under pressure. If there is a loss of air pressure, such as when a sprinkler fuses, the valve will open, flooding the system with water.

In both the dry pipe and preaction systems, if the capacity of the piping exceeds 500 gallons (1893 L), the system must be equipped with an accelerator or exhauster. These devices rapidly remove the air from the system, reducing the time necessary for the water to fill the piping.

Preaction systems can be used when a warning is desired before actual water discharge, such as in computer rooms. In some installations, both the sprinkler and the preaction sensing device must actuate for the system to operate. In other words, the system will not discharge water when a sprinkler is damaged or fused unless the secondary sensing device controlling the valve also detects a fire. This safeguard is used where incidents of vandalism require a backup alarm, such as in parking garages.

Another type of sprinkler system is the **deluge system**. With this system there is no water in the sprinkler piping, and all sprinklers (or applicators) are open. When the system is activated, a detector-operated valve (which is normally closed) opens, releasing water that fills the piping and discharges through the open sprinklers (or applicators) into the protected area. Deluge systems protect areas with high-challenge fire potential, such as flammable liquids, conveyors moving combustible commodities, and transformers. They may also be installed in aircraft hangars as combination water/foam systems. Deluge systems can also be used in many special applications, such as fixed water spray systems **FIGURE 7-4**. Such applications are used to protect aboveground liquefied petroleum gas tanks from fires, dilute flammable liquids, disperse flammable gases, and protect tanks from exposure fires.

Deluge systems are also equipped with manually operated override valves that can be used if the detection devices fail to operate. Many times, the manual override is located at the deluge valve, but it could be located elsewhere. Fire fighters should know where these control valves are located and how to manually activate the system. An exterior sprinkler system, designed as a water curtain to protect a building from exposure fires, may also be available. Some of these systems require manual actuation, and it may also be necessary to supply or augment the water supply to the exposure protection system. This is yet another reason for preincident planning. Many deluge and exterior sprinkler systems have operating instructions printed near the manual control valve.

The most common operational error at properties protected by automatic sprinkler systems is shutting down the system prematurely. Fire fighters working at the scene must be sure the fire is totally under control before shutting down the system. When the system is shut down, a fire fighter should be assigned to stand by the valve in case it is necessary to quickly reopen the valve. This fire fighter should be equipped with a radio or other reliable means of communication.

In many cases the sprinkler system will control the fire but may not completely extinguish fires that are shielded from direct water contact **FIGURE 7-5**. In a warehouse, commodities may be stored in shelves that block the spray pattern and do not allow the sprinkler system to apply water directly on the burning stock. Handlines must be in place, ready to extinguish any remaining fires, and the fire fighter who is assigned to the sprinkler valve must stand by, ready to reactivate the system should it become necessary.

When managing fire department operations at a property protected by an automatic sprinkler system, the IC should keep in mind the potential to deprive the sprinkler system of water. A properly designed, installed, and maintained sprinkler system should have a calculated water requirement that includes enough water to support hose streams. However, the IC must be extremely careful to avoid depleting the sprinkler system of the water pressure and volume necessary to properly support operating sprinklers. The best practice is to connect attack hose lines to a water supply that is separate from the one supplying the sprinkler system. If a decision must be made between using hose lines and properly supplying the sprinkler system, it is usually best to supply the sprinkler system.

FIGURE 7-4 Fixed water spray system.
Courtesy of Veltre Engineering.

augment the sprinkler system or supply apparatus on the scene. The local water department may also be able to bring additional water pumps or water systems online to provide additional volume and/or pressure to the area.

In rare cases, usually where sprinkler piping is damaged, it may be necessary to shut down the system so that enough water is available for manual suppression. If a sprinkler system is having difficulty controlling the fire, it is much better to have hose lines at strategic locations and secure additional offsite water supplies than to shut down the system if the sprinkler piping is still in place.

Following are general guidelines that could be incorporated into department SOPs for operations at protected properties.

Working at a Sprinkler-Protected Building with No Signs of Fire

Listed here are tasks that should be performed at buildings that are equipped with a sprinkler system when no signs of fire or system operation are evident from the outside. It is assumed that there was some report of an unusual condition or a fire alarm system was activated that caused the fire department to be notified.

Gaining Entry

If a key to the building is contained in a lockbox, it may not be necessary to force entry. Likewise, if someone who can provide access is responding and will arrive within a few minutes of the fire department's arrival, it may be advisable to wait for assistance, provided there is no indication that a fire is in progress. Conversely, if there will be a considerable delay in entering the building, it may not be appropriate to wait for assistance. The potential fire or water damage generally outweighs the damage done by forcible entry. When it is necessary to force entry, crews should do so with consideration to property damage. Many times, ladders can be used to gain entry through upper story windows. Guidance concerning forcible entry should be outlined in SOPs. Preincident plans should identify the location of lockboxes, identify if keys are available, and include information concerning potential entry locations.

FIGURE 7-5 Burning materials shielded by obstructions.
© Jones & Bartlett Learning.

If it becomes obvious that the sprinkler system is not achieving the desired result because of damaged piping or related problems, then the IC may elect to redirect water from the sprinkler system to support fire department hose lines.

Efforts should be made to obtain additional water supplies from sources that will not affect the sprinkler system operation. This can be done by identifying nearby water mains that are not part of the system supplying the building. Water can then be pumped from remote water mains to the incident to either

Checking the Main Control Valve

A fire fighter equipped with a radio should be sent to the sprinkler system riser (main control valve). Depending on available staffing, the fire fighter may be assigned to remain at the valve throughout the entire operation. This fire fighter determines whether the system is flowing and checks the valve to ensure that it is in the fully open position. One method of checking for water flow, if the riser is not equipped with a device such as a water motor gong, is to place an ear against the riser and listen for water flowing through the pipe. If this area has the potential to become a hazard area, then the buddy system would apply, with two fire fighters assigned to the control valve.

Several types of main control valves are used to open and close sprinkler systems. The two most commonly used are the outside stem and yolk (OS&Y) valve **FIGURE 7-6** and the post indicator valve (PIV) **FIGURE 7-7**. In preparing preincident plans, the IC should note the sprinkler valve type, location, and operation. In most situations, the sprinkler valve will be locked in the open position. It may be necessary to cut or break the lock to shut down the system.

Checking the Fire Pumps

When the sprinkler system water supply is augmented by a fire pump(s), a fire fighter (or two fire fighters if conditions warrant) equipped with a radio should be assigned to check the fire pumps. If the main pump is operating, there is a good chance that the system is discharging water, either by accident or onto a fire. If a fire is detected but the pumps are not operating, the fire fighter can manually start the pumps at the direction of the IC. It is a poor practice to rely on remote annunciator panels to determine whether the fire pumps are running. An actual physical check of the pumps is needed.

The fire fighter who is assigned to the main control valve may also be in a position to monitor the fire pump. In many cases the fire pumps are located near the main control valve. This is an important position that should not be relegated to a part-time task, requiring the fire fighter to walk a considerable distance between the riser and pumps. A general good rule is to separate the tasks if the pumps cannot be seen or heard by someone standing at the main control valve.

FIGURE 7-6 Outside stem and yolk (OS&Y) valve.
© Jones & Bartlett Learning.

FIGURE 7-7 Post indicator valve (PIV).
Courtesy of Ben Klaene.

Checking the Building for Fire and/or Sprinkler Operation

Fire fighters should be assigned to conduct a systematic check of the entire building. If there was a reason to notify the fire department because of an unusual condition, then it must be verified that nothing is amiss. There is a tendency to approach a sprinkler alarm with no visible sign of fire as a nonemergency situation, making entry without full protective clothing and the equipment needed to handle a fire situation. This is an unsafe procedure that could result in serious injury if a working fire is encountered while checking the building.

Supplying the Fire Department Connection

A pumper connected to an adequate offsite water supply should connect two 2½-in. (64-mm) or two 3-in. (76-mm) hose lines (large-diameter hose if the system is so equipped) to the FDC. The water supply for this pumper should be large-diameter hose, two 2½-in. (64-mm) or two 3-in. (76-mm) hose lines, or a direct hydrant connection. A single 2½-in. (64-mm) or 3-in. (76-mm) supply line is inadequate.

During preincident planning, determine the location and flow of offsite water sources. Department SOPs should assign an engine company the task of supplying water to the FDC.

Large buildings or complexes could have several FDCs supplying different sprinkler or standpipe systems. FDCs are often interconnected; supplying any of the FDCs provides supplemental water to the entire building or complex **FIGURE 7-8**. However, other buildings or complexes are equipped with separate FDCs supplying a standpipe and/or sprinkler system protecting a single area or building within a complex. Preincident plans must address this issue. If the intakes are not interconnected, the area covered by each intake should be specified on the preincident plan and a marking system used to identify the area protected by each FDC.

The fire department should have access to plans for new structures within their jurisdiction and be provided an opportunity to have input regarding the best location for FDCs, fire hydrants, and other fire protection equipment. Response personnel reviewing these plans should make recommendations that coincide with current SOPs and facilitate effective operations. For example, most codes require the FDC to be placed at the front of the building, but do not necessarily address how close to the front of the building. Many department SOPs direct the first-arriving engine company driver to secure a water supply and be prepared to pump into the FDC while the officer and remaining crew members make entry. The fire department intake shown in **FIGURE 7-9** is on the main road at the front of the building but well

FIGURE 7-8 Fire department connection with signage showing area covered by system.
Courtesy of Ben Klaene.

FIGURE 7-9 Remote placement of a fire department connection.
© Jones & Bartlett Learning.

away from the entrance to the protected building. Following an SOP that places the pumper near the FDC would result in the apparatus and driver being located a considerable distance from the building. The driver could not be an effective member of a two-out team, and the apparatus, hose, and tools would be located far from the building. This situation would require a change in the SOP, or special provisions would need to be made in the preincident plan for this specific property.

Working at a Sprinkler-Protected Building with Evidence of Fire Showing from the Exterior

If there are signs of a fire and/or the sprinkler system is operating, the main objective is to support the system while ensuring that occupants are safe.

Gaining Entry

Again, gain entry to the building using the minimum force necessary. However, with a fire in progress or a system operating, the time spent gaining entry will increase the risk to occupants and will cause additional fire, water, and smoke damage; therefore, a more forceful entry is justified.

Checking the Main Control Valve and Fire Pump

Fire fighters should be assigned to the main valve and pump even when there is no visible sign of fire. However, when there is a confirmed fire, their primary responsibilities are to ensure continued operation of the system and to provide rapid shutdown when appropriate. These positions are critical and should be staffed throughout the operation.

Large systems will be equipped with control valves on portions of the system. These valves are not as prone to accidental closing as main control valves, but fire crews should check them as soon as possible.

Supplying the Fire Department Connection

When there is a confirmed working fire, an engine company must supply water by pumping into the FDCs, and the pump operator is required to remain at the apparatus. The department SOPs should identify minimum water supplies and pump pressures. Most departments provide a pressure of 150 pounds per square inch (psi) (1034 kilopascals [kPa]) to the FDCs. It is important to avoid surges when initially charging the hoses connected to the FDC, as many sprinkler systems are designed and tested to a maximum 175 psi (1207 kPa). As was previously noted, the preincident plan modifies SOPs whenever necessary according to the characteristics of the specific property. If a pressure higher than specified in the SOPs is required to adequately supply the sprinkler system or the standard pressure is likely to cause damage, this should be noted on the preincident plan.

When the building is supplied by a public water system, a fire hydrant will normally be located close by. The pumper assigned to supply the FDC has several options:

- Use soft suction to connect to the fire hydrant and hand lay hose to the FDCs (usually two 50-ft (15-m) sections from pumper to each side of a Siamese FDC) or a single section of large-diameter hose to a large-diameter hose connection.
- Forward lay large-diameter hose and connect to the FDC.
- Place one pumper at the hydrant and another close to the FDC.

Each of these layouts has advantages and disadvantages. Placing a pumper near the building may result in the driver and apparatus being within the collapse or falling glass zone, but doing so does slightly reduce the friction loss. Using two pumpers requires that two pump operators remain at their apparatus, reducing initial staffing. However, placing a second pumper at the hydrant provides redundancy should one apparatus experience mechanical problems.

The typical older FDC will be two female 2½-in. (64-mm) intakes **FIGURE 7-10**. Two 2½-in. (64-mm) or preferably two 3-in. (76-mm) hose lines should be connected to the intakes.

FIGURE 7-10 Supplying a fire department connection.
© Jones & Bartlett Learning.

FIGURE 7-11 Large-diameter hose intake to fire department connection.
Courtesy of Ben Klaene.

Large-diameter hose 4-in. (101-mm) or 5-in. (127-mm) connections are now being substituted for 2½-in. (64 mm) intakes **FIGURE 7-11**. Large-diameter hose provides large quantities of water and greatly reduces friction loss. However, there are several potential problems to address when using large-diameter hose to supply a sprinkler or standpipe system:

- The weight of the hose and water bearing on the intake is significant.
- Kinks are more likely.
- The unisex connection can accidentally disconnect (requires safety latch).
- There is no redundancy. (If a single section of large-diameter hose ruptures, the total supply to the FDC is lost.)
- Standard large-diameter hose is generally tested at a lower pressure than 2½-in. (64-mm) or 3-in. (76-mm) hose. However, large-diameter hose with higher pressure ratings is available.

Large-diameter hose is typically tested to 200 psi (1379 kPa). This rating is adequate for most sprinkler operations, but higher pressures may be required for standpipe operations, as discussed later in this chapter.

Department training should include overcoming potential difficulties in connecting to the FDC, such as:

- Damaged threads
- Threads different from those used by the jurisdiction
- Debris and other obstructions (missing intake caps are a clue)
- Damage to intakes or female connections

On occasion, male threads could be encountered at what is believed to be an FDC **FIGURE 7-12**. Fire and building codes require female connections on

FIGURE 7-12 Male threads, intake, test header, or wall hydrant?
Courtesy of Ben Klaene.

standpipes. Male threads are commonly used to test the system (test header) or as a wall hydrant. Preincident plans should be developed for all buildings protected by a fire protection system. Fire companies should routinely familiarize themselves with FDCs throughout their response area and check intakes for damage, operability, or unusual features.

Letting the System Do Its Job

It is better to shut down a sprinkler system too late than too early. The sprinkler system should be permitted to operate until the IC is certain that the fire is under control. When the sprinkler system is shut down, the only fire remaining should be small spot fires. Hose lines should be in place to complete extinguishment.

Backing Up the System

Prepare for an offensive attack and overhaul by strategically positioning hose lines. The hose lines should be staffed by fire fighters in full protective clothing, including self-contained breathing apparatus (SCBA). Hose lines should only be operated to perform rescue operations and limit fire spread or to complete overhaul operations after the sprinkler system has been shut down. However, in the rare case that the sprinkler system is ineffective due to damaged piping, malfunction, or inadequacy, hose lines may take priority.

Ventilating

When it is safe to do so, crews should make ventilation openings above the fire. Proper ventilation will channel the fire and limit its extension. The cooling effect of sprinkler water can inhibit upward smoke movement, possibly making it more difficult to ventilate. Recommended ventilation tactics should be part of the preincident plan.

Some large-area, sprinkler-protected buildings are equipped with automatic or manual roof vents and draft curtains that are designed to limit fire spread. Overhaul will most certainly require ventilation, usually positive- or negative-pressure mechanical ventilation. Fire fighters should not shut down the system to locate the fire; instead they should ventilate.

Performing Property Conservation Tasks

When possible, property conservation tasks should be accomplished while extinguishment is in progress. Extinguishment takes priority, but water damage often dictates that the IC summon enough assistance to start property conservation simultaneously.

Placing the System Back in Service

If fire crews are properly trained to do so, they should place the sprinkler system back in service by replacing fused sprinklers and reopening valves. Most codes require that a supply of spare sprinkler heads, of the type used in the system, be kept on the premises for this purpose. When replacing sprinkler heads, it is critical that the proper sprinkler head be used. In some buildings, the same type of sprinkler head will be used throughout the building. In other buildings, pendant, upright, sidewall, and other types of sprinkler heads may be used. Sprinkler heads designed to operate at different temperatures may also be in use. The sprinkler head being replaced must be of the same type and temperature rating as the existing sprinkler head. If the system is too complex or sprinkler heads are not available, the owner or manager should restore the system or have a licensed sprinkler contractor restore the system. If the system is equipped with division control valves, restoring the unaffected part of the system may be possible even if sprinklers are not replaced in the fire area. Some department SOPs prohibit reactivating the system, owing to possible legal implications; others believe that it is important to place the system back in service as soon as possible.

The property owner should contact a licensed sprinkler contractor to inspect the system after it has operated during a fire. There is a possibility that sprinklers or piping could be damaged. Department policy and guidance on this subject should be included in the department's SOPs.

Working at a Property Protected by a Deluge System

The tasks required at a sprinkler-protected building are basically the same for wet, dry, or preaction systems. The deluge system presents at least one additional consideration: manual operation of the deluge valve. Many of these systems are located outside of buildings. Usually, the hazard protected by deluge systems can create extreme risks for fire fighters who are attempting to manually suppress the fire. System operation will be obvious, thus negating the need for the thorough investigation required in sprinkler-protected buildings. Following are general steps to be taken at these properties. However, preincident planning is the key to a successful operation.

> **NOTE**
>
> Deluge systems often protect highly flammable materials or other high hazard materials. Preincident planning is essential as well as identifying when it is unsafe to approach the area being protected by a deluge sprinkler system.

Checking the Control Valve and Fire Pump

Just as with the wet and dry pipe sprinkler systems, it is important to ensure that valves are open and that the pumps are operating properly.

Operating the Deluge Valve

It is possible, though improbable, that a fire would be in progress in an area protected by a deluge system that failed to operate properly. If the system is needed for fire control, the deluge valve should be activated manually *provided it is in a safe location*.

However, it is more likely that an exposure fire would threaten an area protected by a deluge system. It may be possible to cover these protected exposures with the deluge system by operating the deluge valve if operation of the system does not create a safety hazard or cause additional damage (see Figure 7-4).

Consideration must be given to the water supply requirements for these systems when they are being used for exposure protection. A deluge system may well deplete a private water supply system. In the case of a transformer fire, it is best to stay out of secured areas and allow the system to do its job.

Checking Interlocks

Deluge systems often trip interlocking devices when activated. For example, system operation may deenergize electric transformers, shut down conveyor belts, or shut off a fuel supply. In most cases, there is a means of manually activating the interlock. If these interlocks failed to operate and it is possible to safely shut down a fuel supply, conveyor belts, or other processes that contribute to the spread of fire, this should be done. However, transformers should not be shut down manually. Usually, the deluge system will control the fire even if the interlocks do not function. Sometimes, it may be advisable to wait for plant personnel to shut off fuel supplies or perform other related activities.

Letting the System Do Its Job

As with the wet and dry pipe sprinkler systems, it is better to shut down the deluge system too late rather than too early. A determination must be made that the fire is completely under control before shutting down the system. Deluge systems will be flowing large quantities of water, increasing the temptation to shut down prematurely. Remember, even with hose lines in place, it may be impossible to safely and effectively apply the quantity of water necessary to control the fire if the system is shut down too soon. Many facilities that are protected by deluge systems are such that total loss has already occurred in the area of operation; the deluge system is merely preventing extension. For example, a transformer that catches fire, setting off a deluge system, has probably already sustained the maximum fire loss. In this case, the deluge system is usually designed to protect exposures.

Backing Up the System

Hose lines staffed by fully protected fire fighters are required at strategic locations for fires such as those involving conveyors. However, manually operated hose lines create a substantial safety hazard if the deluge system is protecting high-voltage transformers. The fire officer must evaluate this situation carefully, knowing that charged hose lines offer a dangerous temptation. Most of the time, the protected transformer could completely burn out without endangering lives or additional property.

Working at a Building Equipped with a Standpipe System

A standpipe system is not an automatic fire suppression system. Unlike sprinkler systems, standpipe systems cannot operate without human intervention. A properly operating and maintained building standpipe system is helpful in conducting offensive attacks. In high-rise structures, it may well be impossible to conduct a safe and effective interior operation on upper stories if the standpipe system is inoperative.

NFPA 14, *Standard for the Installation of Standpipe and Hose Systems*, lists six major types of standpipe systems (NFPA 14, 2019):

- *Automatic dry*. In this system, piping is filled with pressurized air or nitrogen but is connected to a water supply that automatically admits water into the system when a discharge is opened.
- *Automatic wet*. In this system, piping is filled with water, and water is provided when the discharge is opened. This is the most common standpipe system.
- *Semiautomatic dry*. In this system, a dry standpipe admits water into the system piping upon activation of a remote-control device located at a hose connection.
- *Manual dry*. This system is filled with air and does not have a water supply; the system relies entirely on water provided via the FDC to supply hose lines.
- *Manual wet*. This system is filled with water connected to a water supply that maintains water in the system, but it is not capable of providing water for firefighting purposes unless it is supplied by a fire department pumper.
- *Combined system*. This system supplies both sprinklers and standpipes.
- *Wet standpipe system*. A standpipe system having piping containing water at all times.

NFPA 14 also defines three types of standpipes (Class I, Class II, and Class III) as follows:

- The Class I standpipe system provides a 2½-in. (64-mm) hose connection to supply water for fire department use.
- The Class II standpipe system provides 1½-in. (38-mm) hose connections and hose stations to supply water for use primarily by trained personnel or by the fire department during initial response. The hose station may be provided with hose as small as 1-in. (25-mm) diameter.

- The Class III standpipe system provides 1½-in. (38-mm) hose stations to supply water for use by trained personnel and 2½-in. (64-mm) hose connections to supply a larger volume of water for use by the fire department.

The standpipe can be independent from or connected to the sprinkler system (combined system), which may have different flow requirements than the three listed classes of standpipes. As mentioned previously, it is important to evaluate the effect of diverting water from the sprinkler system to manually operated hose streams when connected to a combined system. Expect high-rise buildings to be equipped with standpipes, but other buildings may also have standpipes. Most new "big box" mercantile properties will have a drop from the sprinkler system with a hose connection.

Code requirements change. In most cases the standpipe system will meet the requirements of the code in effect when the building was originally built. One major change that occurred was the minimum pressure requirement at the standpipe outlet. Older codes generally specify 65 psi (448 kPa) at the standpipe outlet, whereas modern codes usually require 100 psi (689 kPa). Pressure and flow regulators have changed in design. Given the different types and classes of systems, as well as variations in pressure and design, it is imperative that preincident plan information be available for all buildings equipped with a standpipe system. At a minimum, preincident plan information about the standpipe should include:

- Standpipe location (Many buildings have a standpipe in only one stairway; it is imperative to select the proper stairway for fire attack.)
- Location of FDC and hydrant closest to the FDC
- Special hydraulic considerations (maximum pressure, internal pump operating pressure, pressure zones, unusual pipe configuration)
- Location of fire pump and valves
- Maximum and minimum pressures
- Pressure-limiting valves (location of special tools, how to adjust/defeat)
- Pressure at standpipe outlets
- Maximum flow

When the standpipe system is connected to a fire pump, it is important to determine whether the building's fire pump is providing sufficient pressure and volume to support firefighting operations. If not, it may be necessary for the fire department to either supplement the water supply or provide all of the water needed through the FDC.

Checking Fire Pumps and Main Control Valves

As with a sprinkler system, it is important to check that fire pumps are operating and the main control valve is open. A properly operating system is crucial when a fire occurs in the upper stories of a high-rise building.

Supplying Fire Department Connections

The water supply requirements for standpipe operations are much the same as those for a sprinkler-protected building: two 2½-in. (64-mm) or 3-in. (76-mm) hose lines connected to the FDC (or large-diameter hose when the connection is so equipped), with the pumper being supplied from a hydrant by large-diameter hose or soft suction connection. Some FDCs have more than two intakes. In most cases, this indicates that a flow greater than 500 gallons per minute (GPM) (1892 liters per minute [L/min]) may be required. It is best to supply all intake connections to the system being used.

Some standpipe systems do not have a fixed water supply and rely entirely on fire department pumpers to provide water to the system. Manual wet and dry standpipes require water to be supplied through the FDC. Manual dry standpipe systems are sometimes unserviceable due to valves that have been left open, caps that are missing, pipe damage, or other problems. With a manual dry system, these problems might not be known until the fire department attempts to use the system. An automatic or wet manual standpipe system will flow water if damaged or otherwise compromised, bringing immediate attention to the problem, unless the main control valve has been shut down.

Pressure requirements for a standpipe system include elevation loss, friction loss in the hose, friction loss in the standpipe piping system, and nozzle pressure. Standard fire-ground hydraulic calculations should be adequate, with a slight allowance for system piping. In most operations, an allowance of 10 to 15 psi (69 to 103 kPa) of friction loss is adequate for the standpipe system piping. It is not necessary to determine the exact friction loss in the standpipe system when a fire occurs. However, in preincident planning it should be noted whether additional information regarding friction loss in the system piping is available, especially if it is appreciably more than 15 psi (103 kPa). Communications between the pump operator and the company on the fire floor are critical so that pump pressure can be regulated to meet rate of flow requirements. Some departments carry a pressure regulator with gauge as part of the standpipe equipment. Doing so is highly recommended because

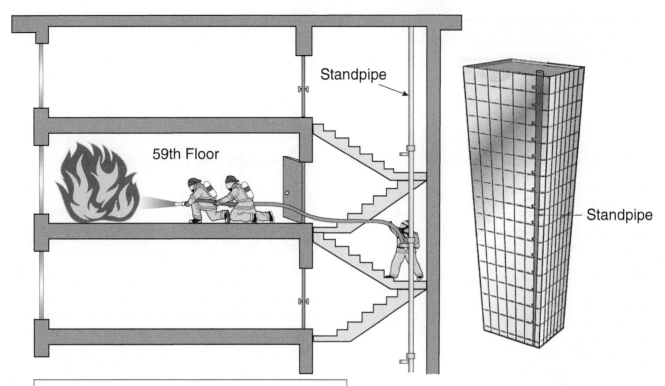

Elevation loss = 255 psi (587.6 kPa)
Nozzle pressure = 50 psi (115.2 kPa)
Friction loss in 150 ft (45.7 m) of 2 1/2 in. (64 mm) hose = 37 psi (85.3 kPa)
Friction loss in standpipe = 15 psi (34.6 kPa)
Total pressure required = 358 psi (824.9 kPa)

FIGURE 7-13 Calculating pressure needed to supply standpipe operations.
© Jones & Bartlett Learning.

it allows the fire attack crew to control the outlet pressure at the point of use. When using 2½-in. (64-mm) smooth-bore nozzles, controlling the nozzle pressure can be critical, as overpressurization can result in an extreme nozzle reaction force.

In large high-rise buildings, there is greater reliance on interior pumps. The laws of hydraulics tell us that there is 0.434 psi of **backpressure** for every 1 ft of elevation (3 kPa of backpressure for every 305 mm of elevation).

For example, in the 55-story high-rise building with a fire on the 50th floor as shown in **FIGURE 7-13**, each floor is 12 ft (3 m) high, yielding a backpressure of 255 psi (1765 kPa) at the base of the building (calculated as follows: 0.434 × 12 × 49; the 50th floor is 49 full stories above ground). In this example, a pump pressure of 358 psi (2468 kPa) is required at the FDC, not including the friction loss in the hose from the pumper to the FDC. This exceeds the maximum test pressure for most hose. The pumper supplying the FDC could augment the water supply, but there is a great reliance on the interior pumping system. A building of this height would probably be designed with multiple pressure zones or pumps in series.

Providing Standpipe Equipment

It is poor practice to use the hose lines that are preconnected to the building's standpipe system. Preconnected hose supplied in a hose cabinet is often unlined, smaller in diameter, and seldom tested or properly maintained. Fire fighters should bring their own hose, nozzles, and adapters into the building.

At a minimum, standpipe equipment should consist of the following:

1. First-arriving engine company
   - Three 50-ft (15-m) lengths of 2½-in. (64-mm) hose
   - One 50-ft (15-m) section of 1¾-in. (44-mm) hose (If staffing is insufficient, the section of 1¾-in. [44-mm] hose can be brought to the fire floor by the second-arriving engine or the truck company.)
   - Smooth-bore, broken-stream, or combination nozzle capable of providing sufficient flow at low pressures
   - Spanner wrench
   - 30-degree standpipe outlet elbow
   - Pressure gauge
   - Set of 2½-in. (64-mm) male and female adapters
   - Standpipe hand wheel

- 18-in. (46-cm) pipe wrench
- Channellock and/or vise-grip pliers
- Set of Allen wrenches
- Screwdriver
- Wire brush
- Door chocks
- 2½-in. (64-mm) gated wye

2. Second-arriving engine company
   - Three 50-ft (15-m) lengths of 2½-in. (64-mm) hose
   - 1¼-in. (32-mm) or 1⅛-in. (29-mm) smooth-bore nozzle (If a combination nozzle is preferred, it must be capable of providing sufficient flow at low pressures and equipped with an adapter that allows it to be converted to a smooth bore.)

3. Truck company
   - Thermal imaging camera
   - Search and rescue equipment, including search rope as well as chalk (preferably sidewalk chalk), tags, or other marking materials to identify areas searched
- Forcible entry, ventilation, and salvage equipment, as required
- Extra air cylinders

Whenever possible these first three companies should work together as a task force or division on the fire floor. Even with the standard four-person staffing, the first-arriving engine company probably will not be able to safely and effectively begin operations on the fire floor of a multistory building. SOPs generally assign the apparatus operator to supply the FDC, leaving a three-person crew to proceed to the fire floor, connect hose at the standpipe outlet on the floor below the fire, and advance a hose line up the stairway through a door to the fire floor. In most cases, it will be necessary to advance hose through one or more doorways or around other obstacles to attack the fire. Maneuvering a hose line through multiple obstacles often requires fire fighters to be stationed at doorways or places where the hose has to make a sharp bend. Placing a fire fighter at the door leading to the hallway and rooms allows the door to be controlled to limit ventilation. With a four-person crew and the apparatus operator outside, even if a two-person crew can reach the fire area, only one person remains to staff the initial rapid intervention crew for fires above grade level or in large buildings.

Many departments use 1¾-in. (44-mm) or 2-in. (51-mm) hose lines as the initial attack hose line from a standpipe. A 150-ft (30-m) length of 1¾-in. (44-mm) hose with a 7/8-in. (22-mm) or 15/16-in. (24-mm) solid stream nozzle or with a low-constant-flow nozzle should provide a flow capable of containing most fires in buildings with compartments that do not require a flow exceeding 100 GPM (379 L/min). Using the formula of volume divided by 100 to calculate the size of the area that could be extinguished with 100 GPM (379 L/min) computes to an enclosure that does not exceed 10,000 ft^3 (283 m^3). Seldom will rooms in residential buildings exceed this size. However, there have been situations when very low pressures were encountered at the standpipe outlet or where fires involved compartments exceeding 10,000 ft^3 (283 m^3). In these situations, larger-diameter hose reduces the friction loss and allows the use of larger-volume nozzles, thus increasing the available rate of flow at low pressures.

Likewise, the minimum 150 ft (46 m) of hose should be enough to reach the seat of the fire, even when it becomes necessary to maneuver around multiple obstructions or doorways to reach the fire area. The 1¾-in. (44-mm) hose line is much easier to advance than larger-diameter hose lines and is particularly beneficial when there is limited staffing on the fire floor. However, larger-diameter hose increases the chance of successfully extinguishing the fire; therefore, it should be the primary attack hose line. The recommended standpipe equipment listed previously allows the attack team to add a 50-ft (15-m) section of 1¾-in. (44-mm) hose at the nozzle tip. At 100 GPM (379 L/min), the one additional section of 1¾-in. (44-mm) hose will add approximately 5 psi (34 kPa) of friction loss to the hose layout.

What diameter and length of hose are best? This question has to be answered by each fire department. If standpipes are spaced such that 150 ft (30 m) of hose is sufficient to reach all areas in the standpipe-equipped buildings in your jurisdiction, three sections of hose may be the best choice. If your department responds to multiple buildings with low-pressure standpipes or with large compartments, the availability of 2½-in. (64-mm) hose is a must.

Preincident plans should be consulted in making decisions regarding hose diameter and length. Preincident plans could also identify the need for additional, specialized equipment. Rate of flow determinations are as essential in standpipe operations as they are in any other manual fire suppression operation. (These are described in Chapter 8, *Offensive Operations*.)

FIGURE 7-14 shows standpipe equipment that would be placed in two quick-opening carrying cases with each fire fighter carrying a hose pack. This arrangement allows fire fighters to share the equipment load and provides an efficient means of carrying the standpipe equipment into the building and deploying it at the standpipe outlet. It is extremely important to

FIGURE 7-14 A. Standpipe equipment. **B.** Fire fighters with standpipe equipment.
Courtesy of Ben Klaene.

select the correct standpipe equipment and hose for your jurisdiction. It is equally important to train on the proper use of this equipment under conditions likely to be found in buildings equipped with standpipes. For example, if 2½-in. (64-mm) hose is selected for use in standpipe operations, training should simulate safely advancing a charged hose line with the minimum crew size.

Some buildings have pressure or flow reduction valves in the standpipe system. These valves can cause problems for fire fighters using the system, as was experienced at the Meridian Plaza fire in Philadelphia that killed three fire fighters (Klem, 1991). In this case, pressure-reducing valves were improperly set to reduce the pressure to less than 60 psi (414 kPa) at the valve. Steps have been taken to eliminate this problem, but good preincident planning and routine inspections will reduce the possibility of this occurring. Many pressure- or flow-reducing valves are field adjustable. Some field adjustments are as simple as removing a fitting that reduces the size of the discharge opening. Other valves are more difficult to

adjust and may require special tools to regulate the pressure. Instructions for increasing the flow and/or pressure should be included in preincident plans, and members must be shown how to change the pressure setting during training sessions. The preincident plan survey should include arrangements to obtain special standpipe adjustment tools. The pressure-regulating tool should be added to the standpipe equipment before ascending to the fire floor. Pressure-reducing valves must be installed and maintained properly to ensure that they can provide the required volume and pressure for firefighting operations. With proper preincident planning and maintenance, it should not be necessary to field-adjust pressure-reducing valves.

Automatic nozzles do not provide good flows at pressures below their design parameters. Some automatic nozzles require a nozzle pressure of 100 psi (689 kPa) and will not provide any flow at very low pressures. These nozzles should not be used when connected to a standpipe. Conversely, smooth-bore nozzles can produce an adequate stream at low discharge pressures and are often the best choice for standpipe operations. A 2½-in. (64-mm) hose line with 50 psi (345 kPa) of nozzle pressure and a 1¼-in. (32-mm) tip will flow approximately 326 GPM (1234 L/min). At 30 psi (207 kPa) of nozzle pressure, a 1¼-in. (32-mm) smooth-bore nozzle will flow approximately 253 GPM (958 L/min).

At the other extreme, it is possible to find standpipe discharge pressures above 130 psi (896 kPa). A 2½-in. (64-mm) hose line with some variable-stream nozzles will produce a large flow at this pressure. The 1¼-in. (32-mm) smooth-bore nozzle is not designed to operate at such high pressures. After discounting a friction loss of 55 psi (379 kPa), the 1¼-in. (32-mm) tip will be flowing nearly 400 GPM (1514 L/min), with a nozzle reaction force of nearly 220 psi (1517 kPa). In this situation, an in-line regulator/gauge placed on the standpipe outlet would be an invaluable tool. The solution is having the right tools at the right place, as well as knowing how and when to use them.

Nozzle manufacturers are now designing nozzles for standpipe use that can be changed from variable to solid bore without changing the tip. When deciding on standpipe hose and nozzles, it is important to determine the actual flow capabilities at the pressures available at standpipe discharges. Some nozzles are significantly more efficient than others. When changing from a variable-stream to a smooth-bore nozzle, it is important to recognize that there may be a significant increase in flow with a resultant increase in nozzle reaction. (See Chapter 8, *Offensive Operations*, for more information regarding rate of flow and nozzle reaction.)

Department training should include the use of the various types of nozzles at low and high pressures. It is essential that members of the department be familiar with and operate standpipe equipment under conditions similar to those found in your community and at the various pressures anticipated. These training sessions should be in full personal protective equipment, including standard-issue gloves. Many tasks that are easily performed in street clothing with clear visibility are difficult to carry out under fire conditions wearing full fire gear.

Departments should also conduct flow tests (see Chapter 8, *Offensive Operations*) to determine the actual flow using various hose layouts at expected pressures. The time to safely experiment and learn how to properly use firefighting equipment is during fire department training sessions. Trial and error on the fire ground can produce deadly results for both civilians and fire fighters.

Connecting to the Standpipe Discharge

Most departments have standing orders requiring connection to the standpipe one floor below the fire. Excess hose in the stairway may prove to be a problem. Hose can be deployed up a stairway to the floor above the fire floor before fire fighters enter the fire floor. This allows fire fighters to enter the fire floor pulling the hose *down* the steps rather than up. Using this method keeps excess hose out of the way. When using this tactic, it is important to lay the hose up the steps before making entry to avoid being above the fire when the door to the fire floor is opened.

Recognize that fire hose in the stairway will impede occupant egress and that smoke conditions in the stairway may place occupants at greater risk. Whenever possible, it is best to dedicate a stairway for fire operations and other stairs for occupant evacuation. In many cases, only one stairway has standpipe outlets, making it the obvious choice as the fire operations stairway. Every effort should be made to keep occupants out of the fire operations stairway. See Chapter 12, *High-Rise Buildings*, for a more in-depth discussion regarding protecting occupant egress in high-rise buildings.

In conducting preincident planning, the standpipe outlets must be identified and special operational requirements noted. Similar to buildings equipped with sprinkler systems, the preincident plan may identify operational needs that are different from department SOPs.

Nonwater-Based Extinguishing Systems

In addition to water-based systems, there are many extinguishing systems that use different agents. Following is a list of fire suppression agents and types of systems that use different extinguishing agents, mostly nonwater-based extinguishing agents.

- Foam
  - Surface application
  - Subsurface application
  - Deluge
- Halon (and other clean agents)
  - Total flooding
  - Local application
- Carbon dioxide
  - Total flooding
  - Local application
  - Extended discharge
  - Handlines
- Dry chemical
  - Local application
  - Handlines
- Other inerting systems (using inert gases to extinguish or contain a fire)

An extensive variety of systems and system components can be used, depending on the hazard being protected. Each of these specialized systems will influence the development of a strategic plan. Therefore, fire officers should have a good working knowledge of all systems. In most cases, the systems will probably have activated before the fire department's arrival. In atypical situations, the IC may need to order that the system be activated manually.

The proper use of in-place suppression devices is yet another reason for preincident planning. The preincident plan should include a drawing showing the location of system components (risers, shut-offs, pumps, and agent supply containers).

Some systems can create hazards for the occupants or fire fighters. For instance, carbon dioxide can suffocate anyone inside the protected enclosure. Dry chemical agents can cause physical harm and, at the very least, obscure vision.

Foam Systems

High-expansion foam systems are designed to protect buildings; however, these systems are rare. They are designed to fill an area, such as a basement, with foam, thereby smothering the fire.

Low-expansion foam systems are usually found at properties storing large quantities of flammable and combustible liquids. Some low-expansion foam systems are automatic; others require fire department support. Even fully automatic foam systems have provisions for manual operation.

Refineries and petroleum storage depots normally protect aboveground storage tanks with a foam system. A **foam house** will be located on the property or nearby, containing additional quantities of foam agent and a means of manually operating the system. System operation will be different at each facility, and the responding fire department should be familiar with the hazards being protected and the operation of the foam system. During the preincident planning survey, consider the proximity of the foam house to the flammable or combustible liquid hazard being protected. If the foam house is located too close to the hazard, it may be unsafe to enter the area when the fire is in close proximity to the foam house.

Fire departments must train on operating the foam system but should also insist that written operating instructions be posted within the foam house. Foam is the number one defense against flammable liquid fires but is nearly useless on pressurized liquids or gases. Remember, for the automatic system, let the system do its job and support it with additional water supply and foam as needed. For the manual system, gain access to the foam house and operate the system as required.

At bulk storage facilities, the quantity of foam required to suppress a fire may be large. The quantity should be determined either by the property owner or during preincident planning. If the required volume is not immediately available onsite, then part of the preincident plan should be directed toward protecting exposures until additional quantities of foam can be obtained. Without the required quantity of foam onsite, it may be counterproductive to launch a fire attack that will ultimately fail. Furthermore, foam will not suppress a three-dimensional fire, such as may occur with leaking fuel.

Class A foam increases the effectiveness of water and adheres well to exposed structures. Therefore, at incidents where these are important operational objectives, the use of Class A foam should be considered. Class A foam systems are rare.

Carbon Dioxide Systems

There are two types of automatic carbon dioxide systems: total flooding and local application. In addition, some occupancies will provide carbon dioxide hose reels for manual application of the agent. Carbon dioxide is used in areas where preventing water damage is a prime objective or where the extinguishing agent is more effective than water or dry chemical agents. Carbon dioxide is not a conductor of electricity and is also used near electrical equipment. Carbon dioxide systems rely on detectors for their activation. Storage of carbon dioxide is limited; therefore, the system has a limited discharge time and coverage.

Total flooding systems are dependent on agent containment for a period of time (soak time) to be effective. If ventilation systems are not shut down or the compartment is opened, the carbon dioxide will quickly dissipate and the fire may rekindle. Carbon dioxide extinguishes by depleting the oxygen supply, and a carbon dioxide system generally requires carbon dioxide concentrations ranging from 34% to 75%. A room flooded with carbon dioxide may appear normal while actually being oxygen deficient. Carbon dioxide is extremely cold when discharged; it is also heavier than air and may accumulate in low or remote locations. Fire fighters entering a room where a carbon dioxide system has discharged must wear SCBA.

Halon and Other Clean Agents

Large computer installations are protected by sprinklers because of the reliability factor. Clean agent systems are also used because of their ability to react quickly and to suppress a fire in its beginning stages, without damaging sensitive equipment. The quick extinguishing capabilities of Halon have long been recognized in explosion-suppression systems in which deflagrations are actually suppressed before pressure builds up.

Halogenated agents were greeted by many as the extinguishing agent of the future. They were considered the ultimate agents for all situations because they are easily applied, believed to be nontoxic and nondamaging, and thought to leave the compartment safe for human habitation during discharge. The excitement that surrounded the use of halogenated agents in the early days has disappeared. We now know that halogenated agents can damage equipment, and although some do not cause immediate death, a health risk may be associated with staying in the compartment after the agent has discharged. Halon is not considered toxic, but the products of decomposition (hydrogen fluoride, hydrogen chloride, and hydrogen bromide) are harmful to humans and might damage some electronic components. Halon 1301 is a known ozone-depleting chemical and is no longer produced; however, reclaimed Halon 1301 is available.

In addition to causing environmental harm by destroying the ozone layer, Halon and other clean agents are expensive. What was once thought to be a miracle agent is now on its way to oblivion. Other clean agents are now being used, and halogenated agents are being phased out.

Fixed clean agent systems rely on smoke detectors to activate the system. They have a limited supply of agent that must be discharged into a confined room or area. Most clean agents do not require suffocating quantities to extinguish the fire; a low concentration is usually sufficient. However, to be effective, the room must remain closed, and the ventilation system must automatically shut down on discharge.

Clean agents are generally considered nontoxic to humans in the concentrations that are found in computer rooms and other installations. This is a time to be skeptical of current beliefs. Insist that fire fighters entering a room where a discharge has occurred wear SCBA until the room has been completely ventilated. Given the fact that it is nearly impossible to determine complete ventilation, it is best that fire fighters continue wearing the SCBA as long as they remain in a compartment where Halon or other clean agents have been discharged.

Dry and Wet Chemical Systems

Dry chemical systems are used in a number of different applications. One of the most common is in restaurant kitchen hoods FIGURE 7-15. Because of the ability of dry chemical systems to suppress fires in cooking appliances, ductwork, and other related areas, it is widely used for this purpose. Dry chemical systems are used in many other applications, such as dip tanks and gasoline-dispensing facilities. However, kitchen hood systems are the predominant application.

A kitchen hood system can be activated either automatically or manually at a pull station. For automatic operation, fusible links are usually located over the area being protected and/or in the ductwork. In the event of a fire, the link will fuse, releasing the tension on a cable that, in turn, will cause the dry chemical agent to be discharged through a series of nozzles.

Wet chemical agent systems, similar in design to dry chemical systems, are also found in kitchen hood applications. Wet chemical agents react with hot grease or oil to form a foam blanket that suppresses the release of combustible vapors. Property owners often prefer wet agent systems because cleaning up after the system discharges is much easier than with dry chemical agents.

As with the other specialized systems, it is likely that these systems will have already discharged before the fire department's arrival.

Working in Areas Protected by Total Flooding Carbon Dioxide or Clean Agent Systems

Letting the System Do Its Job

If the system is controlling the fire, maintain the chemical concentration by keeping the doors closed. Unless employees working in the kitchen failed to escape, there is no need to enter the area if the fire is being controlled. Entering the area will allow the carbon dioxide or clean agent to dissipate, thereby reducing its effectiveness. Unlike sprinkler and standpipe systems, carbon dioxide and clean agent systems have a limited supply of extinguishing agent.

FIGURE 7-15 Kitchen hood systems predominantly use dry chemical agents, though wet chemical agents can also be found in kitchen hood systems.
© WStudio/Shutterstock.

Final Extinguishment and Rescue

If it becomes necessary to enter the room to perform a rescue or for final extinguishment, members must wear full protective clothing and must don their face pieces before entering. After a carbon dioxide release, the area may appear to be completely clear yet pose a serious hazard due to a lack of oxygen. Clean agents may also deplete the oxygen levels, and they may pose a threat because of corrosive decomposition gases. Overhaul operations must be completed, especially where Class A materials are involved. Otherwise, rekindles may occur as the agent concentration is diluted over time.

Manual Activation

These systems will generally be equipped with a manual actuation device that can be operated in the event of fire where automatic detection/activation sensors fail. Some systems are equipped with an abort switch, which can be held to prevent agent discharge. Sometimes the abort switch is located outside the protected area with no provisions to determine what is going on inside the protected area. Employees may decide to keep the agent from discharging for a number of reasons. The first-arriving company must determine whether preventing system discharge is justified.

Checking Interlocks

Fire suppression systems often trip interlocking devices when they are activated. At a minimum, there will be provisions for shutting down the ventilation system when using carbon dioxide and clean agents. Carbon dioxide systems and many clean agent systems have preaction alarms interlocked to the system, allowing occupants time to escape. In most cases there is a means of manually activating the interlock. If ventilation systems are operating, it is imperative that they be closed, especially with carbon dioxide total flood application. However, if the ventilation system has been running for a period of time after discharge of either carbon dioxide or clean agents, then the concentration of agent within the room may be insufficient to suppress a fire, at which time manual hose streams may be necessary.

Checking Agent Supply

Occasionally, a cylinder or other container holding clean agents or carbon dioxide may be accidentally shut off and not reopened. There may be extra containers of agent that can be connected to the system once the original supply has been depleted. A fire fighter should be assigned the task of checking supply valves, or if extra supplies of agent are available, an entire company may be needed to connect the additional supply. However, this task may require specialized knowledge, and by the time an assessment is made, the fire may have grown too large for the system to control. Manual suppression may be required. The IC should be prepared for this potential eventuality.

System Restoration

System restoration will, by necessity, be left to the property owner and/or a contractor who is capable of recharging and resetting the system.

Working in Areas Protected by Local-Application Carbon Dioxide, Clean Agents, Dry Chemical, or Other Special Extinguishing Agents

Letting the System Do Its Job

Make sure valves are fully open and do not interfere with system operation.

Checking the Interlocks

Interlocks may shut off fuel supplies or deenergize equipment. Manual operation of the interlocks may be possible, or employees may be able to shut down equipment as needed.

Manual Activation

Support the system by activating manual devices when necessary. These systems are generally protecting Class B or Class C hazards and do not depend on an enclosure.

Backing Up the System

Be prepared with backup equipment, such as hose lines, foam lines, or portable extinguishers as required to augment the system and/or complete overhaul.

As mentioned, in kitchen hood systems it is common for dry chemical systems to also protect the ductwork leading from the cooking appliances to where it exhausts outside of the building. It is important to inspect the entire ductwork and exhaust system to verify that the fire did not spread beyond the cooking appliance.

System Restoration

System restoration should be left to the property owner or a contractor who is capable of recharging and resetting the system.

Responses to Building Fire Alarm Systems

False alarms transmitted by fire suppression or fire alarm systems are common. In some areas they have become so frequent that special responses are sent to fire alarm activations. Although the number of fire alarms can be a nuisance, it is important to remember that a percentage of these alarms will be for actual fires. The high number of false alarms causes apathy, and apathy lulls fire forces into complacency. Then, when least expected, the inevitable occurs and lives and property are lost—lives and property that could have been saved had the alarm been heeded.

False alarms are often the result of poor system maintenance or improper installation. Unfortunately, no one has all of the answers to this problem. Jurisdictions have assessed penalties on property owners who have an excessive number of false alarms, but this may cause the property owner not to call when the alarm sounds or to shut down the system. For good reason, departments insist on a rapid call when there is an emergency. If the property owner investigates the alarm before calling the fire department, the call for assistance will be delayed.

Particularly difficult are automatic alarms sounding inside a tightly secured building. These situations often present the IC with a dilemma. Forcing entry can cause significant damage; however, failing to enter the building may place people or property at risk.

Fallacy	Fact
Alarms from automatic alarm systems are always false.	Automatic alarm systems are often activated when there is no fire. Different systems have different degrees of accuracy. Treating automatic alarms as false alarms can result in disastrous outcomes.

To assist fire department personnel in gaining entry after business hours, the building owner may be willing to provide a lockbox with emergency access keys or make similar arrangements. However, if a decision is made to force entry, it should be done in a manner that causes as little damage as possible.

Direction should be provided to the owner through code adoption, local ordinance, official notification from the fire chief or fire marshal. Furthermore, the fire department should work with the building owner to identify entry options in advance and include them in the preincident plan.

Wrap UP

CHAPTER SUMMARY

- If the building is equipped with an automatic fire suppression system, the primary tactic is to support the system and let it do its job.
- Preincident planning is essential for any building that is protected by an automatic fire suppression system.
- With few exceptions, a properly designed, maintained, and operated sprinkler system will prevent large loss of life and property.
- To avoid depriving the sprinkler system of an adequate water supply, fire department personnel should not take water from the same source to supply hose lines.
- The main objective when working in a building protected by an automatic fire protection system is to support the system and let the system do its job.
- When working in a building protected by a sprinkler system, back up the system with hose lines and ventilate.
- Hazards protected by deluge systems may pose additional dangers to fire fighters.
- A properly operating and maintained building standpipe system may be helpful in conducting offensive attacks, but it is not an automatic system; it must be manually operated.
- Five major types of standpipe systems exist: automatic wet and dry, semiautomatic dry, and manual wet and dry systems. Automatic wet standpipes are most common and most reliable.
- When working from a standpipe system, check for proper operation of the main control valve and fire pumps, and supply the fire department connection (FDC).
- Fire departments should provide equipment needed for standpipe operations along with standard operating procedures (SOPs) for standpipe use.
- Preincident planning of buildings with nonwater-based fire suppression is essential, as improper use of and/or exposure to some agents can be harmful to fire fighters.
- Unless there are victims inside, do not enter an area where a carbon dioxide or clean agent system has discharged if the fire is being controlled.
- If it becomes necessary to enter a room protected by a carbon dioxide or clean agent system to perform a rescue or for final extinguishment, members must wear full protective clothing and must don their self-contained breathing apparatus (SCBA) face pieces before entering.
- The fire department should not assume that any alarm is false, even if multiple false alarms have occurred from the same system.

KEY TERMS

backpressure Also known as elevation pressure, the pressure required to overcome the weight of water in a piping or hose system. Each vertical foot of water in a pipe, hose, or tank exerts a pressure of 0.434 psi (3 kPa) at the base.

Class A foam Foam for use on fires in Class A fuels such as vegetation, wood, cloth, paper, rubber, and some plastics.

deluge system A sprinkler system in which all sprinklers or applicators are open. When an initiating device, such as a smoke detector or heat detector, is activated, the deluge valve opens and water discharges from all of the open sprinklers simultaneously.

draft curtains Walls designed to limit horizontal spread of the fire that extend partially down (usually no more than 20% of the height of the compartment) from the underside of the roof.

dry pipe sprinkler system A system in which the pipes are normally filled with compressed air or nitrogen. When a sprinkler is activated, it releases the air from the system, which opens a valve so the pipes can fill with water.

foam house A fixed facility consisting of an enclosure housing a foam concentrate supply tank, a foam solution proportioning system, a pump, and sometimes an extra supply of foam concentrate that can be added to the proportioning system.

preaction sprinkler system A dry sprinkler system that uses a deluge valve instead of a dry pipe valve and requires activation of a secondary device before the pipes fill with water.

standpipe system A piping system with discharge outlets at various locations; in high-rise buildings an outlet will normally be located in the stairway on each floor level. Most are connected to a water source and the pressure is boosted by a fire pump.

wet pipe sprinkler system A sprinkler system in which the pipes are normally filled with water.

SUGGESTED ACTIVITIES

1. A grease fire occurs in the kitchen area of the Joe's Rib House restaurant shown in Figure 7-15. This is a popular restaurant that has a maximum seating capacity of 200. Twenty to 25 employees are also in the establishment when fully staffed and open for business. It is Saturday evening December 20th at 7:00 PM, and the restaurant is near full capacity. Assume the role of incident commander arriving at the same time as Engine 1 and Truck 1 from your fire station.

   Upon arrival, the restaurant manager approaches you reporting a grease fire in the kitchen at the rear of the restaurant. The owner of the restaurant adds that it appears the kitchen fire system has extinguished the fire, and the cooks are in the kitchen checking it out.

   Smoke is visible at the door leading to the kitchen and there is light smoke at the ceiling level of the dining area. A few restaurant patrons are leaving the building, but most are still at their tables. **TABLE 7-1** shows the Bentomville Fire Department response.

TABLE 7-1 Fire Department Fire Companies, Fire Officers, and Emergency Medical Resources

	Engine Companies	Truck Companies All Equipped with 100-ft (30.5-m) Aerial Ladder	Other Units
1st Alarm	Engine 1 Engine 2 Engine 3 Engine 4	Truck 1 Truck 2	District Chief 1 Heavy Rescue 1 Paramedic 1
2nd Alarm	Engine 5 Engine 6 Engine 7 Engine 8	Truck 3 Truck 4	District Chief 2 Operations Assistant Chief 1 Safety officer Paramedic 2
3rd Alarm	Engine 9 Engine 10 Engine 11	Truck 5 Truck 6	Fire chief Assistant Chiefs 2, 3, 4, and 5 District Chief 3 Heavy Rescue 2 Rehab unit
4th Alarm	Engine 12 Engine 13 Engine 14	Truck 7 Truck 8	District Chief 5
5th Alarm	Engine 15 Engine 16 Engine 17	Truck 9 Truck 10	Heavy Rescue 3

© Jones & Bartlett Learning.

All engine, truck, and heavy rescue companies are staffed with an officer, driver, and two fire fighters. Medic units are staffed with two paramedics. Dispatch sends a full one-alarm response. Assume the role of incident commander (District 1) arriving simultaneously with Engine 1, Truck 1, and Heavy Rescue 1.

A. List your concerns.

B. Develop an incident action plan.

C. Assign first-alarm companies to specific tasks.

D. Would you order additional alarms or special units? Explain your decision and assign additional units if requested.

E. Develop a NIMS chart.

2. Using the information provided in this chapter and in Chapter 2, *Procedures, Preincident Planning, and Size-Up*, develop a preincident plan for a sprinkler-protected property near you.

3. Compute the pump discharge pressure needed to adequately supply a standpipe system on the top floor of the tallest building in your response district, including areas covered by mutual/automatic aid agreements.

4. Compare standpipe equipment used by your jurisdiction to the suggested list of equipment in this chapter. Justify your present standpipe hose and nozzle combination or recommend a different hose size or nozzle type.

5. Using the information provided in this chapter and in Chapter 2, *Procedures, Preincident Planning, and Size-Up*, develop a preincident plan for a facility equipped with a Class B foam system.

REFERENCES

Hall, John R., and National Fire Protection Association. 2012. *NFPA's U.S. Experience with Sprinklers Boston, MA, March 2012, NFPA Fire Analysis and Research*. Quincy, MA: NFPA.

Klem, Thomas J. 1991. *One Meridian Plaza, Philadelphia, PA, February 23, 1991: Three Fire Fighter Fatalities* Fire Investigation Report. Quincy, MA: National Fire Protection Association.

National Fire Protection Association. 2019. *NFPA 13: Standard for the Installation of Sprinkler Systems*. Quincy, MA: NFPA.

National Fire Protection Association. 2019. *NFPA 13D: Standard for the Installation of Sprinkler Systems in One- and Two-Family Dwellings and Manufactured Homes*. Quincy, MA: NFPA.

National Fire Protection Association. 2019. *NFPA 13R: Standard for the Installation of Sprinkler Systems in Low-Rise Residential Occupancies*. Quincy, MA: NFPA.

National Fire Protection Association, 2019. *NFPA 14: Standard for the Installation of Standpipe and Hose Systems*. Quincy, MA: NFPA.

National Fire Protection Association. 2020. *NFPA 13E: Recommended Practices for Fire Department Operations in Properties Protected by Sprinkler and Standpipe Systems*. Quincy, MA: NFPA.

National Fire Protection Association. 2020. *NFPA 25: Standard for the Inspection, Testing, and Maintenance of Water-Based Fire Protection Systems*. Quincy, MA: NFPA.

CHAPTER 8

Offensive Operations

LEARNING OBJECTIVES

- Compare an offensive fire attack to a defensive fire attack, explaining the basics of each type of attack and identifying the rationale for each strategy.
- List conditions and situations that would warrant a change from an offensive to defensive attack.
- Estimating the size of the apartments and hallways in the pre-plan drawing shown in Figure 8-3, evaluate the rate of flow requirements for areas in the building.
- Describe trial-and-error methods of calculating rate of flow.
- List the advantages and disadvantages of a trial and error approach to a working fire situation in the apartment building shown in Figure 8-3.
- Describe "softening the target" and how it is different from an indirect attack.
- Analyze rate of flow requirements using Royer/Nelson (V/100), National Fire Academy (A/3), and sprinkler calculations.
- Explain why sprinkle systems generally require a smaller rate of flow to extinguish a fire than is required by hose lines.
- Define and contrast ventilation-controlled and fuel-controlled fires.
- Describe "area of involvement" and how it applies to rate of flow calculations.
- Explain the process of softening the target, including when and why this tactic should be used.

- Discuss the advantages of using the Royer/Nelson (V/100) rate of flow formula.
- Explain why a fire attack meeting or exceeding the calculated rate of flow could fail to extinguish the fire.
- Explain the relationship between nozzle type, rate of flow, and nozzle reaction force.
- Describe extinguishment of ordinary combustibles by inhibiting pyrolysis.
- Define internal and external exposures.
- List factors to consider when evaluating external exposures.
- Describe the purpose of a backup hose line and how it can be used to protect fire fighters attacking the fire.
- Evaluate water supply requirements based on rate of flow and other factors.
- Examine the relationship of and proper use of ventilation during offensive extinguishment operations.
- Discuss apparatus management and the factors used to determine the number of apparatus needed at an offensive operation.
- Compute and compare the rate of flow for various areas using A/3 and V/100.
- Evaluate the available flow from standard preconnected hose lines, and determine when the rate of flow for a structure should be preincident planned.
- Estimate the number and size of hose lines needed to apply a calculated rate of flow.
- Discuss the advantages and disadvantages of using a 2½-in. (64-mm) hose line in a typical house.
- Identify the recommended hose line sizes when flowing 150 GPM (568 L/min), 250 GPM (946 L/min), and 350 GPM (1325 L/min).
- Explain what is meant by a transitional attack and why it is used.
- Assess staffing requirements for an offensive attack based on rate of flow and life-safety factors.
- Develop a list of hose lines needed at a well involved structure fire on the third floor of an 80 ft × 80 ft (24 m × 24 m) five-story commercial building during working hours. There are employees on each floor. There is a 36-ft (11-m) wide road at the front on the brick building with occupied five-story commercial buildings on each side.
- Assess the probability of an imminent life-threatening situation.
- Using a fire scenario, assess the total water supply available and apparatus needs in terms of required fire flow.
- Given fire conditions and location, determine the ventilation possibilities and choose the best ventilation method(s).
- Evaluate the flow available from a standpipe system and standard fire department standpipe equipment based on a calculated rate of flow.
- Examine and evaluate various attack positions in a multistory building.
- Discuss factors involved in choosing an offensive strategy.

Courtesy of the Peabody Fire Department.

Case Study

On December 23, 2011, at 12:55 PM, two engine companies, a truck company, and a deputy chief were dispatched to a report of fire in an apartment building. Fire companies on the initial alarm were staffed with a lieutenant, driver, and fire fighter. The deputy chief was first on the scene and assumed the role of incident commander (IC). He observed heavy smoke on the second floor, side Alpha, of a three-story apartment building. He established command, declared a working fire, and entered the building, encountering fire in Apartment 3 on the second floor. The IC then instructed the first-arriving engine company to attack the fire in Apartment 3. The engine company advanced a 1¾-in. (44-mm) hose to the second floor while the truck company made their way to the third floor. After a delay, the lieutenant and fire fighter on the first-arriving engine company entered Apartment 3, which flashed. The lieutenant, with his helmet on fire, was pulled out, but the location of the fire fighter was unknown. The lieutenant returned to Apartment 3 with a thermal imaging camera but was unable to locate the fire fighter. A personal alert safety system (PASS) was sounding, but still fire fighters were unable to find the fire fighter. The officer of a later-arriving company finally located the fire fighter, who succumbed to a cardiac arrhythmia.

1. How does staffing affect safety during an offensive fire attack?

2. What is the value in "softening the target" before entering a fire area?

3. Why should an initial rapid intervention crew (IRIC) be established prior to making entry to the fire compartment?

Data from: NIOSH Fire Fighter Fatality Investigation Summary [F2011-31], April 25, 2013, "Career Fire Fighter Dies during Fire-Fighting Operations at a Multi-family Residential Structure Fire—Massachusetts, https://www.cdc.gov/niosh/fire/pdfs/face201131.pdf, Accessed February 14, 2020.

Introduction

An offensive fire attack is the preferred strategy whenever conditions and resources permit an interior attack. A defensive decision limits operation to the exterior, generally resulting in a larger property loss and limiting rescue options. The offensive versus defensive decision is based on staffing available to conduct an interior attack, water supply, ventilation, and most importantly a risk-versus-benefit analysis. Fire fighters should not enter a building that is in imminent danger of collapse or when fire conditions do not permit safe entry. Likewise, resource capabilities in terms of staffing, apparatus, and water must be able to meet incident requirements for a safe and effective operation. Rate of flow, as described in this chapter, is a major factor in determining if resources are adequate. The objective of an offensive fire attack is to apply enough water directly to the burning fuel to achieve extinguishment, thus providing a safer work environment for fire fighters, facilitating rescues, and reducing property damage. Chapter 5, *Fire Fighter Safety*, addresses fire fighter safety, and Chapter 6, *Life Safety*, describes rescue tactics. Offensive operations are the most effective way to save lives and property, but are potentially the most dangerous strategy. The offensive/defensive decision is critical and the basis of the entire incident action plan. The ability to extinguish the fire is critical to a safe and effective offensive strategy.

The essential question to be answered when addressing extinguishment is how many gallons per minute (liters per minute) are required to extinguish a given fire with properly placed hose lines. Calculating the rate of flow allows the incident commander (IC) to match the number and size of fire lines to flow requirements. In most cases, applying water directly to the burning fuel completes the extinguishment process. However, large offensive fire operations require the IC to consider many variables to successfully save lives and extinguish the fire.

> **NOTE**
>
> An offensive attack is preferred when conditions and resources permit interior operations.

Calculating Rate of Flow

Notable authors of fire tactics textbooks have addressed rate of flow in different ways. In his book, *Firefighting Principles and Practices*, William Clark used a derivation of the fire compartment volume in cubic feet divided by 100 (V/100) This is referred to as the Royer/Nelson method; Royer and Nelson developed the V/100 formula at Iowa State University. Clark provides examples using V/100 to determine the rate of flow in gallons per minute (Clark, 1991). In *Fire Command*, Alan Brunacini stated,

> "When the IC is able to apply more water than the fire can match with heat, we win. Until the IC reaches this level, the fire will continue to burn and eventually win (if the IC cannot overpower it)" (Brunacini, 2002).

Others in the fire service take a trial-and-error approach.

John Coleman notes that a hose stream's extinguishing capability will generally be determined in about 30 seconds. He further recommends a trial-and-error approach: "If you are at the top of a stairway and have a line directing water at a well-involved second floor or attic and you don't darken down the fire within 30 seconds or so, get more water" (Coleman, 1997).

> **NOTE**
>
> Rate of flow calculations as discussed in this text apply only to offensive attacks in ordinary combustible (Class A) materials.

This text takes an approach similar to Clark's methodology. However, trial-and-error methods can also be useful. With trial-and-error methods, fire fighters start with their favorite preconnected hose line, advance the hose line into the building, and, in many cases, make progress on the fire. Successes can reinforce bad habits, warding off any discussion of a more scientific approach. However, when trial-and-error fire fighters fail to make progress with their favorite hose line, they simply add more lines of the same kind. When they are forced out of a building after taking a severe beating during an unsuccessful effort, there is always an excuse. Seldom do they recognize their failure to flow enough water to overpower the fire. The continued use of initial attack hose lines when they are clearly overpowered by the fire is a sign of poor training and a lack of fire-ground discipline.

Many refer to this failure to recognize the need for larger fire streams as a "residential mentality." One method of avoiding the problem of always using the same size fire line is to write standard operating procedures (SOPs) requiring the use of 2½-in. (64-mm) attack hose lines when confronted with a fire in a large commercial building. This protocol provides a higher rate of flow but is less efficient than using a preplanned rate of flow. Some large commercial buildings are divided into small compartments where one or two 1¾-in. (44-mm) hose lines would be sufficient and easier to maneuver. For example, office buildings or strip malls are often subdivided into small individual compartments. Other offices and strip malls have large open spaces where the flow from multiple 2½-in. (64-mm) hose lines would be needed for an advanced fire.

The success rate of the trial-and-error method is highly dependent on the flow rate of the standard preconnected hose line. The larger the flow from the preconnected line, the higher the success rate. With the advent of 1¾-in., 2-in., and 2½-in. (44-, 51-, and 64-mm) preconnected attack hose lines, success rates have improved dramatically from the days of 1-in. (25-mm) or booster line preference.

The trial-and-error method is probably the most commonly used method of determining flow rate. Unfortunately, many fire officers fail to recognize the need for larger flows when the situation arises.

NOTE

Placing a third small-diameter hose line in the same compartment is generally a tactical error.

All of the methods of calculating rate of flow requirements have some imprecision built into them, because it is not possible to take into account every variable that will be encountered at a fire scene. Simply put, how much water is needed to overwhelm the fire? Should operations begin with 2½-in. (64-mm) hose lines? Should the second hose line be a 2½-in. (64-mm) hose? Applying rate of flow principles should answer these questions.

Fallacy	Fact
Rate of flow calculations are used to determine the exact amount of water needed and then to apply a prescribed flow in gallons per minute.	Rate of flow calculations yield a rough approximation of the required flow. They assist the IC in deciding the number and size of fire lines needed for an offensive operation.

Three rate of flow calculation methods are in general use, and all have certain advantages and disadvantages. They are described and compared here.

- Royer/Nelson:

$$\frac{\text{Volume in ft}^3}{100} = \text{GPM}$$

$$\frac{\text{Volume in m}^3}{0.748} = \text{L/min}$$

- National Fire Academy:

$$\frac{\text{Area in ft}^2}{3} = \text{GPM}$$

$$\frac{\text{Area in m}^2}{0.074} = \text{L/min}$$

- Sprinkler calculations: Fire load–specific calculations

NOTE

The purpose of rate of flow calculations is to determine the size and number of hose lines needed when a fire progresses beyond the normal room-and-contents residential fire when an interior, offensive attack is still practical.

The Royer/Nelson Formula

Keith Royer and Bill Nelson of Iowa State University developed fire flow calculations for structural firefighting based on test fires. They hypothesized that most fuels release 535 British thermal units (BTUs, or 564 kilojoules) per pound when combined with 1 ft³ (0.0283 m³) of oxygen. Thus, the fuel load and fire type are not as important as the room volume and can be ignored because fires are **ventilation controlled**. This proposition is somewhat in conflict with standards developed by fire protection engineers in determining sprinkler flow calculations. Sprinkler flows are adjusted based on the size and type of fuel.

The Royer/Nelson formula is based on the premise that the best rate of application is one that results in control of the fire within 30 seconds of effective application (80% efficiency). Iowa State University conducted many test fires to evaluate fire behavior; the Royer/Nelson volume formula was a result of observations and measurements taken at actual test fires. Royer and Nelson calculated the rate of flow in gallons per minute as the volume of the fire area in cubic feet (V) divided by 100, resulting in the formula V/100, as illustrated in **FIGURE 8-1**. This is the formula adapted in Clark's text, *Firefighting Principles and Practices*.

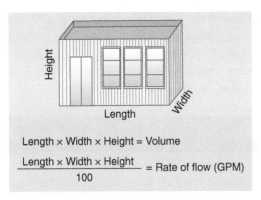

FIGURE 8-1 The Royer/Nelson rate of flow formula.
© Jones & Bartlett Learning.

The major assumption of Royer and Nelson—that structure fires are primarily ventilation controlled—is true for many fires. This assertion was further authenticated in the recent Underwriters Laboratories (UL) research comparing flashover in legacy furnishing to contemporary furnishings. Under conditions of the UL experiments in small residential compartments, the fire remained ventilation limited even when the compartment was well ventilated (Kerber, 2010). However, as the volume of the compartment increases, the fire may be **fuel controlled**, thus negating the major assumption behind the Royer/Nelson theory. These large-area fires are the ones most likely to require larger hose lines and the application of rate of flow principles. Ventilation-controlled fires require less water than a free-burning fire. Therefore, the Royer/Nelson formula may understate the needs in a well-ventilated large-area fire.

Gross underestimation is most likely to occur when fires reach a point beyond the capability of an interior attack and other factors such as the reach of fire streams, water supply, pump capacity, and staffing become more important. Therefore, the Royer/Nelson formula (V/100) will be valid for most fires where an interior attack is advisable.

Although Clark advocates the V/100 formula, he acknowledges that the progression to flashover is based primarily on the ratio of the surface area of the fuel to the size of the enclosure. If other factors such as fuel type and fuel load are held constant and the volume of the enclosure is increased, the time to flashover will also increase.

The process leading to flashover and backdraft is described in basic fire fighter training manuals, such as *Fundamentals of Fire Fighter Skills* (IAFC, NFPA, 2019). Although the conditions necessary for backdraft (an oxygen-deficient atmosphere) and resultant explosive effect are much different than flashover, the process leading to each event is similar. The major difference is the oxygen-starved condition necessary for backdraft and the oxygen-sufficient condition required for flashover. The progression to flashover begins with a fuel being heated until vapors being generated (via pyrolysis) ignite. The fire continues to burn and grow to a point of oxygen depletion. This results in a ventilation-controlled fire, followed by a vented or completely involved fire. If the fire is allowed to smolder at high heat levels rather than venting, the potential for backdraft exists. Too many variables occur in actual fires to accurately predict the progress of an ignition-phase fire to the fully developed phase. However, the effect of volume is fairly predictable. A large-fire area has more oxygen and therefore remains a fuel-controlled fire for a longer period. The heating of all combustibles is delayed as heat rises and reaches equilibrium in large-area fires. All of these factors lead to the conclusion that *no rate of flow calculation is completely accurate.*

The suggestion by UL mentioned previously for additional research into rate of flow calculations is absolutely warranted given the fact that the last field-expedient flow research was conducted over 50 years ago using equipment and tactics that are no longer used. However, to be useful, the new formula must be tested on large-volume fire areas and be easily applied. A formula requiring difficult and complex mathematical calculations would be less useful, and rate of flow calculations are not needed for small fire areas where standard attack hose lines can easily control the fire.

The National Fire Academy Formula

Rate of flow calculations are used to determine whether the fire is beyond the capacity of the standard attack and to determine how many and what size hose lines are needed for an interior attack. Most departments begin an interior attack using one 1¾-in. (44-mm) hose line with a required backup hose line at least as large as the initial attack hose line. The backup hose line can be used to augment flow if necessary. Therefore, the normal attack for most fire departments would be equal to the flow from two standard preconnected hose lines. Thus, the flow delivered using SOPs will vary depending on the size of the backup hose line and the flow produced by the standard preconnected hose line. With this in mind, the more conservative and probably less accurate rate of flow calculation introduced by the U.S. National Fire Academy cannot be fully discounted. Test fires, such as those used to develop the Royer/Nelson formula, were not used as a basis for the **National Fire Academy formula**. Therefore, the validity of the formula is yet to be proven.

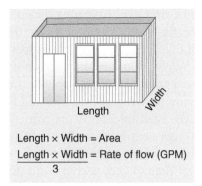

FIGURE 8-2 The National Fire Academy rate of flow formula.
© Jones & Bartlett Learning.

The National Fire Academy developed the A/3 formula for use in their fire tactics courses **FIGURE 8-2**. Many experts believe that calculating the volume at the scene of a structure fire is not practical, thus justifying the simpler area calculations. An article by Royer and Nelson recognizes that the V/100 rate of flow formula is primarily a planning tool (Royer and Nelson, 1959).

The A/3 and V/100 formulas can yield very different results. Typically, the A/3 formula requires more water than the V/100 formula. The A/3 formula generally overstates the need, but as the ceiling height increases, the A/3 formula becomes less conservative.

Large-volume fires are the ones that require the application of a rate of flow formula. Any mathematical calculation on the fire ground is difficult, but experience can be gained through practice by estimating volume or area during training. These estimates can later be compared to the actual area and volume.

Given the difficulty in calculating the rate of flow at the fire scene, including the rate of flow as part of the preincident plan information is strongly recommended when the flow exceeds the capabilities of the two standard initial attack hose lines used by the department. For example, many departments attach an automatic nozzle to four sections of 1¾-in. (44-mm) hose with a pump discharge pressure of 150 pounds per square inch (psi) (1034 kilopascals [kPa]) as their initial attack hose line. Depending on the hose and nozzle configuration, this 1¾-in. (44-mm) initial attack configuration yields a flow of approximately 125 GPM (473 L/min). The standard backup hose line is the same as the initial attack hose line. Therefore, this department would perform a preincident plan for any building with a compartment where the rate of flow exceeds 250 GPM (946 L/min). The rate of flow would then be listed for applicable compartments within the structure—in this case 25,000 ft^3 (708 m^3). If one 2-in. (51-mm) attack hose line with a 1-in. (25-mm) smooth-bore nozzle (approximate flow of 240 GPM [908 L/min]) were used as the initial attack hose line with a matching backup hose line, then a preincident plan would be suggested for any building with a compartment requiring more than a 480 GPM (1817 L/min) rate of flow, or 48,000 ft^3 (1359 m^3).

> **NOTE**
>
> Rate of flow should be included in preincident plans when the size of the largest compartment exceeds the rate of flow established by the department's SOPs for the initial attack, including provisions for a backup hose line.

Sprinkler Rate of Flow Calculations

Both the Royer/Nelson and National Fire Academy formulas ignore the fuel load and fuel type. This represents a weakness in these formulas.

For this reason, **sprinkler system calculations** may prove useful in preincident planning. There are several sources of sprinkler calculations, including NFPA documents and Factory Mutual Data Sheets.

Fires go through a series of stages from the ignition phase to flashover. These stages are characterized by being fuel dependent in the early stages and then oxygen dependent in later stages. A well-involved fire in a vented or large open area will be controlled by both fuel and oxygen. Therefore, the IC must consider the weaknesses of the rate of flow calculations being applied.

A fire in an area containing a small fire load will generally require less water than one in a similar area with a large fire load unless the fire is totally ventilation controlled in a tightly enclosed space. Once this space is vented, either by the fire or by fire fighters, the fire will react to the amount and type of available fuel. Remember, rate of flow is being used to determine the size and number of hose lines needed and whether the fire is beyond the capabilities of available resources. The IC decides when it is time to use larger, more powerful streams. The IC also uses rate of flow calculations to determine whether there are extenuating circumstances (i.e., reasons other than rate of flow) when operations are unsuccessful. If two 1¾-in. (44-mm) lines are not controlling a fire with a calculated rate of flow of 100 GPM (379 L/min), the problem is probably a failure to effectively apply water. Proper ventilation often will make the fire more visible and allow a fire fighter to get closer to the fire, thus improving the chances of successful direct application. However, ventilation will also increase fire intensity by increasing the air supply. By first softening the target, the fire intensity is reduced, allowing coordinated ventilation to control the flow path while effectively applying water onto the burning fuel.

Estimating the Size of the Largest Area

Rate of flow calculations are based on the area or volume of the compartment(s) on fire. There is some difference of opinion among practitioners of rate of flow regarding whether to make allowances for compartmentation. Some advocate applying rate of flow to the entire area or volume on fire; others see each individual compartment as a separate fire area.

This text recommends calculating each room as a separate fire area. This has been a point of discussion over the years, but Royer and Nelson's research clearly indicates that each enclosure should be calculated separately.

In a discussion with Keith Royer, he emphasized that flows should be determined considering each room capable of holding heat as a separate fire area. Royer elaborated by stating that he was talking not just about areas enclosed by fire walls and fire doors, but about any enclosure, even if the door is missing. This is an extremely important distinction. Few large areas are undivided, and it is, in fact, easier to fight fires in a series of smaller compartments than to do battle in one large, undivided area. Fires completely involving a large, undivided area often require a defensive attack.

In looking at the plan view for the apartment building in **FIGURE 8-3**, those who advocate calculating the entire area would apply the rate of flow formula to the entire space. Some would recommend adding each involved floor to the total. The more rational approach would calculate each individual space as a separate fire area. If each individual space is calculated separately, it becomes possible to move down the hallway, then into each apartment, extinguishing the fire in one room at a time. Instead of calculating the entire area, only the flow needed for the largest area (the open dining and living room in this case) is required.

Fallacy	Fact
The entire floor volume is calculated to determine the rate of flow.	Calculate the flow needed for the largest single area on fire. Each compartment (room) can be handled as a separate fire, and the fire can be extinguished in one room at a time.

Practical limitations apply to this logic, as it probably would not be possible to enter the floor if the entire floor were on fire. This would probably lead to a defensive attack where the offensive rate of flow formulas do not apply. If several rooms are on fire, it may be advisable to use multiple lines to expedite extinguishment by simultaneously attacking more than one room at a time.

> **NOTE**
>
> Preincident plan; know your buildings.

Estimating the size of the largest involved area is difficult on the fire scene. The IC does not always have an exact fire location or a detailed floor plan.

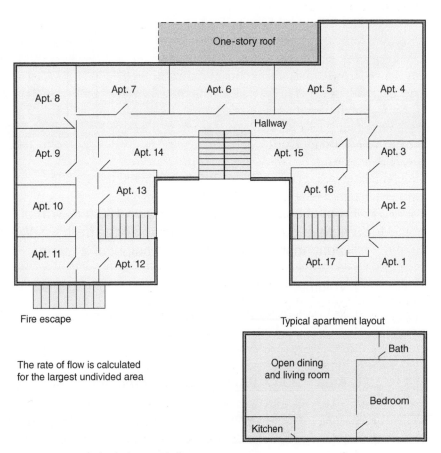

FIGURE 8-3 Calculating each fire compartment as a separate fire area.
© Jones & Bartlett Learning.

Preincident planning the rate of flow where it exceeds the capabilities of two standard attack hose lines is highly recommended. When confronted with a fire exceeding the rate of flow for two standard preconnected hose lines with no preincident-planned rate of flow data available, the IC will be forced to estimate the size of the involved compartment to determine the rate of flow requirement or else revert to trial-and-error methods. Fire fighters making the initial attack may be able to give the IC an estimated compartment size. Also, for a large, undivided building, the IC can estimate the length, width, and height by comparing them to sections of fire hose or to the height of a pumper **FIGURE 8-4**. However, these methods are much more difficult and less reliable than consulting a preincident plan that includes rate of flow calculations. Departments that fail to preincident plan buildings requiring a large rate of flow place their ICs at a considerable disadvantage.

Either of these calculations (A/3 or V/100) is possible at the fire scene, and each calculation only provides a rough approximation. Both calculations assume that the entire area is undivided. If the area is subdivided, the required rate of flow will be less and possibly within the capabilities of 1¾-in. (44-mm) hose lines if the compartments are small. If the building is not subdivided, the large rate of flow calculated by either method would indicate that once a fire involves a considerable portion of the building, an interior attack would probably be unsuccessful even if fire fighters could safely enter. This undivided structure most certainly would *not* be a situation in which additional 1¾-in. (44-mm) lines should be used. If the building is undivided and involves a considerable area, a fire in this structure will require interior 2½-in. (64-mm) hose lines or interior master stream appliances.

Given that calculating rate of flow is not an exact science, most fire officers would estimate, round off, and shortcut the math. For example, a building that is 32 × 32 ft (9.8 × 9.8 m) would become 30 × 30 ft (9 × 9 m). Dividing the first 30 ft by 3 yields 10 × 30 ft, yielding 300 GPM (10 × 30 for the A/3 rate of flow calculation). The metric calculation yields 1136 L/min.

Front of building approximately two 100 ft (30 m) of hose length = 200 ft (61 m)
Width of building approximately half the front = 100 ft (30 m)
Height approximately twice the height of a pumper = 25 ft (8 m)

Volume = 200 × 100 × 25 ft = 500,000 ft³ (14,640 m³)

$$V/100 = \frac{500{,}000}{100} = 5000 \text{ GPM } (18{,}925 \text{ L/min})$$

Area = 200 × 100 = 20,000 ft² (1858 m²)

$$A/3 = \frac{20{,}000}{3} = 6666.67 \text{ or } 6700 \text{ GPM } (25{,}360 \text{ L/min})$$

FIGURE 8-4 Field-expedient area and volume estimations.
© Jones & Bartlett Learning.

This A/3 calculation results in a rate of flow that exceeds the standard 1¾-in. (44-mm) preconnected hose line. Multiple 1¾-in. (44-mm) hose lines may be sufficient. Advancing a 2½-in. (64-mm) hose line as a backup hose line would exceed the 300 GPM (1136 L/min) requirement for this fire.

By taking this concept of matching the hose line to the rate of flow one step further, it is possible to determine in advance the number and size of hose lines needed to combat fires in dwellings, apartment buildings, and small businesses. This is a form of pre-incident planning—that is, preincident planning by general occupancy type. Examining a typical dwelling or apartment building will provide the fire officer with general rate of flow information that is very useful in initiating an offensive attack in these properties. For example, calculating the rate of flow for a 20- × 15-ft (6.1- × 4.6-m) living room with a 10-ft (3-m) ceiling yields:

$$A/3 = 20 \times 15 \text{ ft} = 300 \text{ ft}^2 \text{ divided by } 3$$
$$= 100 \text{ GPM } (379 \text{ L/min})$$

$$V/100 = \frac{20 \times 15 \times 10 \text{ ft}}{100} = 30 \text{ GPM } (117 \text{ L/min})$$

For this situation, 1½-in. (38-mm) or 1¾-in. (44-mm) lines are acceptable.

Calculating the rate of flow for a 20- × 25-ft (6.1- × 7.6-m) basement recreation area with a 10-ft (3-m) ceiling yields:

$$A/3 = 20 \times 25 \text{ ft} = 500 \text{ ft}^2 \text{ divided by } 3$$
$$= 166 \text{ GPM } (628 \text{ L/min})$$

$$V/100 = \frac{20 \times 25 \times 10 \text{ ft}}{100} = 50 \text{ GPM } (189 \text{ L/min})$$

A 1¾-in. (44-mm) hose line is acceptable only when using the V/100 formula. These smaller areas within residential properties tend to reaffirm the lower V/100 rate of flow calculations. When discussing rate of flow in terms of the advantages of 1¾-in. (44-mm) and larger hose lines, fire fighters who preferred using 1-in. (25-mm) booster hose ("red line") often challenged the need for 1¾-in. (44-mm) hose, stating that they have successfully extinguished many fires with the "red line." Most fires occur in dwellings and many in very small rooms within a dwelling. A 10- × 12- × 8-ft, or 960-ft³ (3- × 3.7- × 2.4-m, or 26.6-m³), bedroom is common in a home. Dividing the volume (960 ft³ [26.6 m³]) by 100 (or in metric, 0.748) yields a rate of flow of 9.6 GPM (35.6 L/min). The "red line" easily produces this flow. Calculating the same room using the A/3 formula computes to 120 ft² divided by 3 (or in metric, 11 m² divided by 0.074), which yields 40 GPM (150 L/min). The maximum flow from the low-volume nozzles typically attached to the "red line" is 40 GPM (182 L/min). Thus, the successful encounters with the "red line" tend to verify the V/100 formula. However, anything less than a 1¾-in. (44-mm) line for a structure fire is unacceptable. The larger the flow from the initial hose line, the greater the chance of immediate success. However, larger 2½-in. (64-mm) hose lines are very difficult to maneuver inside the small rooms found in most residential occupancies.

The preceding basement recreation area example represents a fairly large single area for a dwelling. Given that the National Fire Academy figures tend to overstate the case and there should always be a backup hose line, it can be concluded with a fair degree of confidence that fires within one- and two-family dwellings are normally within the flow capabilities of a 1¾-in. (44-mm) or 2-in. (51-mm) hose line with backup hose line. Only extraordinarily large, open-layout dwellings would require rate of flow calculations and subsequent preincident planning.

Apartment buildings can also be included in this category, as they normally have smaller undivided areas, with two exceptions. Apartment buildings with common attics or basements can, and do, progress to large-area fires requiring additional water. Larger apartment buildings and complexes may also have common areas (e.g., recreation rooms, dining rooms) requiring a large rate of flow. Apartment complexes may be preplanned for reasons other than rate of flow. For example, maps showing locations of buildings and streets may be needed to find specific addresses within an apartment complex made up of multiple buildings.

Estimating the Percentage of Area on Fire

The Royer/Nelson V/100 formula is based on fuel consumption that is ventilation controlled by the volume of the enclosure and *does not* reduce the rate of flow by the percentage of volume of the enclosure involved in fire. The National Fire Academy rate of flow formula is based on the area of involvement rather than the volume of the enclosure; therefore, a reduction by the area of involvement is recommended. This difference has to do with the assumption made by Royer and Nelson that the fire is ventilation controlled.

Incident Summary

Cincinnati Residential Fire

On January 31, 2008, Cincinnati fire fighters were dispatched to a residential fire **FIGURE IS8-1**. Two engine companies arrived on scene simultaneously. One company began an offensive attack using a standard preconnected 1¾-in. (44-mm) hose line. The second company initially knocked down fire threatening an exposure on the Delta side. The second company then made entry with a 1¾-in. (44-mm) hose line. One company attacked the fire in the foyer and hallway, while the other company extinguished the fire in two rooms on the first floor. They were able to extinguish the main body of fire and prevent extension to the second floor with two 1¾-in. (44-mm) hose lines due to the limited-volume rooms.

FIGURE IS8-1 Working fire in a two-story residential building with attic.
Courtesy of Denny Baker, Cincinnati Fire Department.

The National Fire Academy recommends a percentage-of-involvement modifier in applying the A/3 rate of flow. For example, if one-fourth (25%) of the floor area is involved, divide the total flow by 4 (or multiply by 0.25). This modifier appears to be reasonable. It takes less water to extinguish 100 ft^2 (9 m^2) of fire than to extinguish 400 ft^2 (37 m^2) of fire.

It is important to point out that the total V/100 flow will not always be required even though a percentage of involvement is not calculated. During the early stages of a fire in a large area, the fire will be fuel controlled, and less than the total calculated flow will be needed. Most fires can be controlled with less than 1 GPM (4 L/min) during the ignition phase.

In most cases, the first company advancing to the fire area will provide the reconnaissance needed to determine the fire area in terms of percentage of involvement and compartmentation. However, heavy smoke conditions may make it difficult or impossible to accurately determine the percentage of involvement.

Advancing small-diameter preconnected hose lines (1¾ in. or 2 in. [44 mm or 51 mm]) is faster than advancing 2½-in. (64-mm) hose. Unless the size of the fire is obvious, the first company advancing to the fire area may decide to use the standard preconnected hose line to size up the situation from the interior. It is important to get companies into all areas exposed to the fire as soon as possible to determine the life-safety and extinguishment needs. Some departments specify the use of a 2½-in. (64-mm) hose line as the primary attack hose line for commercial buildings, which increases the chance of success when confronted with a large-volume fire. The use of 2½-in. (64-mm) preconnected hose lines is highly recommended, as preconnected lines reduce the time needed to deploy larger streams when a large rate of flow is required for extinguishment.

The differences between the V/100 and the A/3 formulas are less significant if the percentage involvement is considered. This text recommends the V/100 for preincident planning, but if A/3 calculation is used at the incident scene (because of a lack of preincident planning), a percentage involvement modifier should be used. Because the purpose of calculating rate of flow is to determine the size and number of hose lines needed, it is reasonable to apply either formula. However, V/100 is based on a substantial body of research and is preferred for that reason.

FIGURE 8-5 Auto parts store example.
© Jones & Bartlett Learning.

Comparing Rate of Flow Calculations

A three-story building that is occupied as an auto parts store will be used as an example. The office and sales area are on the first floor, as shown in **FIGURE 8-5**. The second floor is a storage area for metal automobile parts in cardboard boxes. The third floor is used to store automobile and truck tires, on end. The building is 40 × 60 ft (12.2 × 18.3 m) with 15-ft (4.6-m) ceilings on the first floor and 20-ft (6.1-m) ceilings on the second and third floors.

Royer/Nelson (V/100) rate of flow calculations would be as follows:

$$\text{First floor} = \frac{40 \times 60 \times 15 \text{ ft}}{100} = 360 \text{ GPM (1362 L/min)}$$

This GPM is reduced by half for the first floor, as the first floor is subdivided into two 30- × 40-ft (9.1- × 12.2-m) areas, resulting in a flow of 180 GPM (681 L/min) for the first floor.

Second and third floors:

$$\frac{40 \times 60 \times 20 \text{ ft}}{100} = 480 \text{ GPM (1817 L/min)}$$

No reduction is made on the upper floors, as the second and third floors are not subdivided into smaller compartments.

The A/3 formula would calculate the same floors as follows:

$$\text{First floor: } \frac{40 \times 30 \text{ ft}}{3} = 400 \text{ GPM (1514 L/min)}$$

Second and third floors:

$$\frac{40 \times 60 \text{ ft}}{3} = 800 \text{ GPM (3028 L/min)}$$

It is also possible to determine the required rate of flow by using sprinkler calculations. A number of variables would apply in making these calculations, such as the building type, the number of floors, the occupancy type, the commodity inside the structure, and the storage configuration of the commodity. These sprinkler requirements are addressed in NFPA 13, *Standard for the Installation of Sprinkler Systems* (NFPA 13, 2019).

Properties that are already equipped with a sprinkler system should have a plate on the riser with all of the water flow requirements stamped into the plate. This information could be useful for comparing the two methods outlined earlier with the engineered water requirements.

At first, sprinkler calculations may seem complicated, but with the proper training, they are fairly easy to apply. However, sprinkler calculations will require additional time and a degree of judgment regarding the commodity's classification. Sprinkler rate of flow calculations are most accurate, as they are based on actual fire experience and consider the important factors of area of involvement and fuel load.

NFPA 1142, *Standard on Water Supplies for Suburban and Rural Fire Fighting* (NFPA 1142, 2017), identifies occupancies where extra water supplies may be needed. This list could be used as a reference for deciding when to consult the NFPA 13 commodity storage series of standards for extra hazard rates of flow.

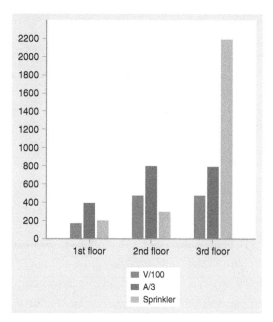

FIGURE 8-6 Comparing rate of flow calculations: Actual ceiling heights from the auto parts store example.
© Jones & Bartlett Learning.

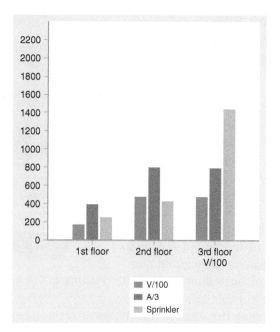

FIGURE 8-7 Comparing rate of flow calculations: 30-ft (9.1-m) ceilings.
© Jones & Bartlett Learning.

FIGURE 8-6 and **FIGURE 8-7** compare the rate of flow calculations discussed in this chapter. Figure 8-6 uses the actual ceiling height from the auto parts store example, Figure 8-5. Figure 8-7 shows how A/3 and V/100 calculations are nearly the same as the ceiling height increases to 30 ft (9.1 m). Both figures calculate the sprinkler rate of flow for the rubber tire storage on the third floor using extra hazard sprinkler tables found in the NFPA 13. Sprinkler calculations require a degree of engineering judgment, but the much higher calculated flow shows how high-hazard fuel storage can affect the rate of flow.

In the 30-ft (9.1-m) ceiling height example, the calculated flow using A/3 is still slightly higher, but V/100 and A/3 calculations are much closer. Mathematically, the rates of flow would be equal for a $33\frac{1}{3}$-ft (10.1-m) ceiling. Ceilings of this height would be common in warehouse or mercantile occupancies but extremely rare on the upper floors of a high-rise.

Sprinkler rate of flow calculations are not suitable for on-scene use unless the preincident plan identifies flow requirements. When a preincident plan includes a pre-calculated rate of flow, no field calculation is necessary.

Sprinkler calculations apply to rate of flow, but the sprinkler system is far more efficient than manual operations. Sprinklers are more effective in "getting the wet stuff directly on the red stuff," and they prewet the exposed fuel much as fire fighters with hose lines wet exposures. Engineers designing sprinklers allow a margin of safety.

A close look at the Royer/Nelson and National Fire Academy rate of flow calculation systems shows that the National Fire Academy system yields a higher rate of flow in most situations. Underestimating the flow will allow the fire to continue to make progress while fire fighters engage in a hazardous and futile battle. By contrast, if the need is overestimated, the fire is extinguished in less time with a reduction in the hazard to the fire fighters. Because all field-estimated rate of flow predictions are prone to error, overestimation is best. However, on closer examination it can be seen that gross overestimations could cause a delay in starting an aggressive interior attack or possibly lead to a defensive or nonattack decision because of a perceived lack of resources.

An article by C. Bruce Edwards in *Fire Engineering* points out that a rate of flow exceeding the minimum will extinguish the fire in less time and with less total water (Edwards, 1992). Thus, when calculating rate of flow at the incident scene, if in doubt, err on the side of higher rates of flow.

Remember that the National Fire Academy recommends a percentage-of-involvement modifier. If 25% of the auto parts store were involved in fire, the calculations would be as follows:

$$A/3 = \frac{40 \times 60 \text{ ft}}{3}$$
$$= 800 \text{ GPM} \times 0.25 = 200 \text{ GPM} (757 \text{ L/min})$$

V/100 remains the same at:

$$\frac{40 \times 60 \times 20 \text{ ft}}{100} = 480 \text{ GPM} (1817 \text{ L/min})$$

The V/100 formula is modified only by the size of the compartment, not by the percentage of involvement, due to the assumption that fuels will produce a given heat quantity per volume of air regardless of fuel size and type.

The rate of flow for sprinkler systems is usually stated as a flow rate (e.g., 0.25 GPM/ft^2 [10.19 L/min/m^2]) over a specific area, such as over the most hydraulically remote area (e.g., 3000 ft^2 [279 m^2]). This estimate generally reduces the flow requirements in large undivided spaces and is based on an assumption that the sprinkler system will operate early in the fire to limit fire extension. This is a valid assumption for a sprinkler system but not for manual firefighting. Therefore, sprinkler flow densities should be applied to the entire compartment when used to determine the rate of flow using hose lines. The sprinkler flow rate example above (0.25 GPM/ft^2 [10.19 L/min/m^2] over the most hydraulically remote 3000 ft^2 [279 m^2]) would be calculated as 0.25 GPM [0.95 L/min] × the area (length × width) for the entire compartment.

Which Rate of Flow Calculation Is Best?

As the preceding discussion shows, there is merit in each of the rate of flow calculation methods. There are also assumptions and limitations for each. The best approach would build on the relative strengths and weaknesses of the various methods. With this in mind, this text recommends that all buildings requiring a rate of flow that exceeds the flow of two standard preconnected hose lines should be preincident planned applying the V/100 formula.

If the fuel load is heavy in terms of total volume and/or fuel type, sprinkler calculations are preferred, with the larger of the two calculations (V/100 or sprinkler) being used to determine rate of flow. With the rate of flow calculated in advance, the IC can more accurately determine the size and number of fire lines needed.

For purposes of discussion, assume that the standard preconnected hose line is four sections of 1¾-in. (44-mm) hose with an automatic nozzle and a predetermined pump discharge pressure of 150 psi (1034 kPa). Furthermore, assume the flow rate for this standard preconnected line is 125 GPM (473 L/min). SOPs should require a backup hose line at least as large as, or larger than, the initial hose line, whenever there is a working structure fire. Therefore, the preincident plan rate of flow threshold for this department would be 250 GPM (946 L/min). Applying V/100 in reverse order would indicate that the volume to be preincident planned in this case would be 25,000 ft^3 (708 m^3) or larger. For example, a 25 × 50-ft (7.6- × 15.2-m) room with 10-ft (3-m) ceilings would equal 25,000 ft^3 (708 m^3). Very few homes will have a room this large.

Returning to the auto parts store example (see Figure 8-5), the second and third floors are 48,000 ft^3 (1360 m^3), because they are undivided and have higher ceilings. The first floor is 18,000 ft^3 (510 m^3). Considerations for the calculations in this building include the following:

- This building would not require a preincident planned rate of flow for the first floor.
- V/100 would be used to calculate the ordinary hazard on the undivided second floor.
- Sprinkler calculations would be recommended for the extra hazard tire storage on the third floor.
- If there is no preincident plan for this building, after arrival on the scene, the IC may elect to use either the A/3 or V/100 formula.

Rate of flow is but one of many reasons to preincident plan a building. (See Chapter 2, *Procedures, Preincident Planning, and Size-Up*, for details on preincident planning.) When confronted with a fire in an area that is beyond the control of two standard preconnected hose lines and in which the fire flow has not been preplanned, the A/3 formula may be easier to apply. However, this places the IC at a distinct disadvantage, as the building size must be estimated, and calculations performed under stressful fire-ground conditions might not be accurate.

Whether rate of flow is calculated in advance or on the fire ground, the IC should modify the formula on the basis of trial and error. If fire fighters on the fire floor report that they have enough hose lines, they probably do (trial and error). Of course, there are times when fire fighters fail to see hidden or extended fires, but this is not where rate of flow calculations are useful. Likewise, if fire fighters report that the calculated 500-GPM (1893-L/min) rate of flow is not getting the job done, the IC should consider the possibility that water is not being effectively applied to the burning materials. In such cases, a change of tactics or stream placement may be needed. If this part of the operation is not being supervised by a division or group leader, staffing a tactical-level management unit to supervise this critical function may improve extinguishment operations. This also could be a case where a lack of "truck work" is hampering the operation. Proper and coordinated venting of the area may allow attack teams to better apply water directly to the seat of the fire.

Royer and Nelson assume an 80% efficiency in getting water into the immediate fire area. If the nozzle team is attempting to cool the area by softening the target, thus making a close approach tolerable and allowing the engine company crew to get water directly onto the burning material, but these efforts are failing, then the 80% efficiency assumption is not valid. Royer and Nelson also assume the flow to be sufficient to knock down the fire in 30 seconds. This assumption supports John Coleman's assertion that progress should be evident within 30 seconds and that the attack teams should be making steady progress on the fire if the rate of flow is sufficient and being properly applied.

Once again, using the auto parts store as an example (see Figure 8-5), assume that the responding department uses 125-GPM (473-L/min) initial attack hose lines (1¾-in. [44-mm] hose line with an automatic nozzle). Furthermore, assume that the first attack hose line will be backed up with another attack hose line of the same size or larger. Looking at each floor, we can apply the rate of flow in terms of the number and size of fire lines needed to control a fire.

First Floor

The first-arriving engine company advancing a 1¾-in. (44-mm) hose line into the front or rear of the building would be well positioned to extinguish the fire with a single 1¾-in. (44-mm) hose line, as the rate of flow of 180 GPM (681 L/min) is nearly met.

The second- or third-arriving engine company would back up the initial attack, with another 1¾-in. (44-mm) hose line, providing enough water to easily overpower the fire (approximately 250 GPM [946 L/min] for a 180-GPM [681-L/min] fire).

Second Floor

The first-arriving engine company would advance a 1¾-in. (44-mm) hose line into the building and have little chance of success if the fire involved a substantial portion of the floor area. The recommended rate of flow of 480 GPM (1817 L/min) is far from being met with the 125-GPM (473-L/min) attack hose line.

If the first-arriving engine company reported anything other than extinguishment with the 1¾-in. (44-mm) hose line, the IC should order a 2½-in. (64-mm) hose line as the second line. Such a line can generate a flow of up to 350 GPM (1325 L/min). The 125 GPM (473 L/min) from the first hose line combined with a maximum 350 GPM (1325 L/min) from the 2½-in. (64-mm) hose line yields a total flow of 475 GPM (1798 L/min). This rate of flow nearly meets the 480-GPM (1817-L/min) requirements and has a good chance of success. If more water is needed, the IC could order an additional hose line or, in consultation with the first-arriving crew, increase the pressure in the 1¾-in. (44-mm) line. The IC could also opt to replace the automatic nozzle with a 7/8-in. (22-mm) solid stream and reduce the pressure while increasing the flow.

Third Floor

The initial 1¾-in. (44-mm) attack hose line would be effective only if the fire were contained to a relatively small area. To extinguish the fire on the third floor, 2½-in. (64-mm) lines or a master stream would be needed. Preincident plan information could prove crucial in this situation. Given the limited access to the third floor and the flow requirements for a fully involved fire, the department may decide that fires beyond a specific level of progression (e.g., beyond the capabilities of two 1¾-in. [44-mm] lines) would result in a quick and limited primary search followed by a defensive attack.

When conditions permit, there is merit to initiating an attack on the fire using the standard preconnected hose lines, thus applying a trial-and-error method. It is often impossible to determine the area involved in fire or the size of the fire from the exterior. A crew advancing a 1¾-in. (44-mm) hose line may be in position to begin their initial attack while performing an interior size-up and evacuating occupants. However, when the fire is beyond the resources of the available units, evacuating occupants followed by a defensive attack is the only option. If the rate of flow is preincident planned, it is much easier to determine when the fire is beyond the department's capabilities to initiate a successful offensive attack.

> **NOTE**
>
> It is possible to have a small fire in a large building, but it is not possible to have a large fire inside a small building.

Selecting the Attack Hose Size

Many of the V/100 calculated flows will be less than the flow capacity of a 1-in. (25-mm) hose line, but given the variables involved in calculating rate of flow and the unknown factors in evaluating the fire from the exterior, booster line hose is clearly inappropriate for structural firefighting. And, contrary to what many believe, the energy expended in stretching a booster line may be greater than that for a 1¾-in. (44-mm) line.

The use of 1¾-in. (44-mm) attack hose lines is recommended as a minimum. The flow rate for this size hose line will be adequate to extinguish most fires when the prescribed backup hose line follows the attack hose line. The amount of work effort for 1½-in. (38-mm) hose and 1¾-in. (44-mm) hose is similar, so why not go for the extra flow?

Preconnected hose length, pump discharge pressures, and hose diameter are highly dependent on the area being protected. If the department's jurisdiction consists of single-family dwellings, anything more than a 1¾-in. (44-mm) hose line is probably unnecessary in most offensive situations. If these dwellings sit well back from the street, a 250-ft (76-m) or longer hose line may be needed. When the average hose lay exceeds 250 ft (76 m), it would be wise to use 2-in. (51-mm) or larger preconnected lines to compensate for the additional friction loss. Some departments connect one to four sections of 2½-in. (64-mm) hose to the discharge and then connect sections of 1¾-in. (44-mm) hose to the 2½-in. (64-mm) hose to reduce friction loss in long hose lays while providing the mobility of a 1¾-in. (44-mm) attack.

A preconnected 2½-in. (64-mm) line with nozzle offers another tool in the arsenal and should be considered when the department responds to commercial buildings. Having a preconnected 2½-in. (64-mm) attack hose line can also play a role in reducing the "residential mindset." Local conditions would dictate the length of the preconnected 2½-in. (64-mm) line.

Occupancy should be considered when choosing an attack hose line. As noted previously, residential fires can usually be extinguished with a 1¾-in. (44-mm) handline, and these should be considered "residential" lines. Conversely, 2½-in. (64-mm) hose lines are often required in commercial occupancies due to their large open areas (larger volume/area) as well as larger fuel load, and these should be considered "commercial" lines. When making the decision regarding which line to stretch, let the occupancy make the call.

As the hose size increases to 2½-in. (64-mm), mobility decreases. Using a larger hose size is a trade-off of mobility for an increase in flow. Handling a 2½-in. (64-mm) hose line in a small area is often unnecessary and extremely difficult. If the fire requires entering small areas to achieve extinguishment, 2½-in. (64-mm) hose lines may be a liability. Conversely, for fires in large, open areas, the 2½-in. (64-mm) hose line is often the line of choice.

It is possible to use building features, such as fire doors, to hold the fire. However, this is not an ideal tactic when initiating an offensive fire attack. Either you advance on the fire or it advances on you. Remember that the V/100 rate of flow is based on extinguishment in 30 seconds. Some fires will take longer, but steady progress should be possible if the rate of flow meets or exceeds the requirement. If water is applied at less than the required rate of flow, extinguishment will be due to either burning all available fuel or a lack of oxygen. In most cases, either of these situations is less desirable than applying the amount of water necessary to reduce the fuel temperature.

Charts from manufacturers sometimes show flow rates for variable-stream nozzles based on pump discharge pressure and the length of hose attached (e.g., 150 GPM [568 L/min] at a 200-psi [1379-kPa] pump discharge pressure through 200 ft [61 m] of 1¾-in. (44-mm) hose). Other manufacturers of variable-stream nozzles base flow rates on nozzle pressure. Smooth-bore nozzles are rated using nozzle pressure; therefore, comparisons are based on nozzle pressure as well. Nozzle pressure is the pump discharge pressure—or in the case of standpipes, the outlet pressure—minus friction loss in the hose and minus pressure loss due to elevation. Pressure loss due to elevation is 0.434 psi (3 kPa) per foot of elevation. In **FIGURE 8-8**, the friction loss for the hose is an estimate.

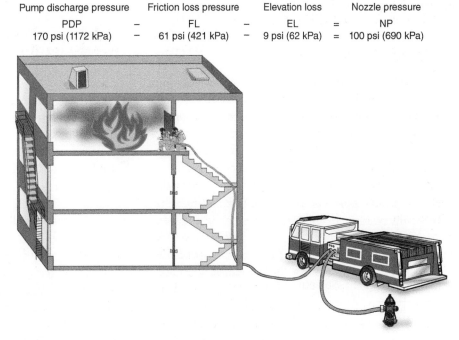

Pump discharge pressure		Friction loss pressure		Elevation loss		Nozzle pressure
PDP	−	FL	−	EL	=	NP
170 psi (1172 kPa)	−	61 psi (421 kPa)	−	9 psi (62 kPa)	=	100 psi (690 kPa)

FIGURE 8-8 Calculating pump discharge pressure.
© Jones & Bartlett Learning.

The nozzle is operating on the third floor; therefore, if each floor is 10 ft (3 m) in height, the pressure loss due to elevation is 0.434 multiplied by 20 = 8.68 psi (60 kPa). Many departments simply use a simplified calculation for elevation loss, such as 5 psi (34.5 kPa) per floor. As can be seen in Figure 8-8, the difference between nozzle pressure and pump discharge pressure can be significant.

Most of the loss in the Figure 8-8 example is due to friction loss in the 1¾-in. (44-mm) hose. Using 2½-in. (64-mm) hose would reduce the friction loss by approximately 75%, yielding a much higher flow rate. However, this higher flow rate would not be necessary in the small compartment shown in Figure 8-8. Calculating friction loss is a bit more challenging than elevation, because friction loss in hose varies by size, hose type, manufacturer, age, and condition. It is best to use calibrated pressure gauges to determine the actual friction loss in the hose carried on department apparatus. Using the manufacturer's friction loss information is a second alternative. Information provided on general friction loss tables, such as in **TABLE 8-1**, is a less desirable method and could be significantly different than the actual friction loss. Notice the difference in friction loss between 1½-in. (38-mm) hose and other hose sizes at 150 GPM (568 L/min). Many departments now use 2½-in. (64-mm) hose off standpipes to reduce the friction loss and improve the flow rate. Reducing the friction loss can dramatically improve flow through the nozzle when standpipe outlet pressures are low. However, 2½-in. (64-mm) hose can be difficult to maneuver through small rooms, such as those found in residential and some office occupancies.

TABLE 8-2 shows the flow rate for various types of nozzles "provided by manufacturers." The flows for automatic nozzles were derived from several manufacturers' data and are provided as examples. Prior to 1998, combination nozzles were rated at 100 psi (689 kPa) of nozzle pressure. Today fire departments choose from a wide variety of combination nozzles. Nozzles are available that offer adjustable flow rates while others limit the options to either low or high pressure. Combination variable-stream and solid-stream nozzles, as well as variable-stream nozzles that are designed to operate at very low pressures, are also being used. Many older nozzle styles will shut down at low pressure. In addition to variable-stream nozzles, there are smooth-bore and broken/aspirated nozzles.

TABLE 8-1 Approximate Friction Loss per 100 Feet (30 m) of Hose

Flow in GPM (L/min)	1½-in. (38-mm) Hose, in psi (kPa)	1¾-in. (45-mm) Hose, in psi (kPa)	2-in. (50-mm) Hose, in psi (kPa)	2½-in. (65-mm) Hose, in psi (kPa)
40 (151)	4.5 (31)	3 (21)	1 (7)	Negligible
60 (227)	10 (69)	5 (34)	2.5 (17)	Negligible
100 (379)	25 (172)	12 (83)	6 (41)	3 (21)
125 (473)	37 (255)	21 (145)	10 (69)	4 (28)
150 (568)	54 (372)	26 (179)	13.5 (93)	6 (41)
175 (662)	Not recommended	34 (234)	18 (124)	8 (55)
200 (757)	Not recommended	45 (310)	24 (165)	10 (69)
250 (946)	Not recommended	70 (483)	37.5 (259)	15 (103)
300 (1136)	Not recommended	95 (655)	54 (372)	21 (145)
350 (1325)	Not recommended	Not recommended	78 (538)	28 (193)

© Jones & Bartlett Learning.

TABLE 8-2 Nozzle Flows

Nozzle Type	Nozzle Pressure in psi (kPa)	Flow in GPM (L/min)	Approximate Friction Loss for 150 ft (46 m) of 1¾-in. (44-mm) Hose, in psi (kPa)	Required Pump Discharge Pressure for 1¾-in. (44-mm) Hose, in psi (kPA)	Approximate Friction Loss for 150 ft (46 m) of 2½-in. (64-mm) Hose, in psi (kPa)	Required Pump Discharge Pressure for 2½-in. (64-mm) Hose, in psi (kPA)
Standard automatic nozzle	30 (207)	Not rated				
	40 (276)	Not rated				
	50 (345)	50 (189)	6 (41)	56 (386)	Negligible	50 (189)
	75 (517)	93 (352)	18 (124)	93 (641)	4 (28)	79 (545)
	100 (689)	256 (969)	110 (758)	210 (1448)	23 (159)	123 (848)
Low-pressure automatic nozzle	30 (207)	110 (416)	23 (159)	53 (365)	5 (34)	35 (241)
	40 (276)	172 (651)	50 (345)	90 (621)	12 (83)	52 (359)
	50 (345)	193 (731)	63 (434)	113 (779)	14 (97)	64 (441)
	75 (517)	237 (897)	95 (655)	170 (1172)	20 (138)	95 (655)
	100 (689)	273 (1033)	122 (841)	222 (1531)	26 (179)	126 (869)
Smooth bore 7/8-in. (22-mm)	30 (207)	123 (466)	32 (221)	62 (427)	5 (34)	35 (241)
	40 (276)	142 (538)	37 (255)	67 (462)	9 (62)	49 (338)
	50 (345)	159 (602)	44 (303)	94 (648)	10 (69)	60 (417)
	75 (517)	195 (738)	65 (448)	140 (965)	16 (110)	91 (627)
	100 (689)	225 (852)	86 (593)	186 (1282)	18 (124)	118 (814)
Smooth bore 15/16-in. (24-mm)	30 (207)	142 (538)	37 (255)	67 (462)	9 (62)	39 (269)
	40 (276)	165 (625)	47 (324)	87 (600)	11 (76)	51 (352)
	50 (345)	184 (697)	59 (407)	109 (752)	14 (97)	64 (441)
	75 (517)	225 (852)	86 (593)	161 (1110)	18 (124)	93 (641)
	100 (689)	260 (984)	113 (779)	213 (1469)	23 (159)	123 (848)
Smooth bore 1¼-in. (32-mm)	20 (138)	206 (780)			15 (103)	35 (241)
	30 (207)	253 (958)			23 (159)	53 (365)
	40 (276)	292 (1105)			28 (193)	68 (469)
	50 (345)	326 (1234)			37 (255)	87 (600)
	75 (517)	400* (1514)			54 (372)	129 (889)
Vindicator® heavy attack	25 (172)	175 (662)	51 (352)	75 (517)	12 (83)	37 (255)
	40 (276)	210 (795)	75 (517)	115 (793)	16 (110)	56 (386)
	50 (345)	250 (946)	105 (724)	155 (1069)	23 (159)	73 (503)
	75 (517)	345 (1306)			42 (290)	117 (807)
	100 (689)	440* (1666)			66 (455)	166 (1145)

* Exceeds recommended maximum flow of 350 GPM (1325 L/min) for a handheld hose line.
© Jones & Bartlett Learning.

Incident Summary

Kmart Warehouse Fire

A fire began in the fully sprinkler-protected Kmart warehouse in Falls Township, Pennsylvania, where aerosol cans containing petroleum-based liquid fell from a pallet after a forklift entered the aisle **FIGURE IS8-2A**. A flammable liquid vapor released from the damaged aerosol cans ignited. The fire quickly spread to involve large areas of the warehouse. The aerosol cans became flaming missiles penetrating through conveyor openings in the fire wall. The fire overwhelmed the sprinkler system, making manual extinguishment necessary. The fire department had preincident planned the building and decided that the best tactic would be to confine the fire to a single quadrant **FIGURE IS8-2B**.

FIGURE IS8-2A The fire at the Kmart warehouse in Falls Township, Pennsylvania.
Bill Benton/AP/Shutterstock.

Calculating the rate of flow shows that a large fire in any quadrant would be beyond the capabilities of the fire department; therefore, using the building's construction features to contain the fire was a good plan. The height of the structure was 30 ft (9.1 m). Quadrant A was the smallest compartment at 215,600 ft^2 (20,030 m^2). If the fire flow were calculated for this building, it would be as follows:

$$A/3 = \frac{215,600 \text{ ft}^2}{3} = 71,867 \text{ GPM (272,017 L/min)}$$

$$V/100 = \frac{6,468,000 \text{ ft}^3}{100}$$

$$= 64,680 \text{ GPM (244,814 L/min)}$$

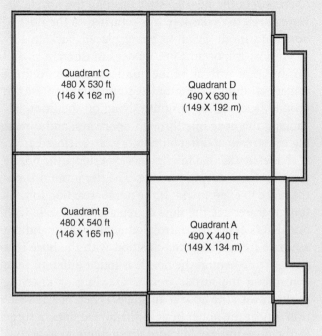

FIGURE IS8-2B Preincident plan of Kmart warehouse.
© Jones & Bartlett Learning.

These calculations are for the smallest area. The fire actually began in the larger Quadrant B. Could any fire department apply 65,000 or 70,000 GPM (246,052 to 264,979 L/min) in an offensive mode?

Could fire streams operated from entry points reach the minimum required distance of 440 ft (134 m) to the far wall of any compartment?

Adding to the problem was an unprotected steel truss roof structure. Fire departments must carefully compare the risk versus the benefit before committing to an interior operation that will ultimately prove futile. Even if the percentage modifier allowed in A/3 rate of flow calculations were used and only 1 percent of the area was heavily involved in fire, the value of property-saving operations would be questionable.

Data from: Best, Richard, 1983, "$100 Million Fire in KMart Distribution Center," *NFPA Fire Journal*, March 1983, pages 36–42; rate of flow comparisons by authors.

In choosing a nozzle it is important to identify features that will improve fire-ground operations. Problems associated with nozzle malfunctions increase with more complex nozzle designs. This is especially true if there are a wide variety of nozzles in use by the department. Do you really need to be able to adjust the flow at the nozzle? Is it important to change from a high- to a low-pressure setting?

One of the advantages of a preconnected hose line is establishing a standard pump discharge pressure, which eliminates the need to calculate hydraulics at the incident scene. However, it will be necessary to add or subtract elevation pressure to maintain a predetermined nozzle pressure. In Figure 8-8, a standard pump discharge pressure might be 150 psi (1034 kPa). With the hose line operating on the third floor, it would be necessary to add 9 or 10 psi (62 or 69 kPa) to the pump discharge pressure.

Data were also gathered from fire departments that tested nozzles. It is important to note that actual test results varied, with some nozzles of the same type and manufacture providing greater or lesser flows. For new nozzles, the flow rates were generally within 10% of the manufacturers' data. However, through actual field testing, departments discovered that improperly operating nozzles produced lower than expected flows, and hose of the same size yielded higher or lower friction loss. Both of these factors affected flow rates. Flow rates will vary according to nozzle pressure, pump discharge pressure, and length of the hose lay. Flow rates will also vary between manufacturers and even for different nozzles produced by the same manufacturer. Table 8-2 provides general comparisons for a variety of nozzles.

Probably the most important point to be made is that departments should flow test their nozzles on a regular basis to ensure they are operating properly and to confirm the flow that can be expected from the hose/nozzle configurations available on their apparatus. If any part of the apparatus, hose, or nozzle configuration is changed, flow testing should be conducted. Every department should perform flow tests on their hose and nozzle combinations using a calibrated flow meter. To do this, first place a flow meter in a hose line between a hydrant and the intake side of the pumper. The flow meter should not be connected directly to the hydrant. After making sure all drains are closed and the tank to pump line is shut, advance the standard attack hose line. Place a pressure gauge in the hose line behind the nozzle and another 100 ft (30 m) back from the nozzle. Increase the pump discharge pressure to the pressure specified in the department's SOPs. The pressure gauges will measure the actual friction loss in the hose, and the flow meter

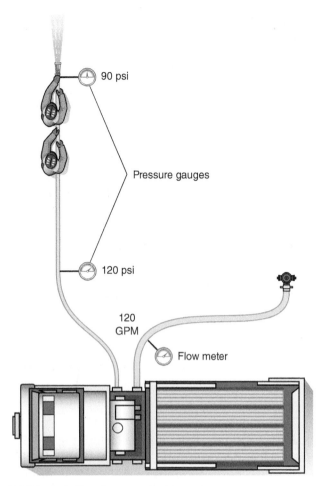

FIGURE 8-9 Flow/pressure test layout.
© Jones & Bartlett Learning.

will provide the actual flow for your hose and nozzle combination **FIGURE 8-9**. This is also a good time to try various hose layouts, including standpipe hose and nozzles. When testing standpipe hose and nozzle combinations, it is important to consider the possibility of extremely low pressures. Determine the rate of flow if the pressure at the standpipe discharge is only 40 psi (276 kPa) or 30 psi (207 kPa).

Most manufacturers do not recommend 1¾-in. (44-mm) hose layouts longer than 250 ft (76 m) because of excessive friction loss in 1¾-in. (44-mm) hose at high flows. If longer hose lines are needed, larger-diameter hose should be used to lengthen the hose line to reduce the friction loss and increase the flow. If mobility is important, the last section of the hose line could be reduced to 1¾-in. (44-mm) hose. Try different nozzles with different lengths and sizes of hose to determine the best combinations.

Training is essential regardless of the hose/nozzle combination your department chooses. If the nozzle has flow or low-pressure selection devices, be sure fire fighters are completely familiar with how to operate these devices. If the nozzle has a flush mode, it is important that fire fighters be trained on proper operation of the nozzle. After flushing the nozzle, always check to make sure it is returned to the normal operating mode prior to being placed back in service. Most important, fire fighters must train in full turnout gear with all of the nozzles available to them under conditions that simulate those found on the fire ground, such as low visibility, lying on the floor, advancing the hose line through doorways, and directing the stream at different points in an enclosed area.

It is essential that fire fighters experience the nozzle reaction force during training. The fire ground is not the place to experiment. If the nozzle reaction force is too great, reduce the flow or replace the nozzle with one that is easier to control. Remember that handling a nozzle from a standing position with the hose in a straight line behind the nozzle is much different than controlling the nozzle from a crawling or kneeling position with the hose at an angle. When nozzle reaction is near the upper limit for handheld lines, a surge in the hose line could cause fire fighters to lose control. Always allow a margin of safety. Nozzle reaction force usually increases as the flow and nozzle pressure increase. However, some nozzles are designed to reduce nozzle reaction force or are equipped with devices, such as pistol grips, that assist in controlling the hose line. When purchasing nozzles, compare flow rates and nozzle reaction force in order to select a nozzle that produces the desired flow and can be safely handled under actual fire-ground conditions.

> **NOTE**
>
> Pump discharge pressure is generally set at 150 to 200 psi (1034 to 1379 kPa) for automatic nozzles. Pump discharge pressure can be directly read at the apparatus pump panel, which simplifies pump operations. Elevation loss and friction loss in the apparatus piping and hose reduce the pressure reaching the nozzle. The friction loss in a pumper and given length of hose varies, thus the flow will be different when operating from different apparatus and when using different hose.

As noted previously, a handheld hose line can generate a flow of up to 350 GPM (1325 L/min). It is possible to control a hose line flowing considerably more than 350 GPM (1325 L/min) by forming a hose loop. However, forming hose loops to counteract the nozzle reaction force is not practical inside most buildings.

Pressures available at standpipes can be significantly lower than those provided by a fire department pumper. The pressure can be different depending on the building's elevation. Some high-rise buildings will have pressure- or flow-reducing valves on the standpipe. (See Chapter 7, *Fire Protection Systems,* for details.) These valves are designed to maintain predetermined pressures throughout the building. However, there have been occasions when these valves were improperly set, resulting in very low pressures at the standpipe outlet. At the Meridian Plaza fire in Philadelphia, Pennsylvania, improperly set pressure-reducing valves created a major problem (see Chapter 12, *High-Rise Buildings*).

NFPA 14, *Standard for the Installation of Standpipe and Hose Systems,* requires that a sign be posted, usually at the main control valve, stating the location of the two most hydraulically remote hose connections and the designed flow and pressure at those connections (NFPA 14, 2019). It is important to note the standpipe pressure and flow on the building preplan. If pressure- or flow-reducing valves are used, note the pressure and flow at locations so equipped. Without pressure- or flow-reducing valves, the most hydraulically remote outlets should present the worst-case pressure and flow scenario, with other outlets in the building providing higher pressure and flow capability. Standpipe systems must be inspected and tested on a regular basis. During preincident planning, it is important to note the most recent residual test pressures.

Standpipe outlets with a residual pressure exceeding 100 psi (689 kPa) or a static pressure exceeding 175 psi (1207 kPa) are required to be equipped with pressure-reducing valves. If a pressure higher than this is needed for fire department operations, the pressure-reducing valve should be adjusted. For most situations, the maximum outlet pressure to a flowing hose line will be 100 psi (689 kPa). Some older standpipe systems are required to provide only 65 psi (448 kPa). There may be older standpipe systems with even lower pressures due to different code requirements at the time of installation and/or reduction in interior pipe diameter due to corrosion or other factors.

Some automatic nozzles are designed to operate at a 100-psi (689-kPa) nozzle pressure (nozzle pressure will be less than the pressure at the standpipe due to friction loss). A nozzle pressure below 100 psi (689 kPa) will result in a reduced—and possibly ineffective—fire flow in a nozzle designed to operate at a 100-psi (689-kPa) nozzle pressure. If the pressure is too low, some automatic nozzles will shut down. The low-pressure nozzles listed in Table 8-2 were designed to operate at lower pressures. Solid-stream nozzles will provide some flow at almost any pressure; however, at very low pressures, the stream will be ineffective, and at very high pressures, it will be difficult to safely handle the attack hose line. The flow increases in proportion to the nozzle pressure. Some automatic nozzles can be fitted with a smooth-bore tip, while others have a built-in solid-bore nozzle. When selecting a nozzle/hose combination for standpipe operations, consider the pressure at standpipe discharges and select standpipe hose and nozzles based on the lowest pressure. If there are large pressure differences in standpipes or buildings that require a large rate of flow, it may be necessary to provide different nozzles and/or hose for these buildings. A pressure gauge with a control valve is included in the recommended standpipe equipment list. (See Chapter 7, *Fire Protection Systems*, for the full list.) This device allows the fire fighter at the standpipe outlet to reduce high standpipe outlet pressures. Changing standpipe hose from 1¾-in. (44-mm) to 2-in. (51-mm) or 2½-in. (64-mm) will result in a reduced nozzle pressure and a larger flow.

Aerial apparatus can be used as portable standpipes by placing the aerial apparatus in position at a window on the floor where a hose line is needed and attaching hose to fittings at the elevating platform or top of the aerial. This method of providing an improvised standpipe is especially effective when the aerial device has a prepiped waterway. This tactic does not place hose through the doorway from the stairs; thus, it reduces smoke infiltration from the hallway into the stairway. Apparatus pumps supply the pressure, thereby eliminating potential standpipe pressure problems. This tactic can be used to advance hose inside a building that is not standpipe equipped. When the aerial ladder or bucket is positioned at an ideal location close to the building, the placement of the portable standpipe outlet will normally be limited to the eighth floor, which is typically the maximum reach of a 110-ft (34-m) aerial ladder. One major disadvantage in using the aerial device for hose lines is that it is no longer available for removing occupants or fire fighters from the building.

Another option is using a rope to advance a fire line to an upper floor. This is a more time-consuming and complex tactic, as it typically requires several fire fighters to advance the hose line, and the hose may need to be supported at intermediate levels. Also, when using a rope to advance a hose line because the stairs are unsafe, the attack team members are at a disadvantage, because they do not have the stairway or aerial device available as a means of escape. A rope can be used as a means of escape; however, fire fighters may not have the equipment, training, or experience necessary to perform a safe and effective rope escape from an upper floor. An alternative is to use a rope to advance the hose line to the floor below the fire, then advance up the stairway to the fire floor, similar to a standpipe operation, if the stairway is safe to use.

When it is possible to conduct an offensive attack, the normal first action is to place the standard attack hose line in position, applying water to soften the target whenever possible then applying water directly on the burning material. Preincident planning or the on-scene size-up may indicate the initial use of 2½-in. (64-mm) hose.

Assuming that the first-arriving company will place a 1¾-in. (44-mm) or larger standard attack hose line in position, the trial-and-error method is generally applied. If this initial hose line does not allow the attack team to make immediate progress, more water is needed. The second crew may also lay the standard attack hose line as a backup hose line. Many times, department SOPs specify that the first-arriving engine company attack the fire and the second crew lay a backup hose line. Quite often, the IC is still in the process of developing an incident action plan as these preestablished procedures are implemented.

A mistake is made when the IC calls for another 1¾-in. (44-mm) standard attack hose line to combat a fire that is not being controlled by two 1¾-in. (44-mm) lines. When more than two 1¾-in. (44-mm) hose lines are needed to control the main body of fire, there is an obvious need for 2½-in. (64-mm) hose lines if an offensive attack is to continue. This text recommends the use of 2½-in. (64-mm) lines when the flow from two standard preconnected lines does not equal or exceed the calculated rate of flow. The reason for using 2½-in. (64-mm) hose during an offensive operation is to apply large quantities of water to the fire. The nozzle pressure and tip size will determine the actual flow from a 2½-in. (64-mm) hose stream. Fire departments should establish SOPs based on the nozzle types available and the flow requirement challenges in the community. The use of 2½-in. (64-mm) lines inside a building is an arduous task and requires the hose crew to gain experience through training under simulated fire conditions. As the flow increases, the nozzle reaction force will also increase.

Selecting the Nozzle Type

A discussion of the best type of nozzle is sure to result in a lively dialogue. Obviously, those from the indirect application school prefer finely divided fog (vapor) streams and the variable-stream nozzle that is capable of providing this pattern. In reality, there is little need for fog streams during offensive structural firefighting. There are times when a fog stream may be appropriate, but these are mostly when the fire stream is being used for exposure protection or for Class B fires.

As Table 8-2 indicates, the variable-stream nozzle operated in the straight-stream position is capable of delivering as much water as solid-stream nozzles typically attached to 1¾-in. (44-m) hose. From an empirical point of view, the solid core of water produced by the solid-stream nozzle would appear to deliver more water with better penetrating power than a straight stream or large droplet stream. A smooth-bore nozzle will deliver a fire stream even at very low pressures. Nozzle manufacturers have conducted technical tests measuring stream force at various distances that have disputed some of the claims made by smooth-bore advocates. Actual data from independent testing sources is sparse, so the controversy continues.

Stream force is an issue, as it affects the distance the stream will carry. In large undivided areas, stream reach can mean the difference between success and failure. Likewise, force allows the nozzle crew to access hidden fires behind drywall and other lightweight building materials by penetrating the ceiling or wall with a powerful, compact stream.

In addition, sprinkler research has proven that large-orifice/large-drop sprinklers have better extinguishing qualities. The distribution pattern from these heads creates larger water droplets that are better at penetrating the fire plume and reaching the burning fuel. This research appears to favor the solid- or straight-stream nozzle.

It is possible to have the best of both worlds by attaching an automatic variable-stream nozzle to the standard preconnected hose lines but with a 1½-in. (38-mm) fitting above the shut-off. These nozzles can be changed to a solid-bore stream by removing the variable-stream tip. This arrangement allows the attack crew to quickly change the nozzle without shutting down the hose line at the pump. Some nozzles are designed with a built-in smooth-bore feature just above the shut-off. Be sure to measure the flow in each mode before deciding on a nozzle capable of a variable and solid stream. Fire fighters should be trained in changing nozzle tips and understand the potential difference in nozzle reaction force.

Selecting the Method of Attack

There are several types of fire attack. The most obvious types are offensive and defensive. This chapter explains different tactics used to control a fire when using an offensive attack.

Chapter 9, *Defensive Operations,* describes defensive tactics. Generally speaking, an offensive attack involves applying water from an interior position or an exterior position in close proximity to the building. An offensive fire attack can be further described as a direct attack or an indirect attack.

Indirect Attack

The indirect attack has very limited application in structural firefighting. However, there are a few possibilities where an indirect attack may be warranted. Because the steam produced by this attack will be dangerous to fire fighters and occupants, it is obvious that this is a poor choice in occupied areas, including areas occupied by fire fighters. Tightly enclosed areas aboard ships are the best application of this attack, but shipboard fires are beyond the scope of this text. Nonetheless, picturing the tightly enclosed unoccupied hold of a ship provides some indication of when this attack may be appropriate.

In conducting an indirect attack, the idea is to keep ventilation to a minimum while introducing a wide-angle fog stream directed at the ceiling through the smallest possible opening, such as a partially closed door or window FIGURE 8-10. This approach will result in maximum steam production disruption of heat layers and an extremely humid environment. For this reason, an indirect attack is usually considered inappropriate when there is any possibility that occupants remain in the building. This tactic may prove useful in unoccupied basements, attics, and storage areas. Directing a piercing, cellar, or distributor nozzle into unvented areas has been used with success. Occasionally, a flat roof is covered by a peaked roof (rain roof) making it especially difficult to access the space between the roofs. Limiting openings to a small hole made to insert a piercing or larger nozzle may be the solution to this difficult fire situation. UL has conducted research into controlling vented and unvented attic fires. Some of this research involved making a small opening in the ceiling to apply water into an unvented attic. Although the indirect attack has extremely limited applications, fire fighters should know and understand every tactical tool.

FIGURE 8-10 An indirect attack.
© Jones & Bartlett Learning.

Direct Attack

The direct attack is preferred in the vast majority of situations where conditions permit an offensive attack. UL's horizontal and vertical ventilation research as well as live studies on Governors Island proved that softening the target and controlling ventilation result in a safer and more effective approach when confronted with a working residential fire where an offensive attack is warranted.

Applying water to the fire for 15 seconds in a small residential fire prior to making entry or just as entry is made softens the target, reducing heat. Larger rooms require water to be discharged for a slightly longer period. Controlling the door by opening it just enough to make entry, and minimizing the opening while advancing the hose line, limits air intake, retarding fire growth and allowing fire fighters to advance on the fire for a safer and more effective offensive attack. This tactic is unproven in large, undivided areas such as those found in many commercial occupancies.

Proper ventilation timed with the fire attack can significantly improve conditions in the fire area. Proper ventilation is critical to prevent flashover and make interior conditions tenable for victims and fire fighters conducting an offensive attack under heavy fire conditions. In **FIGURE 8-11**, notice that the window near the fire is opened as the attack begins, which provides a ready flow path away from the fire fighters entering the room to attack the fire. A roof vent over the fire area would also provide a flow path away from fire fighters entering the door. However, to be effective, a roof vent would require opening the ceiling between the attic and the fire area to allow smoke and heat to escape.

This is an effective vent, but it would place the attic in the flow path, possibly extending fire into the attic.

Ventilation is not always under the control of fire forces. During moderate weather conditions, windows and other openings are often left opened by occupants. In other cases, the heat produced by the fire causes a window to fail or burns through part of the structure creating an unplanned vent.

Softening the target tactic is highly recommended. This tactic is based on credible and exhaustive research by UL involving house fires and live fires conducted at Governors Island (Kerber, 2013).

Many departments are now applying the softening the target tactic. Undoubtedly, this tactic has many

FIGURE 8-11 Softening the target and controlling air flow through the door followed by a direct attack and ventilation. (The second illustration shows an indirect attack.)
© Jones & Bartlett Learning.

advantages, not the least of which is increased safety for fire fighters conducting an offensive fire attack. In most cases, a company officer will direct the initial attack. The research fires were all advanced fires in houses. The softening the target principle has wider application but does not apply to every fire.

Some examples where softening the target may *not* be applicable include, but are not limited to:

- A fire that is inaccessible from the outside at ground level (e.g., upper floor of a high-rise building)
- A compartmented area with no outside access point (e.g., center room in a large office building)
- A small fire that fails to gain intensity (e.g., food on the stove, electrical fire, clothes dryer fire)

The International Association of Fire Chiefs published a "Firefighter Safety Call to Action" in reaction to the UL research (IAFC, 2013). In part, it states the following:

> Water doesn't push fire or threaten trapped occupants:
> - Water should be applied to a fire as soon as possible and from the safest location because research has proven it reduces thermal temperatures.
> - Simply put, if you see fire, put water on it immediately. This greatly increases civilian and fire fighter survivability as well as property conservation.

If written as an SOP this could be misinterpreted to say: "Water should be applied to a fire as soon as possible from the safest location." This would miss the critical point: "*if you see fire.*" Following such an SOP could result in fire fighters opening doors or breaking windows and discharging water into an area within a building where no fire is visible from the outside or from a door opening. The UL research clearly applies to a fire that is rapidly growing toward flashover or one that is in temporary decay waiting for a breath of fresh air. In the absence of visible fire or heavy and/or pushing smoke conditions, the soften the target tactic does not apply.

Several studies by credible sources advocate this method of attack under certain conditions. However, we cannot advocate or denounce this method of attack that undoubtedly has merit, but also may be ineffective under certain circumstances.

There is a reasonable expectation that this tactic could be extended to larger compartments; however, the results may not be the same. Larger compartments have more air space and thus more available oxygen. Larger-volume fires are less likely to go into decay phase as they are less likely to become ventilation limited. Fire growth necessary to involve a large compartment could take considerably longer than rooms typically found in residential occupancies, depending on fuel load, volume of the compartment, and other factors. It is also possible to have a small fire in a large open area where the fire remains confined to a limited area prior to the arrival of fire forces.

Large commercial buildings are more likely to be protected by automatic sprinkler systems. If the system is properly designed and maintained, there is little chance of the fire spreading beyond a relatively small area.

For purposes of this discussion, either the area on fire is not protected by a sprinkler system or the sprinkler system fails to control the fire and the fire has sufficient air and fuel to develop into an uncontrolled large-area fire.

Large-volume commercial buildings tend to have a variety of fuels, some of which result in smaller, slower-burning fires, while others have fuel loads that greatly exceed those found in a residential property. The possibility of being able to soften the target from an exterior position or a position just inside the building is less likely in large commercial buildings. The size and complexity of many large commercial buildings may result in this tactic being less effective and more difficult to apply.

Consider the effects of softening the target in the buildings previously used to explain rate of flow calculations in this chapter. The apartment building in Figure 8-3 is a residential property with fairly small enclosures. It is a large building with small compartments (rooms). The height and configuration of the building may not be conducive to softening the target from the exterior in all areas and levels of the building. However, water can be applied from interior hallways into an apartment to soften the target. This same approach should be effective in a high-rise building except in large open areas, such as in the open-layout office used in the National Institute of Standards and Technology (NIST) high-rise staffing report (Averill et al., 2013) **FIGURE 8-12**.

The supermarket shown in Figure 8-4 has limited access from the exterior. Typically, this mercantile occupancy would have front windows and doors, but no direct access on the other three sides of the building. The inside of the building is probably divided into several areas, including a rear loading dock and storage areas, pharmacy, and refrigerated storage. However, there will be a large open store area that accounts for

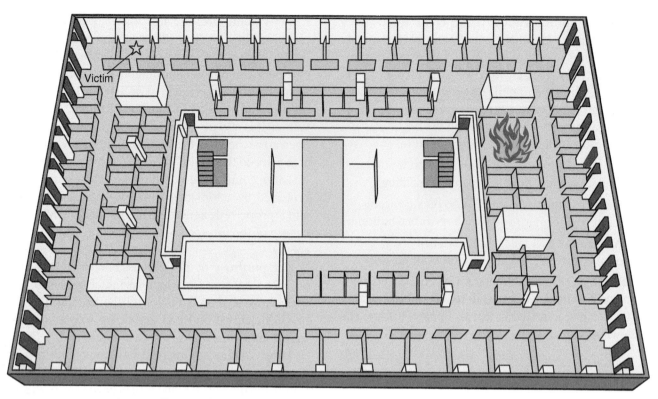

FIGURE 8-12 Open-layout office in a high-rise (from the NIST high-rise report).
Reproduced from: NIST High-Rise Staffing Report- NIST Report on High-Rise Fireground Field Experiments, NIST Technical Study Note 1797, Jason Averill, Lori Moore-Merrill, 2013.

the majority of the 500,000 ft^3 (14,158 m^3) in this example. Aisles make it difficult to apply water from a distance, and the fire could be a considerable distance from the front door. The wide-open store area provides sufficient air to support combustion for an extended period, and there could be many customers needing assistance in finding their way out of the building.

Thankfully, nearly all of these occupancies are sprinkler protected; unfortunately, sprinklers are sometimes poorly designed, poorly maintained, or simply shut down.

The auto parts store in Figure 8-5 presents an interesting scenario in terms of softening the target. The two first floor, 40- × 30-ft (12.2- × 9.1-m) compartments are not much larger than some of the biggest rooms in residential properties. There is a direct entry into each of these areas and the fire could go into decay. Unless there is a considerable quantity of flammable or combustible liquids in the area, softening the target could be an effective tactic on the first floor of the auto parts store.

The second floor provides a potential fire volume (48,000 ft^3 [1359 m^3]) that would exceed the volume of any of the fires used by UL in their residential experiments. The limited fuel supply (metal parts in cardboard boxes) would most likely produce a less intense and slower-growth fire than the modern furnishings used by UL in their research. Obstructions in the form of rack storage may hinder efforts to apply water to the fire prior to making entry or from a position just inside the door leading to this area. The efficacy of softening the target would be uncertain in this case.

The effectiveness of any offensive fire attack is questionable on the third floor of the auto parts store once the fire gains significant headway. The size of the compartment (48,000 ft^3 [1359 m^3]) combined with the extra hazard fuel load has the potential to limit fire attack to a defensive mode once the fire extends to involve a major portion of this floor.

The risk in making entry at the Kmart fire (see Kmart Incident Summary) is extreme and the benefit is near zero. The only credible advice is to stay out of the building and collapse zone.

It is essential that fire departments take advantage of the information gained through valid research. However, it is not possible to conduct live fire research for every building type, especially large buildings. As practitioners, it is up to the fire service to extend the usefulness of research by sharing positive and negative fire-ground experiences.

Final extinguishment involves applying water directly on the burning material, thereby extinguishing the fire by reducing the temperature of the burning and superheated fuel. Class A fuels must be vaporized by heating the solid materials until combustible vapors are emitted; this process is known as **pyrolysis**. By applying water directly to the burning fuel, the temperature of the fuel is reduced, thus reducing or eliminating pyrolysis. Cooling ordinary combustibles to shut down the pyrolysis process is much like shutting off the gas supply at a flammable gas fire.

When fire fighters encounter very high heat conditions, indicative of an impending flashover, directing a stream overhead (pulsing) is advised, but this should be done using short, intermittent blasts with a straight or solid stream aimed at the ceiling in a pulsing fashion before actually entering the fire compartment. If done properly, pulsing will reduce the probability of flashover while not seriously disrupting the heat balance.

Figure 8-11 illustrates softening the target at the entry point. Another way to soften the target is to apply water to the fire from an exterior opening. This approach is sometimes referred to as a *transitional attack*. The transitional attack involves directing a stream from the exterior into a room that is involved in fire where the fire has self-vented **FIGURE 8-13**. Just as with the softening the target attack in Figure 8-11, the fire stream is applied for a short time (usually 15 seconds or less) to knock down a heavy volume of fire, making entry possible. The fire stream should be a handheld solid or straight stream, not a fog stream or master stream. A fog stream could have a negative impact on venting and affect the flow path. Rapidly moving a solid or straight stream from side to side or up and down had a similar negative effect (Kerber, 2013).

From a safety standpoint, the officer should, before entering an area of extreme heat, conduct a risk-versus-benefit analysis. If there is a high potential for saving lives, the continued advancement may be justified. If the goal is keeping the fire out of the common hallway or stairs, it may be better to use the door to the apartment to contain the fire while protecting the stairs **FIGURE 8-14**. Each situation is different, and no single approach is correct for every incident.

FIGURE 8-13 Transitional attack: defensive to offensive.
Courtesy of Rick McClure.

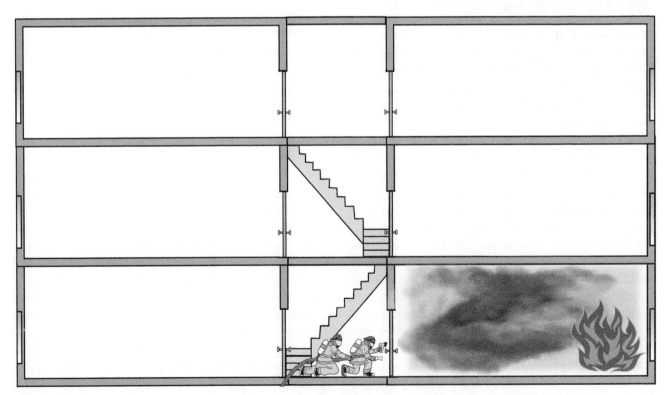

FIGURE 8-14 Fire contained by building features.
© Jones & Bartlett Learning.

FIGURE 8-15 Deflecting water off the ceiling: modified direct attack.
© Jones & Bartlett Learning.

There are times when the fire is shielded by nonburning materials, such as in a warehouse with rows of high-stacked commodities. If possible, the hose should be moved to a position that allows a direct attack. If this is not possible, then deflecting a solid or straight stream off of the ceiling and allowing the droplets to fall on the fuel is an alternative **FIGURE 8-15**. The objective is to get the water directly on the fuel. This tactic should be considered a derivation of a direct attack.

Estimating the Number of Attack Hose Lines

Previous examples show that a single 1¾-in. (44-mm) hose line could probably extinguish most fires within a dwelling. It is good practice to lay a backup hose line to protect egress routes and bolster the attack. Even trial-and-error methods become more successful when provisions are made to double the size of the attack. Although the number of lines needed will be based on flow requirements, there must always be a reserve. Simply stated, enough lines must be placed in service to apply the necessary rate of flow and protect egress routes for interior fire fighters, should conditions deteriorate.

In addition, the IC is responsible for predicting where the fire is going. Construction features may hold a fire within a given area or allow it to travel horizontally and/or vertically throughout the structure. In the case of an apartment building with a common attic, it is necessary to have truck companies gain access to and check for extension in this hidden space while an engine company provides a hose line to limit fire extension. Just as it is necessary to perform search and rescue operations above the fire, it is also necessary to place hose lines above the fire.

Evaluating Exposures

Estimating the number of hose lines above the fire addresses the need to protect interior exposures. Hose lines are needed to prevent the fire from moving through natural pathways such as concealed spaces, stairs, chutes, and shafts. Fire can also extend vertically up the exterior of the building from windows or other openings on the fire floor(s). All of these potential routes of fire extension within the building are classified as **internal exposures** and need to be checked and protected with a hose line as required.

Fire can also extend to adjacent buildings (**external exposures**). Ventilation crews must be aware of this potential, because improper ventilation can spread the fire to exposed buildings—just as proper ventilation can modify a flow path, directing fire and smoke away from exposures.

Nearby buildings and building occupants may need to be protected once the fire breaks out of the containment of the building of origin. Evaluating the need to defend nearby buildings with additional hose streams involves many factors, including the following:

- Life hazard in the exposure
- Proximity to the fire building
- Wind direction
- Height of exposure compared to the level of fire involvement in the fire building
- Hazard presented by the exposed occupancy (e.g., explosive or chemical storage)

The threat to human life is always the first consideration. In the case of exposure buildings, it is important to keep fire from entering or otherwise threatening people in the nearby buildings. Often a protective stream can be placed between the fire building and exposure to limit the hazard **FIGURE 8-16**.

When evaluating exposures, be sure to consider the location of fire apparatus and equipment parked near the fire building **FIGURE 8-17**. Protecting exposures can involve hose lines on the exterior or interior of the exposed building. When working on the exterior, master streams are most often the streams of choice.

FIGURE 8-16 Protecting an exposure using a master stream.
© Jones & Bartlett Learning.

FIGURE 8-17 External exposures.
© Jones & Bartlett Learning.

Estimating Backup Needs

Backup hose lines are needed to protect the crew on the initial attack hose line and to provide additional flow when required. Given that a single 1¾-in. (44-mm) hose line is sufficient for most residential fires (in terms of rate of flow), the backup hose line is additional insurance. In larger-area fires, the backup hose line again provides insurance against rate of flow miscalculation and provides additional protection for retreating fire fighters. Backup hose lines should be at least as large as the initial attack hose line. If you started the attack with 2½-in. (64-mm) hose, the backup hose line should be a 2½-in. (64-mm) hose line. If the attack began with 1¾-in. (44-mm) hose, the backup hose line could be 1¾ in. (44-mm), 2 in. (51-mm), or 2½ in. (64-mm).

Although backup hose lines can be used to augment the initial attack, the defined purpose of a backup hose line is to provide flow in addition to what is needed to extinguish the fire, thus providing a measure of protection for fire fighters combating the fire. If the backup hose line is used to augment the flow rate and the fire is not immediately extinguished, another backup hose line is needed.

Estimating the Number of Hose Lines Needed Above the Fire

It is a good practice to get a hose line on the floor immediately above the fire. Where there is a risk of extension to concealed spaces and attics, additional precautionary lines are needed at each of these areas. Backup hose lines may also be needed in other areas above the fire **FIGURE 8-18**.

FIGURE 8-18 Attack and additional hose streams.
© Jones & Bartlett Learning.

Evaluating Other Hose Lines Needed

On occasion, hose lines are needed below the fire to protect egress routes and to protect external exposures. The total number of lines needed is as follows:

- Hose line(s) required to meet the rate of flow in the immediate fire area
- Backup hose line(s) positioned in or near the immediate fire area
- Hose line(s) to protect egress routes
- Hose line(s) to protect internal exposures (e.g., attic where vertical travel is likely, upper floors, concealed spaces)
- Other backup hose line(s) as needed
- Hose line(s) or master stream(s) to protect external exposures

Estimating Water Supply Needs

Prior to the widespread use of large-diameter hose, the officer assigned to the first-arriving unit in an area with hydrants had to decide whether to rely on a marginal water supply by forward laying a single 2½-in. (64-mm) or 3-in. (76-mm) supply line or use a reverse lay with a direct hydrant connection, which could result in placing the pumper a considerable distance from the fire building. Large-diameter hose eliminates this dilemma and has proven to be an effective tool in providing an adequate water supply at the fire scene FIGURE 8-19. In an area with hydrants, providing at least two water supplies for a working structure fire is considered good practice. In areas with a marginal or nonexistent water supply, estimating and providing an adequate water supply can be a substantial challenge. Fortunately, many of the fires requiring large fire flows occur in areas with reasonably good water supplies. When a significant water supply challenge exists in an area with a limited water supply, operational options should be outlined in SOPs and in the preincident plan.

Attacking a fire with anything less than a water supply from a source of water that is capable of supplying flow requirements over an extended period of time is a gamble. Attack pumper tactics (operating with water from the apparatus tank) work well in many situations, especially in fighting a fire in a small, single-family dwelling. However, relying solely on water from the apparatus water tank can be dangerous.

FIGURE 8-19 Large-diameter hose supply.
Courtesy of David J. Jones, Cincinnati, Ohio.

Many things can happen that will cause fire fighters to lose their water supply. When 1¾-in. (44-mm) lines are used, even a 1000-gallon (3785-L) water tank can supply water for less than 10 minutes. Often, this water discharge time is sufficient. However, if fire fighters need to attack fire in multiple compartments or they are not able to rapidly control the fire for any reason, they may deplete their water supply before exhausting their SCBA air supply.

If an attack pumper concept is implemented, its use must be well understood by all members, and it should be limited to fires in relatively small structures. Where the water supply is limited or nonexistent, consider using two pumpers or other means, such as a water tender, or drafting to augment the water supply to the pumper.

> **NOTE**
>
> A 1000-gallon (3785-L) booster tank, compared to a 500-gallon (1893-L) tank, will add over 2 tons (1814 kg) to the weight of the pumper, which could be problematic, especially in hilly terrain.

Some large urban areas with almost unlimited water supplies will have areas containing small ancillary buildings, such as a shed on a golf course located a considerable distance from a hydrant. In many cases, a long, complex relay is planned to deliver water to these properties. Rather than relying on relay operations, it might be better to use multiple attack pumpers to deliver the needed water. If this is the plan it should be included in the preincident plan.

It is important to remember that a large, effective flow of water for a short period of time will require less water and be more effective than a lesser flow over a longer period. For example, consider a 25- × 16-ft (7.6- × 4.9-m) room with 10-ft (3-m) ceilings in a single-family dwelling. The required rate of flow is as follows:

$$\frac{25 \times 16 \times 10 \text{ ft}}{100} = 40 \text{ GPM (151 L/min)}$$

If the apparatus booster tank contains 500 gallons (1893 L), a 1-in. (25-mm) booster line could discharge 20 GPM (76 L/min) for 25 minutes, or longer than the air supply in a 30-minute SCBA air cylinder. Even a 45-minute SCBA could be depleted as the crew would probably be "on-air" before actually discharging water. The entire 500 gallons (1893 L) would likely be exhausted, and the fire would not be extinguished because the rate of flow was insufficient to suppress the fire.

If 150 GPM (568 L/min) were discharged, the fire would be extinguished in less than a minute, with less than 150 gallons (568 L) of water. In the article by Edwards from *Fire Engineering*, his calculations would indicate extinguishment in about 2 seconds, requiring only 5 gallons (19 L) of water at nearly four times the required flow (Edwards, 1992). Seldom is water applied at maximum efficiency, but the fact remains that the higher-flowing 1¾-in. (44-mm) hose line would quickly extinguish the fire.

The need for a continuous supply of water having been established, standard guidelines should be followed. The use of a single, unpumped 2½-in. (64-mm) supply line is discouraged. From a water supply standpoint, this single line is much better than total reliance on the apparatus water tank, used with the attack pumper, and may be acceptable in residential fires. However, double 2½-in. (64-mm) or 3-in. (76-mm) lines are far superior to the single unpumped 2½-in. (64-mm) or 3-in. (76-mm) supply line. Furthermore, in using 2½-in. (64-mm) or 3-in. (76-mm) supply lines, both the volume and pressure can be improved by assigning a second pumper to relay the water from the hydrant. However, pumping into large-diameter supply hose (4 to 5 in. [101 to 127 mm]) during offensive operations is rarely necessary.

Fallacy	Fact
When the water supply is limited, ration the water so that it will last until a tanker shuttle, relay, or other reliable source of water can be supplied.	A rate of flow below the required flow will be ineffective and waste water. A rate of flow meeting or exceeding the flow requirement will be effective and use less water.

More often, it would be better to position the extra relay pumper in a position to provide a separate water supply or assign it to staging. Only in situations where flows in excess of 1000 GPM (3785 L/min) are anticipated from a single pumper or for extremely long relay lines (over 700 ft [213 m]) is it sometimes necessary to place another pumper in a large-diameter hose water supply layout. Furthermore, once the pumper is placed in operation, the apparatus operator must remain at the pump panel. Thus, unnecessary use of apparatus also results in a reduction of available staffing.

Just as a backup hose line is needed inside structures, obtaining a second source of water is a good practice. This second source of water is essential at working fires, especially when operating off dead-end mains or limited flow hydrants.

Water supply is not enhanced by securing a second source of water off the same dead-end main. Remember, for offensive operations, a minimum water supply layout must be capable of supplying the calculated rate of flow plus backup hose lines and other lines used to protect exposures. Having a second water supply provides a backup source if the initial pumper or hose layout is incapacitated. There have been many instances when a failure in the water supply resulted in a controlled fire getting out of control.

Water supply should be addressed in department SOPs. The initial water supply is crucial. Companies responding must be given clear instructions regarding the water supply. See Chapter 2, *Procedures, Preincident Planning, and Size-Up,* for additional information regarding preplanned water supplies.

Fallacy	Fact
Every pumper should be supplied with water.	There should be at least two separate sources of water whenever possible. In initiating an offensive attack, the total water supply should exceed the calculated rate of flow plus additional lines.

Estimating Ventilation Needs

Venting is most often used to assist in extinguishment efforts by controlling the flow path and making conditions tenable for fire fighters and victims. In Chapter 6, *Life Safety,* venting is also discussed as a life-safety tactic. Timing is crucial when the venting tactic is intended to support extinguishment. The ideal scenario calls for making the ventilation opening just as the attack team enters the fire area. Vertical and horizontal vent placement should be as close to the fire as safety permits and located to direct smoke and heat away from fire fighters making entry. The drawings in **FIGURE 8-20** illustrate a step-by-step progression in a properly timed roof vent.

Fire fighters on the roof are endangered due to the vent opening blocking their egress path and fire fighters inside the room below are endangered due to the roof vent redirecting the fire toward their position. Improperly placed or poorly timed ventilation can hinder extinguishment by spreading the fire into unburned areas.

FIGURE 8-21 shows an improperly placed vent made after fire fighters enter the building to

FIGURE 8-20 Proper venting for extinguishment.
© Jones & Bartlett Learning.

FIGURE 8-21 Improperly placed vent.
© Jones & Bartlett Learning.

extinguish the fire. This vent location would cut off the pathway to the escape ladder on the roof and cause the fire to spread toward the attack team inside the building.

Horizontal and vertical venting should be timed and reversible (open the window rather than breaking it) whenever possible. Proper ventilation allows fire fighters to approach the fire and apply water directly to the burning materials. It also provides oxygen, which intensifies the burning rate, particularly if the fire is ventilation controlled. Therefore, when venting is being carried out to support extinguishment, attack crews with charged lines should be positioned before venting or before opening a door leading directly to the fire area. As entry is made, the door should be controlled to reduce the air flowing into the fire area. Partially close the door, allowing room for hose. The UL flow path study verifies this tactic (Kerber, 2013).

Improper vent placement can channel the fire into concealed spaces or otherwise extend the fire. Remember, regardless of our best intentions, fire and the products of combustion are bound by the laws of physics. When a door leading from the hallway to the room that is on fire is opened, smoke and heat will vent into the hallway unless another vent path is provided. If vertical ventilation is provided via the stairway, expect the fire to enter the stairs unless extinguishment efforts are immediately successful and an alternate flow path established.

If vent openings can be controlled, positive-pressure venting virtually ensures the vent timing, as opening the door to enter the fire compartment allows the pressure to enter the area and exit through an existing or created vent opening. As with natural ventilation, it is imperative that a hose line be ready to immediately attack the fire.

Fires in small compartments are easily ventilated using positive pressure. Research and field use have proven the effectiveness of properly applied positive-pressure ventilation in residential and other small buildings. The effectiveness of positive-pressure venting in larger and more complex buildings was in question until NIST conducted extensive positive-pressure ventilation research in larger buildings. NIST studied the results of large-scale fires in an unused high school as well as high-rise buildings. The results of the school experiments proved that positive-pressure ventilation can be effective in large, complex buildings (Kerber and Madrzykowski, 2008).

Of particular interest were live fire experiments in a 340,078-ft^3 (9630-m^3) school gymnasium. As expected, this large open compartment took longer to fill with smoke and in most cases did not become ventilation limited. Positive-pressure ventilation was effective in reducing heat and clearing smoke during fires involving the moderate fuel loads expected in a gymnasium.

A larger fuel load was introduced to study how effective positive-pressure venting might be under a heavy fuel load. During this experiment, 160 pallets were added to the wooden bleachers and foam padding usually found in a gymnasium. This fuel load resembles the fuel load found in many high-hazard storage facilities. This fire was allowed to fully develop. Positive pressure combined with vertical ventilation resulted in flashover, with the temperature reaching 1290°F (700°C) throughout the gym. This result would be invariably fatal to any remaining occupants and fire fighters inside the building. However, it was noted that conditions inside the gym prior to positive-pressure ventilation and flashover were already lethal. The NIST experiments proved the usefulness of positive-pressure ventilation in large areas, but also revealed limitations in large building compartments containing large fuel loads.

Positive-pressure ventilation, like all other ventilation, can also have negative effects. For example, if the fire has entered large or vented concealed spaces, the vent openings will be uncontrolled and positive-pressure venting is likely to promote fire spread. Consider the effect of positive-pressure ventilation if there is an opening between the room on fire and the attic. The positive pressure could extend the flow path into the attic area and greatly accelerate the spread of fire into the attic. The application of positive-pressure ventilation in high-rise buildings is addressed in Chapter 12, *High-Rise Buildings*.

Ventilation is an important tactic. The IC makes decisions regarding what type of ventilation is best and coordinates hose team and vent crew activities. Fire officers should be completely familiar with the various ventilation methods.

Venting is normally considered an offensive tactic; however, the **trench cut** is sometimes referred to as a defensive tactic. Because the trench cut normally involves interior operations, it falls under the definition of offensive operations used in this text. According to another definition, defensive can mean establishing a fire break and stopping the fire spread at that point. In this sense, the trench cut is defensive, as the objective is to protect interior exposures. The trench cut provides a fire break, separating the burning area from the uninvolved area. To be successful, it must be made well ahead of the fire and effectively separate the burned from the unburned. It takes considerable time and effort to cut a trench completely across a roof. It is imperative that this tactic be planned with sufficient lead time.

Trench cuts can be used in any large structure and are particularly suited to U-shaped apartment buildings and garden apartments. The largest trench cut operation on record was at Tinker Air Force Base in Oklahoma (Goodbread, 1985), where several long trenches were cut in a roof covering 2,531,965 ft^2 (235,227 m^2) for a fire operation spanning 3 days.

Calculating Staffing Needs

Staffing requirements are many and varied. Hose and ladder crews, rescue crews, rapid intervention crews (RICs), incident management system positions, and a tactical reserve are examples of assignments that must be adequately staffed at the scene of a working structure fire. In addition, the IC needs to consider the total extinguishment staffing requirements based on rate of flow, backup hose lines, placement of lines above the fire, and establishment of a secondary water supply. NFPA 1710, *Standard for the Organization and Deployment of Fire Suppression Operations, Emergency Medical Operations, and Special Operations to the Public by Career Fire Departments*, defines tasks and minimum staffing for the initial response to a 2000-ft^2 (186-m^2) two-story single-family dwelling with no basement and no exposures (NFPA 1710, 2020). These tasks and staffing include the following:

1. Establish command (minimum one fire fighter).
2. Establish a minimum uninterrupted water supply of 400 GPM (1514 L/min) (minimum one fire fighter).
3. Establish an attack and backup hose line flowing at least 300 GPM (1136 L/min) (minimum four fire fighters).
4. Assign one support person per line (minimum two fire fighters).
5. Assign at least one search and rescue team (minimum two fire fighters).
6. Assign at least one vent/ladder team (minimum two fire fighters).
7. Establish an IRIC and sustained rapid intervention crew (minimum two fire fighters).
8. If aerial is in use, assign a person operating the aerial (minimum one fire fighter).

On some occasions, the first-arriving unit(s) handles the fire, and the total response defined in NFPA 1710 is not required. For example, the first-arriving engine and truck companies conduct the initial attack on a mattress fire inside a residential property. The engine extinguishes the fire and provides an uninterrupted water supply. Truck company members conduct a quick search and rescue with one crew while the other crew vents the building.

The minimum staffing requirements outlined in NFPA 1710 were the object of considerable debate within the fire service, with many fire service and community leaders disputing the need for four-person minimum staffing or the total staffing required for a typical two-story single-family detached dwelling. In order to prove or disprove the NFPA 1710 staffing requirements, NIST constructed the 2000 ft^2 (186 m^2) two-story single-family dwelling described in NFPA 1710 then conducted 60 full-scale experiments to determine the impact of crew size and total staffing (Averill, 2010). A panel of fire service technical experts identified 22 tasks that should be completed. In these experiments, crew sizes varied from two to five fire fighters per company. The experiments provided scientific evidence that the smaller the crew size, the longer it takes to complete the tasks necessary to successfully combat the single-family dwelling fire identified in NFPA 1710. Especially important was the fact that, on average, it took the four-person crews 30% less time to complete the primary search than the two-person crews. The average time required to stretch the initial hose line was 87 seconds longer for the two-person crew compared to a four-person crew.

It is virtually impossible to conduct safe and effective operations at structure fires when remote fire companies are staffed with fewer than four fire fighters. When two or more fire department units respond from the same fire station, understaffed units may be able to combine staffing to form one adequately staffed company. When a single understaffed fire company located at a remote station encounters a working structure fire, they are forced to make difficult decisions regarding their personal safety versus their commitment to saving lives and property.

For example, a single-engine company fire station (Engine 1) staffed by two fire fighters protects an outlying part of the jurisdiction. The second-due engine company (Engine 2) is located 7 miles (11.3 km) from Engine 1's quarters. The average travel time between stations is 10 minutes. Continuing with this example, a house fire is reported a few blocks away from Engine 1's station. Engine 1 with two fire fighters arrives at the scene of a working house fire within 2 minutes. Engine 1 has several options, none of which is both safe and effective:

- One fire fighter advances a hose line inside the building while the other operates the pumps. (unsafe and ineffective)
- The pump operator sets up the pumper and both fire fighters make entry. (unsafe)
- Both fire fighters working together forward lay a supply line to the pumper then lay a 1¾-in. (44-mm) line to the front door, but do not make entry until Engine 2 arrives. (ineffective)

There are other possible tactical options, but none would be safe *and* effective. With only two fire fighters at the scene, either one or two fire fighters make entry with no IRIC, or extinguishment is substantially delayed, which could result in the total loss of the building and civilian fatalities if occupants are unable to self-evacuate.

A three-person crew would be better but still would result in either an unsafe and/or ineffective operation. Although NFPA 1500, *Standard on Fire Department Occupational Safety, Health, and Wellness Program*, permits crews to make entry without an IRIC when there is the potential to save lives, failing to staff an IRIC always compromises safety (NFPA 1500, 2018).

The 2000 ft^2 (186 m^2) two-story single-family dwelling with no basement and no exposures is sometimes referred to as a "low-hazard occupancy." NFPA 1710 outlines the minimum safety requirements for this occupancy. Even in this typical two-story residential building, staffing needs could be greater than those listed in the standard.

The typical four-person staffing on a fire company consists of an officer, a driver, and two fire fighters. This minimum safe staffing also requires some compromises in terms of safety and effectiveness when responding from a remote station. Using the same example, but with four fire fighters on the first-arriving engine company, the IRIC by necessity will be involved in accomplishing other necessary tasks. In most cases, the pump operator is assigned as one member of the IRIC. In the event of a mayday, rescuing the interior fire fighters will require the pump operator to abandon the pump panel at the apparatus. In many cases, the fire fighter assigned to connect to a fire hydrant is assigned as the second member of the IRIC. This fire fighter could be located a considerable distance from the fire building or may not be able to complete the water supply assignment.

Working fires in a complex or large building, a building with an extraordinary life-safety hazard, buildings containing an extra hazard fire load, and areas with poor water supplies require additional staffing. As a staffing example, consider a fire on the second floor of the auto parts store example shown in **FIGURE 8-22**.

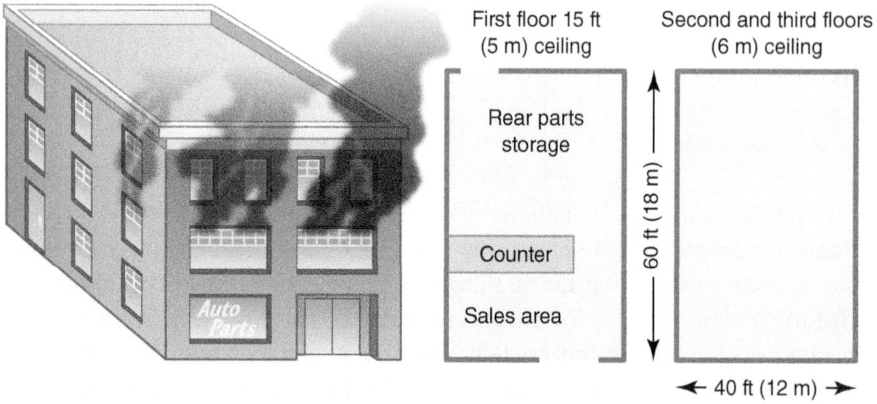

FIGURE 8-22 Auto parts store example.
© Jones & Bartlett Learning.

The nearest exposure is 50 ft to the right. Each fire company is assigned an officer, a driver, and two fire fighters. The standard preconnect is a 200-ft (61-m) 1¾-in. (44-mm) hose line with a 7/8-in. (22-mm) smooth-bore tip. The department's SOP establishes a 65-psi (448-kPa) nozzle pressure, resulting in a flow of 184 GPM (697 L/min).

It is 7:00 AM on Sunday, and the building is believed to be unoccupied. A one-alarm response is dispatched for a reported fire on the second floor **TABLE 8-3**. You are the lieutenant on Engine 1.

- Develop an on-scene report, and describe your initial actions.
- Considering your first actions, assume the role of the District 1 (IC). Assign other first-alarm companies.
- Place companies on an Incident Command System (ICS) chart.
- What conditions would prompt you to request additional alarms?

Engine 1 (First-Arriving Engine Company)

Assume the role of the Lieutenant in charge of Engine 1 (the first arriving company). The building has been preincident planned with an estimated rate of flow of 480 GPM (1817 L/min) based on V/100 calculations (sprinkler calculations are also 480 GPM [1817 L/min] in this case). You take command and give an initial report indicating a working fire on the second floor while working with another member of the crew in advancing a standard preconnected hose line to the second floor **TABLE 8-4**.

As per SOPs, a forward lay of 5-in. (127-mm) hose supplies the initial operation. The apparatus operator remains at the pumper, while the fire fighter assigned to the hydrant hooks up and then advances toward the entry to assist in advancing hose into the building and controlling the door as part of the IRIC.

There is a slight chance that water applied from Engine 1's hose line will have an effect on the fire; however, the flow being applied is well below the 480-GPM (1817-L/min) required rate of flow based on the size of the fire area. This is where trial and error plays a role. If the fire does not involve the entire second floor, it is possible that the current attack will be successful.

TABLE 8-3 Staffing with One Hose Line in Operation

Company	Total Staffing	Staffing at Apparatus	Assignment	Staffing at Assignment	Flow
Engine 1	4	2 (IRIC)	Second floor	2	184 GPM (697 L/min)
Totals	**4**	**2**	**(N/A)**	**2**	**184 GPM (697 L/min)**

© Jones & Bartlett Learning.

TABLE 8-4 Responding Fire Companies, Chief Officers, and Emergency Medical Resources

	Engine Companies	Truck Companies	Other
Initial 1 Alarm Response	1, 2, 3, 4	1, 2	District Chief 1 Heavy Rescue 1
2nd Alarm	5, 6, 7, 8	3, 4	District Chief 2 Heavy Rescue 2 Emergency Medical Services (EMS) 1
3rd Alarm	9, 10, 11, 12	5, 6	District Chief 3 EMS 2 Fire chief Operations assistant chief Safety chief
4th Alarm	13, 14, 15, 16	7, 8	District Chief 4 EMS 3 Personnel assistant chief Fire prevention assistant chief
5th Alarm	17, 18, 19, 20	9, 10	Police District 1 Captain Safety department manager

© Jones & Bartlett Learning.

Otherwise, an additional hose line (preferably a larger diameter hose), will be needed. In any case, there is still a critical need to back up the hose line operating on the fire floor, secure an additional water supply, and get a hose line to the third floor. There is also a critical need to check the floor above the fire and vent.

Engine 2 (Second-Arriving Engine Company)

In the absence of orders, SOPs for this department call for the second-arriving engine company to advance an additional hose line to the fire floor to increase the rate of flow or provide a backup hose line for the first-in attack crew. Otherwise, they are to advance a hose line to the floor above the fire. Realizing that there is a working fire in a preincident-planned building that requires a large rate of flow, the officer of Engine 2 plans to back up Engine 1 by advancing a 2½-in. (64-mm) hose line with a 1⅛-in. (29-mm) smooth-bore nozzle into the building. Engine 2 does not secure a second source of water. The Engine 2 officer radios command to advise that a second source of water has not been secured.

TABLE 8-5 Staffing with Two Hose Lines in Operation

Company	Total Staffing	Staffing at Apparatus	Assignment	Staffing at Assignment	Rapid Intervention Crew	Flow
Engine 1	4	2	Second floor	2	2 (IRIC)	184 GPM (697 L/min)
Engine 2	4	1	Second floor	3	0	305 GPM (1155 L/min)
Totals	8	3	(N/A)	5	2	**489 GPM (1851 L/min)**

© Jones & Bartlett Learning.

The remaining three crew members from Engine 2 advance a 2½-in. (64-mm) hose line off Engine 1's pumper **TABLE 8-5**. Engine 2 is parked out of the way of other arriving apparatus, but in a position to be ready for response to protect exterior exposures, should it become necessary.

There is a good chance that water applied from these two hose lines will be adequate to control the fire, as the flow being applied is slightly above the 480-GPM (1817-L/min) required rate of flow. Had the department opted for a 125-GPM (473-L/min) 1¾-in. (44-mm) nozzle, the flow rate would have been slightly below the needed rate of flow. Again, this is where trial and error plays a role. If the fire is being extinguished with the present attack, the flow is sufficient. Otherwise, an increase in pump discharge pressure to 200 psi (1379 kPa) for the 1¾-in. (44-mm) hose line with automatic nozzle would increase the flow by approximately 50 GPM (189 L/min), thus meeting the required rate of flow when used with the 2½-in. (64-mm) line.

Reconnaissance is the best way to determine the need for more lines to control the fire. In any case, there is still a critical need to back up the hose lines operating on the fire floor, secure an additional water supply, and get a hose line to the third floor.

Truck 1 (First-Arriving Truck Company)

If Truck 1 arrives early, there may be a need for its crew to force entry into the building. Otherwise, the engine company crews will be responsible for this task. The truck company's crew will split up—two fire fighters providing truck duties such as primary search and horizontal ventilation for the engine companies working on the second floor. The other two members of the truck company will work above the fire, conducting the primary search and checking for fire extension on the third floor **TABLE 8-6**.

Chief Officer

The first-arriving chief officer establishes an exterior stationary command post and confers with the company commander of Engine 1, who is serving as the IC **TABLE 8-7**. Engine 1's officer advises the chief officer that slow progress is being made on the fire. After a careful size-up, the chief officer assumes the position of IC and develops the following incident action plan:

- Extinguish the fire.
- Conduct a primary search of all floors. (Companies currently on the second and third floors will be responsible for these floors.)
- Ladder the building to provide alternative egress for fire fighters.
- Assign a complete company as a dedicated RIC—officer, driver, and two fire fighters.
- Advance a backup hose line to the second floor.
- Advance a hose line to the third floor.
- Secure a second water supply.
- Stage an engine and truck company as a tactical reserve.
- Plan for a possible defensive operation.
- Set up a rehabilitation area.
- Provide for medical treatment and transportation.

TABLE 8-6 Staffing with Two Engine Companies and a Truck Company

Company	Total Staffing	Staffing at Apparatus	Assignment	Staffing at Assignment	Rapid Intervention Crew	Flow
Engine 1	4	2	Second floor	2	2 (IRIC)	184 GPM (697 L/min)
Engine 2	4	1	Second floor	3	0	305 GPM (1155 L/min)
Truck 1	4	0	Second floor	2	0	0
Truck 1	(see above)	0	Third floor	2	0	0
Totals	12	3	(N/A)	9	2	489 GPM (1851 L/min)

© Jones & Bartlett Learning.

TABLE 8-7 Staffing with Two Engine Companies, One Truck Company, and a Chief Officer

Company	Total Staffing	Staffing at Apparatus	Assignment	Staffing at Assignment	Rapid Intervention Crew	Flow
Engine 1	4	2	Second floor	2	2	184 GPM (697 L/min)
Engine 2	4	2	Second floor	2	0	305 GPM (1155 L/min)
Truck 1	4	0	Second floor	2	0	0
Truck 1	(see above)	0	Third floor	2	0	0
Chief officer	1	0	Command post	1	0	0
Totals	13	4	(N/A)	10	2	489 GPM (1851 L/min)

© Jones & Bartlett Learning.

This would be a full first-alarm response for some departments, and in many cases mutual aid would be necessary to muster the needed staffing to begin operations. Most urban departments would send a larger response to a fire in a commercial building. Comparing the tasks being performed, the staffing at the scene, and considering the occupancy of a commercial auto parts store, additional staffing would be required. In fact, the IC's incident action plan includes several additional tasks that have not been staffed, and there is no tactical reserve. The two-person minimum is assigned as an IRIC and would not be sufficient. Continuing operations will thus require additional resources and the IC should consider an extra alarm early in the incident.

TABLE 8-8 Staffing with Three Engine Companies, One Truck Company, and a Chief Officer

Company	Total Staffing	Staffing at Apparatus	Assignment	Staffing at Assignment	Rapid Intervention Crew	Flow
Engine 1	4	1	Second floor	2	1	184 GPM (697 L/min)
Engine 2	4	0	Second floor	3	1	305 GPM (1155 L/min)
Truck 1	4	0	Second floor	2	0	0
Truck 1	(see above)	0	Third floor	2	0	0
Chief officer	1	0	Command post	1	0	0
Engine 3	4	1	Third floor	3	0	184 GPM* (697 L/min)
Totals	17	2	(N/A)	13	2	489 GPM* (1851 L/min)

*The flow column reflects the rate of flow required to extinguish the main body of the fire. Engine 3's hose line is on the floor above; it is not added to the total flow calculation. The two water supplies (Engines 1 and 3) can easily handle the total potential flow of 673 GPM (2548 L/min) if the hose line on the third floor is operated.
© Jones & Bartlett Learning.

Engine 3 (Third-Arriving Engine Company)

Engine 3 is instructed to supply its apparatus with a 5-in. (127-mm) supply line and to advance a 1¾-in. (44-mm) hose line with a 7/8-in. (22-mm) smooth-bore tip to the third floor **TABLE 8-8**.

Engine 4

Engine 4 will back up the two lines operating on the fire floor with a 2½-in. (64-mm) hose line laid from Engine 3's apparatus **TABLE 8-9**.

At this stage, the members of Truck 1 should be free to complete a primary search and check for extension on the third floor.

Heavy Rescue, Truck 2, and Engine 5

Truck 2 is assigned as the dedicated RIC with one member assigned as the accountability officer. The fire fighter from Engine 1 can now rejoin his or her company and the driver will continue to operate the apparatus. The heavy rescue will report to command for assignment (assist with search/vent/secure utilities). Truck Company 3 and Engine Company 5 are staged two blocks from the scene.

Two medic units are assigned to rehabilitation, treatment, and transport. Additional staff officers are assigned to fill the positions of safety officer, planning section chief, and Division 2 supervisor **TABLE 8-10**.

While it may be possible to double up on some assignments shown on the various deployment charts, fire fighter safety must never be compromised. In this case the RIC was fully staffed by the first-arriving truck company. Many departments preassign RIC duties (e.g., second- or third-arriving engine or truck company). Doing so ensures a fully staffed RIC but sometimes results in a delay in establishing a RIC. In a large jurisdiction, this fire could be handled by on-duty forces. In smaller jurisdictions, fires such as this would require assistance from mutual aid companies. In some cases, mutual aid assistance would respond automatically to a reported structure fire. Rapid intervention and accountability must be a high priority. Adding additional companies to the initial alarm

TABLE 8-9 Staffing with Four Engine Companies, One Truck Company, and a Chief Officer

Company	Total Staffing	Staffing at Apparatus	Assignment	Staffing at Assignment	Rapid Intervention Crew	Flow
Engine 1	4	2	Second floor	2	2	184 GPM (697 L/min)
Engine 2	4	1	Second floor	3	0	305 GPM (1155 L/min)
Truck 1	4	0	Second floor	2	0	0
Truck 1	(see above)	0	Third floor	2	0	0
Chief officer	1	0	Command post	1	0	0
Engine 3	4	1	Third floor	3	0	184 GPM* (697 L/min)
Engine 4	4	0	Second floor	4	0	305 GPM (1155 L/min)
Totals	**21**	**4**	**(N/A)**	**17**	**2**	**794* GPM (3006 L/min)**

*Although the backup hose line is not assigned to attack the fire, the additional flow is immediately available should the need arise.
© Jones & Bartlett Learning.

with the sole purpose of providing rapid intervention and accountability is recommended. Without an exposure or rescue problem, and with all companies taking appropriate action, this fire required an extra alarm assignment. Seldom is this level of efficiency experienced under the stress and limited information at the incident scene.

Operating a hose line with two fire fighters is an absolute minimum and a very arduous task. NFPA 1710 suggests a minimum of two fire fighters to operate the handline and an additional fire fighter to support and assist with hose deployment for a total of three fire fighters per attack hose line. Two fire fighters can safely operate a 1¾-in. (44-mm) hose stream during practice sessions. During an actual fire, it may be necessary to advance the hose line up stairways, around obstacles, and through multiple doorways, then operate the hose from a crawling or kneeling position, all of which greatly increase the effort needed to successfully place a hose line in position. Consider that Engines 2 and 4 in our scenario advanced 2½-in. (64-mm) lines to the second floor. Not only does assigning all available personnel on a single company to a hose line provide the staffing necessary to properly position and operate the attack hose line, it also maintains company unity and accountability.

On occasion, there may be a need to assign more than one company to a hose line even on departments that assign four people to a company, as was done in the previous auto parts store example. For example, a company is assigned to advance a hose line to the fifth floor of a large warehouse without a standpipe. If the company is supplying its own water, one member must remain at the pumps, several fire fighters will be needed to advance the hose line through the entry door and up the stairway (many warehouse buildings have 20-ft [6.1-m] or higher floors), and the hose line must then be advanced through a door at the fifth floor level and possibly around storage racks and through aisles. In this scenario, multiple companies might be needed to advance the hose line and operate the nozzle that could be flowing as much as 350 GPM (1325 L/min).

TABLE 8-10 Staffing with a Full Complement

Company	Total Staffing	Staffing at Apparatus	Assignment	Staffing at Assignment	Rapid Intervention Crew	Flow
Engine 1	4	1	Second floor	3	0	184 GPM (697 L/min)
Engine 2	4	1	Second floor	3	0	305 GPM (1155 L/min)
Truck 1	4	0	Second floor	2	0	0
Truck 1	(see above)	0	Third floor	2	0	0
Chief officer	1	0	Command post	1	0	0
Engine 3	4	1	Third floor	3	0	184 GPM* (697 L/min)
Engine 4	4	0	Second floor	4	0	305 GPM (1155 L/min)
Rescue 1	4	0	Report to command	4	0	0
Engine 5	4	4	Staged	0	0	0
Truck 2	4	0	RIC	0	4	0
Medic unit	2	0	Treatment/transport	2	0	0
Medic unit	2	0	Rehab	2	0	0
Staff	3	0	Staff	3	0	0
Totals	**40**	**7**	**(N/A)**	**27**	**4**	**794* GPM (3006 L/min)**

*Although the backup hose line is not assigned to attack the fire, the additional flow is immediately available should the need arise.
© Jones & Bartlett Learning.

Determining Apparatus Needs

The apparatus used to deliver staffing to the incident scene is generally more than sufficient to support an offensive operation. Typically, the greater challenge is to properly position on-scene apparatus to best utilize their capabilities. Apparatus that are not needed at the incident scene should be parked out of the way, preferably in positions that allow access to water supplies and are available to support a possible defensive attack. During large-scale incidents, apparatus are often assigned to a staging area. The IC should designate a staging officer to manage and coordinate all companies assigned to the staging area. A staged apparatus has the staffing necessary to function as a unit, such as an engine company, a truck company, or a medic unit. Apparatus without adequate staffing are classified as out of service or parked, rather than staged.

There is a temptation to use on-scene apparatus whether they are needed or not. In the auto parts store example, there is some justification for a third source of water, and a ladder truck could be used to accomplish additional venting or to ladder the building. However, more often than not, during offensive operations, there is a need for additional staffing, not additional apparatus. When companies are staffed with two or three fire personnel, there is an even greater number of unnecessary apparatus on scene. Placing unnecessary apparatus in the immediate fire zone tends to block access that may be needed should a change in tactics be required.

Fallacy	Fact
All nonoperating apparatus at the scene are staged.	Apparatus + staffing = staged. Apparatus − staffing = out of service or parked.

Fallacy	Fact
All apparatus at the scene should be operating.	Use only the apparatus necessary to meet tactical objectives.

Wrap UP

CHAPTER SUMMARY

- Resource capabilities in terms of staffing, apparatus, and water must meet incident requirements for a safe and effective operation.
- Rate of flow is a major factor in determining if resources are adequate.
- Methods for determining rate of flow include trial and error, the Royer/Nelson formula, the National Fire Academy formula, and sprinkler system calculations.
- Each rate of flow approach has strengths and weaknesses, and none is completely accurate, as variables such as fuel load, fuel type, ventilation, and size of the compartment can affect fire dynamics.
- Buildings requiring a rate of flow exceeding the flow of two standard preconnected hose lines should be preincident planned by applying the V/100 formula, with the exception of areas containing an extra hazard fuel load, which require use of sprinkler calculations.
- A 1¾-in. (44-mm) hose line is the minimum recommended size of hose line for an offensive attack at a structure fire.
- Using hose size larger than 1¾ in. (44 mm) is a trade-off of mobility for an increase in flow.
- Nozzle and hose configurations should be flow tested regularly to confirm the flow and proper operation.
- Variable-stream nozzles provide fog or straight streams.
- The number and size of attack hose lines will be based on flow requirements and backup hose lines, plus lines needed to protect means of egress and internal and external exposures.
- Fire can extend within a building to involve other areas or attached structures (internal exposures) or to adjacent structures (external exposures).
- Backup hose lines should be at least as large as the initial attack hose line.
- Each hose line requires at least two fire fighters, but a full company per line is preferred.
- It is good practice to provide at least two water supplies for a working structure fire.
- When a significant water supply challenge exists, options should be outlined in standard operating procedures (SOPs) and in the preincident plan.
- Use of an attack pumper tactic should be limited to fires in relatively small structures.
- A flow of water meeting or exceeding the rate of flow requirements applied for a short period of time will require less total water and be more effective than a lesser flow over a longer period.
- Proper ventilation aids in the extinguishment effort and provides for a safer working environment for fire fighters in the building.
- Staffing needs vary depending on tasks to be accomplished, but certain minimum staffing requirements can be predetermined.
- Safe and efficient positioning of on-scene apparatus is a key consideration.

KEY TERMS

direct attack Firefighting operations involving the application of extinguishing agents directly onto the burning fuel.

external exposures Buildings, vehicles, or other property threatened by fire that are external to the building, vehicle, or property where the fire originated.

fuel-controlled Refers to a fire in which the heat release rate and growth rate are controlled by the characteristics of the fuel, such as quantity, chemistry, and geometry, and in which adequate air for combustion is available.

indirect attack Firefighting operations involving the application of extinguishing agents to reduce the buildup of heat released from a fire without applying the agent directly onto the burning fuel; ventilation is kept to a minimum while a fog stream is directed at the ceiling. This approach is most useful in unoccupied, tightly enclosed spaces.

internal exposures Areas within the structure where the fire originates or within buildings directly connected to the building of origin that were not involved in the initial fire ignition.

National Fire Academy formula A rate of flow calculation that calculates the rate of flow as the area in square feet divided by 3 (A/3).

pyrolysis The chemical decomposition of a compound into one or more other substances by heat alone; the process of heating solid materials until combustible vapors are emitted.

Royer/Nelson formula A rate of flow calculation that calculates the rate of flow as the volume in the fire area in cubic feet divided by 100 (V/100). This formula is based on the assumption that structure fires are primarily ventilation controlled.

sprinkler system calculations Specific rate of flow calculations for sprinkler systems; based on the fuel load. These calculations can be found in various publications, including NFPA documents and Factory Mutual Data Sheets.

trench cut An opening in the roof that extends from bearing wall to bearing wall to prevent horizontal fire spread in a building.

ventilation controlled A fire within a compartment or building that is limited due to a lack of air (oxygen) even though sufficient vapor fuel is available to support continued burning.

SUGGESTED ACTIVITIES

1. On December 23, at 12:55 PM dispatch receives several calls reporting smoke showing on the second floor of an apartment building on Fifth Street near Central Avenue.

    Four engine companies, two truck companies, one heavy rescue company, and a district chief are dispatched to the reported fire. All fire companies are staffed with a company officer (lieutenant or captain), driver, and two fire fighters. **TABLE 8-11** shows the responding fire companies, chief officers, and emergency medical resources.

TABLE 8-11 Responding Fire Companies, Chief Officers, and Emergency Medical Resources

	Engine Companies	Truck Companies	Other
Initial 1 Alarm Response	14, 3, 29, 5	14, 3	District Chief 1 Heavy Rescue 14
2nd Alarm	17, 21, 12, 19	29, 17	District Chief 2 Heavy Rescue 9 EMS 14
3rd Alarm	20, 34, 35, 37	21, 19	District Chief 4 EMS 3 Fire chief Operations assistant chief Safety chief
4th Alarm	23, 32, 9, 2	23, 32	District Chief 5 EMS 17 Personnel assistant chief Fire prevention assistant chief
5th Alarm	18, 7, 36, 39	18, 39	Police District 1 Captain Safety department manager

© Jones & Bartlett Learning.

Assume the role of district chief. You are first on the scene and assume the role of IC. You observe light smoke at the second floor; side Alpha of an apartment building. Part of the building is two stories with two apartments per floor. The two-story part of the building has windows front and rear. The three-story side of the building has front windows, with a blank wall to the rear. The three-story side of the building has three apartments on each floor. This building type is common in the area. You set up command just past the front of the building on side Alpha and transmit a message declaring a "working fire."

Engine 14 and Truck 14 arrive on scene as you establish command. As per SOPs, Engine 14 is laying a 5-in. (127-mm) supply line. Truck 14 positions their apparatus at the far end of the fire building (three-story section) and is raising their aerial ladder to the third floor window farthest from the two-story side of the building. As IC you instruct Engine 14's officer to attack the fire on the second floor via the interior stairs. The officer and two fire fighters on Engine 14 have advanced a 1¾-in. (44-mm) hose to the second floor via the stairway.

Engine 14's driver remains at the apparatus connecting hose lines and operating the pumps. Truck 14 advances to the third floor via the aerial ladder.

Truck 3 is assigned to the rear of the building which is inaccessible via their apparatus due to limited space. They don full fire gear and proceed to the rear of the building. Truck 3 reports that there is heavy smoke and fire visible at the rear of the building FIGURE 8-23. They request

FIGURE 8-23 View of side Charlie during the initial stages of the fire as Engine 14 enters the building from side Alpha and Truck 14 raises the aerial ladder to the third floor on side Alpha.
Courtesy of Bill Strite, Cincinnati, Ohio.

two engine companies hand lay hose lines to the rear of the building.

Engine 5 is assigned as the rapid intervention team (RIT).

After a delay, the lieutenant and two fire fighters on Engine 14 enter Apartment 2 on the second floor, but they are having great difficulty advancing inside the apartment due to heavy fire and smoke. Engine 14 is forced to back out of the second floor apartment, but one fire fighter is missing. The missing fire fighter's PASS began to sound about 10 minutes after making entry, but the lieutenant and the other fire fighter were unable to locate the missing fire fighter. Engine 14's lieutenant declares a mayday as they exit the apartment. They are now reentering the apartment in an attempt to find the missing Engine 14 fire fighter.

A. Was the choice of an offensive attack correct?

B. If you were the IC, would you initiate an offensive or defensive attack on arrival?

C. Explain why you would choose this strategy.

D. At this point in the operation would you continue with an offensive attack or change to a marginal or defensive attack.

E. What indicators would justify changing attack modes?

F. Is there a good reason to continue an offensive operation?

G. Would you request additional alarms?

H. If so, how many alarms and special call units would you request and why?

I. Develop an incident action plan based on your answers to the questions above.

J. Develop an Incident Command System (ICS) organizational chart for the operation after you make adjustments, add companies, or change strategic modes.

2. You are responding to a mixed-occupancy bar and residential fire. Assume the role of the first-arriving engine at the fire shown in **FIGURE 8-24**. It is 7:00 AM on Sunday, August 1. The weather is warm and very humid.

FIGURE 8-24 Sides Alpha and Delta of a mixed-occupancy building—bar and residential.
Courtesy of Bill Strite, Cincinnati, Ohio.

TABLE 8-12 Responding Fire Companies, Chief Officers, and Emergency Medical Resources

	Engine Companies	Truck Companies	Other
Initial 1 Alarm Response	14, 3, 29, 5	14, 3	District Chief 1 Heavy Rescue 14
2nd Alarm	17, 21, 12, 19	29, 17	District Chief 2 Heavy Rescue 9 EMS 14
3rd Alarm	20, 34, 35, 37	21, 19	District Chief 4 EMS 3 Fire chief Operations assistant chief Safety chief
4th Alarm	23, 32, 9, 2	23, 32	District Chief 5 EMS 17 Personnel assistant chief Fire prevention assistant chief
5th Alarm	18, 7, 36, 39	18, 39	Police District 1 Captain Safety department manager

© Jones & Bartlett Learning.

TABLE 8-12 shows the responding fire companies, chief officers, and emergency medical resources.

All companies are staffed with an officer, a driver, and two fire fighters. The first-arriving engine and truck companies from your fire station are also on the scene.

The building is three stories with a basement approximately 40 × 40 ft (12.2 × 12.2 m) at the base.

- Side Alpha: Bar on first floor with apartments on floors two and three
- Side Bravo: Fenced-in parking lot.
- Side Charlie: Fire escape and rear entry/egress to the bar and full basement that is partially above grade.
- Side Delta: Access door to stairway leading to the three apartments on the second floor, two third-floor apartments, and a second means of egress from the bar.

A. Based on a risk-versus-benefit analysis, would you make entry to attack the fire with the staffing on the first-arriving fire company in your department? If your staffing is less than four fire fighters per company, would this staffing level affect your offensive/defensive decision?

B. Is your preconnected hose line adequate to extinguish a fire in any area of this building?

C. Based on the conditions shown in Figure 8-24, where would you place the first attack hose line and why?

D. Assuming you decide on an offensive attack, where would you make entry and how would you reach the fire area to attack the fire?

E. Would it be possible to apply the softening the target tactic?

F. Would door control be effective in this scenario?

G. Would you consider a vent, entry, isolate, search (VEIS) tactic? If so, where would you make entry?

H. You are in charge of the first-arriving truck company. Assuming the first-arriving engine company took the actions you recommended, what tasks would your company perform (in order of priority)?

I. As the first-arriving chief officer, develop and implement an incident action plan, assuming the first-arriving engine and truck companies took the actions you recommended.

J. Develop a National Incident Management System (NIMS) organizational chart for this fire showing all on-scene companies.

3. Assume the role of the first-arriving district chief (IC) at the fire shown in **FIGURE 8-25**. It is 1:00 PM on Monday, July 11. The weather is hot and humid with a light wind from the east. The building is on the north side of the street. The occupant is a costume manufacturer. The fire building is 70 × 90 ft (21.3 × 27.4 m). The three-story attached building on side Delta (28 × 40 ft [8.5 × 12.2 m]) is also owned by the costume company and used as office and

FIGURE 8-25 Fire in a commercial building.
Courtesy of Denny Baker, Cincinnati Fire Department.

sales space. A two-story single-family home is located 20 ft (6.1 m) to the rear. An 80 × 80 ft (24.4- × 24.4-m) one-story manufacturing business is located on side Bravo.

All companies are staffed with an officer, a driver, and two fire fighters. The first-arriving engine and truck companies from your fire station are also on the scene.

A. Would you begin operations in the offensive or defensive mode?

B. What is your rationale for an offensive or defensive attack?

Assume you choose an offensive attack:

C. If you deploy an interior attack hose line, what size line and what nozzle would you order.

D. What is the probability that the costume manufacturing building, the three-story attached office/display building, and two-story single-family home to the rear are occupied at this time?

E. Would you ventilate the manufacturing building? If yes, what kind of ventilation would be best and where would you vent?

F. What would you direct the truck company to do?

G. Develop and implement an incident action plan, assuming the first engine company took the actions you ordered.

H. Assign units to implement your incident action plan.

I. How many units and of what kind would you maintain as a tactical reserve?

J. Develop a NIMS organizational chart for this fire showing all on-scene companies.

4. Assume the role of fire chief in a community of 50,000 people. There are a few stores, churches, restaurants, and schools. However, the community is primarily made up of single-family detached dwellings. No buildings are more than three stories in height. Mutual aid contracts are in effect, but all mutual aid departments are 20 miles or farther from your community. The city is in a financial emergency and must reduce its budget. The fire department is being asked to reduce staffing at the three single-engine fire stations from four to three fire fighters per station. A private ambulance service provides EMS for the community, with the fire department responding as first responders. You are the only paid administrative staff person.

A. Develop a response justifying four-person staffing for your department.

B. Assume the city refuses your proposal and stands firm on reducing staffing. Recommend an alternative that would result in the same reduction in staffing, but provide a safer and more reliable fire and EMS service for the community.

REFERENCES

Averill, Jason D. 2010. *Report on Residential Fireground Field Experiments*. NIST Technical Note 1661. Gaithersburg, MD: National Institute of Standards and Technology.

Averill, Jason D., Lori Moore-Merrell, Raymond T. Ranellone, Jr., et al. 2013. *Report on High-Rise Fireground Field Experiments*. Gaithersburg, MD: National Institute of Standards and Technology. https://nvlpubs.nist.gov/nistpubs/TechnicalNotes/NIST.TN.1797.pdf.

Brunacini, Alan V. 2002. *Fire Command*, Second edition. Quincy, MA: National Fire Protection Association.

Clark, William E. 1991. *Firefighting Principles and Practices*, Second edition. Saddle Brook, NJ: PennWell.

Coleman, John. 1997. *Incident Management for the Street-Smart Fire Officer*. Saddle Brook, NJ: PennWell.

Edwards, C. Bruce. 1992. "Critical fire rate." *Fire Engineering*. September: 97–9.

Goodbread, J. C. July 1985 and August 1985. *Fire in Building 3001*. Quincy, MA: National Fire Protection Association.

International Association of Fire Chiefs, Safety, Health and Survival Section. December 23, 2013. "Firefighter Safety Call to Action." http://www.iafc.org/Media/articlePR.cfm?ItemNumber=7299.

International Association of Fire Chiefs, National Fire Protection Association. 2019. *Fundamentals of Fire Fighter Skills*, Fourth edition. Burlington, MA: Jones and Bartlett Learning.

Kerber, Stephen. 2013. *Study of the Effectiveness of Fire Service Vertical Ventilation and Suppression Tactics in Single Family Homes*. Northbrook, IL: Underwriters Laboratories Firefighter Safety Research Institute.

Kerber, Steve. 2010. *Impact of Ventilation on Fire Behavior in Legacy and Contemporary Residential Construction*. Northbrook, IL: Underwriters Laboratories Firefighter Safety Research Institute.

Kerber, Steve, and Daniel Madrzykowski. 2008. *Evaluating Positive Pressure Ventilation in Large Structures: School Pressure and Fire Experiments*. NIST Technical Note 1498. Gaithersburg, MD: National Institute of Standards and Technology;

National Fire Protection Association. 2017. *NFPA 1142: Standard on Water Supplies for Suburban and Rural Fire Fighting*. Quincy, MA: NFPA.

National Fire Protection Association. 2018. *NFPA 1500: Standard on Fire Department Occupational Safety, Health, and Wellness Program*. Quincy, MA: NFPA.

National Fire Protection Association. 2019. *NFPA 13: Standard for the installation of Sprinkler Systems*. Quincy, MA: NFPA.

National Fire Protection Association. 2019. *NFPA 14: Standard for the Installation of Standpipe and Hose Systems*. Quincy, MA: NFPA.

National Fire Protection Association. 2020. *NFPA 1710: Standard for the Organization and Deployment of Fire Suppression Operations, Emergency Medical Operations, and Special Operations to the Public by Career Fire Departments*. Quincy, MA: NFPA.

Royer, Keith, and Floyd W. (Bill) Nelson. 1959. *Water for Fire Fighting, "Rate of Flow" Formula, Extension Services Bulletin #18*. Ames, IA: Iowa State University.

CHAPTER 9

Defensive Operations

LEARNING OBJECTIVES

- Compare and contrast a defensive fire attack and offensive fire attack, explaining the key differences.
- Explain why an offensive attack is preferred over a defensive attack.
- Enumerate conditions that would lead to a defensive attack.
- List measures that must be taken when changing from an offensive to a defensive attack.
- Describe how collapse zone dimensions are determined.
- Evaluating buildings in a local response area, calculate the minimum distance firefighters should be deployed from a building in danger of collapse.

- Discuss the positive and negative effects of operating a hose stream into a window or roof opening.
- Describe how water should be applied when protecting an exposure from radiant heat.
- Compare and contrast the use of handheld hose streams versus master stream appliances during defensive operations.
- List two ways the water utility company may be able to increase the total water supply at the incident scene.
- The incident commander is ordering six master streams be deployed in a surround-and-drown

tactic. Each master stream is supplied by large diameter hose with pumps operating at maximum capacity. Estimate the total water flow requirement. Determine whether the water supply system in your jurisdiction is capable of supporting this attack.
- Compare and contrast the use of fog versus solid streams during a defensive attack.
- Estimate staffing and apparatus needs when operating master streams.
- List common problems leading to conflagrations.
- Explain tactics used to control a conflagration.
- Discuss why conflagrations are likely to occur immediately after a natural disaster.
- Evaluate the number of conflagrations each year in terms of lives lost and size of the fire. Determine whether the number of conflagrations and related loss of life and property is increasing or decreasing.
- List reasons for a nonattack strategy.
- Given a scenario, calculate the dimensions of the collapse zone.
- Evaluate safety and tactical problems associated with a transition from an offensive attack to a defensive attack.
- Prioritize exposures based on fire conditions, occupancy, and weather factors.
- Develop an incident action plan for a defensive fire and for a conflagration.
- Apply defensive tactics to a defensive fire attack and conflagration.
- Evaluate staffing, water supply, and apparatus needs for a large-scale defensive fire.
- Apply the National Incident Management System (NIMS) to a defensive fire scenario.
- Determine the probability of a conflagration for a specified response area.
- List the six tactical elements that need to be applied when faced with a conflagration.

Courtesy of David J. Jones, Cincinanti, Ohio.

Case Study

On January 8, 2010, fire fighters were called to a fire in an older section of the city containing many large commercial and residential buildings. Upon arrival, they determined that the fire was on the upper two floors of a three-story building of heavy-timber construction. An offensive attack was attempted, but fire conditions required the abandonment of the offensive attack. A defensive strategy was employed to combat the fire in the building of origin and protect three nearby exposure buildings. Additional alarms were sounded, bringing a total response of 85 fire fighters staffing 20 fire companies. After 3 hours, the fire was declared to be under control, but units remained on the scene for more than 13 hours.

1. How does the level of fire involvement affect the offensive/defensive decision?

2. When selecting a strategy, what role do water supply and staffing levels play?

3. What effect does construction type have on structural stability?

Introduction

An **offensive** fire attack is the preferred strategy whenever conditions and resources permit an interior attack. A **defensive** decision limits operations to the exterior, generally resulting in a larger property loss and limiting rescue options. The offensive/defensive decision is based on staffing available to conduct an interior attack, water supply, ventilation, and, most important, a risk-versus-benefit analysis. A defensive operation is used whenever the risk-versus-benefit analysis determines that the risk to fire fighters' lives and safety outweighs any possible benefit that might be achieved through an offensive attack. For example, if a fire occurs in a building that is structurally unsound and the fire does not place any lives at risk, then clearly the operation should be handled defensively. However, situations are not always so clear-cut, and the fire officer must rely on solid fire-ground information, coupled with training and experience, to determine whether the operation should be offensive, defensive, or nonattack. (The nonattack strategy can be used when the building is so far involved that the main concern is to cover the exposures and allow the original fire building to consume all available fuel. This could also be an attack choice when the available water supply will not support an offensive or defensive attack and the available water must be used to protect the exposures.)

> **NOTE**
>
> Rate of flow formulas do not apply to defensive operations. These equations are presented in detail in Chapter 8, *Offensive Operations*.

Classifying as a Defensive Attack

The first-arriving officer must evaluate the situation using the NFPA 1500, *Standard on Fire Department Occupational Safety, Health, and Wellness Program*, risk management process to determine whether to implement an offensive or defensive strategy (NFPA 1500, 2018). This process is described in Chapter 5, *Fire Fighter Safety*. This is a critical decision upon which every other decision is based. The preferred fire attack mode is offensive, and the incident commander (IC) should apply an offensive strategy whenever it is safe to do so.

There are times when initial conditions justify an offensive strategy that ultimately results in a change to a defensive attack. For example, an offensive attack that was at first justified due to life safety may no longer be justified once the building is cleared of occupants. Deteriorating conditions may also require a change from an offensive to a defensive attack.

Each officer who assumes command must reconsider critical size-up factors to determine whether the operation should continue in the present strategic mode. Generally, the longer an uncontrolled fire burns, the more hazardous the building becomes. Therefore, the risk-versus-benefit analysis should be assessed not only during the initial attack or during a change of command, but also at set times and whenever major tactical objectives are achieved.

Changing from an offensive to a defensive attack is a command challenge and potentially hazardous time on the fire ground for several reasons, including the following:

1. The defensive attack should be delayed until the building is completely cleared of fire fighters.
2. When the operation is changed from an offensive to defensive attack it is imperative that a personnel accountability report be conducted prior to commencing the defensive attack. Offensive and defensive attacks must never be conducted in the same building at the same time; however, if warranted, a defensive attack may be conducted in a fire-separated building while an offensive attack is taking place in another building.
3. Fire companies must be reassigned to defensive operational tasks. This can be a major problem if there are a large number of companies operating at the scene when a decision is made to go defensive. Whenever possible, the IC should anticipate the possibility of a move to a defensive attack (plan B) and appoint a planning section chief to develop plan B, then assign officers to relay instructions to companies as they evacuate the building. Firm command and control of the operation is essential during this critical transition.
4. Master stream appliances are primarily used in defensive attacks. Unlike the offensive attack, in which several fire fighters are needed to advance a handheld hose line, the master stream–oriented defensive attack typically requires only one fire fighter to direct the stream and another to operate the apparatus pumps. Once a water supply has been established, unassigned personnel must be managed to avoid dangerous freelancing.

> **NOTE**
>
> Occasionally there are conditions when there is no hope of saving fire fighters inside a building, and/or the risk to fire fighters entering the building will likely result in more fire fighter fatalities, forcing the IC to make the incomprehensible decision to abandon fire fighters and building occupants by ordering a retreat.

Typically, defensive attacks are conducted in the following situations:

- Structural integrity concerns, fire conditions, or other hazards prohibit entry.
- Resource needs outweigh resource capabilities.
- A risk-versus-benefit analysis indicates that the risk is too great compared to what can be saved.

When the IC initiates a defensive fire attack, the objective is to save property that has not already been destroyed and/or to protect the environment. Often, this means sacrificing the building of origin in favor of saving surrounding external exposures. At other times, a defensive attack can knock down the heavy volume of fire that prevents entry, limiting or stopping the fire spread, thus saving the remainder of the building. The overhaul portion of a defensive attack, where part of the building is still standing, can be extremely dangerous.

Defensive fires appear to be a greater challenge and more spectacular, but in reality, they are easier to handle and pose fewer risks to fire fighters if the proper precautions are taken. The fire officer must keep in mind the factors that lead to a defensive decision. Fire fighters should never be needlessly placed in a dangerous environment. When a fire is of such magnitude that the building's structural support system is threatened, the building should be evacuated and a defensive operation initiated.

Establishing a Collapse Zone

When a fire seriously threatens the structural integrity of a building, fire fighters should move to the outside of the collapse zone to avoid life-threatening emergency retreats. **FIGURE 9-1** shows fire fighters in

A

B

FIGURE 9-1 Two sides of a defensive fire attack. **A.** Side Alpha. **B.** Side Delta.
Centers for Disease Control and Prevention, Paid-on-call Fire Fighter Killed by Exterior Wall Collapse during Defensive Operations at a Commercial Structure Fire – Illinois, Death in the Line of Duty . . . A summary of a NIOSH fire fighter fatality investigation. Retrieved from https://www.cdc.gov/niosh/fire/reports/face201115.html

a defensive mode working close to a building showing obvious signs of collapse.

On June 17, 2011, a fire fighter lost his life when he was struck by falling building materials as the building began to collapse. Once a decision is made to conduct a defensive attack everyone, including fire fighters, must immediately retreat from the collapse zone, usually a distance equal to 1½ times the height of the building.

Many theories exist regarding collapse zone distances. Some theorists believe that a building will fall within one-third of its height. Generally, this and other theories are based on sound principles but with questionable assumptions being made about how the collapse will occur. Buildings sometimes collapse within themselves; other times, the walls fall away from the supporting structure as a unit. This type of structural failure results in the collapse zone being equal to the height of the building plus an allowance for debris to scatter **FIGURE 9-2**.

Any collapse zone that extends less than the building's height plus an allowance for debris scatter is a calculated risk. The IC must ask whether the expected benefit is worth the risk.

> **NOTE**
>
> A good method of estimating the collapse zone is to use a distance equal to 1½ times the height of the building.

The ability to apply water from a distance greater than the height of the building may lead to a nonattack strategy for tall buildings that are in danger of collapse. Fire departments should measure the effective water application distance for master streams carried on their apparatus. Collapse zones can be preplanned. The actual height of the building can easily be determined during preincident planning. If the height of the building is not part of a preplan, rough approximations must be made at the time of an incident. For older multistory residential and office buildings, using a factor of 12 ft (3.7 m) per story is reasonably accurate. Modern buildings normally have lower ceilings and are often less than 12 ft (3.7 m) per story, whereas warehouse buildings or special occupancies could have much higher ceilings.

Using the 12 ft (3.7 m) per story estimate, a five-story building would be 60 ft (18.3 m) in height, requiring a collapse zone of approximately 90 ft (1.5 × 60 ft [1.5 × 18.3 m = 27.5 m]) **FIGURE 9-3**. Once this distance has been determined, the question becomes: Can available master streams effectively apply water from a distance of 90 ft (27.5 m)?

Wind conditions must also be considered when estimating the effective distance for a fire stream. Solid-stream nozzles, as compared to straight-stream and fog nozzles, generally maintain better stream continuity for long distances, particularly when the stream is operated in opposition to the wind or when there is a crosswind.

Another limiting factor for apparatus and fire stream placement is the width of the street. If a 90-ft (27.5-m) collapse zone is required and streets are only 40 ft (12.2 m) wide, personnel, apparatus, and equipment placed in the street would be within the collapse zone. Before committing fire fighters and apparatus to a potential collapse zone, the IC must seriously consider the expected benefits. If a decision is made to position personnel and equipment within

FIGURE 9-2 Wall collapse.
Courtesy of St Louis Post Dispatch.

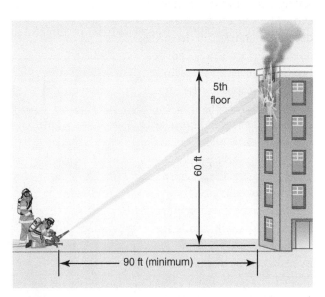

FIGURE 9-3 Safe distance from a wall in danger of collapse.
© Jones & Bartlett Learning.

the potential collapse zone, evaluate the building to determine the safest possible location. For example, positioning companies at the corners of the building is usually safer than positioning for a frontal attack. Anyone who is seriously considering being an IC is strongly encouraged to read texts such as *Brannigan's Building Construction for the Fire Service* to become familiar with the building's structural components (Corbett and Brannigan, 2021).

Evaluating Exposures

Exposures are generally divided into internal and external exposures. Internal exposures deal with fire extending from one area of a building to another within a structure. This is described in greater detail in Chapter 8, *Offensive Operations*. Protecting the interior exposure is only marginally effective in a defensive mode. However, if the building is tightly constructed with closed doors separating areas, there is a possibility that the fire will be contained to the area already involved in fire.

In a defensive operation, the building of origin is often recognized as a total loss; therefore, emphasis is placed on protecting external exposures. The IC should evaluate external exposures in terms of life safety, extinguishment, and property conservation.

The distance between exposed structures and the volume and location of the fire have much to do with prioritizing exposure protection. Radiant heat increases as the size of the flame front increases. Knocking down the main body of fire can reduce the flame front, thus reducing radiant heat energy. Therefore, the most effective way to protect exposures is to extinguish the main volume of fire. However, a risk-versus-benefit analysis is essential. For example, the building in **FIGURE 9-4** appears to be vacant and in poor repair. Thus, it has little value. However, nearby buildings may be high-value structures. If this is the case, extinguishing the fire in this building of origin from a distance could be worth considering.

When inadequate resources, collapse probability, or fire volume makes a direct attack on the fire ineffective, applying water directly on the exposed structures is the best tactic.

The energy levels for radiant heat are inversely proportional to the square of the distance between the heat source and the exposure. Thus the closer the buildings are located to each other, the more radiant heat will be transferred from the fire building to an adjacent structure. Exposure buildings that are higher than the fire building are also at a greater risk.

The fire officer must determine the best method of protecting potential exposures. Available staffing, as well as matching apparatus resources to incident requirements, will greatly affect the IC's options. Wetting the exposure is generally the most effective way to prevent ignition. Radiant heat will travel through transparent materials such as water. Therefore, directing a water stream between buildings is less effective than applying water directly to the exposed building.

Class A foam appears to present possibilities in terms of protecting exposures. The compressed air variety, in particular, tends to cling to exposures and do a better job of prewetting surfaces.

Evaluating a Direct Attack

Hose streams applied directly into the burning structure is an exposure protection tactic that has merit, but the fire intensity and depth may be such that it is impossible to effectively extinguish the fire or knock down the flame front **FIGURE 9-5**. If a direct attack fails to control the fire or the effectiveness of a direct attack is questionable for any reason, first cover nearby exposures and then direct as much water as possible onto the main body of fire from a safe distance.

> **NOTE**
>
> The IC must consider the possibility of fires starting inside exposed buildings due to radiant heat igniting combustibles that are near windows facing the fire building. Radiant heat passes through intact panes of glass.

Sometimes it is possible to place a master stream in a position where it can be operated either directly on the fire or on an exposure without moving the apparatus or master stream appliance **FIGURE 9-6**.

FIGURE 9-4 Fire fighters working within the collapse zone.
Courtesy of David J. Jones, Cincinanti, Ohio.

CHAPTER 9 Defensive Operations   **313**

FIGURE 9-5 Fire fighters operating handheld nozzles from outside the building into the building in a defensive attack.
Courtesy of David Mullis.

Estimating the Number and Type of Master Streams Needed

There are times when master streams are used to augment an offensive attack. However, master streams are used primarily during defensive attacks. Master streams are the tools of choice for defensive operations. A master stream can apply more water from a greater distance and requires fewer personnel to operate compared to a handheld hose line.

FIGURE 9-6 Setting up for direct attack and covering exposures.
© Jones & Bartlett Learning.

Master streams can be operated from ground level, the top of an apparatus, or an elevated position. However, it is important to remember that any tactic that hinders the upward and outward movement of heat and smoke is usually detrimental to the operation. Elevated master streams applying wide-angle fog streams are particularly problematic in terms of interrupting the upward and outward flow path of the heated products of combustion. Sometimes, enough of the building has been destroyed that it is impossible to change the flow path. If the IC considers the building a total loss, directing streams into vent openings may be necessary to protect nearby exposures.

It is extremely difficult for a person to see the effectiveness of a defensive fire stream from behind the nozzle. A fire fighter or officer should step back or to the side to get a better view of the stream and use a thermal imaging camera in order to give directions to the fire fighter who is operating the nozzle.

Once master streams are in position, staffing needs diminish. Personnel inside the fire zone should be kept to a minimum. This is a good time to start rehabilitation. Handheld hose lines have little or no value during defensive attacks on large structures and often result in unsafe tactics. Anything less than a 2½-in. (64-mm) hose line would be ineffective. A common problem arises when fire fighters who are not operating pumps or master streams are standing around at a major fire. They may occupy themselves using handheld hose lines that serve no purpose. It is the IC's responsibility to manage this situation by placing fire fighters who do not have a specific assignment to a safe location. If handheld hose lines were used during an offensive attack that has been abandoned, it is recommended that these hose lines be disconnected from the apparatus during defensive operations to ensure that they are not used inappropriately. However, there are occasions when the fire is being fought defensively in one building and offensively in another. In this case, the use of smaller-diameter hose may be acceptable on the interior of an exposure.

> **NOTE**
> When handheld hose lines are required as part of a defensive attack, reduce fatigue by using hose loops whenever possible.

Estimating Water Supply Needs

Providing water at defensive operations may prove more challenging than for most offensive attacks. Many master stream appliances are capable of flowing 1000 GPM (3785 L/min) or more, thus requiring the total pump capacity of an apparatus. When several of these master stream appliances are operating, the water system can be exhausted even in areas with high-volume and reliable water supplies. Supply hose lines should be large-diameter hose, and there may be a need to relay or shuttle water.

When evaluating the water supply, consider the flow capacity of individual hydrants as well as the total flow capacity of the water system. Some water supply systems are made up of several separate systems that can be connected (cross-tied) when necessary. Typically, the water utility company will open street valves to cross-tie the systems. Cross-tie locations should be identified and preplanned. The water utility company may also be able to bring additional pumps online, increasing the volume or pressure at the fire scene. Consider including water utility representatives on the multiple-alarm call list. If water supplies are not connected (cross-tied) system connections can be accomplished by pumping from one system to the other.

Selecting the Stream Position

The first consideration in placing exterior streams is safety. The second is the ability to apply water to exposures and to the interior of the building. Often, these two considerations are at odds. The best direct application positions are often in the collapse zone.

Unstaffed ground monitors are sometimes the answer to this dilemma. However, the effectiveness of these streams is often limited, which results in sending fire fighters back into the collapse zone to redirect the streams. Elevated streams have a natural advantage when the fire is in the upper stories of the building; however, they require the apparatus to be parked fairly close to the building **FIGURE 9-7**. Likewise, the apparatus-mounted master stream device is easier to place in service but may place the pump operator and apparatus in the collapse zone when directing an attack on tall buildings.

> **NOTE**
> Remember, there is no justification to risk injury to fire fighters during a defensive operation, and there is seldom justification to risk damage to expensive apparatus and equipment.

Selecting the Nozzle Type

Is the solid stream (smooth bore), straight stream, or fog stream best for defensive operations? They all have their place, and it is essential that the fire officer

FIGURE 9-7 Elevated stream on a three-story house.
Courtesy of Bill Strite, Cincinnati, Ohio.

FIGURE 9-8 A fire fighter remotely directing the elevated master stream.
Courtesy of Bill Strite, Cincinnati, Ohio.

know the advantages and disadvantages of all available tools. The solid stream has the greatest reach and penetrating ability and is best suited to situations when the attack is on the main body of the fire in a large structure.

The fog pattern can be gently applied to an exposure, covering a wider area without breaking windows. Straight streams from a variable-stream nozzle are not generally considered equal to solid-stream (smooth-bore) nozzles, but most are capable of penetrating the main body of fire. Using a variable-stream nozzle allows the stream to be used for an attack on the main body of fire, and it can also be repositioned to cover exposures if the apparatus or master stream is properly positioned.

Calculating Staffing Needs

Unlike an offensive attack, in which several fire fighters are needed to advance a handheld hose line, the master stream–oriented defensive attack typically requires only one fire fighter to direct the stream and one to operate the pumper, once a water supply has been established. The officer should stand to the side to direct the stream and provide direction to the fire fighter operating the stream. Unassigned personnel may be rotated through rehabilitation or assigned to staging, making companies available for other responses or to augment the attack.

Determining Apparatus Needs

Offensive attacks are people intensive. Thus during an offensive attack, there may be an excess of apparatus. Defensive attacks are apparatus intensive. Whereas two or more fire fighters are required to advance a 2½-in. (64-mm) hose line flowing 250 GPM (946 L/min) in an offensive attack, one fire fighter can direct a 1000-GPM (3785-L/min) or greater master stream **FIGURE 9-8**.

A master stream that is being supplied from a distant hydrant may require one or more pumpers to supply the necessary water. Furthermore, a large master stream appliance may require one or more pumpers to supply it. However, 1250-GPM (4732-L/min) or 1500-GPM (5678-L/min) pumpers connected to a reliable water supply with large-diameter hose may be able to supply two master streams.

Elevated streams can also require additional pumpers, particularly if the aerial apparatus are not equipped with prepiped waterways and pumps. When water supplies are very limited, tanker shuttles or drafting from distant water sources will require a substantial commitment of personnel, apparatus, and time.

Conflagrations and Group Fires

Conflagrations occur less frequently now than in the past, but they remain a possibility. Fire departments should recognize the extraordinary challenge presented by a conflagration and determine the probability of a conflagration in their response area. Special tactics are needed to resolve a fire incident involving a large geographic area.

Early conflagrations, such as the burning of Rome, are not well documented. More is known about later conflagrations that destroyed large tracts within urban centers. London had conflagrations in 798, 982, 1212, and 1666. Constantinople is the undisputed conflagration champion, with conflagrations recorded in 1729, 1745, 1750, 1756, 1782, 1791, 1798, 1816, 1870, 1908, 1911, 1915, and 1918. In 1607, the first permanent American colony was founded at Jamestown, Virginia; it was destroyed by fire in January of 1608. Plymouth, Massachusetts, was founded in 1620 and was substantially destroyed by fire 3 years later. Boston experienced nine conflagrations before the Revolutionary War. During the Revolutionary War, like all other wars, fire was used as a weapon, destroying many villages in colonial America. Conflagrations continued with great frequency in the New World until the early 1900s. More recently, terrorists used fire as a weapon of mass destruction at both World Trade Center attacks.

Some of the more significant conflagrations to have occurred in North America in the past two centuries are noted in **TABLE 9-1**. The list is not all inclusive. Many group fires and some conflagrations are not investigated by national organizations and, therefore, go unnoticed outside the region where they occur.

The term *conflagration* is often misused. Many people refer to any large fire as a conflagration. It is not unusual to hear a news anchor refer to the conflagration in a downtown tenement building. There are several definitions for conflagration. In an urban setting the term conflagration is used to describe a fire that spreads from building to building over a considerable distance, beyond a natural or artificial barrier (e.g., beyond a city block).

Group fires are similar to conflagrations; however, unlike a conflagration, a group fire is confined within a complex or among adjacent buildings. Some definitions describe a group fire as not extending beyond a complex or confined to the city block of origin. Most group fires have the potential to become conflagrations, but either they are confined early or no exposures are readily available to allow fire spread. Group fires are generally of smaller scale than a conflagration; however, there are exceptions. For instance, the main body of fire at the Santana Row Fire in San Jose, California, was contained within the complex, which would be defined as a group fire even though it did cover a large geographic area. However, this fire does not completely fit the definition of a group fire, as flying brands traveled a considerable distance, igniting buildings well beyond the complex where the fire originated. From a strategic point of view, conflagrations and group fires present similar problems and have similar solutions.

Wildland–urban interface fires that spread from wildland into an urban area and destroy large numbers of buildings are properly defined as conflagrations. The most notable wildland–urban interface fire occurred in Peshtigo, Wisconsin, the same day as the Great Chicago Conflagration—October 7, 1871. The Peshtigo Fire killed 1200 people (although some estimate as many as 2000 fatalities) and destroyed 17 towns. A wildland fire that does not destroy a large number of buildings should not be labeled a conflagration.

The focus of this text is structural firefighting; thus tactics used to combat wildland–urban interface fires are beyond the scope of this text. However, Table 9-1 lists several conflagrations that began as wildfires, because once a wildland fire enters a populated area, many of the tactics described in this text would apply. The discussion here is limited to the tactics used when confronted with a large-area fire in a built-up area.

When conflagrations are studied, common contributing factors are evident, including the following:

1. Closely built structures, especially combustible structures
2. Wood shingle roofs
3. Poor water supplies or fire suppression weaknesses (automatic and manual)
4. Dilapidated structures, especially abandoned buildings in large numbers
5. Large-scale combustible construction or demolition projects

TABLE 9-1 Conflagrations of Significance

Year	Place	Number of Buildings Involved
1835	New York City, NY	700 buildings
1845	Pittsburgh, PA	1000 buildings
1851	San Francisco, CA	2500 buildings
1866	Portland, ME	1500 buildings
1871	Chicago, IL	17,430 buildings (250 fatalities)
1871	Peshtigo, WI	17 towns (approx. 1200 to 2000 fatalities)
1901	Jacksonville, FL	1700 buildings
1904	Baltimore, MD	2500 buildings
1906	San Francisco, CA	28,000 buildings (earthquake)
1908	Chelsea, MA	3500 buildings
1914	Salem, MA	1600 buildings
1916	Paris, TX	1440 buildings
1917	Atlanta, GA	1938 buildings
1918	Minnesota Forest	4000 buildings (559 fatalities)
1922	Ontario Forest	18 townships (44 fatalities)
1922	New Bern, NC	1000 buildings
More Recent Conflagrations and Group Fires		
1961	Bel Air, CA	630 buildings
1973	Chelsea, MA	300 buildings
1979	Houston, TX	25 buildings
1981	Lynn, MA	28 buildings
1982	Anaheim, CA	51 buildings
1983	Dallas, TX	6 buildings (125 apartments)
1991	Oakland/Berkley, CA	2886 buildings (25 fatalities)
2002	San Jose, CA	10 large buildings
2012	New York, NY	130 buildings
2013	Seaside Park, NJ	50 commercial buildings

*Wildland fires are beyond the scope of this publication and are not included in this list.
© Jones & Bartlett Learning.

6. Residential and/or commercial developments near wildlands
7. Built-up areas near high-hazard locations, where a transportation or industrial fire/explosion could quickly involve large numbers of buildings

Response areas with one or more of these seven conditions have an increased probability of a conflagration, especially when experiencing dry and windy weather conditions. Most of the urban conflagrations that occur have two or more of these factors combined with windy weather conditions. However, of the conditions listed, the wood shingle roofing has historically posed the greatest problem and is most often cited as a conflagration factor. Codes allowing untreated wood shingle roofs ignore a real and present danger.

An understanding of how large fires spread is essential in order to effectively apply the tactics involved in controlling a conflagration or group fire. A large flame front created by multiple burning structures creates extremely high radiant heat that rapidly spreads the fire. As the fire gets larger, the flame front expands exponentially, greatly accelerating the rate of fire spread. In addition, wood shingle roofs or other burning materials (flying brands) move upward via convection currents and are then carried by the wind to other areas.

The flying brand hazard is addressed by sending brand patrols into areas downwind, particularly areas containing materials that are likely to ignite. The 2002 Santana Row fire in San Jose, California, serves as an excellent example of the dangers associated with flying brands and wood shingle roofs. At the Santana Row fire, flying brands ignited wood shingle roofs one-half mile away.

Understanding the methods of heat transfer is the first step in developing a conflagration strategy. Once the flame front widens, radiant heat is the primary means of fire extension from building to building and from groups of buildings to other groups of buildings. If the fire is extending along a wide radiant flame front, the primary tactics are narrowing the flame front and setting up primary and secondary lines of defense while protecting exposures. In setting up lines of defense, take advantage of natural fire breaks, such as wide streets.

As always, life safety is the top priority, followed by extinguishment. In large group fires and conflagrations, being proactive and evacuating people in the fire's path long before they are actually threatened is the key to success. Remember that the evacuation area must extend beyond the secondary line of defense, and the police department is typically the best agency to manage the actual evacuation. Isolating the evacuation area from well-intentioned "would be" victims and traffic control are extremely important; thus a wide fire perimeter should be maintained. Establishing a law branch, under police supervision, is generally a good organizational tactic at large-scale fires.

A fully staffed planning section is of great value when commanding a fire that is spread over a wide geographic area; however, many times the initial line of defense is overrun by the fire before planning can be properly staffed.

FIGURE 9-9 shows three lines of defense for a Houston, Texas, apartment complex fire in 1979. The first two lines of defense had to be abandoned before a successful stop was made at the third line of defense. A battle-line strategy established without allowing enough time to position apparatus and equipment results in the IC expending scarce resources in a losing battle.

Incident Summary

Santana Row Conflagration

On Monday, August 19, 2002, a conflagration that began in the Santana Row construction site of San Jose, California, spread via burning embers to the Huff/Moorpark area, which is one-half mile away.

The building of origin that was under construction within the Santana Row complex was a large six-story structure covering 6 acres (24,281 m^2). This fire created burning embers as large as 2 in. × 4 in. (51 mm × 102 mm). High winds carried embers to an apartment complex and townhouses in the Huff/Moorpark area where they ignited wood shake roofs. Some of the roofs had been replaced by composite roofing materials. The damage was largely confined to 10 large apartment and townhouse buildings with wood shake roofs. Buildings with composite roofs were, for the most part, undamaged.

Data from: John Lee Cook, Jr., USFA-TR-153, 1/2004, *Santana Row Development Project Fire, San Jose, California.*

NOTE

When setting up primary and secondary lines of defense for a conflagration, place apparatus so they can be rapidly redeployed if the line of defense must be abandoned.

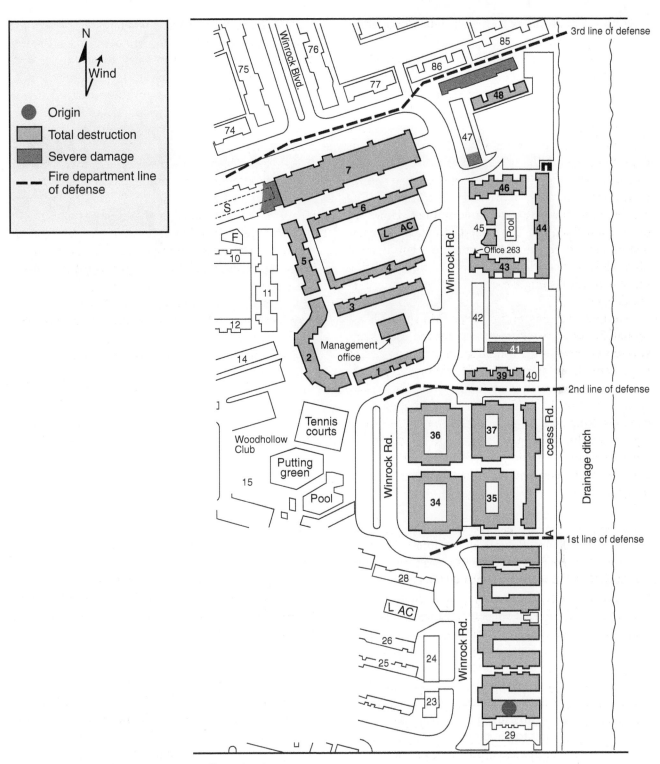

FIGURE 9-9 The Houston, Texas, conflagration in 1979.
© Jones & Bartlett Learning.

The Breezy Point (Queens, New York) conflagration of 2012 occurred during a hurricane. Streets were flooded, and access to the fire area was extremely difficult. Nevertheless, the IC managed to set up fire streams to narrow the fire front using a flanking maneuver, applying a three-pronged attack that limited fire spread (Pfeifer, 2013). Given the flooded conditions, adverse weather, and massive fire, the fact that no lives were lost during this massive conflagration testifies to a well-executed strategy **FIGURE 9-10**.

Halting or preventing a conflagration can seriously challenge any water supply system. The use of water must be prioritized with individual units, realizing the importance of water conservation. In most cases, water

FIGURE 9-10 Aftermath of the 2012 Breezy Point, Queens, conflagration.
© Ramin Talaie/EPA/Shutterstock.

being discharged into a flame front is of little value. The top priorities are maintaining the fire break at the lines of defense and protecting exposures. Once the decision has been made to write off an area, do not waste water trying to save what will surely be lost. Large cities are generally served by several water service areas, and the planning section chief should identify auxiliary water sources well in advance. For instance, when water mains were destroyed in the Marina District fire area during the 1989 San Francisco earthquake, the San Francisco Bay became an auxiliary water source as fireboats relayed water to the shore.

Collapsed structures, accumulations of combustible debris, utility problems (electrical shorts, gas leaks, etc.), and heating and lighting with open flames are only a few of the many reasons that areas struck by a natural disaster are prone to fires and conflagrations. Cincinnati, Ohio, suffered a conflagration in 1937 during a major flood. Petroleum tanks floated off their bases, spreading flammable and combustible liquids on top of the water. The petroleum products were ignited and spread fire to many buildings over a wide area. Fire hydrants and access roads were under water, hampering fire control efforts. Many buildings burned to the water line.

The six tactical elements of a successful conflagration strategy are as follows:

1. Evacuate and rescue people in imminent danger.
2. Evacuate and rescue people in the endangered area to beyond the secondary line of defense.
3. Set up a line of defense in an area with natural or artificial fire breaks.
4. Establish a secondary line of defense.
5. Narrow the flame front.
6. Provide flying brand patrols beyond the line of defense.

These tactics are interrelated and must be accomplished simultaneously or in rapid succession. For example, evacuating people to beyond the secondary line of defense requires knowing the location of the secondary line of defense. Actually protecting the secondary line of defense protects the evacuation.

Classifying as a Nonattack

It is difficult to stand by and watch a building being consumed by fire, but this is sometimes the only safe approach. Nonattack postures are underused because ICs often fail to recognize a total loss.

Often, in nonattack operations, environmental concerns are paramount. In these situations, the IC may assign resources to protect rivers or streams from contaminated runoff water or evacuate downwind neighborhoods. For example, a Sherwin Williams warehouse fire in Dayton, Ohio, threatened the community's water supply (Isner, 1988). A nonattack posture avoided a major environmental catastrophe. If a safe offensive attack is not possible and there is little or nothing to be gained by initiating a defensive attack, consider the nonattack option.

Wrap UP

CHAPTER SUMMARY

- Defensive attacks are used when the building's structural integrity or other hazards prohibit entry, when resource needs outweigh capabilities, or when the risk to fire fighters is too great.
- Solid fire-ground information, training, and experience are used to determine whether to use an offensive, defensive, or nonattack strategy.
- Defensive attacks seek to save surrounding property and/or to protect surrounding exposures.
- The collapse zone is the height of the building plus an allowance for debris scatter, usually estimated as 1½ times the height of the building.
- A master stream can apply more water from a greater distance and requires fewer personnel to operate compared to a handheld hose line.
- Water supply and apparatus positioning are key considerations during a defensive attack.
- A defensive attack is apparatus intensive but generally requires fewer fire fighters than an offensive attack.
- Apparatus at large-scale fires must be positioned for rapid redeployment.
- A successful conflagration strategy includes evacuating civilians, creating fire breaks for a primary and secondary line of defense, narrowing the flame front, and patrolling for fire brands.

KEY TERMS

defensive A type of fire attack in which exterior fire suppression operations are directed at protecting exposures.

offensive A type of fire attack in which fire fighters advance into the fire building with hose lines or other extinguishing agents to overpower the fire.

SUGGESTED ACTIVITIES

1. Residential/commercial building fire: The building in **FIGURE 9-11** is a three-story building and has an occupied basement that is partially underground. It is partially residential (left side of photo) with a commercial building—Global Paper Company (right side of photo). The commercial side can be identified by the arched windows. The businesses on the commercial side contains several stores selling expensive clothing. The residential side of the building has three luxury apartments on the second and third floors and three lower rent apartments in the partial underground section. There is a four-lane road on the left side and a narrow alley on the right side of the building.

   It is December 15 at 8:30 PM. **TABLE 9-2** shows the alarm card that lists the engine companies, truck companies, and other personnel and resources.

   You are the captain of Engine Company 14, and you arrive at the scene at the same time as Truck 14. Per standard operating procedures, you are the temporary IC until a district chief arrives.

   District 1 is out of service at another fire with Engine Companies 3, 5, 17, and 29 as well as Truck Companies 3 and 29. Therefore, you will retain command with a limited crew until District 2 and other second-alarm companies arrive on scene (expected response time 15 minutes).

   Truck 14 has positioned near the center of the building and is raising the aerial ladder preparing to conduct a defensive attack through the roof in the event the offensive attack is unsuccessful. You positioned Engine 14 at the right side of the building and ordered Engine 14 to conduct an interior attack.

   Evaluate the overall initial attack.

   A. If Engine 14 is going to conduct an interior attack, where should they enter the building and where should they begin the attack?

   B. Evaluate the present operation.

FIGURE 9-11 Fire in a large two-story building with an occupied basement.
Courtesy of David J. Jones, Cincinnati, Ohio.

TABLE 9-2 Alarm Card			
	Engine Companies	**Truck Companies**	**Other**
Initial 1-Alarm Response	14, 3, 29, 5	14, 3	District Chief 1 Heavy Rescue 14
2nd Alarm	17, 21, 12, 19	29, 17	District Chief 2 Heavy Rescue 9 Emergency Medical Services (EMS) 14
3rd Alarm	20, 34, 35, 37	21, 19	District Chief 4 EMS 3 Fire chief Operations assistant chief Safety chief
4th Alarm	23, 32, 9, 2	23, 32	District Chief 5 EMS 17 Personnel assistant chief Fire prevention assistant chief
5th Alarm	18, 7, 36, 39	18, 39	Police District 1 Captain Safety department manager

© Jones & Bartlett Learning.

C. Explain your answers to Parts A and B!
D. With the current conditions, should this be an offensive or defensive strategy?
E. Is there a life-safety component to this incident?
F. Would you consider the present operation to be safe and effective?
G. Would you continue with the current operation or are changes necessary?
H. If changes are necessary, explain and justify reassignments.
I. Develop an incident action plan and assign all responding companies.
J. Are responding resources sufficient?
K. If not, request needed assistance and assign incoming companies.
L. When all companies are assigned, develop an Incident Command System (ICS) organizational chart.

2. Use the same alarm card (Table 9-2) for this scenario. Assume the role of the first-arriving battalion chief, District Chief 1, at the fire shown in **FIGURE 9-12**.

It is 5:30 PM on Monday, August 15. The weather is hot and humid with a light wind from the east. The two-building complex is occupied by a costume manufacturer/distributor. The fire appears to be confined to the 70 × 90 ft (21.3 × 27.4 m) building (Bravo [west] side of photo). The three-story (28 × 40 ft [8.5 × 12.2 m]) attached building at the Bravo side is used as office and sales space. A two-story single-family home is located 20 ft (6.1 m) to the rear.

All companies are staffed with an officer, a driver, and two fire fighters. The first-arriving engine and truck companies (Engine 14 and Truck 14) are on scene when you (District Chief 1) arrive.

Engine 14 has secured a 5-in. (127-mm) water supply. The pump operator and the fire fighter assigned to connect to the hydrant remain at the apparatus as the initial rapid intervention crew (IRIC) while the officer and remaining fire fighter are preparing to advance a

FIGURE 9-12 Fire in a one-story commercial building.
Courtesy of Denny Baker, Cincinnati Fire Department.

2½-in. (64-mm) hose line through the door next to the loading dock on side Alpha. You approve!

Truck 14 has divided into two crews. Truck Crew 14A is forcing entry and entering the Bravo side building with search rope and a thermal imaging camera to perform search and rescue. Truck Crew 14B has entered the office/sales building on the Delta side to evacuate the building and check for fire extension.

The remaining first-alarm response arrives. You direct Engine 3 to secure a second source of water via a 5-in. (127-mm) hose line. The pump operator remains at the apparatus while three company members advance a 2½-in. (64-mm) hose line inside the building alongside Engine 14.

You direct the remaining first-alarm response as follows:

- Engine 29 to set up a portable master stream appliance inside the fire building; water is supplied by Engine 14's pumper. Engine 14's captain is assigned as the Division 1 supervisor.
- Engine 5 to lay a 2 ½-in. (64-mm) backup hose line supplied by Engine 3's pumper.
- Truck 3 to evaluate the roof structure and vent the roof near the Bravo side.
- Heavy Rescue 14 to assist in the primary search in the Bravo side building.

As IC, you order an EMS unit.

A. Evaluate the present operation in terms of safety and effectiveness.
  i. Are all necessary tactics being accomplished?
  ii. If not, what else needs to be done? (Add units as needed.)
  iii. Was there an IRIC, and a later rapid intervention crew (RIC)? If not, should there have been?
  iv. Is the water supply adequate?
  v. Should an alternative defensive action plan be developed?

B. Develop an incident action plan and National Incident Management System (NIMS) organizational chart for the operation in progress, adding additional units as you deem necessary.

3. Use the same building, fire, and deployment described in Question 2. Fire conditions are worsening. As you receive the 20-minute notification from dispatch, Truck Crew 14B declares a mayday, stating, "*Mayday, mayday, mayday! We are inside the rear of the building and cannot find our way out! We are running low on air.*"

As the IC, you acknowledge the mayday. They were last assigned to search the rear of the Bravo side building. You decide to begin a transition to a defensive attack but first want to locate and assist Truck Crew 14B to safety.

  A. Explain the deployment and actions that should be taken by the RIC teams.
  B. As the IC, explain your assignments and reassignments of units other than the RIC during the mayday operation.
  C. Restructure your incident action plan and NIMS organizational chart to reflect changes in the operation.

4. The two missing fire fighters are found and rescued. However, fire conditions have changed, as shown in **FIGURE 9-13**.

  A. List the necessary safety measures to be taken as you transition to a defensive attack.
  B. Develop an incident action plan for the defensive attack.
  C. Explain your reassignments, location of apparatus, and use of handheld and master streams. Also describe the water supply necessary to support the defensive operation.
  D. What should be done with the numerous handlines that were used during the offensive attack?
  E. Once the defensive attack is in place, additional personnel will be available (e.g., engine company personnel not operating pumps or fire streams). Describe how you would manage the excess personnel.
  F. Develop a NIMS organizational chart for the defensive operation.

FIGURE 9-13 Defensive fire conditions in a one-story commercial building.
Courtesy of David J. Jones, Cincinnati, Ohio.

5. The fire continues. It is now 8:30 PM (first call occurred at 5:30 PM). Four fire fighters have taken it upon themselves to set up handheld hose lines in the street, using hose lines previously deployed on the interior.

   A. Is the building in danger of collapse?
   B. Are the fire fighters shown in **FIGURE 9-14** within the collapse zone?
   C. Are their hose streams having a positive effect on the outcome?
   D. As the IC, would you add additional hose lines and/or change to master streams?
   E. List any additional changes you would order.
   F. Would additional companies be needed? If so, list the additional companies and other units you need to carry out a safe and effective operation.

FIGURE 9-14 Fire 3 hours into the operation at one-story commercial building.
Courtesy of David J. Jones, Cincinanti, Ohio.

REFERENCES

Corbett, Glenn P., and Francis L. Brannigan. 2021. *Brannigan's Building Construction for the Fire Service*, Sixth edition. Burlington, MA: Jones & Bartlett Learning.

Isner, Michael S. 1988. "$49 Million Loss in Sherwin-Williams Warehouse Fire." *NFPA Fire Journal* 82 (2): 65–73, 93.

National Fire Protection Association. 2018. *NFPA 1500: Standard on Fire Department Occupational Safety, Health, and Wellness Program*. Quincy, MA: NFPA.

Pfeifer, Joseph W. 2013. "Conflagration Hurricane Sandy—NY, NY, Breezy Point, Queens." *Fire Engineering* 5 (166): 61–9.

CHAPTER 10

Property Conservation

LEARNING OBJECTIVES

- List the three tactical priorities, in priority order, and explain how property conservation can be accomplished simultaneously with the life-safety and extinguishment priorities.
- Define primary damage and secondary damage.
- Explain how proper ventilation and forcible entry actually reduce property damage while protecting occupants and fire fighters.
- Given the water pressure and discharge orifice size, estimate the flow from a single and multiple sprinkler heads.
- List and explain six tactics used to reduce water damage.
- Explain why removing property from a building is not generally a good property conservation tactic.
- Calculate the weight of water from a nozzle discharging 350 GPM (1326-L/min) over a 10-minute period.
- Describe the importance of ventilation in property conservation.
- Discuss the importance of overhaul.
- Explain how thermal imaging cameras can be used to reduce overhaul damage, and list the precautions necessary when using thermal imaging cameras to find hidden fires.
- Enumerate safety issues related to overhaul and fire investigation.
- Develop a fire scenario and apply property conservation tactics for a fire controlled by a sprinkler system or an accidental discharge from a sprinkler system.

© Jones & Bartlett Learning.

Case Study

An electrical fire in a four-story apartment building was difficult to overhaul. The fire started in an electrical outlet on the third floor and extended into the wall. The fire department disrupted the power supply to the apartment, opened up the wall, and extinguished the fire. However, the fire had extended into the apartment above and the attic. Even after the wall and ceiling were opened up, some hidden fire was missed. Eventually, all of the fire was found and completely extinguished.

Wires should be traced through the wall space, because heating could occur anywhere along the circuit. Thermal imaging cameras could also be a valuable tool during overhaul operations.

1. What part does a thorough overhaul play in reducing property damage?

2. When trying to locate concealed fire, what are the advantages and limitations in using a thermal imaging camera?

Data from: Frank C. Montanga, Chasing down electrical fires. *Fire Engineering*, July 1999, pp. 83-86

Introduction

Proficient fire officers realize the importance of property conservation and recognize that property conservation tactics can substantially reduce property loss. Although property conservation is the lowest of the three operational priorities, being listed as the third priority does not mean that it must wait until all life-safety and extinguishment tactics are completed. Extinguishment is the ultimate property conservation tactic. Once staffing permits, the incident commander (IC) should simultaneously attend to all three operational priorities during offensive operations.

> **NOTE**
> The three tactical priorities are (1) life safety, (2) extinguishment, and (3) property conservation.

Good public relations results from professionalism shown in property conservation efforts. Few citizens (or elected officials) recognize a good extinguishment effort, generally because expertly conducted offensive operations are hidden from view. Property conservation tasks are obvious, however, even to laypeople. Efforts directed at saving valuable property show the citizens that you care about them and their property. Remember, fire departments are expected to save both people and property. In a residential fire, the occupants' personal property—property that is often irreplaceable—is being protected. The occupants will appreciate close attention to salvaging what can be saved.

> **NOTE**
> The golden rule of property conservation is "Do unto the property of others as you would have them do unto yours."

Fire departments often receive unfair criticism when they conduct forcible entry, ventilation, and overhaul operations. The untrained observer does not understand why it is necessary to cut a hole in the roof or tear walls apart to check for extension. To prevent unfair criticism, the department must educate the media, the public, and politicians about the purpose and importance of these tactics. Most important, the IC should explain the overhaul process to the owner and occupants. Although most secondary damage is essential, training should emphasize keeping the damage to a minimum while stopping the forward progress of the fire and preventing a rekindle.

Property conservation is prioritized behind life safety and extinguishment, but property conservation activities need not be delayed until the fire is extinguished or until the primary and secondary searches are complete. Priority activities are assigned first, and property conservation is delayed when staffing is not sufficient to attend to all of the life-safety and extinguishment tasks. However, proficient ICs include staffing for property conservation in their request for assistance or reassign forces to property conservation activities early in the battle.

Fallacy	Fact
Property conservation must be delayed until all life-safety and extinguishment activities are complete.	Property conservation is prioritized behind life safety and extinguishment. Be certain that resources are assigned to accomplish these priorities first. When staffing is adequate, property conservation can and should be conducted simultaneously.

The purpose of this chapter is to explain the primary components of property conservation operations. However, this is not an exhaustive reference on how to complete the various tasks involved in property conservation. Other sources of information, such as *Fundamentals of Fire Fighter Skills*, are dedicated to the actual tasks involved in protecting property (IAFC and NFPA, 2019).

In commercial and industrial occupancies, employees may be extremely helpful in locating valves, drains, and mechanical controls. Large industrial complexes sometimes have trained fire brigades. Many fire brigades are dedicated to property conservation.

Classifying as an Offensive Attack

Property conservation is generally limited to offensive attacks. If fire crews have been withdrawn from the building because of a lack of resources or because of deteriorating fire and building conditions, property conservation is no longer an issue. It makes no sense to send fire fighters into a building to place salvage covers or otherwise protect property when the entire building is at serious risk of being destroyed by fire. Remember, fire fighters should *never* be placed at risk attempting to save what is already lost, or ultimately will be lost.

Estimating Indirect Damage

A balance must be struck between doing necessary damage to halt the loss being caused by the fire and minimizing the damage caused by fire-ground activities. Categorizing damage helps in understanding this relationship.

Primary damage is caused by the fire and products of combustion. Secondary damage results from fire-ground activities or the operation of a fire protection system. Primary damage will consume the entire property if fire fighters fail to force entry or ventilate. On the other hand, if the building is washed off its foundation, nothing has been saved. In a fire, both primary and secondary damage will occur. Using caution and good judgment, the IC must do everything necessary to limit all types of property damage. However, fire fighters should never be placed at grave risk to save property.

Fire protection systems often cause secondary damage. When these systems are present, it is important to allow the system control the fire and to not shut down the system prematurely. (This was emphasized in Chapter 7, *Fire Protection Systems*.) Smoke and steam reduce visibility, making it difficult to determine the proper time to close valves and shut down systems. Therefore, when the IC decides to shut down a system, a fire fighter should remain at the control valve so the system can be charged again if needed.

Sprinkler systems have an excellent record of extinguishing or holding fires in check with minimal water damage. On occasion, system piping or sprinkler heads are damaged, causing an unwanted or accidental water flow. Although this situation is not a fire, the IC needs to be prepared to mitigate it promptly.

Flows from sprinkler heads can vary depending on the size of the orifice opening, the type of sprinkler head, and the water pressure in the system.

Whether the head was accidentally damaged or opened automatically because of a fire, the flow of water will continue until action is taken to stop it. A 30-GPM (113.6-L/min) head flowing during a 10-minute response time will discharge 300 gallons (1136 L) of water. Once the fire is under control, it is necessary to stop the flow of water, either by shutting the system down at the riser, by closing division valves, or by using sprinkler stops.

Other fire-suppression agents can also harm equipment and stock. The discharge of a dry chemical system, for example, will require considerable clean-up and can damage certain types of equipment, such as computers, telephone equipment, and other sensitive electronics. Generally, clean-up is the responsibility of the property owner, not the fire department.

Evaluating Water Damage

A fallacy exists regarding secondary damage and rate of flow. Many believe that larger-than-needed hose streams cause unnecessary water damage. In reality, the opposite is true. Larger-than-needed hose streams, if properly applied, will result in less water damage, as the fire is quickly extinguished and the time that water is flowing is reduced. Conversely, choosing a fire line that is too small to quickly extinguish the fire will result in both increased water and fire damage, as water will be applied for an extended time without effectively extinguishing the fire.

Life-safety and extinguishment activities concentrate on the fire floor and floors above, whereas efforts to control water damage are concentrated on the fire floor and floors below. The following actions reduce water damage:

1. Promptly extinguishing the fire, and avoiding wash downs. In addition to reducing water damage, prompt extinguishment activities also reduce fire and smoke damage.
2. Stopping the flow of water from sprinkler systems. Promptly shutting down hose lines is one method of stopping the flow of water into the building. In the case of sprinkler systems or other fixed fire suppression systems, it is important to locate control valves. Using sprinkler stops or controlling the water at the main or divisional control valve is the primary way to reduce the flow from an automatic sprinkler system. Be careful not to shut a system down prematurely, as this will result in additional water and fire damage.
3. Containing runoff. Containing the water before it has a chance to run off or migrate to another location is another property conservation tactic. Catchalls that are made with salvage covers, or containers (e.g., drums, buckets), are placed under a ceiling leak below the fire floor. Also, pumps and water vacuums can be effective in property conservation. Preventing further migration of water will also reduce damage.
4. Channeling the water into drains or chutes or otherwise channeling water out of the building. Fire fighters have devised unique ways of channeling water out of the building. Makeshift chutes are sometimes fashioned using ladders, pike poles, and salvage covers. These chutes are then directed out a window, down a stairway, or to a drain.

Fire debris will tend to clog floor drains; it is important to keep these drains clear. If floors do not have drains, a common way to provide one is removing a toilet, creating a drain through the 4-in. (102-mm) opening in the floor. If this is done carefully, the only damage to the toilet will be to the seal and floor bolts. As with any drain, however, it is critical to ensure that it does not become clogged with debris. Removing water from the floor is not only important to property conservation; it is also a safety issue. A large buildup of water on a floor can cause structural failure. The IC or incident safety officer must determine whether it is safe to conduct interior operations once a large quantity of water has been applied. This consideration is particularly important after a long defensive attack when advancing into the building for final extinguishment and overhaul. This amount of water can add substantial weight to an already compromised structure. It is also important to ensure that any standing water in areas such as basements is not hiding hazards, such as openings, changes in elevation, or steps. If it is necessary to send fire personnel into a basement or other confined space that contains standing water, it is critically important to ensure that the electric power has been shut down to avoid any possibility of electrocution.

The nature of the commodities that are stored in a building can affect the weight loading on the floor. If absorbent materials are being stored in the fire area, the weight of the water absorbed will remain for a considerable time rather than running off. In addition, the absorbent materials may expand, placing pressure against outside walls, which can also affect the structural integrity of the building. Water weighs 8.33 pounds per gallon (1 kg/L). When multiple master stream appliances are flowing water into a structure, the weight of the water can rapidly increase the collapse potential. Depending on conditions, the water may flow out of the building or be contained. **TABLE 10-1** assumes that materials stored in the building or fire debris is containing the water being discharged from large master stream appliances **FIGURE 10-1**.

It is difficult to estimate the amount of water remaining inside a building. Water streaming out of a building may represent only a fraction of the total amount of water being applied.

5. Covering valuable property. It is important for the IC, as well as all other members on the scene, to be aware of the need for property conservation. By placing salvage covers on exposed property, an alert fire fighter can often prevent water damage to valuable property. When using salvage covers, crews should group furniture and other items together to protect them from flowing water. This will allow several items to be protected by a single cover. Whenever possible, materials being protected should be raised several inches off the floor using available objects. In an industrial setting, wooden or plastic pallets are often available. In residential properties, place salvageable items on top of a table, bed, or couch.

6. Moving or removing valuable property. Water flows downward through a building, following the path of least resistance. Stairs, elevator shafts, and drains provide paths of least resistance. At times, water flow is

TABLE 10-1 Weight of Water from Master Streams

Number of Master Streams at 1000 GPM (3785 L/min) Each	Weight Added Each Minute at 8.33 lb/gal (1 kg/L)	Total Weight at 30 Minutes	Total Weight at 60 Minutes
1	8330 lb (3778 kg)	249,900 lb (113,353 kg)	499,800 lb (226,705 kg)
3	24,990 lb (11,335 kg)	749,700 lb (340,058 kg)	1,499,400 lb (680,116 kg)
10	83,300 lb (37,784 kg)	2,499,000 lb (1,133,527 kg)	4,998,000 lb (2,267,055 kg)

© Jones & Bartlett Learning.

FIGURE 10-1 Multiple master streams in use.
Courtesy of Bill Strite, Cincinnati, Ohio.

blocked by debris, or the volume simply overwhelms the natural flow out of the building. Even when water is flowing through the ceiling, the path of least resistance will be followed. This is generally around openings in the ceiling, such as light fixtures. If a ceiling is holding water runoff for a period of time, the weight of the water is likely to cause the ceiling to collapse. This ceiling collapse can injure fire fighters assigned to the area and increase damage to the contents below. Using a pike pole to drain the ceiling, being careful to avoid electrical equipment, can alleviate this problem.

As soon as possible, property should be moved away from natural flow paths, grouped together, and covered with salvage covers.

On occasion, property is completely removed from the building by fire crews to save it from water damage. This is a very labor-intensive way to conduct property conservation operations and has limited application. If materials are moved outside, they must then be protected from inclement weather, vandalism, or theft.

Evaluating Smoke Damage

Smoke can cause considerable property damage. In fact, smoke damage often greatly exceeds water damage. Smoke generally follows an upward path but can cause damage below the fire as well. In previous chapters, ventilation was discussed in terms of life safety and extinguishment. However, ventilation is also an important property conservation tactic. Blowers and fans are generally used to remove smoke during property conservation operations. Of course, ventilation tactics used to remove smoke for life-safety and extinguishment purposes also help to prevent property damage.

Calculating Staffing Needs

Once critical life-safety and extinguishment positions are staffed and a tactical reserve has been established, remaining personnel can be assigned to property conservation. The earlier property conservation is addressed, the more successful it will be. Property conservation that is delayed too long will be of little value, as the exposed materials will already be damaged by water and/or smoke.

Property conservation staffing needs vary greatly. For most offensive operations, at least one crew should be assigned to start property conservation activities on the floor below the fire. Several crews may need to be assigned to property conservation duties, depending on the amount of damage, the size of the area to be protected, and the type of property that needs to be protected.

Evaluating Property Conservation Needs

Property conservation activities should not be delayed until other priorities have been completed. This concept needs to be emphasized. It is easy to become overwhelmed with fire attack activities and neglect property conservation until the fire is completely extinguished. Implementing a planning section greatly assists the IC in defining necessary property conservation activity. Staffing and equipment are assigned to life safety and extinguishment first, but if property conservation tasks still remain to be assigned, there is a need to call for additional resources.

Crews generally view property conservation as an undesirable assignment, preferring to be involved in life-safety and extinguishment activities. Companies generally do not freelance into property conservation tasks. The ability to simultaneously conduct life-safety, extinguishment, and property conservation tasks is a sure sign that the IC is effectively managing the incident.

Evaluating Overhaul Needs

Overhaul is extremely important; some tactical priority models list it as a separate priority. Overhaul is the completion of the extinguishment priority. Overhaul can result in what might appear to be additional damage to the property, but this damage is warranted to prevent further primary damage. The use of a thermal imaging camera can substantially reduce damage caused by opening walls and ceilings to locate hotspots **FIGURE 10-2**. However, it is important that members using the thermal imaging camera be well trained in using the device and interpreting readings. If a thermal imaging camera is used to examine walls and ceilings, and any doubt remains regarding the presence of hidden fire, the wall and/or ceiling must be opened.

FIGURE 10-2 Checking for fire extension using a thermal imaging camera.
© Jones & Bartlett Learning.

The purpose of overhaul is to ensure that the fire is completely extinguished and that the building is safe for personnel such as investigators or the property owner to reenter. Permitting people to reenter a building only to have fire rekindle is a serious failure on the part of the IC. Overhaul is an activity that is appealing to few fire fighters, which sometimes leads to a decision to simply wet down everything in sight. This "wash down" is not effective in getting to hidden fires and often results in unnecessary property damage.

There is a tendency for fire scene discipline to deteriorate during the overhaul phase. Fire fighters are often injured as they rush to complete this distasteful chore so they can return to the fire station or home. As the smoke clears, many fire fighters take the opportunity to remove their breathing apparatus and other items of protective clothing. Studies indicate that toxic gas levels remain present during overhaul operations. (See Chapter 5, *Fire Fighter Safety*.) If the IC does not maintain discipline, overhaul can become an uncontrolled, dangerous activity.

To keep property damage to a minimum and to ensure that the fire is completely extinguished, overhaul must be accomplished in a well-planned, systematic manner. A good overhaul technique is to assign a fully protected crew inside as a fire watch while the building is being ventilated. This will allow other fire fighters on the scene to rest, thereby reducing the potential for injuries. If it is practical, keep crews in an outside or sheltered rehabilitation area for at least 15 minutes. After the building is well ventilated, the rested fire fighters can return to a much safer environment to complete the overhaul.

Full protective clothing, including breathing apparatus, must be worn during overhaul activities **FIGURE 10-3**. Opening walls and ceilings will expose smoky, toxic, smoldering debris. In addition, ceilings and walls may contain asbestos or other materials that are capable of harming the unprotected fire fighter. Members operating inside the structure, including fire investigators, cannot be permitted to remove their face pieces.

Many departments use carbon monoxide meters or multiple gas meters to determine when it is safe to remove the breathing apparatus face piece. Carbon monoxide is usually the most prevalent fire gas, and low readings after a fire indicate some measure of smoke and toxic gas removal. However, there are many other inhalation hazards that are not read by carbon monoxide or multiple gas detectors. The best practice is to remain "on air" throughout the overhaul phase of the operation.

CHAPTER 10 Property Conservation 335

FIGURE 10-3 Overhaul in full personal protective equipment.
© Glen E. Ellman.

A Word about Fire Investigation

Improper overhaul can destroy any hope of properly conducting a fire scene investigation. A determination of what areas are to be left undisturbed should be made while fire suppression forces are resting between fire control and overhaul. The IC should attempt to make a preliminary determination regarding what areas will be of interest to the investigators. When a fire investigator is on the scene during overhaul operations, this person is an invaluable resource to the IC. However, proper and complete overhaul should always be performed to ensure that the fire is completely extinguished.

Wrap UP

CHAPTER SUMMARY

- Property conservation is the third priority but can be performed simultaneously with life safety and extinguishment when staffing permits.
- Property conservation is limited to offensive attacks.
- Ventilation and forcible entry generally cause less property damage than delaying extinguishment.
- Covering or moving valuable property can help prevent water damage.
- Smoke damage generally occurs above the fire but can also damage areas below the fire.
- Proper ventilation is a life-safety, extinguishment, and property conservation tactic.
- Staffing and equipment are assigned to life safety and extinguishment first, but if property conservation tasks still remain to be assigned, call for additional resources.
- To ensure complete extinguishment and minimize property damage, overhaul must be accomplished in a well-planned, systematic manner.

KEY TERMS

primary damage Damage caused by the products of combustion.

secondary damage Damage caused by fire-ground activities or the operation of a fire suppression system.

SUGGESTED ACTIVITIES

1. It is 2:00 AM. The automatic fire alarm indicates a fire on the fourth and fifth floors of an insurance office **FIGURE 10-4**. Five companies (three engine companies, two truck companies, and a district chief are dispatched to the fire.

    You arrive as the officer of Truck 2, the second-arriving truck. Engine 1 reports a fire on the fourth floor that has been extinguished by the sprinkler system. Truck 1 is assisting Engine 1. Engine 2 is assigned to the fifth floor to check for fire extension and reports moderate smoke, but no signs of fire.

FIGURE 10-4 An insurance claims office subdivided using mobile partitions (cubicles).
© Jude Lazaro/Shutterstock.

    The IC assigns you (Truck 2) to the third floor with a primary task of conducting property conservation activities. When you arrive on the third floor, there is light smoke but no sign of an active fire. However, water is flowing through the suspended ceiling tiles.

- **A.** Provide a status report to the IC, including a request for additional assistance if needed.
- **B.** Explain measures you would take to protect the office equipment, computers, and written records from water and smoke damage.
- **C.** Water is collecting on the floor. How would you remove water from the third floor without causing additional damage to the third floor and floors below?
- **D.** There is an odor of smoke on the third floor. Describe the best methods to reduce smoke damage.
- **E.** Are the assigned fire companies sufficient? If not, what additional response would be needed?

2. A fire has occurred in the store shown in **FIGURE 10-5**. It appears that the sprinkler system has the fire under control but is still operating.
   - **A.** Assume the role of IC and assign companies to the various tasks necessary to limit damage and prevent a rekindle.
   - **B.** Describe how you would reduce property damage to merchandise not already damaged by the fire.
   - **C.** Explain the use of a thermal imaging camera to find hotspots and how it would be used in overhaul at this fire.

FIGURE 10-5 Overhaul and salvage at a mercantile occupancy.
Courtesy of Ben Klaene.

REFERENCE

International Association of Fire Chiefs, National Fire Protection Association. 2019. *Fundamentals of Fire Fighter Skills*, Fourth edition. Burlington, MA: Jones & Bartlett Learning.

CHAPTER 11

The Role of Occupancy

LEARNING OBJECTIVES

- Classify occupancy risks using an occupancy factor matrix.
- Given the dimensions, number of floors, and occupant load, determine the maximum number of people who could be in a building.
- Define assembly occupancy and provide examples of different types of assembly occupancies.
- Evaluate and compare the six occupancy types listed in the assembly occupancy factor matrix. Which of the six listed occupancies would you consider most hazardous to occupants and fire fighters? Justify your selection.
- Compare and contrast the risk to fire fighters when fighting a fire in an assembly occupancy compared to a residential occupancy.
- Explain why fires in churches often result in a large loss of property.
- Define educational occupancy and compare the life hazards in elementary schools and high schools.
- List some of the fire safety problems associated with college housing units.
- Select a fire from Table 11-2: Notable Fires in Assembly Occupancies. Develop a short presentation with emphasis on lessons learned regarding tactics and strategies.
- Define healthcare occupancy and explain how evacuations in healthcare occupancies are different from evacuations in most other occupancies.
- Compare and contrast occupants in hospitals, nursing homes, and limited care facilities, and

describe how the occupant characteristics in each affect life safety during a structure fire.
- Define residential board and care occupancy and compare these facilities to nursing homes.
- Define detention and correctional occupancy and explain the special challenges associated with combating a fire in a large correctional facility.
- Define residential occupancy and compare various types of residential buildings in terms of life safety.
- Evaluate and discuss civilian fire deaths and fire fighter line-of-duty death rates in residential occupancies.
- Define mercantile occupancy and compare older-style shopping centers to enclosed malls and lifestyle centers in terms of life safety and extinguishment.
- Examine life safety and extinguishment problems related to "big-box" stores.
- Explain how search and rescue procedures in a large commercial structure differ from search and rescue in a residential building.
- Define business occupancy.
- Compare and contrast business and residential occupancies in terms of the risk to fire fighters during fire-ground operations.
- Define storage occupancy and evaluate the effect of fuel load on manual firefighting.
- Describe the problems associated with changing the commodities stored in a sprinkler-protected storage occupancy.
- Compare and contrast storage and residential occupancies in terms of the risk to fire fighters during fire-ground operations.
- Define industrial occupancy and the effect of hazardous materials related to life safety and extinguishment.
- Compare and contrast industrial and residential occupancies in terms of the risk to fire fighters during fire-ground operations.
- List the various types of residential buildings, then compare and contrast the threat to occupants in the various residential (permanent and transient) occupancies.
- Define multiple, mixed, and separated occupancies, and explain the difference between a mixed and separated occupancy.
- Explain the increased hazard to occupants in multi-story buildings where the first floor is occupied by stores and shops, with apartments above.
- Describe the fire hazards associated with buildings under construction, renovation, or demolition.

NIOSH. (July 8, 2015). "4 Career Fire Fighters Killed and 16 Fire Fighters Injured at Commercial Structure Fire – Texas." https://www.cdc.gov/niosh/fire/pdfs/face201316.pdf.

Case Study

Five fire fighters were killed as the result of the May 31, 2013 commercial structure fire in Texas when the roof of a restaurant collapsed on them during firefighting operations. Fatalities and injuries were as follows:

- 35-year-old career captain
- 41-year-old career engineer operator
- 29-year-old career fire fighter
- 24-year-old career fire fighter
- 16 fire fighters injured; one died in 2017 from complications of severe injuries

Upon arrival, the captain of Engine 51 radioed his size-up stating they had a "working fire" in the restaurant with heavy smoke showing. An offensive attack was initiated in the restaurant using a 2½-in. (64-mm) preconnected hose line.

The district chief arrived on scene and established command. He ordered Engine 51 out of the building because Engine 51 was down to a quarter tank of water.

Engine 68 arrived on scene and laid two 4-in. (101-mm) supply hose lines from Engine 51 to a hydrant east of the fire building. Now that Engine 51 had an established water supply, they reentered the building. Engine 68 backed up Engine 51 on the 2½-in. (64-mm) hose line.

Engine 82 (fourth-due engine company) was pulling a 1¾-in. (44-mm) hose line to the front doorway that Engine 51 had entered, when the collapse occurred.

The roof collapsed 12 minutes after Engine 51 arrived on scene and 15 minutes and 29 seconds after the initial dispatch. The fire fighter from Engine 51 was at the front doorway and was pushed out of the building by the collapse. The captain from Engine 82 called a "mayday."

The rapid intervention team (RIT) operations were initiated by Engine 60. During the RIT operations, a secondary wall collapse occurred, injuring several members of the rescue group. Due to the tremendous efforts of the rescue group, a successful RIT operation was conducted.

The captain of Engine 68 was located and removed from the structure by the RIT and transported to a local hospital. The engineer operator from Engine 51 was removed from the structure by the RIT and later died at a local hospital. The search continued for the captain of Engine 51 and the two fire fighters from Engine 68.

Approximately 2 hours after the collapse, the body of the captain from Engine 51 was located on top of the restaurant roof debris. The two fire fighters from Engine 68 were discovered underneath the restaurant roof debris. The officer and two fire fighters were pronounced dead at the scene.

The photo shown below depicts an aerial view of the structure with the restaurant located on the front left (side Alpha–Bravo), motel entrance in the middle (side Alpha), and a sports bar on the right (side Alpha–Delta). Many mercantile occupancies have large open areas that can result in a large internal fire.

Source: NIOSH Fire Fighter Fatality Investigation Summary [F2013-16], July 8, 2015, "4 Career Fire Fighters Killed and 16 Fire Fighters Injured at Commercial Structure Fire—Texas," https://www.cdc.gov/niosh/fire/pdfs/face201316.pdf, Revised April 5, 2017, to update the Executive Summary and the Cause of Death, Accessed March 16, 2020.

NIOSH. (July 8, 2015). "4 Career Fire Fighters Killed and 16 Fire Fighters Injured at Commercial Structure Fire – Texas." https://www.cdc.gov/niosh/fire/pdfs/face201316.pdf

Introduction

As a part of the size-up process, the incident commander (IC) must consider the building's occupancy type. Once the occupancy type is known, the IC can apply a strategy applicable to the occupancy and can begin determining the life-safety, extinguishment, and property conservation tactics required to bring the incident to a successful conclusion. This information will assist the IC in determining the danger to occupants and risk to fire fighters. FIGURE 11-1 lists the occupancy types discussed in this chapter. Residential, industrial, and assembly occupancies all require different strategies. Time of day and day of week are also critical factors. This chapter reviews the process of evaluating occupancies and formulating safe and effective fire-ground strategies.

Some occupancies are dangerous places not only for the occupants but also for fire fighters. For example, fire fighters working at a manufacturing occupancy fire are 8.6 times more likely to be fatally injured than when working at a residential occupancy (Figure 11-1).

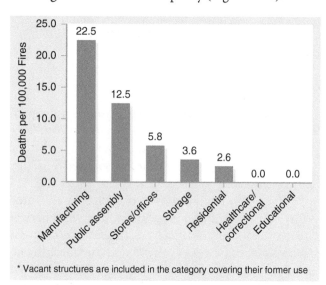

FIGURE 11-1 Fire fighter fire-ground fatalities by occupancy type, 2013 to 2017.
Modified from: Fahy, Rita F. *Firefighter Fatalities in the United States*. 2019. National Fire Protection Association.

Classifying the Occupancy Type

The function or use of a building has much to do with life safety. Codes and standards are written with a focus on the special hazards presented by the occupancy. Some of the major factors related to occupancy type are as follows:

- Time factors. When are people present in the structure? (places of worship, schools, night clubs)
- Occupant load and total number of occupants (single family dwelling vs. large apartments or nursing homes)
- Awareness. Are occupants sleeping or mentally impaired? (hospital, group home)
- Age of the occupants (nursing home)
- Mobility of occupants (hospitals, rehab facilities)
- Familiarity. Do occupants know the building layout? (hotel, place of assembly)
- Leadership. Will there be an organized evacuation? (educational facility or home vs. place of assembly)

Occupant load is defined in NFPA 101, 2018 Edition as:

> **3.3.170.2** Occupant Load. The total number of persons that might occupy a building or portion thereof at any one time. (SAF-MEA).

Codes usually state occupant load as the number of square feet (ft^2) per person. In some assembly occupancies, the maximum load is 7 ft^2 (0.65 m^2) per person (occupant load factor). In other words, 100 people could occupy a 700-ft^2 (65-m^2) area. Some occupancies have different maximum loads. Most codes require the maximum number of people allowed in a given area to be posted in assembly occupancies and sometimes in other occupancies. If the area is not posted, a rough approximation of the number of people who could safely occupy an area can be determined by dividing the area in ft^2 by the occupant load factor. Occupancy factors depend on the occupancy use and load factors, and may be very different from building to building. The ability to evacuate without assistance will be rated as part of the discussion of each major occupancy type. A scale of 1 to 10 will be used to indicate whether each factor is positive (10), neutral (5), or negative (1) for the specific occupancy type. For example, an awareness rating of 10 would mean that all occupants within this occupancy would be expected to be fully awake and mentally unimpaired. An awareness rating of 1 would be given to an occupancy where occupants are totally unaware, such as people who are under general anesthesia. The numeric ratings are general in nature, and buildings of the same occupancy classification may have different ratings. For example, some ambulatory care facilities anesthetize patients, which would result in a very negative awareness rating, while other ambulatory care facilities do not anesthetize or use medications that reduce patient awareness.

The following occupancy types will be addressed in this chapter:

- Assembly occupancies
  - Places of worship

- Eating and drinking establishments
- Sports arenas
- Convention centers
- Theaters
■ Educational occupancies
  - Elementary schools
  - Middle, junior high, or high schools
■ Healthcare occupancies
  - Hospitals
  - Nursing homes
  - Limited care facilities
  - Ambulatory care facilities
■ Residential board and care occupancies
■ Detention and correctional occupancies
■ Residential occupancies
  - One- and two-family dwellings
  - Apartment buildings
  - Dormitories
  - Hotels/motels
■ Mercantile occupancies
  - Shopping centers
  - Enclosed shopping malls
  - Lifestyle centers
  - Big-box stores
  - Multilevel department stores
■ Business occupancies
■ Storage occupancies
■ Industrial occupancies
■ Multiple occupancies
■ Buildings under construction or demolition
■ Renovated buildings

Occupancy and suboccupancy types will be discussed in terms of the operational priority list: life safety, extinguishment, and property conservation.

Codes are promulgated to avoid the repetition of past mistakes. However, strengthening codes is not enough. Each of the major loss-of-life fires described in this chapter contains lessons for the fire officer. In many of these cases, existing codes were ignored, resulting in a large loss of life. The various occupancies are used to discuss common problems and solutions

NOTE

The philosopher and poet George Santayana said, "Those who cannot remember the past are condemned to repeat it." As practitioners of fire strategy and tactics, it is imperative that we learn from the past.

but also to offer history lessons. The U.S. Fire Administration (USFA), National Institute for Occupational Safety and Health (NIOSH), and National Institute of Standards and Technology (NIST) produce technical reports that are available for recent large loss-of-life fires. Periodicals such as *NFPA Journal* and *Fire Engineering* have offered technical reports to the fire service for over 100 years.

Assembly Occupancies

NFPA 101, *Life Safety Code*, 3.3.196.2, defines an assembly occupancy as an occupancy (1) used for the gathering of 50 or more persons for deliberation, worship, entertainment, eating, drinking, amusement, awaiting transportation, or similar uses; or (2) used as a special amusement building, regardless of occupant load (NFPA 101, 2018).

The discussion here is limited to larger places of assembly and tactical difficulties related to the occupancy. Included under the **assembly occupancy** category are the following:

■ Places of worship
■ Eating and drinking establishments
■ Sports arenas
■ Convention centers
■ Theaters

An occupancy factor matrix for assembly occupancies is shown in **TABLE 11-1**. Note that occupant familiarity and occupant load tend to be negative factors, but mobility and awareness tend to be positive factors for most assembly occupancies.

Places of assembly are not occupied at all times. However, when they are occupied, large numbers of people gather in a relatively small area. An open-air assembly occupancy of concentrated use (meaning no fixed seating) with an occupied space of 100 ft × 100 ft equals 10,000 ft^2, divided by 7 equals 1428 attendees, would be allowed pursuant to NFPA 101, 2018 Edition. The primary challenge in managing an incident in an occupied assembly building is removing people to the outside or to a place of safe refuge.

The fuel load will vary by the assembly classification. The concern is not only for the total fuel load; flame spread rate is also critical. Decorative materials with high flame spread rates were significant contributing factors in the large loss of life at the Cocoanut Grove, Beverly Hills Supper Club, and The Station nightclub eating and drinking establishment fires (discussed later). Fire control efforts must place a high priority on keeping the fire out of the exits. Aside from the main tactical consideration of removing victims during times of peak occupancy,

TABLE 11-1 Occupancy Factor Matrix: Assembly Occupancies

Assembly Occupancies	Mobility of Occupants	Age of Occupants	Leadership	Awareness	Occupant Load	Familiarity of Occupants
Places of worship	8	5	9	10	2	5
Eating and drinking establishments	8	7	5	4	2	3
Sports arenas	8	7	6	7	2	2
Convention centers	8	6	5	10	2	1
Theaters	8	6	5	10	2	2

1 = extremely negative factor; 10 = very positive factor.
© Jones & Bartlett Learning.

responders must remain mindful that these structures are unattended for long periods when they are not in use, allowing fires to gain considerable headway before discovery. The probability of total involvement on arrival in these situations dramatically affects fire fighter safety. Fires in places of assembly are less common than home fires, accounting for approximately 3 percent of all structure fires (NFPA, 2014-2018), but the potential for a large loss-of-life fire in these buildings requires preincident planning and special tactical considerations.

Places of Worship

Places of worship often use open flames as part of religious services and tend to be overcrowded on religious holidays. Places of worship are also left unattended for long periods, and many fires occur when these buildings are unoccupied. However, places of worship are occupied at times other than during religious services. Meetings, educational programs, fundraisers, and social events take place within the building or attached rooms. Some places of worship provide a wide array of social services, such as child care, elderly programs, and shelter for the homeless. These services require supplies that add to the fuel load. Collecting clothing and food for poor or homeless people in the community is a common practice. These supplies—food and clothing—are often stored in places that are not separated from the main building, often in the exit way, and can pose a severe fire threat.

In the case of old Gothic-style churches, fires quickly involve large concealed spaces such as roofs and bell towers **FIGURE 11-2**. At the time of the 2019 fire, Notre Dame Cathedral was being renovated. The fire originated in the church steeple. Church steeples typically contain large, heavy bells that present a potential collapse hazard, making interior operations hazardous. Furthermore, many churches are attached to other buildings, such as rectories or schools, which can create an exposure problem.

Fire department notification for the Notre Dame fire was delayed by approximately 30 minutes, partially due to the local fire alarm system that did not automatically transmit the fire alarm to the fire department. It was reported that 400 or more fire fighters fought the fire in this historic church.

Older Gothic-style churches typically have large timbers supporting the roof structure **FIGURE 11-3**, whereas modern churches often have truss roof structures **FIGURE 11-4**. Recent fire fighter fatalities have occurred due to truss roof collapses when fire fighters were working on the interior. (See Chapter 5, *Fire Fighter Safety*, for a detailed analysis.) The roof structure should be noted during preincident planning.

The large open spaces in places of worship can result in a large-volume fire that is extremely difficult to extinguish once the fire gains considerable headway. In addition to these traditional church structures, religious services are sometimes held in buildings that were not originally designed as churches.

Statistically, fires and loss of civilian lives are rare in places of worship. However, there is a potential for a significant loss of life when large numbers of people gather for religious services, particularly on religious holidays. Most fires in these occupancies take place when services are not being held, resulting in an extended alarm time that allows the fire to gain considerable headway prior to the fire department's response.

FIGURE 11-2 Notre Dame Cathedral Fire.
REUTERS/Benoit Tessier.

FIGURE 11-3 Gothic-style church.
Courtesy of Ben Klaene.

FIGURE 11-4 Modern church.
Source: NIOSH report F2011-14, January 18, 2012.

Many times, fires that occur when a church is unoccupied are not reported until fire breaks through the roof.

Fires in places of worship and funeral homes account for 0.4% of all structure fires and 0.8% of structure fire property loss (NFPA, 2014-2018). These occupancies tend to have priceless artifacts, stained glass windows, and other high-value property. Unfortunately, large fire volume and structural damage/collapse may render property conservation efforts ineffective. Large church fires usually dictate a defensive operation and often result in full or partial building collapse.

Eating and Drinking Establishments

Overcrowding is a common problem in eating and drinking establishments. If a fire occurs while the occupancy is overcrowded, exits may quickly become blocked. Many tragic fires have occurred in these occupancies during the past 100 years; most notable are the ones listed in **TABLE 11-2**. The largest loss of life in an eating or drinking establishment in the United States was in the Cocoanut Grove fire in Boston, Massachusetts, where nearly 500 people were killed, many

TABLE 11-2 Notable Fires in Assembly Occupancies

Fire	Year	Deaths
Iroquois Theater, Chicago, IL	1903	602
Rhoads Opera House, Boyerton, PA	1908	170
Rhythm Club, Natchez, MS	1940	207
Cocoanut Grove, Boston, MA	1942	492
Ringling Brothers and Barnum and Bailey Circus, Hartford, CT	1944	168
Indiana State Fairground, Indianapolis, IN	1963	75
Beverly Hills Supper Club, Southgate, KY	1977	165
Happy Land Social Club, New York, NY	1990	87
The Station Nightclub, West Warwick, RI	2003	100

© Jones & Bartlett Learning.

of them service men on their way to fight in World War II.

An analysis of eating and drinking establishment fires reveals the killing potential of fires in assembly occupancies. Occupants will often be mobile, but they may be impaired because of alcohol consumption. Fire probability is fairly high, owing to cooking, and highly combustible decorations and furnishings. Fire fighters must quickly confine the fire while protecting the exits.

Most patrons in assembly occupancies will attempt to exit the building using the same route they used to enter the building. Although alternative exits are available, they are often underutilized and overcrowding occurs at the main entrance. Fire fighters should direct people to alternative exits while keeping the crowd at the main entrance moving. It is essential to identify alternative exits, including windows and other openings that could be used to provide emergency egress, during preincident planning for eating and drinking establishments. Preincident plans should also consider the possibility of using forcible-entry tools to create alternative exits or to widen existing exits during an emergency evacuation.

Several large loss-of-life fires have occurred outside of the United States, including a January 28, 2013, fire in Santa Maria, Brazil, that killed 238. The fire in Brazil and The Station nightclub fire were ignited when entertainers used pyrotechnics inside the occupancy.

Many times, occupants escaping the building stop as soon as they reach safety, thereby unintentionally blocking the exit for others. Someone should be stationed at exit discharges to keep people moving away from the building. This is best done by fire fighters, but police, security personnel, or property management can be assigned this task when fire department staffing does not permit assigning fire fighters. However, consideration must be given to fire and smoke conditions at the exit discharges. If heavy smoke or fire is present at the exit discharge, only fire fighters will be able to facilitate movement at the exit.

Panic has long been blamed for many deaths at nightclub fires. In reality, panic is a scapegoat. Often the real problem is a lack of proper fire protection features. Extensive human behavior research was done after the Beverly Hills Supper Club Fire **FIGURE 11-5** (Best, 1978). The findings indicated that people actually assisted one another as long as options were available. Only in desperation, when no chance of escape was available, did social norms break down. Studies have consistently shown altruistic behaviors among occupants in emergencies.

People are known to respect role models under emergency conditions; children will follow their parents, and workers will look to the manager. At the Beverly Hills Supper Club, patrons listened to their server

FIGURE 11-5 Beverly Hills Supper Club, Southgate, Kentucky, 1977.
Rf/AP/Shutterstock.

while ignoring the bus person and others. Fire fighters most certainly represent an authoritative role model. People should be expected to follow any reasonable directions given by fire officials during an emergency. The challenge is not persuading the occupants to follow orders; the challenge is formulating clear, concise, and correct instructions.

As mentioned previously, extinguishment, the second tactical priority, often is the most effective means of accomplishing the number one tactical priority of life safety. Extinguishing or confining the fire to areas away from exit pathways will do more to save lives than individual rescues when there are large numbers of victims and the fire is controllable. The first-arriving company at a major fire in an occupied eating or drinking establishment is faced with a critical life-safety decision: Should efforts be directed toward removing endangered occupants or should the operation begin with an offensive fire attack? If the fire can be controlled quickly and many occupants are in danger, extinguishment is generally the best choice.

Sports Arenas

In 1985, 56 people died and 300 were injured in a Bradford, England, stadium fire (Klem, 1986). The 77-year-old grandstand was an outside facility that resembled wooden grandstands of the past. Problems would be greatly magnified in an inside facility holding nearly 100,000 people, such as the huge, domed stadiums of today.

Fortunately, newer facilities limit the amount of combustible material; nonetheless, there is still plenty to burn, largely in the form of plastics. Stadium owners may point to their facility's modern construction, stating that "it can't happen here." However, that type of thinking is a recipe for disaster. These occupancies pose an extreme life hazard threat, and a close

Incident Summary

The Station Nightclub Fire, Rhode Island

At The Station nightclub in West Warwick, Rhode Island, on February 27, 2003, pyrotechnics used as part of a stage act ignited a highly flammable polyurethane foam insulation material used on the walls and ceiling around the stage, resulting in a rapidly advancing fire in a crowded nightclub. At first, occupants failed to realize the need to evacuate, but 30 seconds after ignition the band stopped playing and occupants began to evacuate. Smoke quickly filled the building, and guests began stacking up at the main entrance approximately 100 seconds after ignition. Some occupants escaped via windows at the front of the building. Flame was visible at the front of the building 312 seconds after ignition **FIGURE IS11-1**. The foam covering and crowded conditions provided little time for evacuation, resulting in 100 civilian fire deaths.

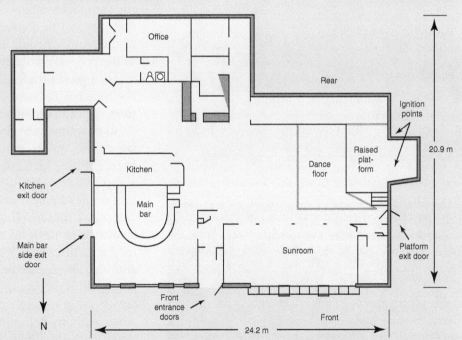

FIGURE IS11-1 The floor plan of The Station nightclub showing exit locations; ignition points are also shown.
Source: Grosshandler, W., Bryner, N., Madrzykowski, D., Kuntz, K. *NIST NCSTAR 2: Vol. I, Report of the Technical Investigation of the Station Nightclub Fire*. National Institute of Standards and Technology, U.S. Department of Commerce; 2005.

Data from: William Grosshandler, Nelson Bryner, Daniel Madrzykowski, Kenneth Kuntz, *NIST NCSTAR 2: Vol. I, Report of the Technical Investigation of The Station Nightclub Fire*, National Institute of Standards and Technology, U.S. Department of Commerce, 2005.

examination of evacuation tactics is a must. The logistics associated with evacuating 10,000 or 100,000 people are complex. When conducting preincident planning at these facilities, the fire officer should keep in mind that fire is not the only type of incident that may require a rapid evacuation. Today, any place that attracts large crowds is a potential target for terrorists.

A fire and explosion occurred in Indianapolis, Indiana, in 1963, killing 74 people at the Indiana State Fairgrounds Coliseum. A situation of this nature should be handled like a nightclub fire: protecting egress while directing and facilitating evacuation. Of course, a stadium or coliseum fire presents a significant life-safety hazard. On the positive side, open-air arenas allow smoke and toxic gases to dissipate. Even coliseums and covered stadiums have large open-air spaces that will allow more time for evacuation. Newer facilities, such as the sports arena shown in **FIGURE 11-6**, are equipped with sprinklers, which greatly improve life safety. Sports arenas are multipurpose facilities—they are also used for a variety of nonsports functions.

A

B

FIGURE 11-6 Modern sports arena. **A.** Exterior. **B.** Interior.
Courtesy of Ben Klaene.

Figure 11-6B shows an inside sports arena being used for a graduation celebration. Typically, sports fans would occupy the stadium seats and a small number of athletes and coaches would occupy the playing floor. However, nonsporting events often result in a larger number of people in this building.

Convention Centers

The 1967 McCormick Place convention center fire in Chicago, Illinois, caused $52 million in damages, the largest property loss in U.S. history at that time. Since the McCormick Place fire, automatic fire sprinklers have become commonplace in convention center construction. Sprinklers greatly diminish the potential loss of life and property. With the exception of the September 11th attacks, large loss-of-life fires are unheard of in fully sprinkler-protected buildings when the sprinkler system is installed and maintained in compliance with nationally recognized codes and standards.

However, sprinkler systems are not the total solution; conditions such as blocked sprinkler heads, closed valves, or a fuel load exceeding the design capacity can render the system ineffective. Also, event organizers may be reluctant to cancel events when the sprinkler system is not operating. Fire companies that respond to convention centers should be aware of the event schedule and conduct onsite inspections of the property whenever large events are planned. It is also a good idea for fire companies to participate in the code enforcement process for buildings located in their response districts.

Improvements in fire protection features allow for larger and more complex convention center buildings. The new McCormick Place convention center is a good example. Finding and accessing the fire location is a key element of a preincident plan in any large, complex structure. Under normal conditions, it is a challenge to navigate throughout these large multilevel structures even with a building diagram in hand and interior signage pointing the way. Adding to the confusion is the facility's ability to change the interior layout using movable dividing walls. Locating a fire when visibility is limited and with people rushing to evacuate the buildings is extremely difficult.

The main entrance is not always the best route to the fire area. Contingency plans are a must. Part of the contingency plan should address staging later-arriving first-alarm fire companies at a predetermined location so they can be deployed based on information gained by the first-arriving units. The preincident plan should delineate the best access locations for each major building area. In many cases, security personnel will meet the first-arriving unit at a designated location to provide incident information and to direct fire fighters to the alarm location. However, fire conditions or injuries to security personnel require contingency planning.

Large quantities of combustibles may be brought into convention centers, and hazardous demonstrations may be performed. Because of the combination of large fuel loads and large numbers of people, suppression efforts can be difficult. As previously mentioned, it is important to ensure that the building's fuel load does not exceed the designed capacity of the sprinkler system. If a fire does overwhelm the sprinkler system and involves a large area, the rate-of-flow requirement may be beyond the fire department's resources and capabilities. In this case, extinguishment may not be a viable option. Fire departments must be prepared to combat a large fire combined with an extraordinary rescue operation in these occupancies. Tactical considerations for convention centers, such as protecting egress routes, are similar to those for nightclubs and similar public assembly properties.

Theaters

Movie theaters built during the first half of the 20th century were built to accommodate live shows or to present both live shows and movies. The construction of these facilities was similar to that of the Iroquois Theater in Chicago, where a 1903 fire resulted in 603 fatalities (mostly children). Some older theaters have now been subdivided into several smaller units.

Modern movie theaters are typically housed in very large buildings. Generally, these are one story, but there is a movement toward mega-theaters that have multiple stories. These buildings are usually subdivided into several smaller theaters **FIGURE 11-7**.

The current trend in movie theaters toward smaller theaters within larger buildings should make the fire fighter's job less difficult. However, the interior layout is sometimes very complex. Tactics are much the same as for the eating and drinking establishments, except that there are generally fewer people in an individual movie theater unit. The modern concept of having numerous but smaller theaters located in a single large building also increases the probability that extinguishment efforts will be effective. In addition to the multiple, smaller theaters allowing for a reduced rate of flow, the tendency for everyone to exit via the main entrance is also reduced in these buildings, as alternative exits should be well marked and easily identified by patrons. Many newer theaters have plainly marked exits at several locations within the individual theater and throughout the building complex. Furthermore, modern movie theaters are often equipped with sprinkler systems.

While movie theaters are getting smaller, venues for live entertainment are not. The fire during a live stage show at the Iroquois Theater taught building designers some important lessons, but it was not until recently that theaters were required to be constructed with sprinkler systems. If an older theater has not been retrofitted with a fire sprinkler system and its exit design updated, it could be just as dangerous as the Iroquois Theater was in 1903. Today, many live performances, formerly staged in theaters, are presented in large arenas, in stadiums, or outdoors using temporary stages and seating.

FIGURE 11-7 Modern movie theater. **A.** Exterior. **B.** Interior.
A: Courtesy of Ben Klaene; B: © iStock/Thinkstock.

Educational Occupancies

NFPA 101, *Life Safety Code*, defines an **educational occupancy** as follows (NFPA 101, 2018):

> **3.3.196.6 Educational Occupancy.** An occupancy used for educational purposes through the twelfth grade by six or more persons for 4 or more hours per day or more than 12 hours per week.

Post-secondary institutions such as college classrooms are not technically considered educational occupancies. However, there are similarities between college and other classrooms in terms of life safety, extinguishment, and property conservation. Therefore, the discussion of college classrooms is included here. For purposes of this comparison, three types of educational occupancies will be discussed:

- Elementary schools
- Middle, junior high, or high schools
- Colleges and universities

An occupancy factor matrix for educational occupancies is shown in **TABLE 11-3**. Mobility factors tend to be very good in school settings. In the case of elementary and high schools, most of the occupants would be mobile. At the college level there will be an older population, but this should not pose a problem. However, younger elementary school students may not be capable of taking independent action, thus the lower age ratings for elementary school occupancies in the educational matrix. A high degree of leadership is provided by classroom teachers, especially at the elementary school level. At the high school and college levels, students typically move from area to area and from building to building; therefore, there are times when they are not under the direct control of a teacher.

The Federal Emergency Management Agency (FEMA) conducted a study of National Fire Incident Reporting System (NFIRS) school fires from 2009 to 2011 (FEMA, 2014). Students at all levels should be aware of fire conditions and be able to hear the evacuation alarm. Fire drills held on a regular basis in elementary and high schools increase awareness and better establish teacher-to-student leadership. The occupant load factor for elementary and high school classrooms is 20 net ft^2/person (NFPA 101, 2018). There are fewer people in the same relative area in elementary and high school settings compared to an assembly occupancy, but there are more people in these settings than in many other occupancies.

The NFPA 2017 report, *Structure Fires in Educational Properties*, by Richard Campbell, provides more information.

Findings of the FEMA study include the following (FEMA, 2014):

- An estimated 4000 school building fires were reported by United States fire departments each year and caused an estimated 75 injuries and $66.1 million in property loss.
- Fatalities resulting from school building fires were rare.
- There was a general increase in school building fires toward the beginning and end of the academic year.
- The three leading causes of school building fires were as follows:
  - Cooking (42 percent)
  - Intentional action (24 percent)
  - Heating (10 percent)

TABLE 11-3 Occupancy Factor Matrix: Educational Occupancies

Educational Occupancies	Mobility of Occupants	Age of Occupants	Leadership	Awareness	Occupant Load	Familiarity of Occupants
Elementary schools	10	6	10	10	4	10
Middle, junior high, or high schools	10	10	7	10	4	10
Colleges/universities	8	8	6	10	5	5

1 = extremely negative factor; 10 = very positive factor.

- The leading area of fire origin in nonconfined school building fires was the bathroom at 25 percent.
- In 75 percent of school building fires, the fire spread was limited to the object of origin.
- Smoke alarms were reported as present in 66 percent of nonconfined school building fires.

Elementary Schools

Most tragic fires involving educational occupancies occurred in primary or elementary schools, mainly because younger children are less likely to take appropriate action on their own. All other things being equal, the younger the student, the greater the danger.

Some notable school fires are listed in **TABLE 11-4**. **FIGURE 11-8** shows the scene at Our Lady of Angels School in Chicago in 1958.

Only recently have schools been protected with automatic sprinkler systems. Very few older school buildings are sprinkler protected. Placing large numbers of children within a small, nonsprinklered area creates the potential for a tragedy. In addition to not being sprinkler protected, older schools tend to be three or more stories high **FIGURE 11-9**.

The saving factors in elementary schools are that children become disciplined through frequent fire drills, and teachers are natural role models that the children follow in an emergency. In the event of a fire in a school, it is necessary to coordinate operations with the school principal to determine the evacuation status and the location of students, teachers, and staff members who are missing. If rescue efforts are necessary, fire fighters can be directed to the probable

FIGURE 11-8 Our Lady of Angels fire, Chicago, Illinois, 1958.
© AP Photos.

location for trapped students, teachers, and staff on the basis of this information.

Fire department personnel should observe school fire drills whenever possible. Observing a school fire drill provides invaluable training for fire fighters, who should know the prescribed location for students and the school contact person. Being present during a fire drill also provides an opportunity to evaluate accountability and evacuation procedures.

Elementary and high schools maintain visitor logs. Overall security measures, such as the installation of electronic locking systems, have improved in recent years due to multiple "active shooter" incidents. The NFPA has promulgated NFPA 3000 NFPA 3000™ (PS), *Standard for an Active Shooter/Hostile Event Response (ASHER) Program*, providing measures to protect students from active shooters (NFPA 3000™ (PS), 2018). It is important that the fire department review new security measures to determine what effect security has on evacuation and fire department access to the building and the fire department's role during an active shooter incident.

Knowing how long it takes to totally evacuate the school in case of fire is important, but not nearly as important as knowing whether students are under control and whether the accountability system is reliable. Determining whether the school evacuation plan is consistent with fire department operations is also important. Sometimes students are placed in areas that could obstruct or slow fire department access. Other times the school contact person is located in an area that is not in the line of travel for arriving fire units. Minor changes in the school evacuation plan can significantly improve fire department response.

TABLE 11-4 Notable Fires in Educational Occupancies

Fire	Year	Deaths
Lakeview Grammar School, Collinwood, OH	1908	175
Cleveland School, Beulah, SC	1923	77
Consolidated School, New London, TX	1937	294
Our Lady of Angels, Chicago, IL	1958	95
Star Elementary School, Spencer, OK	1982	7

© Jones & Bartlett Learning.

Many jurisdictions conduct unannounced fire drills that are timed and otherwise evaluated by the fire department.

In older schools (see Figure 11-9A), heavy reliance is placed on the use of hallways and stairways. As with any building, the stairs in multistory school buildings must always be controlled by the fire department during evacuations. Evacuating trapped victims through the interior stairs is typically the most effective and safest option. Some newer one-story schools have doors from each classroom leading directly to the outside, greatly reducing the life hazard (see Figure 11-9B). Few schools are protected by sprinkler systems. As with other occupancy types, sprinklers provide a high degree of safety for students and staff (see Figure 11-9C).

When absolutely necessary, aerial ladders, elevated platforms, and ground ladders can be used to evacuate a multistory school. Ladders allow pupils to form a steady evacuation stream with teachers assisting from the inside. However, fire department ladders should be used only when fire fighters cannot gain control of the interior stairs. Establishing an occupant accountability system that ensures complete evacuation and accounting for all students, teachers, and staff is essential. It is also important to make a sound evaluation as to whether the immediate priority is rescue or fire control.

A

B

C

FIGURE 11-9 Buildings used as schools. **A.** Multistory, older high school. **B.** One-story school. **C.** Modern school with sprinkler protection.
Courtesy of Ben Klaene.

As mentioned previously, many school buildings pose special firefighting challenges. Fire extension should be expected in nonsprinklered school buildings. Where mobile buildings are used as temporary classrooms, complete destruction may occur in a short time. Many schools add to the fire control problem by storing school supplies, clothing, furniture, and other materials within the school, sometimes under stairwells and in corridors. Such storage practices greatly increase the fuel load and the threat to life safety.

Middle, Junior High, and High Schools

Fire drills are required in all kindergarten through grade 12 schools. As mentioned previously, the student accountability process may not be as efficient at the high school level where students often move to different classrooms during the school day.

Fire fighters are inclined to think schools are occupied only during school hours. When determining the probability of a school being occupied, it is important to remember that teachers, maintenance workers, and administrative personnel may occupy the school outside of normal classroom hours. There are also many sporting and social events that occur at schools outside the normal school day. At high schools and colleges, there will usually be more of these events than at the elementary school level. The fire department should obtain a list of scheduled events and include event information with the preincident plan. A basketball court or auditorium will likely be classified as an assembly occupancy within the educational occupancy. High schools may also have evening classes or permit citizens to use computers, gym equipment, and other school resources outside of normal school hours.

Colleges and Universities

Colleges and universities usually have much larger campuses compared to elementary or high schools, with many more evening and weekend activities. The adult population is better able to take independent action to evacuate a building; however, students move from building to building at different time intervals and are not under close supervision. Furthermore, typically there is no occupant accountability system, and universities may be situated on a large and complex campus, requiring a map to find specific buildings. Most campuses will have one or more large classroom facilities that hold 200 or more students. These classrooms resemble an assembly occupancy and are classified as such. Any college classroom with 50 or more students is considered an assembly occupancy.

Fires in classrooms, except for large college classrooms or large laboratory and shop areas, will generally be within the flow capabilities of standard preconnected hose lines. Rate of flow should be precalculated for any large undivided areas within the school property.

Schools contain library collections, computers, and other high-value contents. Therefore, property conservation tactics should be carefully thought out and included in the preincident plan.

Healthcare Occupancies

NFPA 101, *Life Safety Code*, defines a **healthcare occupancy** as follows (NFPA 101, 2018):

> **3.3.196.7 Health Care Occupancy.** An occupancy used to provide medical or other treatment or care simultaneously to four or more patients on an inpatient basis, where such patients are mostly incapable of self-preservation due to age, physical or mental disability, or because of security measures not under the occupants' control.

Four types of healthcare facilities will be discussed:

- Hospitals
- Nursing homes
- Limited care facilities
- Ambulatory health care facilities

An occupancy factor matrix for healthcare occupancies is shown in **TABLE 11-5**. Mobility is a major concern in hospitals and nursing homes. Some patients are considered ambulatory in these facilities, but ambulatory and mobile are not the same. An ambulatory person is able to move but may move slowly and require a walker or cane. Many patients in hospitals and nursing homes are nonambulatory, meaning they are unable to self-evacuate and will need assistance if it is necessary to move to a place of safe refuge. The nursing staff generally provides this assistance, but the staff available may not be sufficient if large numbers of people need assistance. The products of combustion may force the nursing staff to evacuate. Occupants in limited care facilities are usually more mobile than in a nursing home, but many move slowly or need walkers, canes, wheelchairs, or other types of assistance.

Ambulatory health care facilities treat illnesses and injuries on an outpatient basis. However, some of these facilities perform surgical or testing procedures that require the patient to be anesthetized or restrained. If this is the case, anesthetized and restrained

TABLE 11-5 Occupancy Factor Matrix: Healthcare Occupancies						
Healthcare Occupancies	Mobility of Occupants	Age of Occupants	Leadership	Awareness	Occupant Load	Familiarity of Occupants
Hospitals	2	4	10	3	7	2
Nursing homes	2	1	8	3	7	7
Limited care facilities	4	2	7	5	8	9
Ambulatory care facilities	6	5	6	5	7	2

1 = extremely negative factor; 10 = very positive factor.
© Jones & Bartlett Learning.

patients will be immobile. Anesthetized patients have limited or no cognitive ability. Part of the preincident planning process should include notation of ambulatory care facilities where patients are anesthetized or restrained.

Residents of nursing homes and limited care facilities are typically much older than the general population. The nursing staff provides leadership and assistance during an emergency. In a hospital or nursing home, patients will look to their nurses for direction in an emergency. Limited care facilities will generally have staff personnel who can provide leadership, but patients are more independent. Patients at ambulatory care facilities may not recognize the staff, and conscious patients will be more likely to take independent action in an emergency. If patients are restrained in any way, a notation should be made on the preincident plan showing the location of the restraining devices and, if necessary, instructions on releasing the restraints.

A nursing home, on the other hand, is home to the occupant; thus, most patients will be familiar with the layout of the building. However when under stress, residents may not remember details of the building, and some patients may not be able to follow directions. Many patients in these facilities are disoriented due to dementia or Alzheimer's disease and may resist following instructions or be combative. Patients and visitors in hospitals and ambulatory care facilities may be unfamiliar with the building.

Large numbers of people located in a single building always represent life-safety concerns. Physically or mentally challenged people present additional difficulties for rescuers. Fire fighters should expect facility personnel to assist with evacuations; this should be reinforced through training. In the hospital or nursing home, nurses and attendants will be needed to care for patients.

These occupancies often require fire forces to rely on people who are not trained to work under fire conditions. Healthcare facilities use places of safe refuge within the facility during evacuations. Hospital and nursing home personnel sometimes move patients through fire doors to safe areas. Seldom are hospital or nursing home patients evacuated to the outside. Codes require doors and hallways to be wide enough to allow the patients to be moved while remaining in bed. Moving people on beds with life support equipment is difficult, to say the least; therefore, the defend-in-place strategy is the preferred option when conditions permit.

In addition, hospital and nursing home corridors are often congested with computer carts, laundry bins, and food service equipment that partially block the pathway to a place of safe refuge.

FIGURE 11-10 shows incidence of structure fires at healthcare properties by occupancy type.

Hospitals

Building features assist in accomplishing the life-safety mission at hospitals, with building areas being separated by substantial fire barriers and fire doors. Making sure that fire separations are not compromised is an essential part of a hospital fire protection plan. Fire doors are often left open in hospital facilities, allowing smoke and fire to enter common hallways and other patient areas, as was the case in a hospital fire in Petersburg, Virginia, which killed six patients.

The fire department may be needed to assist in moving patients from the fire area to another designated defend-in-place location **FIGURE 11-11**. The IC must make an early determination regarding progress

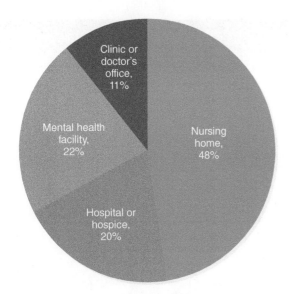

FIGURE 11-10 Structure fires in healthcare facilities by occupancy type.
Source: Campbell, Richard. Structure Fires in Health Care Facilities. National Fire Protection Association Research. Quincy, MA. October 2017. Pg. v. https://www.nfpa.org/News-and-Research/Data-research-and-tools/Building-and-Life-Safety/Fires-in-Health-Care-Facilities. Accessed March 18, 2020.

FIGURE 11-11 Hospital fire.
U.S. Fire Administration, Major Fires Investigation Series.

Hospitals will often have very expensive equipment. Some of this equipment is extremely sensitive to water and smoke damage. Given the high life-safety problem, property conservation could be delayed. However, the IC should call for enough assistance to assign personnel to property conservation activities as soon as possible.

Nursing Homes

Most nursing home residents are unable or only partially able to assist themselves. In some cases, special security provisions such as restraints or exit doors that require a code number to open hinder self-rescue efforts. This special population presents many evacuation and rescue problems, requiring a significant commitment of fire department resources whenever a working fire is encountered in a nursing home.

During the past few decades, as the population has aged, the need for nursing home facilities increased. A larger percentage of the population moved into these occupancies, and large loss-of-life fires became common. Most areas of the United States now have codes requiring sprinklers in nursing homes. In addition, the Centers for Medicare & Medicaid Services requires sprinklers in all existing nursing homes to receive Medicare/Medicaid funding. These measures have greatly reduced the number of large loss-of-life fires in healthcare facilities. Multifatality fires still occur in nursing homes but not where sprinklers have been properly installed and maintained.

> **NOTE**
>
> According to NFPA research (Ahrens, 2017), sprinkler systems greatly improve occupant safety when installed in healthcare occupancies. This NFPA study estimated a reduction in civilian fire deaths at 71 percent when any type of sprinkler system is present and an 88 percent reduction when wet pipe sprinkler systems are installed.

When a properly installed and operating sprinkler system is in place, fires are generally limited to one room, and the primary tactic is to move patients out of smoky areas to places of safe refuge while supporting the sprinkler system. If the nursing home is not sprinkler protected or the sprinkler system is not operating properly, use fire doors to contain the fire whenever possible. Some nursing home doors automatically close when a smoke detector senses a fire or the sprinkler system is activated.

Even when self-closing fire-rated doors separate rooms from hallways, obstructions may prevent the door from closing, which allows fire and smoke to

being made in moving patients to an area of safe refuge and whether the place of safe refuge is defendable. Evacuation to a lower floor or to the outside is a mammoth undertaking and is rarely necessary. The nursing staff should be able to account for patients, but in most cases, they will not be able to account for visitors or doctors. Therefore, a primary search must be conducted.

There are large open areas within hospitals and some of these areas, such as laundries, storage areas, and refuse areas, have a heavy fuel load. Furthermore, the presence of oxygen can increase flame intensity and fire spread. However, patient rooms in hospitals are usually within the flow capacity of a standard preconnected hose line or fire department standpipe hose line. It is essential to consider the possible smoke spread to the area of safe refuge when a fire door is opened to advance a hose line.

spread to the hallway and other rooms. Moving patients out of harm's way is the primary tactic, but extinguishment is essential to defend the place of safe refuge. If the fire is not extinguished, smoke and possibly fire will eventually enter the area of safe refuge, requiring further evacuation and rescue.

Some notable fires in healthcare facilities (nursing homes and hospitals) are listed in **TABLE 11-6**. According to a 2012–2014 study of fires in medical facilities, the death rate per 1000 fires is not significantly higher in nursing homes when compared to other medical facilities; however, the injury rate is more than twice as high in nursing homes (USFA, 2016).

Nursing homes and hospitals make a good case for total sprinkler protection, as is evident in Table 11-6. Prior to 1970, few nursing homes were equipped with total automatic sprinkler protection. The few multifatality fires that have occurred since were in nonsprinklered buildings. Likewise, the majority of hospitals were not sprinkler protected in the early 1970s. According to the NFPA report, "U.S Experience with Sprinklers, July 2017," 67 percent of nursing home or hospitals were protected by sprinkler systems (Ahrens, 2017).

Limited Care Facilities

Limited care facilities are similar to nursing homes, but the occupants do not require continuous nursing care **FIGURE 11-12**. Some limited care facilities provide services that are similar to comprehensive nursing home care, while others establish independent living criteria for occupants. Often a nursing complex will include both residential board and care, and full nursing care. Elderly residents generally move from their homes into the residential board and care facility with a degree of self-reliance. Over time, these residents may require additional nursing services, and eventually occupy a full nursing facility. Most codes now require automatic sprinkler protection for all new nursing homes, but they may not require existing nursing homes to be retrofitted with sprinklers or require sprinkler protection in limited care facilities.

Part of the fire protection plan for a nursing home requires a minimum number of on-duty staff to assist patients in an emergency. Staffing will generally be less at a limited care facility, and there may not be places of safe refuge where a defend-in-place strategy can be implemented (NFPA 101, 2018). Many limited care facilities resemble apartment buildings; therefore, a primary search and quick extinguishment are the principal tactics. Occupants of these facilities, who are mostly elderly, would have great difficulty descending a ladder; therefore, attempting to rescue residents using fire department ladders would be an option of near last resort. Many residential board and care facilities are multistory buildings and occupants often have great

TABLE 11-6 Notable Fires in Hospitals and Nursing Homes*

Fire	Year	Deaths
Hospital, Cleveland, OH	1929	125
Hospital, Effingham, IL	1949	74
Psychiatric facility, Davenport, IA	1950	41
Nursing home, Hillsboro, MO	1952	20
Nursing home, Warrenton, MO	1957	72
Hospital, Hartford, CT	1961	16
Nursing home, Fitchville, OH	1963	63
Nursing home, Marietta, OH	1970	31
Nursing home, Buechel, KY	1971	10
Nursing home, Honesdale, PA	1971	15
Nursing home, Cincinnati, OH	1972	10
Nursing home, Denmark, WI	1972	10
Nursing home, Springfield, IL	1972	11
Nursing home, Philadelphia, PA	1973	11
Nursing home, Wayne, PA	1973	15
Hospital, Osceola, MO	1974	8
Nursing home, Chicago, IL	1976	23
Hospital, Riverside, CA	1986	5
Healthcare center, Memphis, TN	1988	6
Nursing home, Norfolk, VA	1989	12
Hospital, Petersburg, VA	1994	6
Health center, Hartford, CT	2003	16
Nursing home, Nashville, TN	2003	14

*Notable fires refers to fires with five or more deaths.
© Jones & Bartlett Learning.

FIGURE 11-12 Nursing and limited care facility.
Courtesy of Ben Klaene.

difficulty using stairways, hindering their exit from the building. Elevators will only be available to these residents when placed in fire department service and controlled by the fire department.

Many apartment buildings rent exclusively to elderly occupants. These buildings are classified as a residential occupancy and have less stringent fire protection code requirements. However, the firefighting challenges are similar to those in a limited care facility.

Ambulatory Care Facilities

As noted previously, these facilities provide a variety of services. Some are group practice physicians who provide emergency care for minor illnesses and injuries. Others perform outpatient surgery or testing that requires full anesthesia. An emergency care facility where patients are conscious and unrestrained requires the same tactics as an office building. If patients are anesthetized, they will have to be physically removed from the building and provided medical care. If restrained, they will need to be released from the restraining equipment (e.g., MRI tunnel). These facilities should be preincident planned with notations regarding the types of service provided and a special notation if patients are anesthetized or restrained.

Residential Board and Care Occupancies

NFPA 101, *Life Safety Code*, defines a **residential board and care occupancy** as follows (NFPA 101, 2018):

> **3.3.196.12 Residential Board and Care Occupancy.** An occupancy used for lodging and boarding of four or more residents, not related by blood or marriage to the owners or operators, for the purpose of providing personal care services.

TABLE 11-7 shows an occupancy factor matrix for residential board and care occupancies. Awareness is

TABLE 11-7 Occupancy Factor Matrix: Residential Board and Care Occupancies

	Mobility of Occupants	Age of Occupants	Leadership	Awareness	Occupant Load	Familiarity of Occupants
Residential board and care occupancies	5	5	7	3	7	7

1 = extremely negative factor; 10 = very positive factor.
© Jones & Bartlett Learning.

given a rating of three due to the diminished mental and/or physical capacity of many of the residents and resultant problems in evacuating the building. There may be one or more staff members on duty depending on the facility size and the mental/physical status of patients being housed. In most cases, staff members are less qualified that hospital and nursing home personnel. Thus, leadership is a positive factor, but not to the extent found in schools and hospitals.

During the 1970s, mentally challenged patients who were formerly housed in large, fire-resistive structures were moved to board and care facilities. This created a problem similar to the problems experienced in nursing homes in the past. However, the board and care problem has been more difficult to regulate, as fewer patients are kept in a single structure. Staff requirements are less in board and care facilities, and many resemble a residential occupancy. Facilities classified as board and care occupancies in NFPA 101 are identified by a variety of designations such as group homes, rest homes, shelters, halfway houses, and other names.

The board and care fire problem is substantial, and tactics must be planned in advance. It is essential that the fire department identify and inspect board and care facilities, as well as developing preincident plans for occupancies located within their jurisdiction. Tactics should take into account the limited ability of residents to escape on their own. Residents will be categorized as independent, but will usually be mentally and/or physically disadvantaged. Many of these buildings are residential buildings; therefore, the standard preconnected hose line should be sufficient to extinguish most fires that occur. Converted two- or three-story residential buildings typically have only one means of egress from upper floors.

The NFPA has issued several fire investigation reports for fires in board and care facilities. Some of these facilities are occupied by mentally challenged or drug-impaired residents; others are scaled-down facilities for elderly patients who do not require full nursing care.

Some notable board and care fires are listed in **TABLE 11-8**. Widespread use of residential board and care facilities is a relatively new phenomenon; therefore, the multifatality fires shown in Table 11-8 are fairly recent.

The number of multifatality fires in residential board and care facilities seems to be trending downward, as shown in **TABLE 11-9** (Ahrens, 2016). This improvement is due in part to more stringent codes and better code enforcement, but also to increased awareness among fire departments that these occupancies exist in their communities.

TABLE 11-8 Notable Fires in Residential Board and Care Occupancies*

Fire	Year	Deaths
Pleasantville, NJ	1973	10
Connellsville, PA	1979	10
Farmington, MO	1979	25
Washington DC	1979	10
Pioneer, OH	1979	14
Brady Beach, NJ	1980	24
Detroit, MI	1980	5
Keansburg, NJ	1981	31
Pittsburgh, PA	1982	5
Eau Claire, WI	1983	6
Gladstone, MI	1983	5
Worcester, MA	1983	7
Gwinnett County, GA	1983	8
Cincinnati, OH	1983	7
Lexington, KY	1984	5
Southfield, MI (hospice facility)	1985	8
Johnson City, TN	1989	16
Colorado Springs, CO	1991	10
St. Isidore, Quebec, Canada	1991	5
Detroit, MI	1992	10
Mobile, AL	1994	6
Broward County, FL	1994	6
Laurinburg, NC	1996	8
Mississauga, Ontario, Canada	1996	8
Sainte-Geneviève, Quebec, Canada	1997	7
Harveys Lake, PA	1997	10
Arlington, WA	1998	8
Maryville, TN	2004	5
Anderson, MO	2006	11
Mt. Carmel, CA	2011	5

*Notable fires refers to fires with five or more deaths.
© Jones & Bartlett Learning.

TABLE 11-9 Structure Fires in Residential Board and Care Facilities by Year, 2003 to 2013

Year	Total Fires	Civilian Injuries
2003	2020	74
2004	2100	44
2005	2060	55
2006	2250	124
2007	1940	51
2008	1990	55
2009	1730	35
2010	1710	39
2011	1,920	68
2012	1,920	38
2013	1,890	49

Ahrens, Marty. Structure Fires in Residential Board and Care Facilities. Table 1. NFPA Fire Analysis and Research, Quincy, MA. July 2016. https://www.nfpa.org/-/media/Files/News-and-Research/Fire-statistics-and-reports/Building-and-life-safety/osboardandcare.ashx?la=en. Accessed March 18, 2020.

Detention and Correctional Occupancies

NFPA 101, *Life Safety Code*, defines a **detention and correctional occupancy** as follows (NFPA 101, 2018):

3.3.196.5 Detention and Correctional Occupancy. An occupancy used to house one or more persons under varied degrees of restraint or security where such occupants are mostly incapable of self-preservation because of security measures not under the occupants' control.

An occupancy factor matrix for detention and correctional occupancies is shown in **TABLE 11-10**. Residents of nursing homes and hospitals are characterized as having physical and/or mental challenges that hamper self-evacuation, whereas prisoners in detention facilities are generally physically restrained. During a prison fire, inmates are moved to other areas within the confines of the detention property.

The prison staff provides leadership for inmates. Close cooperation between prison staff and the fire department is needed for successful operations in a jail or prison environment. Some detention and correctional occupancies are no more than cells within a police station designed to temporarily hold prisoners. Other correctional facilities are nearly self-sufficient mini-cities with laundries, storage areas, large-scale food preparation areas, and many other facilities included within the prison complex.

During a detention facility fire, maintaining security is extremely important; however, dealing with tight security measures while implementing fire department tactics is sometimes exasperating for responders. Close cooperation between the IC and prison administrators is an absolute necessity.

Prisoners present a physical threat to the fire fighter, therefore close coordination with prison officials is needed to enable fire fighters to operate safely while suppressing the fire. Modern prisons are usually constructed with a fair degree of fire safety in mind, but older, less fire-safe structures are still being used as prisons. Generally, older prisons are not sprinkler protected and have fewer fire protection features. These less-protected prisons and jails complicate operations by permitting the free circulation of smoke and toxic gases as well as allowing rapid fire extension.

Modern and old prisons alike tend to be overcrowded, holding more prisoners than they were designed to house. Any form of overcrowding increases the life hazard.

TABLE 11-10 Occupancy Factor Matrix: Detention and Correctional Occupancies

	Mobility of Occupants	Age of Occupants	Leadership	Awareness	Occupant Load	Familiarity of Occupants
Detention and correctional occupancies	1	6	9	6	7	6

1 = extremely negative factor; 10 = very positive factor.

© Jones & Bartlett Learning.

Detention and correctional structures in Latin America and elsewhere sometimes ignore fire and building codes that are generally enforced in the United States. As a result, at least nine large loss-of-life fires with 10 or more fatalities occurred during the 10-year period from 2003 through 2012. Most notable was a fire in Comayagua, Honduras, on February 14, 2012, that resulted in 361 inmate deaths (Moncada, 2012).

Several notable fires in detention facilities are listed in **TABLE 11-11**.

Preincident planning is an absolute necessity for a detention or correctional facility. Securing a water supply and advancing hose lines may require special precautions. The details of fire department interaction with security are worked out during preincident planning. Questions that must be answered in advance include the following:

- How will fire fighters be provided access to secured areas? This generally involves a security person meeting the fire department at a predesignated location and escorting them while inside the prison. Be sure to detail how security will provide access when smoke or fire conditions require fire fighters to advance ahead of security personnel into areas requiring self-contained breathing apparatus and full protective clothing.
- Are there areas within the prison complex that do not house prisoners, and can they be accessed without assistance from security? Work areas such as laundries may be occupied by prisoners during certain times of the day but may be unoccupied at other times.
- What are the emergency procedures in the event of system malfunctions, an automatic lockdown, or a power failure?
- What are the evacuation routes, and where are the locations of safe refuge for prisoners?

The goal is to conduct a safe and effective operation while avoiding direct contact with the prison population. There may be a need for special rapid intervention crews (RICs) and accountability and rehabilitation procedures during a detention/correctional occupancy fire.

Residential Occupancies

NFPA 101, *Life Safety Code*, defines a **residential occupancy** as follows (NFPA 101, 2018):

> **3.3.196.13 Residential Occupancy.** An occupancy that provides sleeping accommodations for purposes other than health care or detention and correctional.

Residential occupancies include:

- One- and two-family dwellings
- Apartment buildings
- Dormitories
- Hotels/motels
- Lodging or rooming houses (B&Bs, bunk rooms with no more than 16 sleeping accommodations)

An occupancy factor matrix for residential occupancies is shown in **TABLE 11-12**. Note that mobility

TABLE 11-11 Notable Fires in Detention and Correctional Occupancies*

Fire	Year	Deaths
Raymond, MS	1913	35
Columbus, OH	1930	320
Lancaster, SC	1972	11
Sanford, FL	1975	11
Marion, NC	1976	9
Point Pleasant, WV	1976	5
Danbury, CT	1977	5
St. John, New Brunswick, Canada	1977	18
Columbia, TN	1977	42
Leavenworth, KS	1979	7
Biloxi, MS	1982	29
Jersey City, NJ	1982	7
Bakersville, NC	2002	8

*Notable fires refers to fires with five or more deaths.
© Jones & Bartlett Learning.

TABLE 11-12 Occupancy Factor Matrix: Residential Occupancies

Residential Occupancies	Mobility of Occupants	Age of Occupants	Leadership	Awareness	Occupant Load	Familiarity of Occupants
One- and two-family dwellings	5	5	7	4	8	10
Apartment buildings	5	5	5	4	6	9
Dormitories	7	7	5	3	4	8
Hotels and motels	6	5	2	3	4	1

1 = extremely negative factor; 10 = very positive factor.
© Jones & Bartlett Learning.

and age are near average (5), except in the dormitory, where occupants tend to be young adults who are very mobile. There is little leadership in most hotels and motels except when provided by a parent traveling with children. In one- and two-family dwellings, the head of household should be in a strong leadership position, particularly if the family has practiced a home escape plan. Awareness is rated slightly negative in all residential occupancies because occupants are sleeping part of the time that they occupy the building. With the exception of dormitories, hotels, and motels, there are few people in a fairly large area; thus, exit capacity is not generally considered a problem, although the number of exits from upper floors of a home is generally limited. People living in a residence or dorm should be familiar with their surroundings. Hotels and motels are transient in nature; therefore, most occupants will not know the location of alternative means of egress.

Most structure fires, most fatal fires, and most property loss occur in residential buildings. Between 2013 and 2017, fire departments responded to an average of 354,400 home structure fires annually (Ahrens, 2019). Most of the fire deaths occurred in residential occupancies. The "Catastrophic Multiple-Death Fires in 2017" report included 12 fires that occurred in one- and two-family dwellings where five or more fatalities resulted (Badger, 2018). This is not an unusual circumstance; every year several multiple-death fires occur in single-family homes.

The residential occupancy classification includes more than just one- and two-family dwellings; it also includes apartment buildings, hotels, motels, and dormitories. Large loss-of-life fires have occurred frequently in apartment buildings. **TABLE 11-13** lists apartment building fires that resulted in five or more civilian fatalities from 2009 through 2018.

TABLE 11-14 lists some significant residential fires, excluding one- and two-family homes and apartment buildings.

In residential occupancies, as with most occupancy types, there is a direct correlation between life safety and the time of day. In residential occupancies, the hours when residents are sleeping are the most dangerous. Wide acceptance of smoke alarms and detectors has greatly diminished the number of deaths in residential properties. Many high-rise buildings are now constructed for use by senior citizens and family living. Expect to find many people throughout the structure in need of assistance, including the possibility of the convergence cluster phenomenon (Bryan, 1985).

Convergence cluster behavior was studied in depth after the MGM Grand Hotel and Casino Fire in Las Vegas. Many guests attempted to use stairs to exit the building but encountered smoke as they descended. They then sought refuge in a room other than their own. This resulted in several people being in a single room, rather than the expected one to five guests per room. It would be logical to assume that guests had evacuated the floor successfully after finding several rooms empty, but this assumption could have serious, deadly consequences. The convergence cluster phenomenon should reinforce the importance of conducting a complete, systematic search.

TABLE 11-13 Notable Fires in Apartment Buildings, 2009 to 2018*

State	Year	Deaths
Mississippi	2009	9
Illinois	2010	7
Minnesota	2010	6
New York	2010	5
Illinois	2011	6
Washington	2011	5
Mississippi	2011	5
Missouri	2012	5
New Jersey	2012	5
New York	2014	8
Massachusetts	2014	7
Texas	2014	6
South Carolina	2014	5
Maryland	2016	7
New York	2017	13
Michigan	2017	5
South Dakota	2017	5
North Carolina	2018	5
New Jersey	2018	5
Texas	2018	5

*Notable fires refers to fires with five or more deaths.
© Jones & Bartlett Learning.

TABLE 11-14 Notable Fires in Residential Occupancies Other Than One- and Two-Family Dwellings or Apartment Buildings*

Fire	Year	Deaths
Gulf Motel, Houston, TX	1943	54
La Salle Hotel, Chicago, IL	1946	61
Winecoff Hotel, Atlanta, GA	1946	119
MGM Grand Hotel and Casino, Las Vegas, NV	1980	85
Hilton Hotel and Casino, Las Vegas, NV	1981	8
Dupont Plaza Hotel and Casino, San Juan, Puerto Rico	1986	96
Ramada Hotel, Evansville, IN	1992	11
Paxton Hotel, Chicago, IL	1993	20
Fraternity house, Chapel Hill, NC	1996	5
Off-campus student housing, Columbus, OH	2003	5
Residential board and care facility, Nashville, TN	2004	5
Residential hotel, Reno, NV	2006	12
Homeless shelter, Paris, TX	2009	5
Bed and breakfast, New Ulm, MN	2011	6

*Notable fires refers to fires with five or more deaths.
© Jones & Bartlett Learning.

One- and Two-Family Dwellings

One- and two-family dwellings are an important and large part of the fire problem **FIGURE 11-13**. According to the USFA's National Fire Incident Reporting System and NFPA's Annual Fire Experience Survey (NFPA, 2014-2018),

- Each year, from 2014 to 2018, fire departments responded to an estimated 382,399 fires in residential buildings across the nation.

- These fires resulted in an annual average of 2746 deaths; 11,477 injuries; and almost $7.6 billion in property loss.

- The residential building portion of the fire problem is of great national importance, as it accounts for the vast majority of civilian casualties.

- National estimates for 2014 to 2018 show that 97 percent of all fire deaths and 90 percent of all fire injuries occurred in residential buildings.

- In addition, residential building fires accounted for 72 percent of the total dollar loss from all fires.

FIGURE 11-13 Fire at a single-family dwelling.
Courtesy of David J. Jones, Cincinnati, Ohio.

The term *residential buildings* includes what are commonly referred to as "homes," whether they are one- or two-family dwellings or multifamily buildings. It also includes hotels and motels, residential hotels, dormitories, residential board and care, and halfway houses—residences for formerly institutionalized individuals (patients with mental disabilities, drug addiction, or those formerly incarcerated) that are designed to facilitate their readjustment to private life. The term *residential buildings* does not include institutions, such as prisons, nursing homes, juvenile care facilities, or hospitals, even though people may reside in these facilities for short or long periods of time (USFA, 2020).

Fire fighters generally feel confident attacking a fire in a one- or two-family dwelling "bread and butter" fire. However, more fire fighter line-of-duty deaths occur in these properties than any other occupancy; this is primarily due to the large number of fires that occur in these properties. The point that must be emphasized is that no fire should ever be considered routine. Proper precautions must be taken at every structure fire.

One- and two-family dwellings are primarily occupied at night and on weekends, but people can be home and asleep at any time. Whenever it is safe to conduct an offensive attack, a primary and secondary search should be conducted, unless there is a *credible* and *verifiable* report that everyone escaped. Reports that everyone is out of the building should be carefully scrutinized for reliability. When in doubt, conduct a primary and secondary search. When the fire occurs late at night, the primary search should normally begin in bedrooms.

Except for some large open-layout buildings or mansions, fires in one- and two-family dwellings should be within the capacity of one or two preconnected hose lines. As is most often the case, extinguishment is usually the best way to achieve the life-safety objective.

The actual monetary value of the building's salvageable contents will vary, but it is important to remember that in many cases, personal items cannot be replaced and are of great value to the occupants. Property conservation efforts must take into consideration that personal mementos are invaluable and cannot be replaced.

Apartment Buildings

In terms of life safety, extinguishment, and property conservation, apartment buildings are much like one- and two-family dwellings. However, a larger number of people are concentrated in a smaller area and the probability of someone being able to provide a *credible* and *verifiable* report that everyone has successfully escaped is unlikely. The life-safety problem is directly proportional to the number of units in a single building. Larger buildings with more apartment units require a larger staffing commitment to complete the primary and secondary searches **FIGURE 11-14**.

With the large number of people potentially exposed to the fire and the lack of a reliable evacuation status report, extinguishment becomes even more important. Most fires in apartment buildings will require flows that are within the capacity of one or two standard preconnected hose lines. However, some apartment buildings have large open storage or common spaces that may require additional flow.

As is the case in other residential properties, the occupants' personal items will be of great value.

Exposure problems are more likely in large apartment complexes. At a minimum, a preincident plan

FIGURE 11-14 Apartment building fire.
Courtesy of Bill Strite, Cincinnati, Ohio.

Incident Summary

Bremerton, Washington, Four-Story Apartment Complex Fire

In a four-story apartment complex fire in Bremerton, Washington, four residents were killed in a fast-spreading early morning fire. The fire originated in a third-floor apartment that was unoccupied at the time and quickly spread from the apartment door that was left open when the fire was detected. The fire moved into the attic space and then throughout the complex. There were four fire separation walls in the attic space, but openings in these walls allowed the fire to spread. This case shows how an attic fire can create a severe challenge for the fire department in attempting to contain the fire and protect the occupants.

Data from: Edward R. Comeau, Apartment Fire, Bremerton, Washington, NFPA Fire Investigation Report. Quincy, MA: NFPA. 1999. https://www.nfpa.org/-/media/Files/News-and-Research/Resources/Fire-Investigations/Bremerton.ashx?la=en. Accessed February 7, 2020.

should include a map showing the building layout with an address for each unit within the complex, as well as areas requiring a rate of flow exceeding the flow capacity of two standard preconnected hose lines. Preincident plans should also note access problems. Some apartment buildings have back-to-back units with fire separations between units or a center hallway separating units. Apparatus access may be limited to the front side of the building, with no road access to rear units. A serious fire in a rear unit may not be visible from the front of the building, and access to rear apartments may be limited.

If a fire is being attacked in a defensive mode at an apartment complex where buildings are closely spaced, the primary concern is that other buildings will quickly become involved in fire, as was the case in the Houston, Texas, apartment complex conflagration (NFPA, 1980). This event is discussed in Chapter 9, *Defensive Operations*. It is important to get ahead of the fire, for both extinguishment and evacuation purposes, as it is fairly easy to alert and evacuate residents who are not in immediate danger. However, if the fire spreads and threatens the occupants, the rescue effort will be much more complex.

High-rise residential properties would also fall into the category of apartment buildings. These buildings have special perils and require specific tactics. High-rise buildings are discussed in Chapter 12, *High-Rise Buildings*.

Dormitories

College and university dormitory housing is supervised. University policies and jurisdictional codes require a specified level of safety in dormitories. However, there are other types of housing for college students. Students rent housing in apartments or homes that have no affiliation with the college or university. Fraternity and sorority housing may or may not be controlled by the college. College dormitories tend to be larger facilities than off-campus housing; some are high-rise buildings. The number of people in both on- and off-campus structures is fairly high compared to most residential properties. Dormitories tend to provide small living spaces for each student, with common public areas. Many fraternity, sorority, and private houses rented by students are older residential properties that have been modified to hold more tenants than the original design intended.

The occupant load is increased substantially when guests are invited to social events. A 2003 fire in private housing for students near Ohio State University claimed the lives of five students. The three-story building had 12 residential units plus a large number of visiting students. In another deadly fire, three students perished in an off-campus housing unit near Miami University in Oxford, Ohio, in 2006 (Campus Firewatch, 2020).

Off-campus statistical information is difficult to obtain, because many fire reports categorize these fires as apartment occupancies, and many smaller fires go unreported. Approximately 75 percent of college students live in off-campus housing, compared to 25 percent living in dormitories. A dormitory fire at Seton Hall University and a fraternity house fire at North Carolina University prompted officials to require sprinkler protection. The American Fire Sprinkler Association, Campus Firewatch, NFPA, National

Fire Sprinkler Association, USFA, and other organizations and fire safety advocates are working to require sprinkler protection in dormitory, sorority, and fraternity housing, but many on- and off-campus housing units remain unprotected.

Life safety is a critical issue in dormitory and off-campus housing. Alcohol consumption can result in a lack of awareness and reduced mobility. Many victims of fatal campus housing fires have high blood alcohol levels. When a fraternity or sorority house (or other private student housing unit) invites other students to social events, alcohol consumption is often excessive and many of the invited students spend the night at the off-campus housing unit. The fire officer must consider the possibility that more students than expected may occupy an off-campus house and occupants may be in areas other than normal sleeping areas. Many of the multifatality fires in off-campus housing occur after a party. Furthermore, arson is a leading cause of fire in student housing; therefore, a heavily involved fire with multiple victims is not uncommon.

Most rooms in dormitory or off-campus housing will be within the flow capacity of standpipe or preconnected hose streams. Common areas used for study and recreation will be larger, but most will still be within the capabilities of one or two standard hose lines. Irreplaceable personal heirlooms are not as common in dormitories and off-campus student housing as compared to some other residential occupancies; however, as with other residential properties, the importance of personal items should be considered during incident operations. Students may have course assignments or research on a computer that took considerable time and effort to compile.

The U.S. Fire Administration report "Campus Fire Fatalities in Residential Buildings (2000-2015)" listed the following findings (USFA, n.d.):

- During the last 16 academic years from 2000 through 2015, there have been 85 fatal fires in dormitories, fraternities, sororities, and off-campus housing, resulting in 118 fatalities—an average of approximately seven per school year.
- An astonishing 94 percent of fatal campus fires examined took place in off-campus housing.
- Smoke alarms were either missing or had been tampered with (disconnected or battery removed) in 58 percent of fatal campus fires.
- Fire sprinklers were not present in any of the 85 fatal campus fires.
- A disproportionate number of fatal campus fires occurred on the weekend—70 percent on Friday, Saturday, and Sunday.

Incident Summary

University of North Carolina—Chapel Hill Fraternity House Fires

On May 12, 1996, a fire in an off-campus fraternity house at the University of North Carolina at Chapel Hill resulted in five fatalities and $475,000 in property loss **FIGURE IS11-3**. Less than a month later, an arson fire struck another fraternity house. On June 19, 1999, an ordinance was passed requiring all fraternity and sorority houses to be sprinkler protected.

FIGURE IS11-3 Aftermath of fraternity house fire—University of North Carolina at Chapel Hill.
© The News & Observer.

Data from: Michael S. Isner, NFPA Fire Investigation Report, Fraternity House Fire, Chapel Hill NC, May 12, 1996.

- Males were more likely than females to die in campus fires, accounting for 67 percent of all victims.
- Alcohol was a factor in 76 percent of all fatal campus fires—fires where at least one of the students was drinking and, according to reports, legally drunk, which is at or above 0.08 percent blood alcohol concentration (BAC).
- Smoking (29 percent) was the leading cause of fatal fires in campus housing, followed by intentional actions (16 percent), electrical (11 percent), and cooking (9 percent), with 18 percent of the fires classified as "cause undetermined."
- The adage "nothing good happens after midnight" rings true for fatal campus fires, with 73 percent occurring between midnight and 6 AM.
- April was the peak month (13 percent) for fatal fires in campus housing, with January, May, and October at 12 percent each. Predictably, the

lowest number of fires occurred in June, July, and August, when there are fewer students enrolled in classes at colleges and universities.

For more information, read the full report, which is available from USFA, "Campus Fire Fatalities in Residential Buildings (2000–2015)."

Hotels and Motels

Like dormitories and off-campus housing, hotels and motels have a much greater occupant load than the typical home or apartment. Small motels may have doors that lead directly to the outside at grade level or doors leading to a walkway with direct access to ground level. Larger hotel and motel room doors typically lead to an interior hallway with well-marked exits. Motels and hotels are required to post evacuation information inside each unit. Unfortunately, many people fail to read the evacuation instructions or familiarize themselves with exits when they are assigned a room. Occupants in a rush to evacuate will seldom read the instructions in their room, and exit signage may be obscured by smoke at the time of the fire.

Many times, fire apparatus access will be limited to only one or two sides of a hotel or motel; therefore, fire department aerial devices will not be able to reach occupants on the nonaccessible sides of the building. Furthermore, fires in units on the opposite side of back-to-back units or units with an interior hallway may not be visible upon arrival. Expect to find occupants still in their rooms, possibly with occupants from several rooms converged in a single unit (convergence cluster) or disoriented in hallways, in elevator lobbies, on exterior balconies, and in stairways. If the roof is accessible, occupants may try to seek refuge there as well.

When encountering heavy fire and smoke conditions, several fire companies will be needed to conduct search and rescue operations in a large hotel or motel property. The desk clerk may have a list of people occupying the hotel, but the hotel staff will not know which occupants are in the facility or be aware of people who share rooms without notifying hotel management. Most large hotels will have assembly, food, and drinking establishment areas within the building. Because of the potential for multiple rescues and a large rate of flow, it might be necessary to assign multiple companies to these areas. Most of the large loss-of-life fires in hotels and motels occur in high-rise hotels. High-rise tactics are addressed in Chapter 12, *High-Rise Buildings*.

Mercantile Occupancies

NFPA 101, *Life Safety Code*, defines a **mercantile occupancy** as follows (NFPA 101, 2018):

> **3.3.196.9 Mercantile Occupancy.** An occupancy used for the display and sale of merchandise.

An occupancy factor matrix for mercantile occupancies is shown in **TABLE 11-15**. Shoppers and employees should be fully alert, and most will be able to evacuate with little assistance. However, there may be people who need assistance in an emergency, particularly in multistory mercantile buildings. For mercantile occupancies with a sales area on a street floor, the occupant load factor is 30 ft^2/person (NFPA 101, 2018). In many cases the actual number of occupants is found to exceed that.

Time factors are critical in sizing up a mercantile occupancy. During times of peak occupancy, the life hazard can be high and content losses can be significant in these buildings. Property loss is typically the largest problem facing fire forces in mercantile occupancies, but don't be misled; the danger to life may be substantial. Consider the life-safety problem in a crowded department store during the Christmas season.

Mercantile occupancies include shopping centers and malls, individual stores, and shops. Individual stores and shops can range from small shops or convenience stores to mammoth "big-box" stores. Small convenience stores tend to be more congested, with stock in the aisles making it difficult to navigate even under normal conditions. Each type of large store, whether a big-box retailer, multistory department store, shopping center, lifestyle center, or enclosed mall, presents special challenges.

TABLE 11-15 Occupancy Factor Matrix: Mercantile Occupancies						
	Mobility of Occupants	Age of Occupants	Leadership	Awareness	Occupant Load	Familiarity of Occupants
Mercantile occupancies	7	5	3	9	3	2

1 = extremely negative factor; 10 = very positive factor.
© Jones & Bartlett Learning.

Any building with large undivided spaces requires special search and rescue techniques. The right- and left-hand searches used in residential occupancies are not effective in a store that could be hundreds of feet in each direction, with aisles, cross aisles, and various obstructions. These large areas require team searches, which are difficult to coordinate and are labor- and time-intensive. Team search techniques using ropes and thermal imaging cameras will be needed when vision is obscured. However, it takes longer for smoke to fill these large-volume compartments, allowing many evacuations to be completed in a relatively clear atmosphere. If smoke is filling the store, proper venting can restore visibility and may allow occupants to evacuate with little assistance.

Large undivided spaces will require a substantial rate of flow. Extinguishment may not be possible within the store of origin once the fire gains sufficient headway. Rate of flow should be computed during preincident planning for each store and storage area that exceeds the flow capacity of two standard preconnected hose lines. Consider using 2½-in. (64-mm) hose lines or master streams as initial attack hose lines inside large stores.

Shopping Centers

Smaller shopping centers are usually configured as a line of stores (strip malls) with direct front and rear access to each store. Larger shopping centers characteristically have an anchor stores on each end. Anchor stores are typically big-box stores or multistory department stores. Having direct access to the outside makes evacuation less confusing compared to evacuating stores located inside enclosed malls. Most occupants of stores in a shopping center or lifestyle mall will be able to find the main entrance with little difficulty. The rear of the store is often used as a loading dock and storage area, thus rear exits will require occupants to go through unfamiliar surroundings to reach the outside. Rear doors tend to be heavily secured.

Many shopping centers lack sprinkler protection, although sprinkler systems are becoming more commonplace. As stated previously, the rate of flow could be very large, and it may be beyond the fire department's ability to extinguish a fire in a large store. Shopping center stores, other than anchor stores, are attached on both sides, which creates an internal exposure hazard. Exposure hose lines should be deployed to stores on each side of the store of origin as soon as critical life-safety and initial extinguishment assignments are made or if a defensive attack becomes necessary. Fire can penetrate anywhere along the common fire wall, but the roof line and any place with utility penetrations are most likely extension paths. Older strip malls may have a common attic, which could allow the fire to spread horizontally to all attached stores. To deal with this scenario, hose lines must be quickly extended into these false spaces on each side of the fire. Some shopping centers have double fire walls between buildings with no penetrations. Extension is less likely with this higher degree of compartmentation.

Enclosed Shopping Malls

The enclosed shopping mall took shopping convenience one step further by enclosing the total shopping area with entrances to various shops from an interior walkway, thus providing access to shops in a climate-controlled environment. Shopping malls tend to be larger than shopping centers and may have several anchor stores located at the ends and off multiple wings. Shopping malls are usually two or more levels. The layout of these structures is much more confusing than that of shopping centers. It is virtually impossible to find a specific smaller store without a map and/or guidance from building management. Preincident plans should include exterior access areas for fire apparatus, and floor layout drawings showing the location of each store should be carried on the apparatus of the first-arriving units. Primary and alternative access locations for each general area inside the mall should also be noted. If preincident plans are computerized, it is possible to search electronically for specific stores and identify the best exterior access points.

Stores within a shopping mall will be separated by partition walls of noncombustible construction but will be open to the interior walkway area. During business hours, large door openings are designed to provide easy customer access to the store from the mall walkway. At night, metal grates are lowered and secured. These metal grates provide a barrier to criminals but will not stop the forward progress of a fire. Most modern shopping malls are sprinkler protected, greatly reducing the potential fire problem. Without a properly maintained and operational sprinkler system, these properties would present an unmanageable life hazard for occupants and fire fighters.

The size of enclosed shopping malls has increased dramatically. Some of the largest malls cover several blocks and are virtually small cities inside a single building. These larger malls contain many occupancy types, including places of assembly. One such mall in Edmonton, Alberta, Canada, has hundreds of stores, an amusement park, a hotel, and a wave pool inside the structure.

Along with sprinkler protection, smoke control systems are an integral part of the fire safety system

Incident Summary

Shopping Mall Fire in Altoona, Pennsylvania

A fire in Altoona, Pennsylvania, demonstrates the problems associated with shopping malls that are not sprinkler protected. The fire began in an unprotected store and rapidly spread to the mall walkway area and other stores. This shopping mall fire also demonstrates another common fire problem in strip and enclosed malls: common concealed spaces. The fire in Altoona began in a utility closet then spread to the ceiling space, involving a large concealed area before breaking out into the mall area. The result was the total destruction of 15 stores and damage to 37 other stores within the mall in the weeks before Christmas **FIGURE IS11-4**.

Data from: Michael S. Isner, Shopping Mall Fire, Township of Logan, Pennsylvania, December 16, 1994, Fire Investigation Report. Quincy, MA: NFPA.

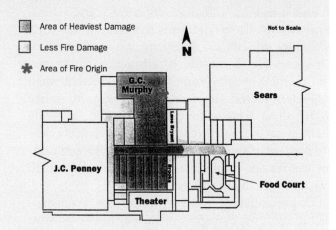

FIGURE IS11-4 Areas of damage from the mall fire in Altoona, Pennsylvania.
Reprinted with permission copyright © 2007, National Fire Protection Association, Quincy, MA. All rights reserved.

within a mall. If not controlled early, fire in one store will likely spread smoke and heat into the walkway area and other stores. Smoke control systems and large open walkways should allow enough time for a successful evacuation.

Lifestyle Centers

A lifestyle center is a type of shopping center laid out as blocks of stores, much like a city street. Large anchor stores are usually surrounded by smaller shops at grade level, and each individual store can be accessed directly from the outside. The same basic tactics can be used as in the shopping center, but lifestyle centers will have fewer stores in each cluster and nearly all will be protected with automatic sprinklers. Finding individual shops and stores will normally be more difficult in the lifestyle mall than in a shopping center. Like the enclosed mall, a drawing showing the location of each store is needed as part of the preincident plan.

There is generally a regular turnover of tenants in shopping centers, enclosed malls, and lifestyle centers. It is important to communicate with building management on a regular basis and update the preincident plan whenever a store's occupant changes.

Big-Box Stores

Many big-box stores are anchor stores at shopping centers, but some are either stand-alone stores or have a few small stores nearby. The typical layout for a big-box store is an open-layout store to the front with an attached storage area to the rear. The big-box store should be sprinkler protected. As long as the sprinkler system is operating properly and store management has not increased the fuel load beyond the capabilities that the sprinkler system was designed to protect, the sprinkler system operational guidelines will be sufficient. These guidelines are presented in Chapter 7, *Fire Protection Systems*. Many of these stores will have standpipe drops from the sprinkler system, but it may be better to lay hose directly from the apparatus to avoid robbing water from the sprinkler system.

The sprinkler system is the only positive factor in these buildings. Variances in construction methods, exit facilities, and size are permitted owing to the sprinkler protection. This is a valid concept provided the sprinkler system is properly maintained and is fully operational. If the sprinkler system is out of service or unable to control the fire, the big-box store is a very dangerous place for fire fighters and occupants.

Even with a properly operating sprinkler system, the height and configuration of storage can require substantial intervention to achieve final extinguishment. For example, consider an indoor lumber yard where sheets of plywood are stacked with no in-rack sprinkler protection, such that the upper stacks of plywood shield the lower racks from overhead sprinkler water. The storage racks are typically metal; therefore, a serious fire in a lower rack could result in a collapse of the entire rack. Furthermore, these stores may contain merchandise of an extra-hazardous nature (e.g., chemicals, compressed gas cylinders, aerosols, racked clothing), which are extremely difficult to extinguish and create a serious life hazard for occupants and fire fighters.

Multilevel Department Stores

As noted previously, multilevel department stores can be part of a shopping center, enclosed mall, or lifestyle center. They can also be stand-alone buildings. Life safety and extinguishment are more difficult in these occupancies because of the multiple stories. Evacuation from upper floors requires the use of stairs or escalators. In many stores there is little separation from floor to floor; thus fire on a lower level is likely to extend to upper floors. Due to the large fuel load, extinguishment will be difficult. Property conservation is also more problematic in a multilevel structure. Property conservation is a major issue in all mercantile occupancies where high-dollar inventories are common and many items are extremely sensitive to smoke and water damage.

Business Occupancies

NFPA 101, *Life Safety Code*, defines a **business occupancy** as follows (NFPA 101, 2018):

> **3.3.196.3 Business Occupancy.** An occupancy used for the transaction of business other than mercantile.

An occupancy factor matrix for business occupancies is shown in **TABLE 11-16**. The occupants of a business occupancy should be awake and aware while occupying the building. Occupant mobility is generally not a major problem, but there may be physically or mentally challenged people in the building who require special assistance. Furthermore, many business occupancies are located within high-rise buildings that may require the use of stairways for evacuation as well as places of safe refuge for occupants with special needs.

In business settings, the people occupying the premises are typically between the ages of 18 and 65 years. Leadership depends on the specific occupancy, but many business occupancies rehearse evacuation plans, particularly in high-rise buildings. Occupant load can be higher than one person per 100 ft^2 (9.3 m^2) which is considered less crowded than most places of assembly, but more densely populated than many other occupancies. Employees in a one- or two-story office building with exits leading directly to the outside will probably be familiar with egress. People working in high-rise buildings usually take elevators to their work areas and may be unfamiliar with stairway locations. The large open areas and complex layouts place fire fighters at greater risk.

Business occupancies have a relatively good history in terms of life safety for occupants. However, the trend toward more and larger high-rise structures increases the probability of a large loss-of-life fire in a business occupancy. For many years, fire professionals predicted that a future high-rise fire would result in the loss of a thousand or more lives. Unfortunately, this prediction became a reality on September 11, 2001; however, the circumstances of the terrorist attack on that day were much different from the fire the experts envisioned. The 2001 attack on the World Trade Center was the deadliest fire in U.S. history, but it was not the first terrorist attack on a high-rise building, nor was it even the first attack on the World Trade Center. The first World Trade Center attack, occurring in 1993 (USFA, 1993), and the 1995 attack on the Alfred P. Murrah Federal Building in Oklahoma City (Comeau and Foley, 1995) were also the result of terrorist activity, and both attacks included fire and explosions as part of the scenario. Historically, fire has been used as a weapon, and most experts agree that high-occupancy buildings will continue to be targets of choice for terrorists. The World Trade Center had approximately 50,000 occupants at the time of the first bombing on February 26, 1993.

TABLE 11-16 Occupancy Factor Matrix: Business Occupancies						
	Mobility of Occupants	Age of Occupants	Leadership	Awareness	Occupant Load	Familiarity of Occupants
Business occupancies	7	7	7	9	5	6

1 = extremely negative factor; 10 = very positive factor.
© Jones & Bartlett Learning.

The Murrah Building is believed to have been occupied by 350 people at the time of the explosion on April 19, 1995. At nine stories, the Murrah Building was small in comparison to many high-rise office buildings. High-rise buildings create special problems and are discussed in detail in Chapter 12, *High-Rise Buildings*.

At night, when most business buildings are unoccupied, fires can progress for a long period if the building is not protected by an alarm and/or suppression system. Most new construction in business occupancies provides built-in automatic fire suppression and alarm systems. There is also a trend toward retrofitting older high-rise buildings with sprinkler systems and other fire safety equipment.

In office occupancies, separation between individual offices is typically poor or nonexistent. Many times, movable partitions (cubicles) are used to separate work stations, allowing heat and smoke to spread quickly throughout a large, undivided area. Furthermore, this layout makes search and rescue operations complex, because the configuration often resembles a maze. Wall separations are sometimes not full fire barriers, as some walls separating offices do not extend beyond the suspended ceiling. The area above the suspended ceiling provides an easy path for fire extension.

Fire control efforts vary widely in business occupancies, depending on the fuel load and the size of the undivided area. Fire control will not always be possible or could be substantially delayed owing to large flow requirements. Business occupancies tend to have a light-to-moderate fuel load. As for other wide-open areas, the rate of flow requirements depends on the volume of the largest undivided compartment. If the sprinkler system does not control a fire in a large open-layout office, heavy-volume solid or straight-stream appliances may be needed for suppression.

Property conservation in business occupancies must emphasize the importance of protecting sensitive electronic equipment and business records.

Storage Occupancies

NFPA 101, *Life Safety Code*, defines a storage occupancy as follows (NFPA 101, 2018):

> **3.3.196.15 Storage Occupancy.** An occupancy used primarily for the storage or sheltering of goods, merchandise, products, or vehicles.

The factors affecting occupant evacuation in **TABLE 11-17** are mainly positive for a storage occupancy. Employees in these occupancies are expected to perform physical labor; thus, most employees will be mobile. Generally speaking, the very young and elderly are at greatest risk during a structure fire. Age should not be an evacuation factor in a storage occupancy with a working-age population. Some storage facilities have an emergency action plan that prescribes internal leadership, while others will lack leadership in the initial stages of an emergency. Employees should be awake and alert, as well as familiar with the building layout and exit facilities.

Occupant load is generally low in storage facilities, especially in automated warehouses. Many modern warehouses are automated, with conveyors or robotics extending into remote and difficult-to-access locations. The increase in online sales is expected to continue with a subsequent increase in the number and size of automated warehouses.

The risk of a fast-moving fire or explosion may be high within the storage occupancy, which could have an immense fuel load. Yet the fire safety record in these occupancies has been good, primarily because of steps taken by responsible owners in protecting their workforce and property from fire.

Storage facilities vary in size from small self-storage modules to mammoth facilities like the Kmart warehouse in Falls Creek, Pennsylvania, or the general merchandise warehouse in New Orleans, Louisiana.

Most one-story modern warehouses are of noncombustible construction with metal truss roof structures **FIGURE 11-15**. Like the big-box stores described

TABLE 11-17 Occupancy Factor Matrix: Storage Occupancies						
	Mobility of Occupants	Age of Occupants	Leadership	Awareness	Occupant Load	Familiarity of Occupants
Storage occupancies	9	8	6	9	9	8

1 = extremely negative factor; 10 = very positive factor.

© Jones & Bartlett Learning.

FIGURE 11-15 Interior of large warehouse.
nd3000/Shutterstock.

in the section "Mercantile Occupancies," these storage buildings can have tremendously high fuel loads, and the roof structures make them prone to early roof collapse. On the positive side, most large modern warehouses are sprinkler protected. Conversely, many older warehouses will be of fire-resistive, heavy-timber, or ordinary construction, though they are often multistory structures with limited egress and confusing layouts. Many of these older warehouse buildings lack sprinkler protection.

Fighting a fire in a large commercial building is considerably more challenging tactically than attacking a fire inside a small residential building. The size and complexity of the building, coupled with the possibility of an extremely high fuel load, make these buildings more hazardous for fire fighters working inside. The large volume of air inside each compartment will provide oxygen for an extended time, allowing the fire to grow rather than become ventilation controlled.

Obviously, escape time will be much longer from a building that may be 500,000 ft^2 (46,500 m^2) or greater as compared to a residential structure, where room dimensions seldom exceed 1000 ft^2 (93 m^2).

A fire in a vacant warehouse in Worcester, Massachusetts, resulted in six fire fighter fatalities. The interior of this multistory warehouse was very confusing. Fire fighters are more likely to be killed in these large commercial structures than in residential fires, and the danger increases exponentially in large, vacant buildings.

In warehouses with an emergency action plan, including provisions for employee accountability, the primary search and rescue will be much easier because the search team can focus on a targeted area, provided the emergency action plan was successfully implemented.

If the property is sprinkler protected, supporting and augmenting the sprinkler system will be the primary extinguishment tactic. Facility management will sometimes increase the fuel load inside a warehouse beyond the design capacity of the sprinkler system. Additional water supplied by fire department pumpers could make the difference between a successful operation and a total loss of the building when the fuel load exceeds the design capacity of the sprinkler system.

When a fire overwhelms the sprinkler system and reaches an advanced stage, extinguishment will be difficult. Every effort should be made to increase the effectiveness of the sprinkler system before attempting to control a major fire with an offensive attack using handheld hose lines.

A common practice in warehouse facilities is to place large stacks of idle pallets in a staging area. This creates a tremendous challenge for the sprinkler system. Only an extra-hazard sprinkler system is capable of handling a fire in large quantities of wooden pallets. When a new occupant leases a storage facility, the new owner/occupant may store a different commodity, possibly exceeding the design capacity of the sprinkler system. For example, a warehouse sprinkler system may be designed to protect metal parts stored in cardboard boxes. If a new owner or occupant were to store rubber tires, flammable liquids, or plastics in this same area, the system, which was designed for the lower-hazard commodity, would be inadequate to extinguish the fire. The result would likely be a total loss.

The Kmart fire in Falls Creek, Pennsylvania, is a good example of the loss potential in a storage occupancy fire (Best, 1983). The Kmart warehouse, covering nearly 1.3 million ft^2 (approximately 12 hectares), was fully sprinkler protected and compartmentalized, yet it was completely destroyed in a $100-million fire. Aerosol cans of carburetor cleaner, which exceeded the design capacity of the sprinkler system, were first ignited at this incident. The burning aerosol cans rocketed from the compartment of origin to other compartments, eventually spreading the fire to all four building quadrants. Had the fire chief failed to see the danger to his personnel, this fire could well have resulted in several fire fighter fatalities. By developing a preincident plan for the structure, the chief recognized the collapse potential and the futility of manual firefighting operations when confronted with an advanced fire.

The rate of flow must be calculated during preincident planning. Standard V/100 calculations (discussed in Chapter 8, *Offensive Operations*) may not be sufficient for storage occupancies. Sprinkler system calculations should be used when dealing with extra-hazard storage. It is important to remember that sprinkler calculations assume the fire will be contained to a limited area due to quick operation of the system, with water discharging directly onto the fire. These assumptions are not valid when manually attacking the fire with handheld hose lines, because sprinklers are much more efficient in applying water directly on the fire. Therefore, the calculated rate of flow for hose lines may result in a flow that is beyond the available water supply and fire department staffing and pumping capacities. With that said, when calculating rate of flow and developing preincident plans, remember that in many cases the fire will be limited to a small area within a large building.

Many firms that suffer a large fire go out of business. Even with insurance, the loss of production and subsequent loss of customers are more than many businesses can overcome. Competent business people apply risk management principles to protect their profitability. One of the most common risk management measures involves installation of suppression systems or other built-in fire protection.

Most warehouses contain large quantities of valuable commodities. Some of these commodities are susceptible to water and smoke damage. Storage in racks or on pallets reduces water damage from runoff. Storage of materials that absorb water, especially if the materials are stored on the floor, will greatly increase the water damage and add significant weight to the floor.

Industrial Occupancies

NFPA 101, *Life Safety Code*, defines an **industrial occupancy** as follows (NFPA 101, 2018):

> **3.3.196.8 Industrial Occupancy.** An occupancy in which products are manufactured or in which processing, assembling, mixing, packaging, finishing, decorating, or repair operations are conducted.

The occupant profile in an industrial setting is much like that of the storage occupancy, with mobile, alert workers who are familiar with their surroundings **TABLE 11-18**. The occupant load is usually higher in an industrial occupancy facility as compared to a storage occupancy. Industrial plants are required to have emergency evacuation plans; therefore, leadership is usually better in the industrial occupancy than in many other occupancies. Emergency action plans often include an accountability process that can assist the IC in facilitating an orderly evacuation and accounting for plant personnel.

Incident Summary

New Orleans Warehouse Fires

In 1996, the New Orleans Fire Department was successful in containing a fire that was beyond the capabilities of the sprinkler system, even though 30 overhead and 17 in-rack sprinkler heads operated. After manually bringing the fire under control, all valves controlling the systems were closed to eliminate additional water damage within the heavily damaged 930,020-ft^2 (86,400-m^2) undivided warehouse. Subsequently, the power was restored to the warehouse area, resulting in a second fire caused by damaged wiring. The second fire resulted in a total loss of the warehouse and contents valued at $280 million (Comeau and Puchovsky, 1996).

TABLE 11-18 Occupancy Factor Matrix: Industrial Occupancies

	Mobility of Occupants	Age of Occupants	Leadership	Awareness	Occupant Load	Familiarity of Occupants
Industrial occupancies	9	8	8	9	7	8

1 = extremely negative factor; 10 = very positive factor.
© Jones & Bartlett Learning.

Recognizing hazards and equipment limitations is paramount during an industrial fire. In the large industrial building or industrial complex, preincident planning is critical to fire fighter safety and efficient operations. However, preincident planning is not enough. Cooperation between plant personnel and fire fighters is essential. In-plant employees are plant experts; they know valve locations, how to navigate the plant, and how to shut down processes. The fire fighter is the fire suppression expert. Successful industrial fire suppression activities involve plant staff assisting fire department personnel. Anything less than a cooperative effort invites total destruction and places fire fighters at unnecessary risk. The number of hazards and the potential for harm dictate that special tactics be developed for each large manufacturing and storage property within a jurisdiction.

Industrial occupancy fires are notable because they tend to cause a large loss of life and/or large financial loss. Some notable large loss-of-life fires in industrial occupancies are listed in **TABLE 11-19**. Several of the multifatality fires in industrial properties involved hazardous materials, such as ammunition, fireworks, explosives, chemicals, and flammables. Table 11-19 lists industrial fires that killed at least six people. There were many additional oil refinery and tank fires that claimed three to five lives that are not listed in the table.

In the industrial setting, the main tactical activity often involves controlling a manufacturing process or otherwise stabilizing the incident to protect life and property. Depending on the type of burning fuel, the use of water may be counterproductive. Using knowledgeable facility personnel as advisors to the IC during an incident is imperative, as they are familiar with the building, systems, and processes and can advise the IC on a course of action to take (or not take) to safely mitigate the problem.

Fire control efforts must be customized to the hazard. Preincident planning, which includes identifying and outlining the responsibilities of plant personnel, is key to successfully controlling fires in industrial occupancies. In addition to the normal salvage activities, property conservation includes stabilizing manufacturing and operating processes. Many industrial facilities have interlocks that sense a drop in pressure in the water supply and/or fire protection system. These interlocks sometimes interrupt production or dump materials being processed. Flowing water from a fire hydrant located on the property may unintentionally interrupt a process, which could result in significant property loss.

Damage or destruction of high-value storage and manufacturing equipment at industrial properties results in many large-dollar losses. A million-dollar

TABLE 11-19 Notable Fires in Industrial Occupancies

Fire	Year	Deaths
R. B. Grover Shoe Factory, Brockton, MA	1905	50
Triangle Shirtwaist, New York City, NY	1911	145
Eddystone Ammunition Corporation, Eddystone, PA	1917	133
Aetna Chemical Company, Oakdale, PA	1918	193
Semet-Solvay, TNT Manufacturing, Split Rock, NY	1918	50
Hercules Powder Company, Kenvil, NJ	1940	52
Elwood Ordnance Plant, Joliet, IL	1942	54
East Ohio Gas, Cleveland, OH	1944	130
Munitions depot, Port Chicago, CA	1944	300
Chemical plant, Pasadena, TX	1989	23
Chicken processing plant, Hamlet, NC	1991	25
Fireworks manufacturing, Osseo, MI	1998	7
Power plant, Dearborn, MI ($1 billion property loss)	1999	6
Automobile insulation manufacturing, Corbin, KY	2003	7
Oil refinery, Texas City, TX	2005	15
Sugar refinery, Wentworth, GA	2008	14
Hydrocarbon refinery, Anacortes, WA	2010	7
Power generating plant, Middletown, CT	2010	6
Grain elevator, Atchison, KS	2011	6
Ammonium nitrate fertilizer plant, West, TX	2013	15*

*Includes nine fire fighters and an emergency medical technician.

© Jones & Bartlett Learning.

fire is commonplace in an industrial occupancy. The $100-million-plus fires at Kmart in Pennsylvania (Goodbread, 1985), Tinker Air Force Base in Oklahoma (Goodbread, 1985), Central Storage and Warehouse in Wisconsin (Isner, 1991), and West Fertilizer Company in Texas (Reuters, 2013) are examples of large-loss fires eclipsed by the billion-dollar fire at a Texas oil refinery in 2005.

Multiple (Mixed or Separated) Occupancy Buildings

NFPA 101, *Life Safety Code*, defines a **multiple occupancy** as follows (NFPA 101, 2018):

> **3.3.196.11 Multiple Occupancy.** A building or structure in which two or more classes of occupancy exist.

Multiple occupancies are further categorized as *mixed* or *separated*, as follows:

> **3.3.196.10 Mixed Occupancy.** A multiple occupancy where the occupancies are intermingled.

> **3.3.196.14 Separated Occupancy.** A multiple occupancy where the occupancies are separated by fire resistance–rated assemblies.

Separated multiple occupancy buildings can be handled as two buildings due to fire-resistive separations. **Mixed multiple occupancy** buildings are more problematic because fire can easily travel from one occupancy to another. Having an unoccupied area in the same building where other areas are occupied can result in a severe life-safety threat. For example, many buildings have a bar, restaurant, and small stores or shops on the first floor with residential units above. (See Chapter 8, Figure 8-24.) Fuel loads in the shops will typically be greater than the fuel loads found in residential occupancies and often include hazardous processes. Furthermore, the shops are usually unoccupied outside of normal business hours, allowing a fire to go undetected in an area below sleeping residents. By the time residents realize there is a fire, egress routes may be untenable and the fire may have already extended into the residential sections of the building.

Buildings Under Construction, Renovation, or Demolition

Buildings under construction, renovation, or demolition are not classified as specific occupancies, but these buildings are vulnerable to fire and warrant special attention. Buildings under construction or undergoing extensive renovation may lack fire protection equipment and structural features designed to impede fire until late in the construction process. Conversely, in buildings being demolished, these same fire protection features and equipment are often the first to go, including disabling the sprinkler system and removing interior walls.

Buildings under construction may not have fire barriers or exterior coverings that retard fire growth, and large quantities of building materials may be stored within the interior framework of a partially constructed building. Like the building under construction, the building being renovated may have large stacks of building materials stored inside the building. When a building is being razed, some interior materials are removed for salvage, then work begins to demolish the outer shell, with large quantities of debris accumulating around and within the structure.

Buildings under construction, renovation, or demolition are typically unoccupied when work crews leave for the day; however, these structures may be illegally occupied by youths or homeless individuals. In either case, after-hours fires in these structures usually gain considerable headway before being reported, allowing the fire time to spread throughout the structure and involve external exposures. Free-burning fires involving wooden framing, combustible building materials, or debris can develop a large fireball with subsequent high levels of radiant heat. These buildings can also produce flying brands that can ignite fires far from the building of origin, as occurred at the Santana Row fire in San Jose, California. (See the incident summary in Chapter 9, *Defensive Operations*.) The radiant heat ignites nearby structures while flying brands can travel a considerable distance beyond the building of origin. This is discussed further in Chapter 9, *Defensive Operations*. The open status of buildings under construction, renovation, or demolition also makes these structures vulnerable to flying brands from nearby fires **FIGURE 11-16**.

Buildings Under Construction

A building that is only partially constructed or one that is partially razed is more likely to collapse than a finished structure. Frequently, scaffolding is placed around buildings under construction, tempting fire fighters to use the scaffolding to gain rapid access to upper floors of the building. However, fire fighters should avoid using scaffolding, because water from extinguishment efforts can undermine the base of the scaffolding and the fire could destroy the scaffolding

FIGURE 11-16 Multistory building with a large volume of fire.
Roman, Jessee. (December 29, 2014) In A Flash. NFPA Journal. https://www.nfpa.org/News-and-Research/Publications-and-media/NFPA-Journal/2015/January-February-2015/News-and-Analysis/In-A-Flash

connections to the building. Either scenario could result in a total collapse of the scaffolding system, injuring or killing fire fighters who were working from the scaffolding at the time of collapse, while leaving fire fighters who used the scaffolding to access upper floors stranded with no means of escape.

Renovated Buildings

Once the renovation process is completed, the occupancy classification may change. Many multistory, downtown buildings that were once offices, stores, schools, or factories have been converted to other occupancy types. Many old school buildings are now apartments, condominiums, and restaurants; old department stores, lofts, and buildings that were built as factories are now offices. These are but a few examples of the kinds of conversions taking place in many urban areas.

In many cases, converted buildings are safer, because building and fire codes require upgrades when occupancy classifications are changed or when a significant portion of the building is renovated. However, these converted occupancies can also create challenges. For instance, interior walls are often removed to create large open spaces to accommodate the new building use. In addition, false ceilings are added to reduce the ceiling height for aesthetic and energy conservation reasons. Heavy dimensional lumber is often replaced with lightweight engineered wood.

The present trend of moving back into the central city core is expected to continue. And, with preservationists using their political clout to protect older buildings from being razed, much of the new residential and commercial space will be located in renovated structures. Many of the potential problems in renovating these older buildings can be avoided by enforcing building and renovation code requirements. Unfortunately, local and state government officials, in their zeal to revitalize the urban core while at the same time appeasing special interest groups, may bow to pressure to approve code waivers. For fire fighters, this means combating fires in buildings that contain many false spaces that can hide fire and provide channels for fire extension, confusing floor plans, and large open areas that will complicate search and rescue efforts, possibly requiring a large rate of flow.

The building under renovation in **FIGURE 11-17** was originally a one-story commercial storefront. After renovation it will be a three-story residence with indoor parking on the first floor. The original building was a Type III, ordinary construction building (masonry exterior walls with a true dimensional lumber roof structure). After renovation, the first floor will

FIGURE 11-17 Building under renovation.
Courtesy of Thomas Lakamp, Cincinnati, Ohio.

still have exterior masonry walls; however, the second and third floors will be constructed using lightweight wood trusses. In addition, a large portion of the dimensional lumber in the original roof assembly has been replaced with lightweight wood trusses to increase the open area on the first floor. Fire fighters responding to a fire in this structure could inaccurately assume this structure is ordinary construction, as it will assimilate into the existing architecture of the area when complete. However, this structure should be considered a lightweight truss constructed building.

General Considerations for Special Occupancy Fires

Special occupancy fires require fire departments to not only preplan for specific properties but also to develop action plans for general use within various types of buildings and occupancies. Many of the fire fighter fatalities that have been recorded over the years were in residential properties, including single-family detached dwellings. However, the size and complexity of business and industrial buildings place the fire fighter at additional risk when combating fires in these structures. Extra precautions and more frequent accountability procedures may be necessary in combating assembly, mercantile, storage, business, and industrial fires.

It is fair to say that fire fighters are at much greater risk of dying when confronted with commercial fires than when working a residential structure fire. Obviously, the threat to a fire fighter's life is substantial at the residential fire, but the fire officer must recognize the even greater potential danger when working at a commercial fire.

Using operational priorities as a basis, fire suppression forces can wage a safe and effective operation at commercial and other special occupancies. Due to the relatively rare incidence of these special occupancy fires, few ICs gain confidence through actual firefighting experience; therefore, it is imperative that they learn through training and from the experiences of others.

Most of the large loss-of-life fires in commercial and other special occupancies occurred in the fairly distant past. The number of civilian fire deaths and line-of-duty deaths for fire fighters has shown much improvement in the recent past. Two important points should be made about these positive statistics:

1. The fire threat to human life is real and present. Maintaining a properly staffed, well-trained fire force is absolutely necessary.
2. Improvements in codes and standards and the widespread use of smoke detectors, fire sprinkler systems, and public education programs are directly related to life-safety improvements.

Estimating the Number of Potential Victims

Determining the total number of occupants is closely related to the type of occupancy. A high-rise office building could have thousands of occupants, compared to one or two occupants in many single-family detached dwellings. Occupant load is also time sensitive. The office building will most likely be fully occupied during normal working hours, while the residential building will most likely be occupied at night.

The staffing requirements for a fire department response to a single-family detached dwelling are usually inadequate for larger building fires because of the additional search and rescue requirements combined with the need for larger fire flows. The more complex the structure, the greater the need for personnel. However, some commercial buildings will be more substantially built and are more likely to be protected by fire suppression systems. These fire-protective features will lower the risk to fire fighters and occupants, provided that the fire safety features designed into the building are maintained and operating properly.

The first step in determining evacuation needs is an evaluation of the time of day, building size, and occupancy type. Classifying these variables will provide a rough estimate of how many people could be in the building; however, it is important to remember that this evaluation is, at best, only an estimate.

Wrap UP

CHAPTER SUMMARY

- Occupancy type, time of day, and day of week are critical factors in determining fire-ground strategy.
- Major factors related to occupancy type are number, mobility, and age of the occupants; leadership in guiding evacuation; occupant awareness; occupant familiarity with the structure; and time of day, week, and year.
- Assembly occupancies include places of worship, eating/drinking establishments, stadiums, convention centers, and theaters where 50 or more persons gather.
- Mobility factors are usually above average in school settings, but younger elementary school students may not be capable of taking independent action to safely evacuate.
- Leadership and preparedness are generally positive factors in educational occupancies.
- Operations in detention/corrections occupancies are complicated by the fact that prisoners cannot self-evacuate and must be kept secure during firefighting operations.
- Preincident planning is an absolute necessity for a detention or correctional facility.
- Residential occupancies include single- and multiple-family dwellings, apartment buildings, dormitories, and hotels/motels.
- Most fires, fatalities, and property loss occur in residential fires.
- In many occupancies, including residential, there is a direct correlation between life safety and the time of day.
- More fire fighter on-duty deaths occur in residential properties than in any other occupancy due to the large number of fires that occur in these properties.
- Most fires in one- and two-family dwellings should be within the capabilities of one or two preconnected hose lines.
- Property loss is typically the largest problem facing fire forces in mercantile occupancies; however, depending on time factors, danger to life safety may be substantial.
- Preincident plans should include exterior access areas for fire apparatus and floor layout drawings showing the location of each store in shopping centers, malls, and lifestyle centers.
- High turnover of tenants in shopping centers and malls makes it important to update preplans on a regular basis.
- Large fuel loads in mercantile occupancies make extinguishment difficult.
- The risk of a fast-moving fire or explosion is high in a storage occupancy, especially if the fuel load exceeds the design capacity of the sprinkler system.
- Most large, modern warehouses are of lightweight construction but sprinkler protected; older warehouses are often of fire-resistive, heavy-timber, or ordinary construction, but often lack sprinkler protection.
- Emergency evacuation and preincident plans are required in most industrial facilities.
- Mixed and multiple occupancies combine the fuel loads and life-safety hazards of two or more occupancy types.
- Structural collapse and/or collapse of construction scaffolding is highly likely in buildings under construction.
- Occupant load is related to building size, type of occupancy, time of day, and day of week.

KEY TERMS

assembly occupancy An occupancy used for the gathering of 50 or more persons for deliberation, worship, entertainment, eating, drinking, amusement, awaiting transportation, or similar uses; or used as a special amusement building, regardless of occupant load.

business occupancy An occupancy used for account and record keeping or the transaction of business other than mercantile.

convergence cluster phenomenon A reaction to fire conditions in which groups take shelter together to provide mutual support.

detention and correctional occupancy An occupancy used to house one or more persons under varied degrees of restraint or security where such occupants are mostly incapable of self-preservation because of security measures not under the occupants' control.

educational occupancy An occupancy used for educational purposes through the 12th grade by six or more persons for 4 or more hours per day or more than 12 hours per week.

healthcare occupancy An occupancy used for purposes of medical or other treatment or care of four or more persons, where such occupants are mostly incapable of self-preservation due to age, physical or mental disability, or security measures not under the occupants' control.

industrial occupancy An occupancy in which products are manufactured or in which processing, assembling, mixing, packaging, finishing, decorating, or repair operations are conducted.

mercantile occupancy An occupancy used for the display and sale of merchandise.

mixed multiple occupancy A multiple occupancy where the occupancies are intermingled.

multiple occupancy A building or structure in which two or more classes of occupancy exist.

occupant load The total number of persons that might occupy a building or portion thereof at any one time.

residential board and care occupancy A building or portion thereof that is used for lodging and boarding of four or more residents, not related by blood or marriage to the owners or operators, for the purpose of providing personal care services.

residential occupancy An occupancy that provides sleeping accommodations for purposes other than health care or detention and correctional.

separated multiple occupancy A multiple occupancy where the occupancies are separated by fire resistance–rated assemblies.

storage occupancy An occupancy used primarily for the storage or sheltering of goods, merchandise, products, vehicles, or animals.

SUGGESTED ACTIVITIES

1. Classify specific occupancies in your jurisdiction using an occupancy factor matrix similar to the general occupancy factor matrices tables used in this chapter.

2. Compute the occupant load in a 100- × 100-ft (30.5- × 30.5-m), 10-story office building assuming 100 ft^2 (9.3 m^2) per person.

3. For this activity, use the plan-view drawing showing the exit doors and ignition points from The Station nightclub incident summary. Change the ignition sequence to a slower-moving fire (one not involving foam plastic materials). When you arrive as the officer in charge of the first-in engine company, there are heavy smoke and fire conditions with people jammed at the main entrance blocking people who are attempting to escape.

   A. What is the best course of action for this first-arriving unit, assuming the company is staffed with an officer, apparatus operator, and two fire fighters? Explain the rationale for your decision.

   B. If you decide to lay a hose line, what would be the best place to enter the building to attack or control the fire?

FIGURE 11-18 Insurance claims office subdivided using movable partitions (cubicles).
Jude Lazaro/ShutterStock, Inc.

    C. If you decide to rescue occupants, explain how you could rescue the most people in the least amount of time.

    D. Assume the role of a battalion chief arriving to take command. Develop an incident action plan and deploy units to implement the plan. Organize units deployed and in reserve using a National Incident Management System (NIMS) chart.

4. Review an evacuation plan for a local elementary or high school in your jurisdiction.

    A. Does the plan clearly spell out the location of the school contact person for a fire emergency? Who are the primary and alternate contact persons?

    B. Are students assembled in areas that interfere with fire department access or operations?

    C. If the school has an active shooter plan, how does it affect fire response?

5. Obtain a list of residential board and care occupancies in your response area. What are the occupant characteristics in these facilities in terms of their ability to escape without assistance?

6. It is 2:00 AM and you are responding as the officer of the first-due engine company to a 30-unit apartment building. Upon arrival, you can see fire in the main stairway, and the fire appears to have originated in the basement storage area. Some residents are already out of the building; others are at windows or on balconies calling for help. What is your best first action? Explain.

7. An insurance company's claims adjustment unit is housed in a 250- × 250-ft (76- × 76-m) building of noncombustible construction. Office space is divided using 4-ft (1.2-m) high cubicles with aisle and cross-aisle access to individual work areas **FIGURE 11-18**. Describe a search method that could be used under conditions of low visibility in this building.

8. A warehouse and residential occupancy are located in the same general area **FIGURE 11-19**. The alarm card for both occupancies is presented in **TABLE 11-20**.

FIGURE 11-19 Large apartment building with large, fully involved fire.
NIOSH. (November 19, 2013). Death in the Line of Duty... A summary of a NIOSH fire fighter fatality investigation. https://www.cdc.gov/niosh/fire/reports/face201213.html
A: Courtesy of David J. Jones, Cincinnati, Ohio; **B:** NIOSH Report 2012.13.

TABLE 11-20 Responding Fire Department Fire Companies, Fire Officers, and Emergency Medical Resources

	Engine Companies	Truck Companies All Equipped with 100-ft (30.5-m) Aerial Ladder	Other Units
1st Alarm	Engine 1 Engine 2 Engine 3	Truck 1 Truck 2	District Chief 1 (No chief officers have an aide) Paramedic 1
2nd Alarm	Engine 4 Engine 5 Engine 6	Truck 3 Truck 4	District Chief 2 Operations assistant chief Safety officer Paramedic 2 Heavy Rescue 1
3rd Alarm	Engine 7 Engine 8 Engine 9	Truck 5	Fire chief Operations assistant chief District Chief 2 Paramedic 3 Heavy Rescue 1
4th Alarm	Engine 10 Engine 11 Engine 12	Truck 6	Fire chief District Chief 4
5th Alarm	Engine 13 Engine 14 Engine 15	Truck 7	

© Jones & Bartlett Learning.

Compare the hazard to fire fighters in this late-night fire in a large warehouse to the fire in the four-story residential building.

Compare and contrast the probability of savable occupants in each building.

A. Would you initiate an offensive or defensive attack on the warehouse fire?

B. Would you initiate an offensive or defensive attack on the residential fire?

C. Compare the hazard to fire fighters in this late-night fire in a large warehouse to the fire in the four-story residential building.

D. Compare and contrast the probability of savable occupants in each building.

Assume the role of the IC:

E. Develop an incident action plan for the residential and warehouse buildings.

F. Is a one alarm sufficient for the residential fire?

G. Is a one alarm sufficient for the warehouse fire?

H. If not, how many alarms would you request for the residential and warehouse fires?

I. Assign each responding unit.

J. Develop an incident command chart for this incident.

REFERENCES

Ahrens, Marty. March 2006. *U.S. Fires in Selected Occupancies 1999 to 2002*. Quincy, MA: NFPA Fire Analysis and Research Division.

Ahrens, Marty. July 2016. *Structure Fires in Residential Board and Care Facilities*. Quincy, MA: NFPA Fire Analysis and Research Division.

Ahrens, Marty. July 2017. "U.S. Experience with Sprinklers." *NFPA Research*. Quincy, MA: NFPA Data and Analytics Division.

Ahrens, Marty. October 2019. "Home Structure Fires." *NFPA Research*. Quincy, MA: National Fire Protection Association.

Badger, Stephen. 2018. "Catastrophic Multiple Death Fires in 2017." *NFPA Journal* September/October.

Best, Richard L. 1978. *Reconstruction of a Tragedy: The Beverly Hills Supper Club Fire, Southgate, Kentucky, May 28, 1977*. Quincy, MA: National Fire Protection Association.

Best, Richard L. 1983. "$100 Million Fire in K-Mart Distribution Center." *NFPA Fire Journal* March: 36–42, 80.

Bryan, John L. 1985. "A Phenomenon of Human Behavior Seen in Selected High-Rise Building Fires: Convergence Clusters." *NFPA Fire Journal* November: 27–30, 86–90.

Campbell, Richard. September 2017. "Structure Fires in Educational Properties." National Fire Protection Association Research; nfpa.org/-/media/Files/News-and-Research/Fire-statistics-and-reports/Building-and-life-safety/oseducation.pdf. Accessed June 12, 2020.

Campbell, Richard. October 2017. "Structure Fires in Health Care Facilities." National Fire Protection Association Research; p. v. https://www.nfpa.org/News-and-Research/Data-research-and-tools/Building-and-Life-Safety/Fires-in-Health-Care-Facilities. Accessed March 18, 2020.

Campus Firewatch. 2020. "Fatal Fire Spreadsheet." http://www.campus-firewatch.com/resources/current-fire-information/fatal-fire-spreadsheet/. Accessed March 19, 2020.

Comeau, Edward R., and Stephen Foley. 1995. "Oklahoma City, April 19, 1995." *NFPA Journal* July/August.

Comeau, Edward R., and Milosh T. Puchovsky. March 1996. "Warehouse Fire, New Orleans, Louisiana." *Fire Investigative Report*. Quincy, MA: National Fire Protection Association.

Evarts, Ben. May 2012. *Structure Fires in Residential Board and Care Facilities*. Quincy, MA: National Fire Protection Association.

Federal Emergency Management Agency. April 2014. "School Building Fires (2009–2011)." *Topical Fire Report Series* 14 (14): 1–16. https://www.usfa.fema.gov/downloads/pdf/statistics/v14i14.pdf. Accessed March 18, 2020.

Federal Emergency Management Agency. June 2017. "Residential Building Fires (2013–2015)." *Topical Fire Report Series* 18 (1): 1–17. https://www.usfa.fema.gov/downloads/pdf/statistics/v18i1.pdf. Accessed March 18, 2020.

Goodbread, J. 1985. "Fire in Building 3001." *NFPA Fire Command* July: 34–7.

Isner, Michael S. 1991. "$100 million fire destroys warehouses." *NFPA Fire Journal* November/December: 37–41.

Klem, Thomas J. May 1986. "Soccer Stadium Fire: Bradford, UK, May 11, 1985." Fire Investigations. *Fire Journal* May. https://www.nfpa.org/-/media/Files/News-and-Research/Resources/Fire-Investigations/fibradford.ashx. Accessed April 15, 2020.

Moncada, Jaime A. 2012. "Lessons of Comayagua." *NFPA Journal* September/October: 50–9.

National Fire Protection Association. 2018. *NFPA 101®: Life Safety Code.* Quincy, MA: NFPA.

National Fire Protection Association. 2018. *NFPA 3000™ (PS): Standard for an Active Shooter/Hostile Event Response (ASHER) Program* Quincy, MA: NFPA.

National Fire Protection Association. Fires by occupancy or Property Type: Number of Fires Reported to Local Fire Departments in the United States by Property Use: 2014-2018 Annual Averages. https://www.nfpa.org/News-and-Research/Data-research-and-tools/US-Fire-Problem/Fires-by-occupancy-or-property-type. Accessed June 25, 2020.

NFPA Fire Investigations Department. February 1980. *Wood Shingle Roofs Fuel Conflagration.* NFPA Fire Command. Boston, MA: NFPA.

Reuters. May 7, 2013. "Ammonium Nitrate Stores Exploded at Texas Plant: State Agency." https://www.reuters.com/article/us-usa-explosion-texas-idUSBRE9460GP20130507. Accessed April 23, 2020.

U.S. Fire Administration (USFA). n.d. "Campus Fire Fatalities in Residential Buildings (2000–2015)." https://www.usfa.fema.gov/downloads/pdf/publications/campus_fire_fatalities_report.pdf. Accessed March 19, 2020.

U.S. Fire Administration (USFA). 1993. "The World Trade Center Bombing: Report and Analysis." *Technical Report Series* [USFA-TR-076]. Emmitsburg, MD: USFA; originally published in *Fire Engineering* in December 1993.

U.S. Fire Administration (USFA). 2016. "Nursing Home Fires (2012–2014)." National Fire Incident Reporting System (NFIRS) 5.0. NFIRS Data Snapshot. Last reviewed July 18, 2016. https://www.usfa.fema.gov/data/statistics/reports/snapshot_nursing_home.html. Accessed April 15, 2020.

U.S. Fire Administration (USFA). 2020. "National Fire Data Center." https://www.usfa.fema.gov/data/statistics/. Accessed January 24, 2020.

CHAPTER 12

High-Rise Buildings

LEARNING OBJECTIVES

- Describe the magnitude of the high-rise fire problem in terms of number of fires, number of fire fatalities, and property loss.
- Define a high-rise building from a fire department perspective.
- Explain elevator recall and the responsibilities of lobby control.
- Explain why using an elevator during a fire emergency is a calculated risk and list the rules for elevator safety.
- Assume the role of incident commander and apply the seven rules of elevator safety to the fire on the fiftieth floor of the 68-story Trump Tower.
- Explain the first rule of elevator safety when considering occupant rescues.
- Explain the limitations of aerial ladders at a high-rise building fire.
- Compare the advantages, disadvantages, and limitations of aerial ladders, elevators, and stairways when rescuing occupants of a high-rise building.
- Apply the seven rules of elevator safety.
- Explain the duties of ground support and calculate the number of fire fighters needed to staff ground support for a fire on the 30th floor where the elevators are unsafe to use.
- Identify negative and positive aspects of using helicopters at a high-rise building fire.
- Compute the approximate pump discharge pressure needed to supply a hose stream operating on the 20th floor.

- Explain when the command post should be located in the lobby, in a fire command room, outside the building, or at a remote location.
- List four occupancy types that are commonly found in high-rise buildings.
- Compare old-style tower construction to modern planar-style high-rise buildings.
- Define and explain the advantages of a smoke-proof tower.
- Discuss methods that can be used to maximize the limited stairway capacity in a high-rise building and how fire department operations can hinder evacuation.
- Describe an emergency voice/alarm communications system (EVACS), and identify the advantages and disadvantages as compared with other notification systems.
- Explain how wind, stack effect, and heat affect smoke movement in a high-rise building.
- Enumerate measures taken to counteract the effects of a wind-driven fire.
- Discuss the importance of extinguishment to life safety in a high-rise fire.
- Explain what is meant by a "wrap-around" fire.
- Describe the hazards involved in ventilating upper floors by removing window glass and how to protect fire fighters and civilians on the street below.
- List pathways for floor-to-floor fire extension in a high-rise building.
- Define "lead time" and extrapolate the estimated lead time for a fire on the 40th floor of a high-rise building when elevators are unavailable.
- List practical forms of non-radio communication that can be used at a high-rise building fire.
- Discuss when interior and exterior staging would be used at a high-rise fire.
- List the duties of the interior staging officer and lobby control.
- Compare the designed "accidental aircraft" impact at the World Trade Center to the actual impact of the aircraft on September 11, 2001.
- Describe the conditions leading to structural collapse at the World Trade Center.
- Describe how a "convergence cluster" could affect search and rescue operations at a high-rise building.
- Using a simulated high-rise fire scenario, size up the incident, develop an incident action plan, assign units to carry out the plan, and develop a National Incident Management System organizational chart.

Anadolu Agency/Getty Images.

Case Study

In May of 2018 a fire on the 50th floor of the 68-story* Trump Tower in New York City claimed one life and caused multiple injuries. The fire originated in an apartment on the 50th floor. The fire alarm sounded an alarm for the 50th floor with no verification of a fire. Because the alarm was received via a fire detection system on the 50th floor with no other call verifying a fire, a single truck company was dispatched to the alarm.

The building engineer responded to the 50th floor. The first FDNY fire company was unable to communicate with the building engineer on arrival. Unable to obtain a status report, the first-arriving company proceeded to the 50th floor where they encountered heavy smoke conditions. A full alarm response was ordered. Another ladder company arrived and proceeded to the 50th floor. The two companies were able to close the door to the apartment on fire to confine the fire.

The windows in the 50th floor apartment failed and the fire caused the windows on the 51st floor to fail, allowing fire to spread to the 51st floor. The high heat in the apartment of origin made it difficult to conduct search and rescue efforts. The occupant of the 50th floor apartment of origin was the only fatality.

A 2½-in. (64-mm) hose line was deployed into the apartment. Three engine companies and two truck companies were staged on the 48th floor, with emergency medical services (EMS) staged on the 47th floor. The total four-alarm response included 16 engine companies and 8 truck companies, as well as rescue and squad companies. The fire was brought under control in approximately 1 hour.

1. What effect does a delay in dispatching a full alarm assignment have on the potential loss of life and the extent of fire involvement?
2. What role does compartmentation play in saving lives? In confining the fire?
3. Why is it important to have companies in staging?

---

*There seems to be some dispute regarding the number of floors in the Trump Tower. Most accounts of the fire list the building as 68 stories high. However, reliable sources state the building actually has 58 stories.

Introduction

The basic tactics and strategic objectives used in high-rise firefighting are essentially the same as those that apply to any other structure fire, but with special considerations because of the height of the building. The differences are great enough to deserve this separate chapter.

A recent high-rise fire report from the NFPA indicates that during the period 2009–2013, ". . . U.S. fire departments responded to an estimated average of 14,500 reported structure fires in high-rise buildings per year. These fires caused an average of 40 civilian deaths, 520 civilian injuries, and $154 million in direct property damage per year" (Ahrens, 2016).

The World Trade Center bombings in 1993 and aerial attack in 2001 are not typical high-rise fire situations, but this is the nature of the high-rise problem. It is not unusual for there to be intervals of several years between major loss-of-life or large property-loss fires in high-rise buildings in the United States; yet, the potential always exists. The overall numbers of fires, civilian fire deaths, and property loss appear to be trending downward. This is probably due to more high-rise buildings being sprinkler protected. In the mid-1970s, most jurisdictions required new high-rise buildings to be sprinkler protected, but few required existing buildings such as Trump Tower to be retrofitted. The World Trade Center was fully sprinkler protected, but the system was severely damaged on impact and the jet fuel provided a fire load beyond the design capacity of the system (NIST, 2005). There is absolutely no doubt that a sprinkler-protected building is a safer building for fire fighters and occupants; however, fire departments should not place total reliance on the sprinkler system. Another factor contributing to fewer fire deaths is the common use of pressurized stairways beginning in the 1970s.

High-rise fires represent an extraordinary challenge to fire departments and are some of the most challenging incidents a fire department will encounter. High-rise buildings can hold thousands of people well above the reach of fire department aerial devices, and the chance of rescuing victims from the exterior is near zero once the fire is above the operational reach of ladders or elevated platforms.

The National Institute of Standards and Technology (NIST) issued its "Report on Residential Fireground Field Experiments" (NIST, 2010) in 2010 by conducting experiments in a two-story house to verify the staffing requirements specified in NFPA 1710, *Standard for the Organization and Deployment of Fire Suppression Operations, Emergency Medical Operations, and Special Operations to the Public by Career Fire Departments*, for a low-hazard occupancy (NFPA 1710, 2020). The results of that study have serious safety implications. These implications are discussed in Chapter 5, *Fire Fighter Safety*. One of the results of the residential study was a recommendation to determine the staffing required for a successful outcome in a high-hazard building.

In 2012, NIST conducted field experiments in a 13-story high-rise with a simulated working fire on the 10th floor in an effort to determine the effect of company-level staffing on successful outcomes during a high-rise building fire. In the NIST "Report on High-Rise Fireground Field Experiments", crew sizes ranging from 3 to 6 were compared (Averill et al., 2013). Deployment based on the number of companies responding on the initial and subsequent alarms was also evaluated. Other variables included sprinkler protection (sprinkler protected versus nonsprinkler protected) and availability of elevators for fire department use (elevators available for use versus unsafe elevators). Each field experiment was conducted multiple times to verify accuracy. This important report is available at the NIST website.

Case histories provide an excellent learning tool. By studying past fires, it is possible to gain experience that may not be gained in any other way. Outside of New York, Los Angeles, Chicago, and several other large cities, few chief officers experience enough working high-rise fires to gain confidence in managing these challenging incidents. Communities with one or two high-rise buildings may never experience a serious high-rise fire, yet the fire department is expected to be prepared if a high-rise fire occurs. For this reason, several high-rise fires are discussed at the end of this chapter. Studying these and other high-rise fire case studies is strongly recommended.

Realistic training sessions applying high-rise standard operating procedures (SOPs), tactics, and tasks are essential to maintain the skills necessary to successfully combat high-rise fires. Many fire companies from the Washington, DC, area departments were involved in the NIST high-rise fire-ground field experiments. The study notes how these departments used the experiment scenarios to improve fire-ground skills and cooperation between the many departments likely to work together at a major high-rise fire in the region.

Training can be as simple as testing a single function such as the ability to apply the National Incident Management System (NIMS) or a tabletop exercise using scenarios such as the RGB high-rise fire scenario presented in the "Suggested Activities" section at the end of this chapter. Full field exercises, where standpipe equipment is actually moved to the fire floor and hose lines are connected to the standpipe and advanced to a simulated fire area, combined with search and rescue activities, are extremely difficult to develop and manage. Maintaining adequate staffing for real fire responses and taking precautions not to cause damage to the structure used for the scenario can be a significant challenge. The NIST high-rise fire-ground field experiments explain many ingenious ways to provide realistic training with a low probability of injury or property damage.

Developing and Revising High-Rise Standard Operating Procedures

Preincident planning and code enforcement can reduce the scope of the high-rise problem. However, special tactics will be needed to control fire forces working in different areas within these large structures while providing the necessary logistical support. Using NIMS and developing high-rise SOPs can do much to ensure successful operations.

A difference of opinion exists regarding the definition of a high-rise building. Most codes define a high-rise building in terms of height and/or number of stories. Fire departments tend to think of a high-rise building as being beyond the reach of the available aerial fire equipment. Because the focus of this text is on fire-ground tactics, the fire department definition is most appropriate. However, do not forget the obvious: an eight-story building will not present the same challenges as an 80-story high-rise building. Logistics and access problems increase with height. The more floors that are located above the fire, the more people are likely to need fire department assistance and the more there is to burn.

High-rise buildings were once found exclusively in larger cities, but today they are commonly found in small and midsized communities as well. In most cases, high-rise buildings in these smaller communities are newer, lower in height, and protected by automatic sprinkler systems. Even if your department does not respond to a high-rise building at present, if urban sprawl continues as expected, it probably will in the future. If you have a mutual aid contract with a jurisdiction that contains high-rise buildings, you will likely be called upon to help in the event of a working high-rise fire. Several larger cities that once contained just a few notable high-rise structures have experienced an incredible growth in high-rise construction. Given the special problems that these buildings present, each department that could reasonably be expected to respond to a high-rise fire should have a high-rise SOP and train accordingly.

Fire Fighter Safety

The risk to fire fighters and occupants increases in proportion to the height of the building and the height of the fire above grade level. Once fire fighters are operating above the reach of aerial devices, the only viable means of egress is the interior stairs; the extra protection afforded by laddering the building is not possible.

Good tactics and fire fighter safety cannot be separated. The tactics that are explained throughout this chapter improve fire fighter safety. All of the standard safety considerations apply to high-rise firefighting. (See Chapter 5, *Fire Fighter Safety*.) However, additional measures need to be taken during a high-rise operation.

Occupants should not use elevators to escape a fire except under special circumstances, and even then, only under fire department supervision. Modern high-rise building elevators have fire department controls. When the alarm system is activated, elevators return to the ground floor and remain there "locked out" for fire department use. An elevator key is required to unlock the elevators. Older buildings may have fire department controls but may not have the feature that automatically returns elevators to the ground level.

A responsibility of **lobby control** is to control, operate, and account for all elevators. Some fire service professionals say that elevators should never be used under fire conditions or suspected fire conditions until their safety can be verified from the fire floor. This may not be practical where fire companies respond to alarms in high-rises several times each day. Requiring fire fighters to ascend 50 flights of stairs to check an odor of smoke is not a productive use of resources. A more reasonable approach is to develop procedures and conduct training to reinforce the safe use of elevators.

Fire Fighter Use of Elevators

A critical variable in high-rise fire operations is the availability of reliable elevators. If fire fighters can safely use the elevators, fire-ground logistics are dramatically improved. When the fire is located many floors above ground level, there is a strong inclination to use the elevators. However, elevators often stall or act erratically under fire conditions. Fire fighters who are trapped in a stalled elevator become part of the problem, as other fire fighters are then needed to rescue the rescuers. Therefore, the department SOPs should address the safe use of elevators, including circumstances when it is unsafe for fire fighters to use them. These procedures should include alternative measures for transporting needed equipment to the fire floor when elevators cannot be safely used.

Fire department service controls and other elevator safety features will vary by age and elevator manufacturer. Building preplans should include information and instructions regarding the safe use of the elevators.

One of the dangers in using elevators is that doors may open on the fire floor, exposing fire fighters on the elevator to smoke and heat before they are in position with hose lines to attack the fire. Once elevator doors open on the fire floor, sensing devices (e.g., electric eyes) may prevent the doors from closing, thus trapping the fire fighters. This is a potentially deadly mistake that should be avoided at all costs. If fire fighters are caught in this situation, they may be able to use an override switch to close the elevator doors. If not, they may be forced to exit the elevator in an attempt to escape the fire floor via the stairway. Similarly, elevators can stall in the shaft at or above the fire floor, also trapping fire fighters in the elevator car. Fire fighters should never, under any circumstances, use an elevator if there is a chance the elevator will travel to or above the fire floor. This is another possibly fatal mistake.

Elevators are equipped with redundant safety systems to prevent them from falling. In many cases, an elevator car has to be nearly destroyed before it will fall. The elevator shaft is a fire-protected enclosure, but a fire of sufficient intensity can invade the shaft, and smoke and toxic gases will most certainly enter elevators that are stalled above the fire floor. This could prove deadly for fire fighters, equipped with personal protective clothing, including self-contained breathing apparatus (SCBA). Civilians without protective gear have even less chance of survival when trapped in an elevator on or above the fire floor. Maintenance and security people should never be taken into the elevator until it has been verified that the elevator is completely safe.

The first rule of elevator safety is to avoid the use of elevators unless they will substantially improve operations. Fires on lower floors do not warrant the use of an elevator unless someone on the fire floor can verify its safe use. The first-arriving companies should use the stairways for fires on lower levels. Some departments prohibit the use of elevators for fires on the seventh floor or below. Department SOPs and preincident plans must address this issue specifically. Once the safety of an elevator has been established, fire fighters can use it under the close supervision of lobby control. Two other factors to be considered are fire separations between the elevator bank and fire location and whether the elevator goes to or above the fire floor.

High-zone/low-zone (split-bank) elevators are common in taller buildings. Some split-bank elevators are divided into several zones. For example, a three-zone bank of elevators would have one bank serving lower floors only, another bank serving only the middle floors, and a third bank providing access to upper floors only. This is important information that should be included in the preincident plan. For example, for a confirmed fire on the 10th floor, it is fairly safe to use elevators that terminate at the eighth floor **FIGURE 12-1**.

When there is a fire separation between the elevators and the fire area, it may be possible to safely use the elevators in a lower building zone and then travel horizontally to the fire via the stairs. This approach is much easier than ascending the stairs from the ground floor. Large high-rise buildings, especially hospitals, often have building zones. Fire walls and fire doors create a horizontal barrier to smoke and heat from the

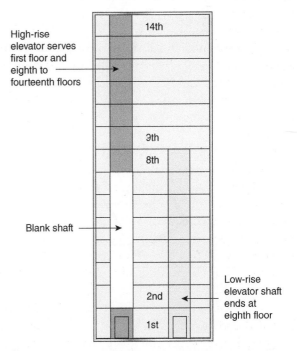

FIGURE 12-1 Split-bank elevators with upper and lower zones.
© Jones & Bartlett Learning.

fire. Hospital patients are usually moved horizontally to another zone, rather than taken down to the ground floor or evacuated from the building. If another building zone is safe for the patients, it should certainly be safe for fire fighters.

Other occupancies also use horizontal separation as a fire protection strategy. The NFPA report on the 1981 Las Vegas, Nevada, Hilton Fire includes a plan view of a floor divided into three horizontal zones (Demers, 1982). These zones provide a possible area of safe refuge for guests and a staging area for fire fighters. This hotel is actually three separate buildings built at three different times **FIGURE 12-2**.

When separate low-zone/high-zone elevators are unavailable or the building is not adequately zoned horizontally and the fire is reported to be on upper floors of the building, extreme caution should be used if the incident commander (IC) decides to use elevators to transport fire fighters.

The following list describes several rules that should be observed in taking the calculated risk of using an elevator when responding to a reported fire area, including alarms from an alarm system. These rules for elevator safety should be considered when writing department high-rise SOPs and rigidly enforced during fire alarm response.

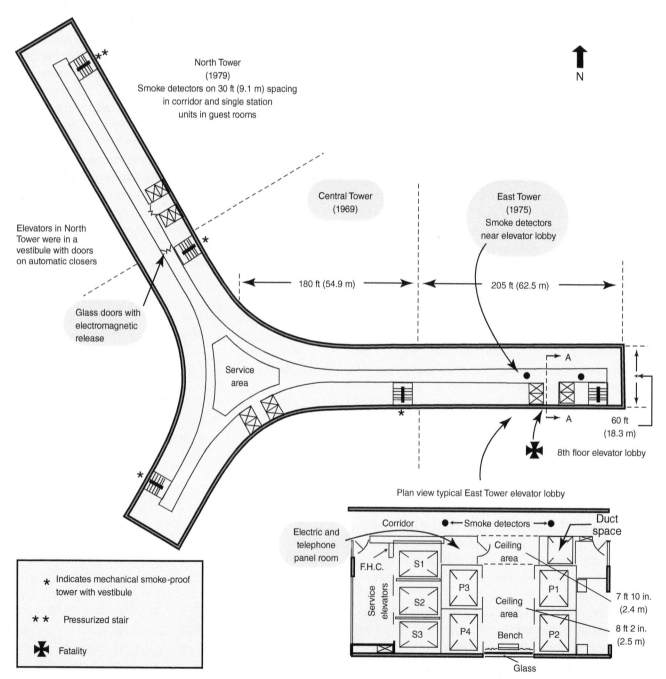

FIGURE 12-2 Illustration of the Las Vegas Hilton floor plan.
© Jones & Bartlett Learning.

1. *Do not use an elevator for a fire on a lower floor level.* Elevators should not be used for fires on lower floors in the building or if there is any doubt about the safety of the elevator.
2. *Never take an elevator directly to the fire floor or above.* This is the cardinal rule of elevator safety and must be rigidly enforced. Department SOPs should state that the elevator should be taken two floors or more below the fire, and then fire fighters should walk up the stairways to the fire floor. This rule also applies to split-bank elevators that do not travel to the fire floor. If split-bank elevators are available, use an elevator that does not travel to or above the fire floor. Again, exit the elevator at least two floors below the fire and use the stairway to reach the fire floor.
3. *Place the elevators under independent (fire department) control.* Keys should be made available to the fire department so that elevators can be placed under independent control. Newer elevators cannot be operated without a key once the fire alarm has sounded. This independent control greatly increases fire fighter and civilian safety, as the elevator will not be responding to calls from occupants.
4. *Control all elevator cars in multiple hoistways.* Doing so not only provides the fire department access to upper floors, but also prevents erratic response to other calls within the building. Controlling elevators is the responsibility of lobby control.
5. *Never overcrowd elevators.* This principle is doubly important for fire fighters, not only because it is unsafe to exceed the weight limit on an elevator, but also because fire fighters may need space to don SCBA or to use tools to force their way out of the elevator.
6. *Wear personal protective clothing, including SCBA, and bring forcible entry tools.* Air supply should be conserved by not donning the SCBA face piece until approaching the fire area or when conditions indicate there is a need to be on air. *Never forget that using an elevator is taking a calculated risk.* Even when every reasonable precaution is taken, what lies at the top of the ascent is unknown. Forcible entry tools must be available to escape the elevator if necessary. Understanding elevator door operations can be extremely valuable if personnel become trapped in an elevator that stalls. If trying to escape a stalled elevator, *always* activate the emergency stop switch. Opening an emergency escape door or ceiling panel should also stop the elevator, as the interlocks are designed to prevent accidental movement when people are trying to escape. However, the emergency stop switch should be activated.
7. *Send equipment rather than fire fighters on elevators.* Many times elevators that are considered unsafe for fire fighters can be used to transport tools and equipment to the interior staging area (usually two or more floors below the fire floor). Fire fighters can then safely ascend the stairs without the burden of heavy tools and equipment. It is not always possible to send an unstaffed elevator to a desired floor. When the fire alarm is activated, it may be necessary to mount the elevator and hold a floor button or elevator key before the elevator will move. In this case, equipment cannot be sent above without a fire fighter. Freight elevators are usually less safe than passenger elevators, but they may be suitable for sending equipment without placing fire fighters in the elevator car.

When ascending in an elevator, it is good practice to stop periodically, possibly every five floors, to check for smoke. This is accomplished by opening the top elevator escape panel, when the elevator is so equipped, and using a flashlight to see whether smoke is present in the shaft. This is also a test of the elevator's ability to stop on demand. It is further recommended that fire fighters stop the elevator three or more floors below the fire floor, step out into the hallway, and check the general floor arrangement to get a feel for the building layout. By examining an uninvolved floor, fire fighters can become familiar with landmarks and can quickly identify the location of secondary exits, floor configurations, and potential ventilation locations. The first floor, mezzanine floors, and second floors are generally not typical floor layouts; therefore, it is better to check an upper floor that will likely (but not always) be similar to the fire floor. Many high-rise office buildings have floor layouts that resemble a maze, increasing the probability of a fire fighter becoming disoriented and lost. Therefore, fire fighters should constantly ask themselves, "Where am I in relation to the stairways?" Crews working away from a hose line should use guide ropes.

The fire officer must always conduct a careful risk-versus-benefit analysis before placing fire fighters in an elevator. Furthermore, there is seldom justification for placing civilians in elevators before the fire is completely extinguished and the smoke is ventilated from the building. SOPs regarding the use of elevators should be kept current, and all fire fighters should be thoroughly familiar with them. These procedures should consider the height of buildings in your

FIGURE 12-3 The north face of the World Trade Center Tower 2 viewed from the corner of West Broadway and Spring Street in Soho, New York City, on September 11, 2001.
Laperruque/Alamy Stock Photo.

jurisdiction, automatic fire suppression equipment, type of occupancy, and the safety considerations outlined in this text.

Code officials in some countries are now including elevators in the civilian evacuation plan. In the United States, the use of elevators by the occupants of high-rise buildings is being examined by NIST and others. At the World Trade Center attack on September 11, 2001 **FIGURE 12-3**, many building occupants successfully escaped using the elevators, although elevators were not part of the evacuation plan. It is estimated that 3000 people safely evacuated the South Tower using the elevators in the 16 minutes between attacks. Of particular concern is the evacuation of handicapped occupants. To allow elevators to be used for evacuation, it is recommended that fire-resistive construction be improved in elevator shafts and lobbies.

Ground Support

Moving equipment via a stairway relay function was formerly named "stairway support" but is now properly termed "ground support." The easiest and quickest way to move cylinders, nozzles, hose, first aid supplies, forcible entry tools, and other materials to the interior staging area in a high-rise building is using the elevator, but what if the elevator is not safe to use? As previously discussed, many times elevators are out of service or unsafe to use. Moving supplies and staff up 10, 20, 30, or more stories is an arduous task. If it is not properly managed, no one will reach the fire floor with the physical stamina necessary to fight the fire. Imagine returning to the apparatus for a fresh air tank from the top floor of the 1454-ft (443-m), 110-story Willis Tower (formerly the Sears Tower) in Chicago. When elevators cannot be used, gaining access to the upper floors of a building will take significant time. Getting fire fighters and equipment to an interior staging area will reduce the time required to support a high-rise fire operation.

Ground support is a high-rise support unit used to move supplies to the interior staging area when using elevators is deemed to be unsafe. Fire fighters assigned to ground support ascend two stories with air cylinders and other equipment, where the next fire fighter picks up the equipment and relays it two additional floors. After moving the equipment up two stories, fire fighters descend two stories empty handed, providing a rest period. During extended operations involving many companies in rescue and suppression activities, it may be necessary to place fire fighters on every floor or possibly recycle air cylinders to grade level for refill.

> **NOTE**
>
> Interior staging is set up in a safe area two or more floors below the fire floor when it is impractical for fire fighters to go outside at ground level to change SCBA cylinders and for rest and recuperation (rehab).

Moving equipment up through the building when the elevators are out of service is a mammoth undertaking; ground support provides a reasonable but slower alternative to using elevators. During the NIST high-rise fire-ground field experiments, it was found that moving the required equipment to the eighth-floor staging area via ground support took 10 minutes longer than moving the same equipment using elevators. Obviously, setting up ground support and moving equipment to a higher floor would require more time and staffing. However, once each person assigned to ground support is in place, they would only move up and down two stories.

Depending on conditions in the stairway, it may be possible to allow members who are assigned to ground support to work without their SCBA and turnout gear. This will preserve their energy and allow them to transport more equipment in less time. For example, experience indicates that wearing rubber boots while ascending the stairs is particularly fatiguing. However, it is essential to always err on the side of caution.

If there is any chance that members assigned to ground support will encounter smoke or other hazardous conditions, they should be required to wear appropriate protective equipment. Even when personal protective equipment is not being used, it should be immediately available. Ground support should be one of the first assignments given when the fire is on the upper floors in a building without elevator service. Ground support should be explained in the department's high-rise SOP.

Life Safety

As search and rescue teams proceed with a systematic search, they must provide status reports to their supervisor and mark areas that have been searched. A simple marking system involves placing a chalk mark "X" on doors to rooms that have been searched and/or indicating that the whole floor has been searched by marking hallway doors or walls opposite the elevator. Sidewalk chalk is preferred as it is easier to handle while wearing gloves. Door hangers or other marking devices are commercially available for this purpose as well. A method of indicating areas searched should be part of the department's high-rise SOP.

Many high-rise buildings lock the doors from the stairs to the hallway, further complicating search and rescue efforts. This was a major factor in the loss of life at the Cook County Administration Building Fire in Chicago, Illinois, where occupants encountered fire on the 12th floor while attempting to escape. Fleeing occupants then attempted to reenter floors above the fire where doors leading to the hallway were locked on the stairway side. Forcible entry tools should be carried by rescue and extinguishment teams to force entry when necessary. When selecting the tools to carry, remember that power saws might not operate because of heavy smoke conditions. Furthermore, using gasoline-powered equipment inside of a building can create other hazards if the area is not adequately ventilated.

Once doors have been opened, it is important to prevent them from relocking behind fire fighters entering the floor. Some departments use a piece of rubber with two holes that covers the doorknob and lock to prevent the door from relocking. For most door types, this method works well. During preincident planning, examine the doors to determine which method would work best for the type(s) of doors in that building. Whatever method you decide to use, be careful not to leave doors propped open, as doing so will most certainly allow smoke to move from the stairs into the hallway or from the fire floor into the stairs. Open doors to hallways also reduce the effectiveness of stairway positive-pressure ventilation and can provide air to a ventilation-limited fire.

The primary search should also include a search of all elevators. Elevators should be brought down to ground level and checked by lobby control. If elevators are stalled or otherwise located above the ground floor, they must be checked for victims. It is essential that all elevators be accounted for and checked for occupants.

Rescuing and Evacuating Occupants

Helicopter Rescues

There have been occasions when helicopters were successfully used to rescue occupants during high-rise fires. However, helicopter rescues can be extremely dangerous and, in most cases, unnecessary. Few instances warrant the use of a helicopter in removing occupants from a roof, and many roofs make poor helicopter landing zones. A few fire departments have developed programs that use helicopters to place fire fighters on roofs, sometimes by having them rappel from the helicopter to the roof. Fire fighters placed on the roof can calm the occupants and control or limit the fire threat, and these fire fighters can descend to areas that are inaccessible from below.

These programs have merit but require constant training on the part of the rappel team and helicopter crews. At last count, New York City had 32 buildings over 600 ft (183 m) in height. New York and other very large cities can justify the training and associated expenses related to a helicopter rescue program. However, the need or justification for such a program is doubtful for a department with a few well-protected high-rise buildings within their response area. It is unlikely that members of such a department would have the expertise or equipment needed to safely operate from helicopters.

Helicopters can sometimes be used for reconnaissance. An aerial view can provide information that is not available from the interior or the exterior at ground level. At the World Trade Center on September 11, 2001, police in helicopters could see signs of an impending structural collapse. Unfortunately, they were unable to communicate this message to the fire department.

Unfortunately, helicopter operations above a burning building can actually create additional risks owing to the thermal updraft. Helicopters flying near a burning building can create high winds that negatively affect operations. Smoke can reduce visibility, and heat generated by the fire can affect lift needed to safely operate the aircraft. Drone technology, on the other hand, can provide reconnaissance data and not place the operator at risk. A drone can be operated at a fraction of the operating cost of a helicopter and can be airborne in minutes.

Remember, as with any tactic, before deciding to use helicopters for reconnaissance, for assisting fire fighters in gaining access, or for rescuing occupants, always conduct a risk-versus-benefit analysis. Is the risk to occupants, fire fighters, and the helicopter crew warranted? Are other less risky options available?

Partial or Sequential Evacuation

A decision to utilize a partial or sequential evacuation can be made in advance of a fire and incorporated into the alarm system. Alternately, the IC may make this decision at the time of the fire. When the fire is small or smoke has entered the stairway, the IC may decide that it would be best to leave occupants in their rooms rather than attempting a complete evacuation. A defend-in-place tactic places a great burden on the IC, as a successful outcome depends on the fire being promptly extinguished. However, many times, occupants are placed in greater danger by attempting to evacuate a building through smoke, using resources that could have been assigned to extinguishment.

If the fire is not quickly controlled, the people who remain in the building will be in great danger as a result of the nonevacuation decision. Much of this decision-making process has to do with fixed fire protection features that are designed into the building. For example, there is less risk associated with a nonevacuation decision if the building is sprinkler protected.

Many departments permit the use of alarm systems that advise only the occupants in affected areas to evacuate the building. Rather than sounding a general alarm throughout the building, the evacuation alarm may sound only on the fire floor and on one floor above and one floor below. Occupants on the other floors may receive a prealarm notification but be advised to remain on their floor awaiting further notification.

This evacuation tactic reduces stairway traffic but also leaves occupants inside the burning building. Use of a partial or sequential evacuation system requires a sophisticated fire alarm system and/or manual control. The IC should have preincident plan information about the operation of the system readily available at the command post.

Relying on building management or security to notify occupants of the need to evacuate is generally a mistake. Maintenance or security people have a propensity to investigate alarms first, thus delaying notification of the fire department. This has been the case in many deadly high-rise fires, such as the First Interstate Bank Building in Los Angeles, California, and One Meridian Plaza in Philadelphia, Pennsylvania. When a fire is discovered, many times in-house employees are not properly trained in how to initiate a safe, orderly evacuation. The answer lies in requiring direct and immediate fire department notification, regardless of the type of internal alarm system, fire protection, or method of evacuation.

Any time a decision is made to leave occupants inside a burning building, the IC is taking a calculated risk with the lives of the remaining occupants. Preincident plan information related to construction and fire protection systems, along with accurate, timely, and continuous status reports, will provide the IC with the information necessary to make the correct decision.

Emergency Voice/Alarm Communications System

Many high-rise buildings are equipped with an emergency voice/alarm communications system (EVACS). The EVACS uses recorded messages to notify occupants of a fire and provides specific directions for reaching places of safe refuge within the building. The EVACS sometimes directs people above the fire to higher floors within the structure, avoiding the need to pass the fire floor. People on the fire floor and the floor below are directed to areas two or three floors below the fire or to evacuate to the outside. When occupants are sent to other floors within the building, the occupants on the receiving floors are also notified of the evacuation. The EVACS notifies people on elevators that the elevator is responding to the ground floor, where they are to exit. By directing occupants away from the immediate fire area soon after the fire is detected, the EVACS helps to solve many high-rise evacuation problems.

Because a fire scenario presents many uncontrolled variables, caution is in order when using any automated system such as EVACS. Stairways that would be perfectly safe if the fire were to occur in one floor area might not be safe in another circumstance. All evacuation variables should be thought out in advance, and a manual backup system must be in place. The EVACS will generally have manual overrides that allow fire fighters or building management to direct the evacuation. Manual directions could include which occupants are to evacuate and which stairway(s) to use. Most of these systems can notify occupants on specific floors or all of the building's occupants. Sequential evacuation may be best to reduce stairway congestion. Stairways should be identified

> ### Incident Summary
>
> #### Cook County Administration Building Fire
>
> On October 17, 2003, at 5:00 PM a fire on the 12th floor of the 37-story Cook County Administration Building in Chicago, Illinois, resulted in six civilian fatalities. The fire was believed to have originated in a closet within a 2629-ft^2 (244-m^2) suite of offices on the 12th floor. Actual fire damage was contained to the office suite due to fire-resistive building features, but smoke and fire migrated to the entire 12th floor. The fire self-vented through eight exterior windows, creating the potential for floor-to-floor fire extension via the exterior.
>
> Building occupants were notified to evacuate the building via the stairways. However, some occupants used the elevators to escape while others used the east and west stairways to evacuate. The fire department advanced a hose line into the 12th floor via the east stairway. When the door to the 12th floor was opened to advance the hose line into the fire area, smoke levels in the east stairway increased dramatically. As far as can be determined, there were 13 occupants above the 12th floor in the east stairway when the fire department began attacking the fire. Doors from the hallways on each floor were locked on the stairway side, trapping these occupants in the stairway above the fire. Seven of the occupants in the stairway found a door that did not completely latch on the 27th floor and were able to leave the east stairway and escape. Six others perished in the east stairway. Fire fighters were faced with an intense fire that they were unable to extinguish from their hallway position. Elevated master streams were used to knock down the fire from the exterior. Interior hose streams were then redeployed to achieve final extinguishment.

by marking both sides of the stairway door using an alphabetic or other easily understood designation. Reliance on the EVACS should be limited to compartmentalized buildings with full sprinkler system coverage. If the sprinkler system fails to control the fire and fire forces are unable to quickly extinguish or confine the fire, all occupants above the fire floor should be evacuated. Furthermore, the IC and lobby control officer should have a thorough knowledge of the EVACS system.

Extinguishment

What makes high-rise firefighting different? In the lower portions of the building, the main difference is the exposure of many floors above the fire to vertical fire spread. Above the eighth floor, most exterior defensive fire control tools are no longer effective. Elevated streams can sometimes reach the fire floor, but the angle of deflection inside the window diminishes with each floor above the maximum height of the appliance. To improve fire stream penetration, it is necessary to move the nozzle farther from the building, thus reducing the angle of deflection. In most cases, the width of the street is the limiting factor in determining how far away the appliance can be placed from the building. If the width of the street is not a limiting factor, the effective reach of the nozzle will determine the maximum distance the nozzle can be placed from the building. At the Cook County Administration Building Fire an elevated master stream was used to control a fire on the 12th floor; however, a delay occurred while the apparatus was repositioned (Independent Commission, 2003). In operating above the eighth floor, it is unlikely that ground-based appliances will deliver effective stream penetration to the fire area **FIGURE 12-4**. Occupants are also beyond the reach of platforms, aerials, and ground ladders; therefore, attempting an exterior rescue would be extremely dangerous for both fire fighters and occupants. In fact, the building's stairways are the best means of egress, even when floors are within the reach of aerials. For these reasons, it is extremely important that fire forces prevent fire extension into the stairways, because stairways are the safest and most effective egress routes.

Generally, the pressure at the standpipe discharge will be lower than the pump discharge pressure used to supply hose streams connected directly to a pumper. Standpipe operations and equipment are discussed in Chapter 7, *Fire Protection Systems*, including pressure- or flow-reducing valves. Departments compensate for the lower pressure by using smooth-bore or low-pressure nozzles. Some departments have purchased hose that has less friction loss or use 2-in. (51-mm) or 2½-in. (64-mm) hose for standpipe operations. Any method of reducing the friction loss in the hose will increase the nozzle pressure and flow.

Taller building standpipes could be equipped with pressure- or flow-reducing valves. If a building is not

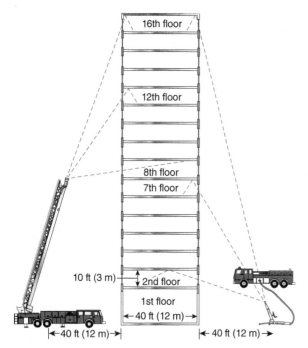

FIGURE 12-4 Stream deflection.
© Jones & Bartlett Learning.

equipped with pressure-reducing valves, the pressure on the lower floors of the building will be higher than the pressure on the top floors. It is also possible to increase the pressure in the standpipe system by pumping into the fire department connection. Pumping into the fire department connection may be an effective means of boosting the pressure and increasing the water supply. Very high buildings have complex water supply systems where pumping into the fire department connection may not increase pressure at the standpipe outlet. In these buildings, fire department pumpers alone cannot supply sufficient pressure for firefighting on upper floors, and the building's internal fire pumps must provide primary pressure. Although the ceiling height of individual floors will vary in high-rise buildings, a fair estimate is allowing 20 ft (6 m) for the first floor and 10 ft (3 m) for each additional floor.

The loss of pressure due to gravity (elevation) is 0.434 pounds per square inch (psi) per foot (3 kilopascals [kPa] per 0.305 m). The pressure loss for a hose line operating at the top floor of a 40-story building (39 floors above grade) with 10-ft (3-m) floors would be calculated as follows:

> First floor   = 20 ft (6 m)
> 38 floors × 10 ft per floor   = 380 ft (116 m)
> Total elevation   = 400 ft (122 m)
> 400 ft × 0.434 psi/ft = 174 psi pressure loss
> (122 m × 3 kPa/0.305 m = 1200 kPa loss)

If the floor height is changed to 12 ft, then the pressure loss would be calculated as follows:

> First floor   = 20 ft (6 m)
> 38 floors × 12 ft/floor   = 456 ft (139 m)
> Total elevation   = 476 ft (145 m)
> 476 ft × 0.434 psi/ft = 207 psi pressure loss
> (145 m × 3 kPa/0.305 m = 1426 kPa loss)

In addition to the pressure loss due to elevation, friction loss in the standpipe system and the hose line, as well as nozzle pressure, must be added. Friction loss will vary depending on the type of hose used, and nozzle pressure will vary depending on the type of nozzle. It is obvious that standard fire department pumpers alone could not support hose streams operating on the upper floors of an ultra-high-rise building.

Pumping into the fire department standpipe connections can assist internal fire pumps and is a good practice. If provisions are not made to provide an external backup supply and the internal pumps fail, the only protection will be the fire-resistive nature of the structure (fire enclosures). In high-rise building fires, when the fire is above the reach of aerial ladders and the building is not protected by an automatic sprinkler system, the fire can be expected to burn out all floors above the fire unless the fire can be brought under control by an offensive attack. Contingency plans for supplying the standpipe and/or reservoirs in case of breakdown, developed in cooperation with building management, could prevent a disaster. These contingency plans should be included in the preincident plan.

Command Post Location

Fire department SOPs often list the lobby as the preferred command post location. Many times, this is true, but on occasion the lobby is a poor choice for a command post. When fire or the products of combustion threaten the lobby, the command post should be located elsewhere. It is often stated that the larger the incident, the farther away the command post should be. (This is discussed in Chapter 1, *Organizing, Coordinating, and Commanding Emergency Incidents*.) The idea is to isolate the IC from disruption so that the command process can be carried out efficiently. When large numbers of people are attempting to exit through the lobby and fire fighters are gathering there in an effort to ascend toward the fire, the lobby is a poor location for the command post. The September 11 World Trade Center Fire provides an example of the lobby being a poor choice for the command post. At

FIGURE 12-5 Command post.
Courtesy of David J. Jones, Cincinnati, Ohio.

this incident, the buildings collapsed. However, even if the structures had not collapsed, the heavy traffic on the first floor of each building would have made the lobby a poor choice for the command post.

Most newer high-rise buildings are equipped with a command center, which is typically near the lobby **FIGURE 12-5**. For most incidents, this is the best command post location, as these command centers often provide good communications and the needed work space for command activities.

Once the command post has been established, its location should be communicated to all responding companies. Some department SOPs use a street name, such as Main Street Command, to notify companies of an exterior command post location. A command post located inside a building may be identified using the building's name, such as Rockefeller Command.

Developing Building-Specific High-Rise Preincident Plans

All high-rise buildings should be preincident planned. Knowing the location of interior command rooms is just one of the many issues that should be addressed in the high-rise preincident plan. A good high-rise preincident plan will address all of the issues covered in a standard building preincident plan, including occupancy type. Chapter 11, *The Role of Occupancy,* describes the importance of occupancy types, and Chapter 2, *Procedures, Preincident Planning and Size-Up,* lists occupancy type as one of the major subjects to be addressed in a preplan. Knowing the building's use is essential to preplanning. Most high-rise buildings are businesses, hotels, apartment buildings, or healthcare occupancies. Many high-rise buildings are mixed occupancies. Most high-rise buildings have lobby and/or shops on the first floor, with stores, offices, apartments or other occupancies on the floors above.

Trump Tower is an example of mixed occupancy in a high-rise building with a lobby on the first floor, offices on the lower floors of the tower, and living apartments on the upper floors. In addition to the standard preincident planning factors, there are many special considerations in a high-rise, including the following:

- Elevator operations and elevator key location
- Access/egress issues such as stairway doors that automatically lock
- Standpipe operations, especially if field-adjustable pressure-reducing valves are present
- Floor layouts for each floor that is different or special (e.g., when tenants occupy several contiguous floors, there may be internal open stairs between floors.)
- Ventilation, such as the use of the heating, ventilation, and air-conditioning (HVAC) system, and window venting considerations, including the presence of windows that can be opened, tempered glass, or special windows (e.g., hurricane resistant)
- Procedures or operations that are unique to the building

High-rise buildings are generally of fire-resistive construction, with older high-rise buildings often being superior to newer buildings in many construction features **FIGURE 12-6**. Older tower buildings have better compartmentation, more fire-resistive components, and generally better exit facilities, however many are not protected by sprinkler systems.

Many newer buildings have pressurized stairways, and they are often fully sprinkler protected. However, there was a period when high-rise buildings were constructed using modern, lightweight construction methods but were not protected with automatic sprinklers. These new-style, nonsprinklered high-rise buildings are likely to be the most problematic. Many high-rise buildings are not protected by sprinkler systems. NFPA statistics quoted in the NIST high-rise report indicate that 41 percent of high-rise office buildings, 25 percent of high-rise healthcare facilities, 45 percent of high-rise hotels, and 54 percent of high-rise apartment buildings are not protected by sprinkler systems.

Since the mid-1970s, codes used in the United States, with one notable exception, required high-rise buildings to be sprinkler protected. In New York City, until a change was made in March of 1999, it was

FIGURE 12-6 Old-style tower construction on left versus new-style planar construction on right.
Courtesy of Ben Klaene.

possible to construct a residential high-rise building without a sprinkler system. Two tragic fires led to a change in the building codes, and currently all new residential high-rise buildings built in New York City are required to have a sprinkler system installed. Also, within the past few years, many major cities such as New York and Chicago have implemented high-rise sprinkler retrofit programs for some existing high-rise buildings. For instance, the city of Chicago adopted an ordinance that required high-rise buildings above 80 ft (24 m) to be retrofitted with sprinklers. However, if residential properties and certain other high-rise buildings that are classified as historical pass a locally developed life-safety evaluation, they may opt out of the retrofit requirement. Trump Tower in New York City is one of many high-rise buildings that is not sprinkler protected.

In many new-style high-rise buildings, where core construction methods are used, fire attack and evacuation tactics are further complicated. In these buildings, stairways are usually located in the center of the building **FIGURE 12-7**. Therefore, if occupants are trapped or if the fire occurs near the core or between the occupants and the core, evacuation and fire attack positions are limited. Preincident plan drawings will prove to be extremely valuable to the IC in developing

CHAPTER 12 High-Rise Buildings

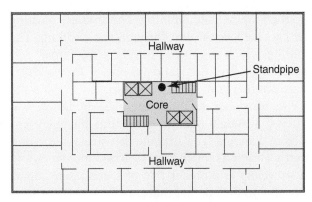

FIGURE 12-7 Central core layout.
© Jones & Bartlett Learning.

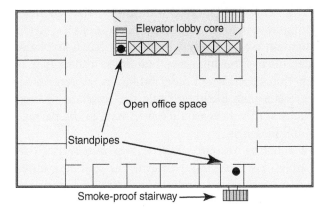

FIGURE 12-8 Side core layout.
© Jones & Bartlett Learning.

evacuation and fire attack plans in these buildings. Some core construction methods move the core away from the center of the building, which is sometimes referred to as side core construction **FIGURE 12-8**.

Interior building configurations can be radically different in buildings that appear to be the same from the exterior. For example, in Figure 12-7, there is an open floor layout with individual offices separated by cubicles, and there are standpipes in the stairway. In Figure 12-8, individual offices are separated by full floor-to-ceiling walls, with the standpipe located in the core area. These seemingly subtle differences can have a major impact on firefighting and rescue operations.

The stairway at the periphery of the side core layout shown in Figure 12-8 is a **smoke-proof tower**. Smoke-proof towers are built as a separate structure, thus reducing the possibility of smoke entering the stairs. Also, the stairs could be pressurized to reduce smoke infiltration. Unless there is a compelling reason to use another stairway, the smoke-proof tower or the pressurized stairway should be the stairway of choice for occupant evacuation. This should be noted in the preincident plan.

Analyzing the Situation Through Size-Up

All of the usual factors regarding size-up apply to high-rise fires. These factors are discussed in Chapter 2, *Procedures, Preincident Planning, and Size-Up*. However, size-up at the scene of a high-rise fire will often be particularly difficult, because visual information may be nonexistent or misleading. Merely finding the fire can be a perplexing chore. A large structure can contain a substantial fire without displaying any external signs. If flame and smoke are visible, it is both a good sign and a bad sign. The good news is that the fire has self-vented, and the IC has some indication of the location and intensity of the fire. The bad news is that the fire has probably gained considerable headway before the arrival of fire fighters and has the potential to be a wind-driven fire.

Observing a high-rise fire from the exterior could result in confusion in directing interior assignments. High-rise buildings often do not have a floor numbered 13, or the mezzanine may not be counted as a floor. There may also be half-floor equipment rooms within the building. A heavy volume of fire that originated at a central core may appear at a single window, while a fire that originates near the periphery may cause several windows to break. Visible evidence gained from the exterior, many floors below the fire, may be unreliable. Occupants may report smoke or odors, but these reports typically come from occupants who are many floors above the fire. Information may be available from building occupants regarding evacuation status, but this information is seldom completely accurate in a large building. Even internal alarms are sometimes misleading, as smoke detectors sense smoke above the fire floor or in other areas where the HVAC is depositing smoke. For these reasons, even with fire and smoke showing outside the building, finding the seat of the fire may require assigning several fire companies to different areas of the building.

Matching the occupancy classification to time factors generally gives the IC a good idea of the life hazard potential. There is a tremendous life-safety difference in a high-rise office building during working hours compared to the same building during nonbusiness hours. There may be people in the building during nonbusiness hours, but the occupant load during normal working hours will be much greater. How many people would you expect to find in an office building during normal working hours? As a general rule, means of egress from business use spaces are sized based on a minimum of one person for every 150 ft^2 (14 m^2) (NFPA 101, 2018). In the Trump Tower, the upper floor apartments would tend to have a higher occupancy at night and the offices below a higher occupancy during the day. An accurate and up-to-date preincident plan most certainly aids in determining the life-safety problem.

The validity of dispatch information varies depending on the alarm type. A report of a smoke detector alarm on the 30th floor could be for a fire several floors below or for a localized problem on the 30th floor. A report of an odor of smoke from a building occupant may be no more reliable than the smoke alarm in determining the location and nature of the problem. However, a report of visible smoke or flames on the 30th floor in Suite 3012 provides the IC a wealth of information.

Due to the frequency of false alarms from smoke detectors, many departments send a limited response when the only report of fire is from an alarm system. This can conserve resources, but it may also lead to complacency among responders who think that fire alarms are less important than traditional, phoned-in fire responses and as a result may not wear full PPE. It is imperative that fire companies treat fire alarms with the same importance and preparedness and resist the temptation of complacency. We should never be surprised to find fire!

It is not unusual for the dispatch center to receive follow-up calls from the public during a high-rise fire. Many of these calls repeat and verify information already received. In many fires, building occupants call dispatch after the initial alarm to report people trapped, to request evacuation instructions, or to report fire conditions in specific areas of the building. This information is critical and should be immediately relayed to the IC. The IC should notify the operations chief and/or divisions, groups, or companies working in these areas and/or deploy companies to evaluate the situation. Communications is often a problem; in some past cases, the IC had vital information but was unable to relay it to units working within the building.

In high-rise buildings, structural stability will typically be very good. The First Interstate Bank Building and Meridian Plaza fires were both in new-style high-rise buildings, and both sustained large volumes of fire over several floors for an extended period of time. While there was structural damage, especially at the Meridian Plaza Fire, neither building collapsed. Furthermore, an old-style high-rise building can be expected to endure even more fire exposure than modern, lightweight construction. The Empire State Building serves as a good example of the structural stability of the older, tower-type high-rise buildings. On July 28, 1945, a bomber crashed into the Empire State Building. The aircraft's burning fuel cascaded down and through the building. The building, which sustained both the crash impact and the resulting fire, still stands as one of America's famous landmarks.

Fire resistance in older high-rise buildings was achieved through massive structural members encased in concrete. The airplane that struck the World Trade Center was much larger, carried more fuel, and was traveling at a higher rate of speed upon impact compared to the bomber that crashed into the Empire State building. Therefore, it is difficult to compare the two incidents. However, one of the major factors leading to collapse at the World Trade Center was aircraft debris removing sprayed-on fire-resistive coatings from steel structural members. This type of sprayed-on insulation was not used in the construction of the older tower-type high-rise buildings, although it is sometimes used when older buildings are renovated. There is a high likelihood that the concrete-covered structural members would have survived the barrage of aircraft debris.

Evaluating resource needs is extremely difficult in a high-rise building. However, the staffing and resources necessary to combat a large fire are significant, as you will see in the analysis of high-rise case studies later in this chapter.

The high-rise decision tree shown in **FIGURE 12-9** considers a few major factors that could lead to the call for additional alarms at the scene of a high-rise fire. This decision tree by no means provides a complete size-up of the potential challenges that a high-rise fire presents. However, it does offer a thought process to assist in determining resource requirements. An examination of the decision tree demonstrates that a large fire, located beyond the reach of ground equipment and with elevators unavailable, will require extensive resources.

Some department SOPs require that the first-arriving company request an additional alarm for any working fire in a high-rise building. There is justification for this precaution, given the life hazard, extinguishment, and property conservation potential. In most instances, additional personnel will be needed to augment the initial attack and to provide logistical support. The actual staffing requirements will vary depending on the circumstances, as demonstrated by the scenarios in this chapter.

Smoke Movement

Weather conditions, together with fire intensity, can have a significant effect on smoke movement within a high-rise structure. While the IC should consider these factors, a degree of unpredictability should be expected regarding the effects of weather and fire intensity on smoke movement.

Heat of the Fire

The intensity and size of the fire will determine the extent to which combustion gases are heated and how high they will rise inside the building. In lower structures there is generally enough heat energy to cause the heated fire gases to rise to the highest level in the structure. In high-rise buildings, the smoke and toxic gases will tend to rise until they reach temperature equilibrium, at which point they will stratify. It is not unusual to have heavy smoke on a midlevel floor and smoke-free floors above. This stratification can endanger occupants who enter a smoke-free stairway and then discover smoke several stories below. Many times, doors leading back into a floor area are locked, forcing the fleeing occupant to seek refuge in the stairway or proceed through the smoke.

Stack Effect

The airtightness of the structure has much to do with stack effect—the vertical airflow caused by temperature differences within and outside the building. The unpredictable behavior of smoke within a high-rise is due, in large part, to stack effect. Many fire departments have been surprised when they tried to ventilate a high-rise building using a technique that was previously effective in the same or a similar building, only to get an entirely different outcome. In some buildings the stack effect is so great that it interferes with the proper operation of the HVAC system. The colder it is outside and the warmer it is inside, the greater the positive stack effect (upward movement). Conversely, the stack effect can be negative (downward) on a warm day within an air-conditioned building. The heat of the fire and stack effect are interdependent. The chances of smoke stratification are less on a cold day than on a warm day. There are formulas for calculating stack effect, but they have little application for field use.

Wind

There is a point within a high-rise structure of sufficient height called the neutral pressure plane (NPP). Below the NPP, air is moving into the building; at the NPP, forces are neutral (air is not moving in or out); above the NPP, air is moving out of the building

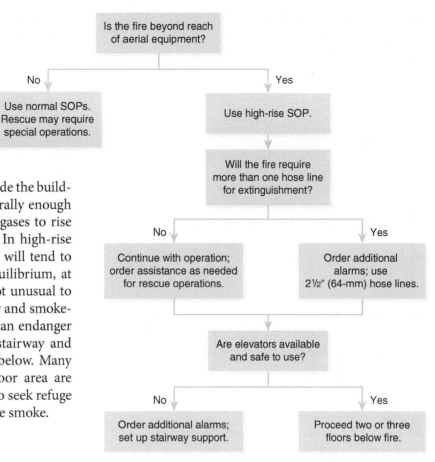

FIGURE 12-9 High-rise decision tree.
© Jones & Bartlett Learning.

FIGURE 12-10 Neutral pressure plane.
© Jones & Bartlett Learning.

FIGURE 12-10. The NPP is affected by heat from the fire and the stack effect. Wind also plays a major role **FIGURE 12-11** (NFPA, 2003).

Ground-level winds are not always a good indication of wind direction and speed at higher elevations. Downtown areas of large cities containing large

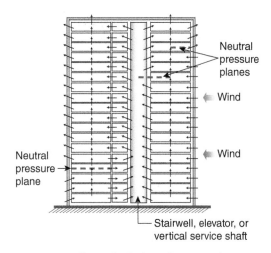

FIGURE 12-11 Effect of wind on the neutral pressure plane.
© Jones & Bartlett Learning.

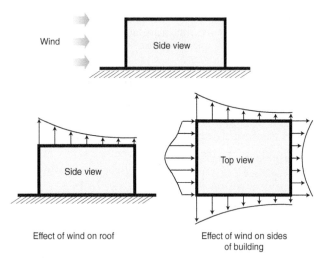

FIGURE 12-12 Air pressure distribution along the four sides and roof of a building.
© Jones & Bartlett Learning.

numbers of high-rise buildings are like giant canyons. Wind entering the high-rise canyon is redirected and becomes turbulent. These wind gusts also prevail high above the ground but possibly in another direction and/or at a higher velocity.

Wind passing over a roof opening has a pulling or Venturi effect. In addition, the wind will push smoke and fire back into the building on the side facing the wind and help to vent smoke out of the building on the opposite side of the building **FIGURE 12-12**.

In reality, predicting the wind factor on the fire ground is more of an art than a science. Wind direction and velocity can change dramatically and without warning, even when atmospheric conditions are not changing significantly.

Wind-Driven Fires

Much research has been done regarding wind-driven fires in high-rise buildings. Several possible tactics have been introduced as a result of this research:

- Controlling windows and doors
- Applying positive-pressure ventilation
- Deploying wind control devices
- Using exterior fire streams

Of these, controlling windows and doors is the most practical.

Controlling Windows and Doors. Even under high-velocity wind conditions, it is possible to protect stairs and hallways by maintaining barriers between pathways to hallways and stairways. The usual pathway to the stairway is through a window opening, and subsequently through one or more doorways leading to the stairs. Any tactic that interferes with this pathway can be effective in retarding fire extension and minimizing the effects of high wind.

Given the unpredictability of the wind and internal conditions such as stack effect, breaking windows to vent the fire is highly discouraged. Nonetheless, the heat of the fire often causes windows to fail, allowing wind to enter. In a residential building or healthcare occupancy there will likely be a fire door leading to a relatively small room or apartment. This door provides a fire barrier that can retard fire growth and delay extension into other areas. Fire fighters must be cognizant of the fact that a virtual "blow torch" fire could erupt when a door is opened to the fire area under high-wind conditions. A properly staffed charged hose line must be in place before opening doors leading to an active fire area, and the door must be controlled. If a wind-driven fire is suspected, drive the point of the Halligan tool through the door. If the fire exits the hole under pressure (like a blow torch), suspect a wind-driven fire on the other side. Do not open the door.

In the case of healthcare occupancies, there will usually be at least one additional fire door in the hallway before reaching the stairway. Multiple doors provide barriers to fire extension into the stairway. In these buildings it may be possible to seek temporary safe refuge in another room if the fire breaks out of containment, although the stairway is a much better place of refuge if it can be safely reached.

Large, undivided office spaces or common areas in other occupancies may have a single fire door between the fire area and the stairway. These areas are particularly hazardous under wind-driven fire conditions, as the wind could well push the fire directly into the stairway when the door to the fire floor is opened and there may not be an enclosed room in which to seek safe refuge. During a high-rise fire, controlling the door leading to the fire area is critical.

Applying Positive-Pressure Ventilation. Positive-pressure ventilation (PPV) was introduced to the fire service in the 1970s and is an effective tool when properly implemented. Like all ventilation tactics, improper venting can have a detrimental effect. Departments that protect high-rise buildings had many questions regarding the effectiveness of PPV in high-rise buildings, particularly on upper floors of taller buildings. A study completed by NIST in 2007 (Kerber, 2007) compared various PPV equipment and techniques, providing answers to many questions regarding PPV tactics at high-rise fires. Effective PPV can clear stairways during a high-rise fire in two ways:

- Push smoke out of the building either through windows or vertical vent openings
- Pressurize the stairway to keep smoke from entering the stairway

Both methods rely on creating a positive pressure in the stairway. These methods can be very effective; however, there are several challenges. Stairways that terminate in a large open space such as a first-floor lobby are more difficult to pressurize due to the large open-air space. Open areas on the first floor are subject to depressurization when doors are opened. Sometimes using larger, more powerful blowers, or deploying blowers in tandem, can overcome this problem. It is far more efficient to pressurize a stairway leading directly to the outside than it is to pressurize a stairway terminating in a large lobby area.

When pressurizing a stairway, open doors leading from hallways to the stairway can result in a loss of stairway pressure. During the NIST PPV experiments, pressure loss was inconsequential when doors were opened just enough to advance a fire hose from the stairway into the hallway. However, when doors were opened wide, as they may be by escaping occupants or during firefighting operations, the drop in stairway pressure was significant. Placing additional blowers operating into the stairway on upper floors below the fire floor can increase stairway pressure and thus increase PPV effectiveness.

High wind conditions can have an adverse effect on PPV methods, particularly if the wind is blowing into the building through an opening such as a window. This leads to a discussion of other wind-driven fire tactics.

Deploying Wind Control Devices. Realizing the probability of a wind-driven fire and the fact that wind-driven fire conditions are exacerbated during a high-rise fire, experiments were conducted to evaluate devices designed to minimize the effect of high winds entering a window opening (Madrzykowski and Kerber, 2009). At present, wind control devices are basically a weighted mat that is placed over a window to keep the wind from entering the building through an existing opening. These devices are deployed from an area above and in line with the fire floor. NIST evaluations determined wind control equipment to be effective in partially blocking the wind entering a window.

Currently, these devices seem to have limited application and are not in widespread use. Questions need to be answered regarding how to best deploy the wind control device.

Using Exterior Fire Streams. Experience in using master streams operated from the exterior of a building in a defensive mode indicates that this is a less than ideal way to combat a fire and should be used only when the fire area cannot be safely entered to conduct an offensive attack. However, pulsing and softening the target tactics are used to knock down heavy volumes of fire before entering the immediate fire area. With this in mind, there are situations when a limited exterior attack may be warranted in a high-rise building. On lower floors of a high-rise building, solid or straight streams operated from an aerial device can be used for this purpose. On floors beyond the limitations of aerial devices, a nozzle with extension pipe could be operated into the fire floor from the floor below. At present, one nozzle manufacturer is marketing a special nozzle designed to be operated from the floor below the fire floor into a window on the fire floor. This nozzle was found to reduce heat in the compartment of origin during the NIST wind-driven fire research.

Questions arise as to how to position nozzles to apply an effective stream on the floor above. In many cases, high-rise buildings have windows that are not meant to be opened or removed. Breaking windows is a questionable tactic. Determining the best window to use from the floor below the fire is difficult, as is the accurate application of the fire stream. Actual field experience using exterior streams operated from the floor below the fire is very limited.

Developing and Implementing an Incident Action Plan

Incident priorities (life safety, extinguishment, and property conservation) remain the same, regardless of the type of structure. Each type of building or occupancy presents a special set of hazards to occupants and property. When people and property occupy buildings that are 110 stories high, the risk to life safety and property is obvious.

The primary rescue tactic in a high-rise building is a well-placed offensive attack. In a high-rise, more than in any other building, quickly confining and extinguishing the fire are critical, as ventilation and evacuation options are limited. Once the fire is extinguished, the toxic products of combustion are no longer being produced and the operation becomes more manageable.

How much water flow will be needed to extinguish a well-involved fire in a large high-rise building? Compartmentation plays a major role in fire flow calculations. Some office occupancies are divided by substantial walls; others have glass panels or portable dividers separating work areas. Furthermore, there are often large, undivided spaces within high-rise buildings. These undivided areas create a high hazard requiring large flows. In the case of the MGM Grand Hotel Fire (shown later in this chapter) the required rate of flow was beyond a reasonable flow capacity, and the size of the area made it impossible for fire streams to reach the opposite interior walls from the access points.

There have been cases in which master stream appliances were brought into upper stories of a high-rise building and operated on the interior, but few standpipes could provide sufficient water to support master streams. To compensate, 5-in. (127-mm) hose was used in the building at One Meridian Plaza. However, even this tactic was unsuccessful in controlling or suppressing the fire.

Using master streams within a building is an unusual tactic, but this could be the only realistic way to control a large-area fire on the upper levels of a high-rise building. Methods have been developed for advancing large-diameter hose up interior stairways for use as a portable standpipe, but the time and staffing required to advance the hose to an upper floor present significant challenges and become unrealistic at higher elevations. Master stream appliances could be supplied by fire department pumpers located at street level, provided that the elevation pressure, friction loss, and required nozzle pressure do not exceed safe operating pressures. If master streams are operated on or into the interior, the structural integrity of the building must be closely monitored. Water from master streams could add a hundreds or thousands of pounds of weight within the structure in a relatively short time. It is also easy to imagine the damage that internal master streams would do to property within the structure.

There are several water distribution appliances that deliver water to the fire floor from the floor below **FIGURE 12-13**. These devices are capable of flowing large amounts of water but are still dependent on the water supply from the standpipes or hose laid in the stairwell. These devices also require a window to be removed from the floor below the fire and on the fire floor. While these nozzles can be effective, they are not usually "Plan A." The floor-below nozzles require a substantial commitment of personnel to deploy, assemble, and get into operation.

Until ventilation is accomplished, the floor area of a fire-resistive building can be dangerous for fire fighters operating hose lines. Like all other energy and

A

B

FIGURE 12-13 Exterior water application nozzles being operated from a lower floor into a floor above.
A: Courtesy of The Hero Pipe; **B:** Courtesy of Mike Terpak.

matter, the fire will follow the path of least resistance. If the floor is not properly ventilated, fire, smoke, and heat will follow pathways through hallways or concealed spaces. This can be extremely dangerous in a closed high-rise building, as the fire can "wrap around," when moving through these spaces and get between fire fighters and their exit route. The floor should be ventilated as soon as possible; at the same time, fire fighters should be aware of the special hazards associated with breaking windows on the upper floors of a high-rise building. Backup hose lines should always be in place to protect exit routes.

In attacking a fire in a high-rise building, as in any structure fire, engine company and truck company operations must be coordinated. Truck company members will be needed to force entry. On reaching the fire floor, a thermal imaging camera should be used to check heat levels in the ceiling area above a suspended ceiling. If there is any doubt regarding fire entry into this false space, remove a ceiling tile under the protection of a hose line to control fire that might intensify as air is introduced to this space. Fire fighters should continue to monitor this space to ensure that the fire is not getting behind the fire attack crew. Suspended ceilings are common concealed spaces and a likely place for a wrap-around fire. This area sometimes contains cables and other equipment that can drop down, creating an entanglement hazard.

The attack hose line is used as the lifeline to safety. Members should stay within range of this protective line, not only for fire suppression purposes but also as a means of finding the stairway in heavy smoke conditions. It is also good practice to place a fire fighter at the stairway opening to the floor. This fire fighter will be needed to help extend the hose, control the door and can direct fire fighters to the exit if necessary. Department SOPs should require the use of lifelines (ropes) in situations where fire fighters are working away from hose lines.

Use of Stairways

Most occupants of an office building are mobile and able to escape on their own, provided that stairways are available for their use. When fire fighters are trying to advance up stairways while occupants are attempting to evacuate, the result is gridlock. What can the fire department or occupants do to alleviate the problem? Many high-rise building managers and fire officials recognize that it is not always the best policy to have all of the building's occupants in the stairways at the same time. The people on the fire floor and the floor immediately above are in the greatest danger and should be among the first groups evacuated. When the stairway is filled with people from other floors, those in greatest danger cannot safely evacuate the building.

Mass evacuations also complicate fire operations, as fire fighters are forced to "swim upstream" in a stairway full of evacuees moving downward. Furthermore, when there is a mass evacuation, occupants above the fire must pass by the fire floor to reach safety. Fire fighters tend to exacerbate this problem by blocking doors open as they progress with fire lines, thereby venting the products of combustion into the stairway. These were contributing factors to the loss of life at the Cook County Administration Building Fire in Chicago, Illinois. There is no easy or, for that matter, best way to deal with this dilemma. Several ideas are presented here that may or may not work in buildings protected by your department.

First, search efforts must be systematic and include a complete primary search of the fire floor and floors above the fire. In searching above the fire, it is important to check all stairways. One way to accomplish this is to enter the floor using one stairway and exit the floor using another stairway. However, the evacuation stairway must not be contaminated with smoke from the fire floor or other floors where smoke has infiltrated the hallways. It is best to assign a crew to check and clear the attack stairway of building occupants. Some important questions that need to be answered include:

- Are people in the stairway?
- Are any stairways filled with smoke? Smoke may stratify; therefore, do not assume that the entire stairway is clear simply because it is clear on an upper floor.
- Are doors from the stairway to each floor level locked?

Forcible entry is often needed, so the search and rescue team must carry proper tools. Forcing doors not only results in damage but also will physically exhaust the team. Master keys, supplied by building management, can be invaluable in gaining access to rooms. It is necessary to check individual rooms to ensure that the occupants on any floor that is endangered by the fire or smoke have escaped.

As previously mentioned, another method of facilitating both evacuation and extinguishment is to designate separate stairways for occupant and fire department use. Fire companies must have control of the evacuation and good communications with occupants to successfully use this tactic. An internal public address system or EVACS can be used to direct occupants to evacuation stairways and away from the fire operations stairway. Even though occupants are directed to use stairways other than the fire operations stairs, the fire operation stairway must also be checked, as people do not always follow instructions. As was noted earlier, smoke-proof towers or pressurized stairs may be the best option for occupants. Controlling the stairs and dedicating one stairway for firefighting also reduces the possibility of opposing hose lines on the fire floor, as it is easier to control the entire attack if it is being made from one entry point. A stairway with standpipe outlets is the usually the best choice for a fire operations stairway.

Ventilation

Ventilation and control of the HVAC system should be accomplished using reversible methods. Breaking windows high above grade level creates a serious hazard below in addition to the possible wind-driven effect mentioned previously. Plate glass windows tend to break into large, irregular pieces of glass, known as shards. These shards of glass create a dangerous situation as they drop to the ground below. Tempered glass breaks into tiny pellets and is less hazardous to people below. Even if windows are pulled into the building or are made of tempered glass, opening a window is much preferred to breaking a window. If a window is opened and the effect is negative, it is possible to reverse the negative effect by closing the window. Unfortunately, most high-rise buildings have sealed windows that cannot be opened. Sometimes fire codes require that a percentage of the windows be operable or made of tempered glass so that manual ventilation is possible. Codes that require vent windows generally require that windows designated for ventilation be marked as such. The location of tempered glass or operable windows should be noted on the preincident plan.

The operation of the HVAC system is usually reversible. If operating the HVAC system does not have the desired effect or spreads the smoke, shut it down.

Interior Exposures

All floors above the fire are exposures. Companies need to check above the fire floor for extension and be equipped to fight the fire. In modern buildings with curtain walls, fire can extend upward, inside the building, near the exterior wall. Stairways and elevator shafts are vertical openings through which fire can spread. Additionally, pipe chases and utilities penetrate floors, creating openings through which fire can spread.

Fires can extend up the exterior of the building via auto-exposure by "lapping" from floor to floor. While upward extension is the main concern, the fire can spread downward through melting expansion joints or burning materials dropping below via the HVAC or by other means. Therefore, areas below the fire must also be checked. The search and rescue team can check floors above the fire while fire fighters who are assigned to property conservation can check floors below the fire.

Property Conservation

As previously stated, the first priorities are life safety and extinguishment. However, property conservation should also be considered early in the incident. High-rise office buildings typically contain valuable contents, such as computers; other office equipment; and important files and documents. High-rise residential buildings contain personal items that cannot be replaced. Often, protecting these valuables is overlooked, owing to the labor-intensive nature of rescue and extinguishment challenges in a high-rise building fire.

When considering property conservation, the height of the building comes into play in an opposite way. In life safety and extinguishment, people and property on and above the fire floor are normally considered to be most at risk, because fire, smoke, and heat will first affect these floors. However, the greatest property conservation exposure is often downward as water flows through curtain walls, electrical fixtures, and other openings, damaging valuable property beneath the fire. In addition to moving and covering valuables, fire fighters should channel the flow of water down stairs or through drains. Like life safety and extinguishment, property conservation operations must be planned and well executed to reduce the loss.

Lead Time

Experienced ICs understand that when an assignment is given, it takes time to see results. Ordering a portable master stream appliance to be deployed on the 10th floor could take a considerable amount of time. The level where the fire occurs has the greatest impact on lead time, especially when elevators are not available. **TABLE 12-1** is based on a study conducted by the Cincinnati Fire Department. The results clearly demonstrate an increased lead time related to the height of the building. For purposes of this study, fire recruits and instructors were in full turnout gear, but not on air. Each five-person company (four recruits and an instructor) was assigned 150 ft (46 m) of 1¾-in. (44-mm) hose, a nozzle, extra SCBA cylinders, and hand tools. The average time to walk up one floor for the first nine floors was 20.8 seconds or a total of 3 minutes 7 seconds to reach the 10th floor level. The last eight floors took 59.0 seconds per floor. On average it took 28 minutes and 52 seconds to reach the top floor of this

TABLE 12-1 Climbing Stairs to the Top of 48-Story Building

Floors	Average Time Per Floor in Seconds
1–10	20.8
11–20	27.8
21–30	33.6
31–40	45.9
41–48	59.0

© Jones & Bartlett Learning.

48-story building. This does not include time needed to connect to the standpipe, place the SCBA in service, and advance on the fire. This building is less than half the height of some high-rise buildings and the participants were in better than average physical condition. Studying high-rise fire reports and conducting tests reveals that fire fighters are sometimes unable to reach the fire floor or need to stop and rest during the ascent.

The NIST report on high-rise fire-ground field experiments compared the time required to place the first attack hose line in position when elevators were available versus the time needed to perform the same task using the stairway to access the 10th floor. In the NIST experiments, it took 2 to 4 additional minutes to reach the fire floor via the stairs. Any delay in extinguishment or rescuing occupants could have serious consequences. A fire on a higher floor would delay operations to a much greater degree, as indicated in Table 12-1.

Achieving a successful outcome when confronted with a working high-rise fire requires that an adequate firefighting force be rapidly assembled. During the NIST high-rise fire-ground field experiments, 66 to 120 fire fighters were assembled to carry out the 38 identified tasks. These experiments did not include property conservation or overhaul. It was found that the total number of fire fighters responding was not as important as the response time for resources allocated to complete critical tasks. Amassing the necessary personnel early in the operation resulted in reduced times to carry out the life-safety and extinguishment objectives. The simulated fire on the 10th floor of an open-layout office required two 2½-in. (64-mm) hose lines for extinguishment. A 2½-in. (64-mm) hose line was also advanced into the 11th floor. Search and rescue operations were conducted on the 10th and 11th floors, with one victim located on each floor. Other tasks, such as ventilation and evacuation of the remainder of the building, were also conducted in priority order.

Crew sizes were three, four, five, and six with high and low responses. During the exercise, the crew size remained the same regardless of how many units responded. For example, a low response of three engine companies and three truck companies each with a crew of three would total 18 fire fighters on the first alarm plus ambulances and battalion chiefs. Each alarm duplicated the first-alarm fire company response with changes in other responses such as higher-ranking chief officers. A high response was four engine companies and four truck companies per alarm. Eventually three-person staffing provided adequate on-scene fire fighters. However, many critical tasks were delayed as more distant companies responded to the scene. Using the worst-case scenario, a nonsprinklered building with elevators unavailable took the low response three-person crew nearly 14 minutes longer to rescue the victim on the fire floor than the low response four-person crew.

The fire dynamics simulator modeled fire growth, and the fractional effective dose scale was used to estimate the impact toxic gases would have on victims. The toxic effects on the victim exposed to fire and toxic gases for 14 additional minutes were significant. Higher levels of staffing (five- and six-persons per company) further improved times to complete critical tasks. This study proved that the crew size makes a dramatic difference in saving occupant lives.

If the IC waits to give assignments until the need is obvious, he or she has waited too long. To be successful, the IC must be able to anticipate needs and assign resources in advance.

Establishing a Wide Fire Zone

Immediately adjacent to the exterior of a high-rise building fire is an unsafe place, particularly if glass is falling. A perimeter should be established to keep civilians out of danger and to provide a safe working area for fire fighters outside the building. Fire fighters who must work within the perimeter should be kept within vehicles or under protective structures. The importance of establishing zones changes with each incident. If there is danger from falling glass, a 200-ft (61-m) perimeter should be enforced around the building. It may be necessary to enlarge the perimeter when high wind conditions indicate the need. As mentioned previously, plate glass falls to the street in large shards and at times will float a considerable distance from the base of the fire building. Opening windows by pulling the glass inward greatly increases safety at high-rise building fires. This technique has been tried with limited success using tape, suction cups, and other methods. When there is a working fire in a high-rise building, fire fighters may be forced to break windows on the upper floors or the fire may self-vent without warning. Therefore, the IC should expect glass to fall at any time and direct that a safe perimeter be established early in the incident.

Applying the National Incident Management System to a High-Rise Fire

The NIMS should be used at every structure fire, and a high-rise building fire is certainly no exception. Proper use of the NIMS is a test of the department's training, preincident planning, and discipline.

A look at the logistical requirements and multiple staging areas makes it apparent that the span of control can be quickly exceeded at a major high-rise fire. In the case of a high-rise fire, the IC is well advised to hand off the operations and/or planning sections early in the incident so that other priorities can be addressed. Record keeping is essential. If the IC has an assigned aide, this person usually acts as the initial planning officer and records situation reports, resource status, and assignments. Units operating on the fire floor and staging area can be managed by a forward rescue/suppression branch director or operations section chief.

Communications

Communication and accountability are simplified by using the Incident Management System. Companies working on the fire floor may change; however, the geographic designation remains the same. For example, if Truck 1 and Engine 2 are assigned to the 15th floor for search and rescue, they would be identified as Division 15. If Truck 1 and Engine 2 are later reassigned to rehab and replaced by Truck 3 and Engine 4, these replacement companies become Division 15. Remember that "division" is the NIMS term used to identify a geographic location or assignment.

Suppose a fire is on the 10th floor of a 20-story building. A rescue and evacuation group is doing the primary search on floors 12 to 20. Fire attack units are on the 10th and 11th floors. **FIGURE 12-14** is a NIMS chart and communications network for this operation. Communications for this fire would be as follows:

- The rescue and evacuation group would be communicating with companies in their group as to the status of various floors.
- Division 10 (fire floor) would be communicating with both the operations section or IC (if operations is not staffed) and the companies assigned to Division 10.
- Division 11 would be doing much the same as Division 10.
- Companies assigned to ground support would be communicating with each other and with the logistics section.
- Operations, if assigned, would be communicating with Divisions 10 and 11; otherwise, they would report to a branch manager or the IC.
- The staging officer would be communicating with the operations section or IC, which would then order the necessary staffing, equipment, and supplies for the staging area.
- Safety, as well as the operations and logistics sections, would be communicating with the IC.
- If staffed, the command staff positions of liaison and information, as well as the planning and administration/finance sections would most likely be at the command post location, communicating face to face with the IC.

Communications is the most frequently cited problem during major emergency operations. Communications can be very complicated within a high-rise building. If separate radio frequencies or hard-wired communications are not provided at the incident scene, a breakdown in communications is likely to occur.

Radio discipline *must* be maintained, and alternative methods should be arranged (e.g., messenger, hardwired telephones. Note that the rescue and evacuation group in Figure 12-14 is communicating via telephone. Hard-wired telephones are abundant in most high-rise buildings; consider using them as a means of emergency communications. In Figure 12-14, for example, the search and rescue group supervisor could be located in a room off the lobby, on a lower floor, or in another building where two or more telephones are readily available. All units working in this group would be given the contact numbers for the group supervisor, which would be used to call in status reports and receive new assignments. Alternate communications must be available because the fire could damage the telephone system. During a large-scale search and rescue operation, there is a need for considerable two-way communications within the search and rescue group. Complicating the communications problem at a high-rise building is the fact that using radios is difficult within many structures. In fact, it is often impossible to receive radio transmissions from inside a structural steel building.

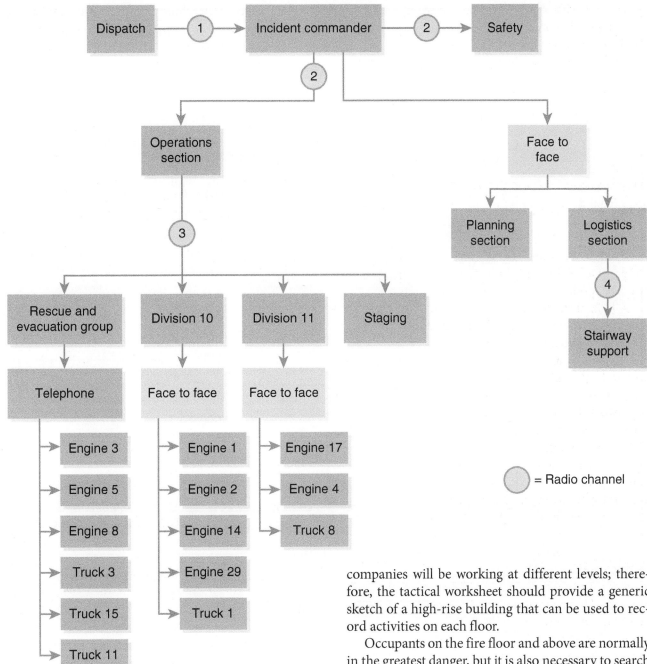

FIGURE 12-14 High-rise communications network with the operations section assigned.
© Jones & Bartlett Learning.

Tactical Worksheets

Fire departments have devised a wide assortment of tactical worksheets to be used in documenting and tracking activities on the fire scene, including some that are specifically designed for high-rise fires. Most worksheets contain checklists to remind the IC of important functions.

A good tactical worksheet will include sections to record the fire location and the location of companies at the incident scene. During high-rise fires, companies will be working at different levels; therefore, the tactical worksheet should provide a generic sketch of a high-rise building that can be used to record activities on each floor.

Occupants on the fire floor and above are normally in the greatest danger, but it is also necessary to search below the fire. A search and rescue group normally works above the fire to conduct a primary search and to assist in evacuation. This group is generally very active, moving from floor to floor. Therefore, it is important that the group leader keep the IC (or operations chief) current on the group's location in the building and the status of the primary search.

The high-rise worksheet in **FIGURE 12-15** lists various high-rise positions described in this chapter and provides space to enter companies working on each level.

The value of an aide is clear. The IC needs to know the location of all companies operating in the building but does not always have time to keep the worksheet current. An aide, acting in a planning capacity, can assist in receiving and recording this information.

Upper Search and Rescue Group		Engine Cos.	Truck Cos.	Others
Floors: From _____ to _____ Supervisor _____				

Floor # _____	Engine Cos.	Truck Cos.	Others	Division # _____ Supervisor _____
Floor above fire Floor # _____	Engine Cos.	Truck Cos.	Others	Division # _____ Supervisor _____
Fire floor Floor # _____	**Engine Cos.**	**Truck Cos.**	**Others**	Division # _____ Supervisor _____
Floor # _____	Engine Cos.	Truck Cos.	Others	Division # _____ Supervisor _____
Staging Floor # _____	Engine Cos.	Truck Cos.	Others	Division # _____ Supervisor _____
Floor # _____	Engine Cos.	Truck Cos.	Others	Division # _____ Supervisor _____

Lower Search and Rescue/Property Conservation Group		Engine Cos.	Truck Cos.	Others
Floors: From _____ to _____ Supervisor _____				

Floor # _____	Engine Cos.	Truck Cos.	Others	Division # _____ Supervisor _____

Lobby control	Stairway support
Supervisor _____ Company _____	Supervisor _____ Companies _____

FIGURE 12-15 High-rise tactical worksheet.
© Jones & Bartlett Learning.

Base (Exterior Staging for High-Rise Fires)

The term *base* is used by wildland fire fighters to identify a location housing reserve equipment and personnel (base camp). This term is not normally used in structural firefighting, but it has utility in managing high-rise fires. The base for high-rise fires is a location where support equipment and personnel are held in reserve on the *exterior* of the building. Depending on interior conditions and the location of the fire, an exterior staging area may be established near the fire perimeter. The reason for this distinction is that the staging area is normally moved inside the structure during a high-rise fire. Unless there is a possibility of moving to an exterior operation, the fire is involving other structures, or is located on a lower floor, it would be unusual to amass a large force on the exterior at a high-rise fire when the fire is on an upper floor. However, some department SOPs maintain a predetermined reserve force in base as a tactical reserve.

Staging (Interior)

In a high-rise situation, most of the reserve force is moved through the lobby and then to the interior staging area. This area is normally two or more floors below the fire. A rehabilitation area may be set up on the same floor as the staging area or on another floor. The concept is the same as in exterior staging—to provide a readily available reserve force. To avoid the confusion of having a different name for the exterior staging area, the interior staging area could be identified as interior staging.

> **NOTE**
>
> It is essential that uncommon terminology be shared not only with fire fighters in departments having jurisdiction but also with mutual aid fire departments that are likely to respond.

Department SOPs should address this issue, using either the suggested terms *base* and *staging* or other terminology that is consistent throughout the region.

The supervisor in charge of the staging area or an assistant can act as the accountability officer for crews that are working out of the staging area. Tracking crew rotations can be a challenging task. If an extended operation becomes necessary, fire fighters will be needed for each position on a hose line. For example, a four-person fire company operating a 2½-in. (64-mm) hose line on the fire floor will exhaust its air supply in approximately 15 to 30 minutes. To continually operate the hose line on the fire floor, this crew will need to be relieved by an available crew in staging.

As the crew leaves the fire floor, it will move to rehab, and another crew will move to the ready position **FIGURE 12-16**.

The duties of the staging officer are as follows:

- Report to the operations section, if it is staffed. Otherwise, interior staging would communicate directly with the IC. The IC could elect to have the interior staging officer communicate directly with the logistics section.
- Maintain records of the companies that are in staging and rehab.
- Maintain a minimum reserve of engine and truck company personnel as established by the IC.

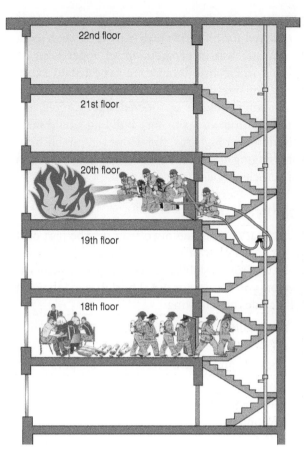

FIGURE 12-16 Fire companies rotating from rehab.
© Jones & Bartlett Learning.

- Request additional resources to maintain the established reserve force.
- Maintain an adequate supply of air cylinders and other equipment as needed.
- Supply first aid equipment and medical services for units that are involved in rescue and suppression.

In studying the NIST report on high-rise fire-ground field experiments (Averill et al., 2013), the importance of staging during a high-rise fire became obvious.

Consider the Trump Tower case study at the beginning of this chapter. The fire was on the 50th floor. The most likely place for staging would be on the 48th floor. If elevators were not safe to use, this would require fire fighters to walk up the stairs to the 48th floor, before being assigned to a task on the 50th floor. Without staging, fire fighters would first walk the steps, probably in full fire gear, and carrying equipment needed for their assignment. Once staging is established, fire fighters are assigned to carry needed equipment to the staging area, sometimes using a relay where fire fighters carry equipment for five to ten floors. Other fire fighters then carry equipment another 5 to 10 floors until the needed equipment reaches the staging area. The staging officer is managing a major logistics and rehabilitation operation.

The concept of an interior staging area should not be reserved for high-rise fires. Any time an extended interior attack is in progress on an upper floor of a building where fire fighters are expected to expend more than one tank of air, interior staging and rehabilitation should be considered. For example, a fire on the top floor of a six-story building would not by definition be a high-rise fire. However, establishing a rehab/staging area on the fourth floor could result in a more efficient operation.

Lobby Control

The primary accountability officer may be part of the lobby control crew or may be co-located in the lobby. Lobby control is established regardless of whether the elevators will be used or not. The duties of lobby control include the following:

- Controlling, operating, and accounting for all elevators
- Locating and controlling all interior stairs
- Directing incoming companies to the proper elevator or stairway
- Consulting with the building engineer
- Controlling or shutting down the HVAC system after consulting with the IC

Controlling the elevators and stairways is the only way to effectively gain access to the upper floors of a high-rise building. Many department SOPs assign these initial lobby control tasks to a member of the first-arriving truck company. Another important lobby control duty is to gain control of the HVAC system. If operated properly, the HVAC system can prevent heat, smoke, and toxic gases from reaching occupants. Conversely, if it is operated incorrectly, it may spread the products of combustion well beyond the immediate fire area.

In some buildings, it is virtually impossible to know what will happen when the HVAC system is operated because of the many unknown variables, such as heat produced by the fire, stack effect, and wind. Furthermore, few building engineers are completely knowledgeable about the operation of the HVAC system. For these reasons, many fire departments require an emergency shutdown switch near the lobby.

High-Rise Case Histories

The One Meridian Plaza Fire

The One Meridian Plaza Fire in Philadelphia, Pennsylvania, occurred at 8:40 PM on Saturday evening, February 23, 1991 **FIGURE 12-17** (Eisner, 1991;

FIGURE 12-17 One Meridian Plaza, Philadelphia, Pennsylvania.
© George Widman/AP Images.

Klem, 1992; Routley, 1991). The building, which had 38 floors, sustained millions of dollars in damage, and nine floors were destroyed. Major building operating systems failed, including main electrical power and emergency generator power. Because the fire occurred outside of normal working hours, the threat to occupants was minimal. However, fire fighters had to withdraw from the building due to the potential for building collapse. This fire resulted in three fire fighter deaths and 24 fire fighters were injured.

Fire fighters in Los Angeles at the First Interstate Bank Building Fire (discussed next) successfully mounted a manual attack on the 16th floor, whereas the One Meridian Plaza Fire was not contained until it reached a sprinkler-protected floor. The difference was the standpipe system. At One Meridian Plaza, pressure-reducing valves reduced the water pressure below the operating pressures needed to properly supply the automatic nozzles being used by the fire department. The codes governing standpipe installations required reducing valves to avoid overpressurization. The flow-reduction valves that were used to meet code requirements were incorrectly set at a pressure below the required pressure and could not be changed at the time of the fire. Manufacturers now provide nozzles that will operate effectively at lower pressures. Smooth-bore nozzles are another option when faced with low pressures.

Fire fighters provided a considerable occupant load in the fire building, as the operation resulted in 316 fire fighters being called to the scene. The building was in total darkness and without elevator service.

Perhaps the One Meridian Plaza Fire would not have gained the same headway had the building been occupied, because the fire probably would have been detected earlier. The reports do not provide the total number of occupants in the building during normal working hours, but the conceivable scenario is alarming. Each floor had 17,000 ft^2 (1579 m^2) of gross floor area. According to the NFPA 101, *Life Safety Code*, the means of egress would need to be sized to accommodate at least one person per every 150 ft^2 (14 m^2) of gross floor area. Based on this typical factor, there could have been 113 people or more per floor (NFPA 101, 2018).

The fire began on the 22nd floor and dropped down to the 21st floor. The top (38th) floor was a mechanical floor, meaning that the occupants on floors 21 through 37 (17 floors) would have been in immediate danger had the building been fully occupied. At 113 people per floor, 17 floors could be occupied by up to 1921 people. The eight floors that were destroyed could have held 1360 people or more. It would be a realistic assumption that the eight floors that were destroyed could have been occupied by 904 people or more.

The One Meridian Plaza Fire most certainly points out the potential for a large loss-of-life fire in a non-sprinklered high-rise building. The building was being retrofitted with sprinklers at the time of the fire. Floors 22 through 29 were not sprinkler protected and were lost to the fire. It is interesting to note that the fire was suppressed on the 30th floor by 10 sprinkler heads operating at the points of penetration. Can the value of sprinkler protection in high-rise buildings be underestimated? This fire serves as a testimonial to the value of automatic sprinkler systems in high-rise buildings.

The First Interstate Bank Building Fire

On May 4, 1988, at 10:25 PM, a fire similar to the One Meridian Plaza Fire occurred in Los Angeles, California, at the 62-story First Interstate Bank Building **FIGURE 12-18** (Klem, 1988; Routley, 1988). At this fire, four floors were completely destroyed and a fifth floor was heavily damaged. The building was

FIGURE 12-18 First Interstate Bank Building, Los Angeles, California.
© Russell Johnson/AP Images.

unoccupied except for security and maintenance personnel. The fire burned floors 12 through 15 and was stopped at the 16th floor. To evaluate the potential loss of life, use NFPA 101, *Life Safety Code*, to calculate the maximum number of people that could legally be in this building, remembering that codes are not always observed. In this case, the fire occurred at 10:25 PM, outside hours of peak occupancy. The fire was on the 12th floor of the 62-story building. Therefore, 51 floors were exposed to the fire and smoke conditions. In this case, the actual number of building occupants was listed as 4000. If the actual occupant load is averaged equally among the floors, as many as 3290 people could have been above the fire during normal working hours.

Comparing the One Meridian Plaza and First Interstate Bank Building Fires

Examining similarities between these two fires enhances the learning experience that would be gained by studying a single fire. **TABLE 12-2** compares the two incidents.

TABLE 12-2 Similarities Between the One Meridian Plaza and First Interstate Bank Fires

	One Meridian Plaza	First Interstate Bank
Deaths	3 fire fighters	1 maintenance person
Occupancy	Office building	Office building
Time of day	8:30 PM, unoccupied	10:25 PM, 40 occupants
Alarm	Delayed, > 4 minutes	Delayed, > 16 minutes
Sprinklers	Partial, being retrofitted	Partial, being retrofitted
Floor of origin	22	12
Area of involvement	8 floors, 17,000 ft²/floor (1579 m²/floor)	5 floors, 17,500 ft²/floor (1626 m²/floor)
Responding fire fighters	316	383

© Jones & Bartlett Learning.

When studying case histories, the effect of delayed alarms should be noted. A 16-minute delay occurred at the First Interstate Bank Building Fire. Had the fire department been notified 16 minutes earlier, there is a high probability that the fire could have been extinguished without exposing fire fighters to prolonged and extreme danger, and with much less damage. Some would speculate that this delay is unlikely when the building is fully occupied.

Francis L. Brannigan, in one of his monthly *Fire Engineering* articles, took exception to this theory. Someone suggested that during normal working hours the employees at the First Interstate Bank Building would have used the occupant-use hose line to extinguish the fire. Brannigan states the following in response:

> I think this answer is unrealistic. On a trading floor, operatives are dealing in multi-million-dollar split-second transactions. I think it would be more like the Bradford Stadium fire in England, where the fans kept alternately watching the fire and the game until they died. In addition, to the typical computer nerd, the thought of water on his computer is anathema (Brannigan, 1998).

There are many documented large-loss fires in occupied buildings where delayed alarms played a significant role. In hotel and office occupancies, security or maintenance personnel may first attempt to investigate the fire before calling the fire department. Sometimes, they become involved in firefighting operations with occupant-use hose lines or fire extinguishers and delay calling the fire department until the fire is out of control.

FIGURE 12-19 Fire at Marco Polo Apartments, Honolulu, Hawaii.
©Audrey McAvoy/AP Images.

A fire on the 26th floor of the Marco Polo apartment building on Kapiolani Boulevard in Honolulu, Hawaii spread to the 28th floor and killed three residents **FIGURE 12-19**. Conditions in the stairway on one side of the building were unsafe and unusable, complicating access and evacuation. There were reports that people were trapped. Fire fighters battled the fire, staging their equipment on the 24th floor.

The Marco Polo was built in 1971 and was not sprinkler protected.

A fire that occurred on January 6, 1995, in North York, Ontario, Canada, verifies Brannigan's assertion that delayed alarms should be anticipated in occupied buildings (NFPA, 1995). It also demonstrates that multifatality high-rise fires are not limited to office and hotel occupancies. This fire, which occurred in a 29-story apartment building, killed six occupants. The occupant in the apartment of origin first attempted to extinguish a couch cushion fire with water. Other occupants smelled smoke and investigated, then assisted in trying to extinguish the fire. Finally, one of the occupants called 911. The door to the apartment of origin was left open, allowing the fire to spread into the hallway and extend into one of the two stairways located at the central core.

Experienced fire officers have speculated that a working fire in a large high-rise building during periods of full occupancy could require the services of 500 or more fire fighters. The fires in Philadelphia and Los Angeles make this estimate appear low, given that both occurred when the building was occupied by only a small staff of security and maintenance personnel.

A last example is the fire at the Strand Condominium building that occurred in New York City in 2014 **FIGURE 12-20**. The fire occurred in an apartment on the 20th floor at 11:00 AM. Residents who called 911 were instructed to remain in their units. Occupants attempted to exit the building in stairwells that became filled with smoke. The building had 41 stories. For safety reasons, elevators were not in use. After the fire, some residents indicated there had been insufficient communication from building management. A 27-year-old male tenant died and four fire fighters were injured.

FIGURE 12-20 Fire on the 20th floor at the Strand Condominium building in New York City in 2014.
© Katherine Bourbeau/AP/Shutterstock.

Terrorist Attacks at the World Trade Center

The two largest high-rise incidents in the United States to date were the result of terrorist attacks on the World Trade Center in New York City **FIGURE 12-21**.

World Trade Center Bombing: February 26, 1993

The first attack involved a truck bomb (Isner, 1995; Manning, 1993). Although this incident was the result of a terrorist's bomb, fire and smoke were the major concerns. A quote from Chief Anthony L. Fusco after the first attack identifies the fire danger and reinforces the contention made throughout this book that the most important life-safety measure is often extinguishment. In his remarks, Chief Fusco states, "The decision to attack the basement fires in the initial stages of the incident was the most important decision of the incident [commander] in that hundreds, maybe thousands of lives were saved owing to timely extinguishment of the fires."

FIGURE 12-21 New York City skyline view including the World Trade Center.
© Jemini Joseph/ShutterStock, Inc.

The first World Trade Center attack should have removed any doubts about the potential for death and injury in high-rise building fires. The World Trade Center was a seven-building complex that included two 110-story towers. The area of each floor was 40,000 ft^2 (3716 m^2). It was estimated at the time of the truck bomb incident that 60,000 people were working in the buildings and there were thousands of visitors. Six deaths and 1042 injuries were reported at this incident. Fifty thousand people were evacuated from the building.

This incident was the equivalent of simultaneous multiple-alarm incidents.

- Five alarms for the Vista Hotel
- Five alarms for Tower 1
- Four alarms for Tower 2
- One alarm for additional resources

Forty-five percent of the department's on-duty resources were dispatched to this incident. In addition, a large contingent of police officers responded, and 174 ambulances treated and transported the injured. Few fire departments in the world have comparable resources. In fact, few fire departments could provide the staffing needed to search the 99 elevators in the towers.

The World Trade Center was sprinkler protected, but the blast destroyed the sprinkler piping in the garage area where the fire occurred. With the sprinkler system out of service in the garage, the fire continued to burn and push smoke to the floors above. Once the fire was extinguished, smoke production stopped. However, the smoke that had already entered the building lingered. As previously mentioned, venting a high-rise is problematic and usually takes considerable time.

World Trade Center Attack: September 11, 2001

The second attack on the World Trade Center resulted in the largest number of civilian and fire fighter fatalities ever recorded in a building fire **FIGURE 12-22**. Lessons are still being learned from this disaster. Although aircraft flying into buildings is not a typical high-rise fire scenario, it occurs frequently enough to deserve consideration. Small aircraft have flown into high-rise buildings on several occasions. A B-25 bomber flew into the Empire State Building in 1945. World Trade Center building designers had actually planned for an accidental aircraft accident involving a Boeing 707 aircraft making a landing approach. World Trade Center structures were designed to withstand the impact and fire from this slightly lighter aircraft

FIGURE 12-22 One of the fires resulting from the September 11, 2001 World Trade Center attack.
Greg Semendinger/Nypd/Shutterstock.

assumed to be low on fuel and traveling at approximately 180 miles per hour (mi/h [290 km/h]). This "planned for" accident would produce much less impact and fire than the Boeing 767s that were nearly full of fuel and traveling at an estimated 470 to 590 mi/h (756 to 950 km/h) when they were purposely flown into the buildings on September 11, 2001.

The 767 airplanes caused considerable structural damage on impact, but the two towers withstood these initial impact forces and remained standing. However, the aircraft debris moving through the building removed sprayed-on fire-resistive coatings that protected steel structural components. The ensuing fire weakened the unprotected steel, resulting in the collapse of Towers 1 and 2, killing 2749 people, including 340 fire fighters. NIST theorizes that the two buildings would not have collapsed had the thermal insulation remained in place (NIST, 2005). It is estimated that Towers 1 and 2 at the World Trade Center were 33% to 50% occupied at the time of the attack. Had they been fully occupied, NIST estimates that it could have taken as long as 3 hours to evacuate the buildings and as many as 14,000 people could have lost their lives.

Prior to the collapse of the two World Trade Center buildings, high-rise buildings had not been known to collapse, even when subjected to intense fires on multiple floors. The Meridian Plaza and Interstate Bank fires discussed earlier in this chapter both withstood severe fires on multiple floors and neither collapsed.

The MGM Grand Fire

During the MGM Grand (now Bally's) Fire in Las Vegas, Nevada, occupants displayed what we now call convergence cluster behavior **FIGURE 12-23** (Best and Demers, 1980; NFPA, 1982). They gathered in certain rooms as groups, gaining a feeling of safety in the presence of others. What implications does this have for the fire department? Searching fire fighters might not find anyone in several rooms or in an entire floor area, while one room may contain far more victims than anticipated.

The occupancy type has a direct bearing on the behavior and awareness level of occupants as well as the population density. The MGM Grand had 3400 registered guests at the time of the fire. The death and injury toll was due to smoke infiltrating the 21-story tower due to a fire that originated in the first floor casino. The death toll of 85 people made this the second most deadly hotel fire in U.S. history (119 died in the Winecoff Hotel in Atlanta, Georgia, in 1946).

FIGURE 12-23 MGM Grand Fire, Las Vegas, Nevada.
© Robert M. Stanzler/Las Vegas Review-Journal/AP Images.

The fire in the MGM Grand started at the rear of the casino and eventually involved the entire 150- × 450-ft (46- × 137-m) casino floor. The ceiling height was not given in the reports, but the ceilings were probably a minimum of 20 ft (6 m) high. Using the V/100 formula for 20-ft (6-m) ceilings yields a rate of flow of 13,500 gallons per minute (GPM [51,103 L/min]). The size of the fire area and the required rate of flow led to the conclusion that this fire was not going to be extinguished by manual means, and there were no sprinklers in the fire area. Therefore, the best tactic was to keep the fire out of the stairways and towers; this was accomplished. However, smoke did spread throughout the structure, killing 85 people.

Wrap UP

CHAPTER SUMMARY

- Tactics and strategic objectives used in high-rise firefighting are the same as those that apply to any other structure fire, but with special considerations because of the height of the building.
- High-rise fires are some of the most challenging incidents a fire department encounters because of the high occupant load and building height/configuration.
- Preincident planning, code enforcement, use of the NIMS, and development of high-rise standard operating procedures improve the chance for successful operations in high-rise buildings.
- Logistics and access problems increase with building height.
- Avoid using elevators unless they will substantially improve operations, and even then, use with extreme caution.
- During a structure fire, never use an elevator to travel to or above the fire floor; exit the elevator at least two floors below the fire.
- All elevators should be under the control of fire fighters.
- When in doubt about the safety of elevators, a possible option is to use the elevator to send equipment (rather than personnel) to the floors above.
- Getting fire fighters and equipment to an interior staging area via stairs is essential when elevators are unsafe or unavailable and returning to ground level is impractical.
- Stairway transport of equipment is done via relay (ground support) to conserve personnel.
- When the fire is on an upper floor of a high-rise building, interior staging should be set up two or more floors below the fire.
- Helicopter rescues are extremely dangerous and usually unnecessary. Partial or sequential evacuation strategies can be determined in advance of a fire and incorporated into the alarm system.
- An emergency voice/alarm communications system (EVACS) is available in some buildings, but occupants do not always follow the instructions given.
- Most exterior defensive fire control tools are ineffective above the eighth floor. Fire forces must prevent fire extension into the stairways, which are the safest, most effective, and sometimes only available egress routes.
- Pressure-reducing valves and fire pumps both affect standpipe hose pressures.
- Due to large numbers of people and other distractions in the lobby, a high-rise lobby may be a poor choice of location for a command post.
- A command post's location should be transmitted to all responders regardless of whether it is inside or away from the building.
- Subtle differences in interior building configurations can have a major impact on firefighting and rescue operations.
- Weather conditions, fire intensity, stack effect, wind, and atmospheric pressure can have a significant effect on smoke movement within a high-rise structure.
- Dedicating a stairway with standpipe outlets for firefighting reduces the possibility of opposing hose lines on the fire floor.
- To avoid exposing people and equipment at ground level to falling hazards, a wide fire perimeter is essential.
- Communications is the most frequently cited problem during major emergency operations.
- Controlling the lobby, elevators, stairways, and HVAC system facilitates both fire fighter access to and successful evacuation of the upper floors of a high-rise building.

KEY TERMS

lobby control A high-rise assignment in which a crew is responsible for managing the stairways; elevators; heating, ventilation, and air-conditioning systems; and related duties.

smoke-proof tower A stairway that is designed to be separated from the building by a landing. This creates a separation that will limit the spread of smoke into the stairway and keep it clear for evacuation.

stack effect The vertical airflow within buildings caused by temperature differences between the building interior and the exterior; depending on conditions, stack effect could be positive or negative, causing smoke to move upward or downward.

SUGGESTED ACTIVITIES

1. This question is based on the Trump Tower Fire (see photo in the "Case Study" section at the beginning of this chapter).
   A. Provide an initial report to dispatch including any request for addition alarms.
   B. List high-priority assignments.
   C. How many and what size hose lines would you deploy to the fire floor and above?
   D. How is extinguishment related to rescue on the upper floors of this high-rise building?
   E. What role does compartmentation play in saving lives and properties?
   F. List evacuation options that are available to occupants.
   G. If an elevator is used, should companies using the elevator travel to the fire floor via elevator? If not, where should they exit the elevator and why?
   H. Develop an incident action plan, including companies needed for each task.
   I. Is there a need for nonfire department resources, e.g., law enforcement or water utility?
   J. Should a staging area be established? If yes, where, how many, and what kind of units (engine company, EMS, etc.) should be assigned to staging?
   K. What equipment is needed in the staging area(s)?
   L. Develop an incident command system (ICS) chart for this incident.

2. This question is based on the Cook County Administration Building Fire.
   A. What evacuation options are available to occupants beyond the reach of aerial devices?
   B. How is extinguishment related to rescue on upper floors of a high-rise building?
   C. What role does compartmentation play in saving lives and properties?

3. This scenario is based on an actual fire that occurred in Kensington, Maryland **TABLE 12-3**.

   It is 5:15 AM. A resident noticed a fire on the ninth floor and went into the main hallway, transmitted the alarm, and tried to extinguish the fire using a fire extinguisher. The building is equipped with a standpipe but not a sprinkler system.

   Assume the role of the district chief responding from a nearby station with an engine and ladder company. Provide an initial report to dispatch, including any request for addition alarms.
   A. List actions that need to be taken immediately.
   B. How many and what size hose lines would you deploy on the fire floor and above?
   C. How is extinguishment related to rescue on upper floors of this high-rise building?
   D. What role does compartmentation play in saving lives and properties?
   E. What evacuation options are available to occupants?

TABLE 12-3 Responding Fire Department Fire Companies, Fire Officers, and Emergency Medical Resources

	Engine Companies	Truck Companies All Equipped with 100-ft (30.5-m) Aerial Ladder	Other Units
1st Alarm	Engine 1 Engine 2 Engine 3 Engine 4	Truck 1 Truck 2	District Chief 1 (you) Paramedic 1
2nd Alarm	Engine 5 Engine 7 Engine 6 Engine 8	Truck 3 Truck 4	District Chief 2 Operations Assistant Chief Safety officer Paramedic 2 Heavy Rescue 1
3rd Alarm	Engine 9 Engine 11 Engine 10 Engine 12	Truck 5 Truck 6	Fire chief Operations assistant chief District Chief 2 Paramedic 2 Heavy Rescue 1

© Jones & Bartlett Learning

- **F.** Should companies assigned to the fire floor and above use the stairs, elevators, or both to reach the fire floor, and why?
- **G.** If an elevator is used, should companies using the elevator travel to the fire floor via elevator? If not, where should they exit the elevator and why?
- **H.** Develop an incident action plan, including companies needed for each task. Also, request nonfire personnel resources as needed (e.g., law enforcement or water utility).
- **I.** Should a staging area be established?
- **J.** If yes, where should staged personnel be located? How many, and what kind of units (engine company, EMS, etc.) should be assign to staging?
- **K.** What equipment is needed in staging?
- **L.** Develop a NIMS chart for this incident.

4. Use the following RGB Management high-rise fire scenario to do the following:
   - **A.** Size up the fire using information from this chapter and the size-up factors discussed in Chapter 2, *Procedures, Preincident Planning, and Size-Up*.
   - **B.** Establish a strategic mode: offensive, defensive, or non-attack.
   - **C.** Outline an incident action plan.
   - **D.** Assign/reassign companies to tactics and tasks necessary to carry out the incident action plan.
   - **E.** Organize the operation using NIMS. Use high-rise positions described in this chapter as necessary, such as ground support, lobby control, staging, and base.

   FIGURE 12-24 is the preincident plan narrative for the RGB Building.

RGB Building
500 Vine Street

Life Safety
Occupancy Type
- Business Occupancy (Office Building) with attached garage
- First floor Lobby, Restaurant, Bank, Gift Shop
- 26th floor storage area, 27th floor mechanical room

Estimated Number of Occupants
- Possibly 9000 weekdays 0700 to 1800 hrs
- 2 Security Personnel at all times
- 5 to 40 Maintenance Personnel outside working hours

Primary and Alternative Egress Routes
- 3 Stairways (see floor plan)
- Doors from stairway to hallways are locked on stairway side (no re-entry to hallways)

Access to Building
- Paved street access on sides A, B and C
- Parking garage on side D
- Building lobby open weekdays 0600 to 2000 hrs.
- Security maintains occupant log at other times

Construction Type
- Fire Resistive, central core high-rise
- 26 stories (no 13th floor), height 316 ft (96 m)
- Basement and Sub-basement
- Area at base (sub-basement to 3rd floor) 144 ft × 185 ft. (44 m × 56 m)
- Area on tower floors (4th to 27th) 60 ft. × 160 ft. (18 m × 49 m)
- Open stairs and escalators between 1st and 2nd floors

Condition
- Inspected every 6 months, minor violations only
- Building well maintained

Live and Dead Loads
- Typical Office Occupancy
- Large fire load on 26th floor storage area
- Large live load in mechanical room on 27th floor

Enclosures and Fire Separations
- First floor lobby open layout
- Floors 3 to 23 mostly open layout with cubicle separations between work stations
- Floors 24 and 25 executive offices with additional security. Subdivided into large offices around periphery, work stations near core.

Age
- Built in 1970

Extinguishment
Probability of Extinguishment
- Most building areas will require multiple 2½" lines for extinguishment

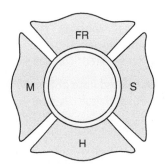

FIGURE 12-24 RGB high-rise preincident plan narrative.
Courtesy of Ben Klaene.

External Exposures
- Wide streets on Sides A, B, and C

Internal Exposures
- Attached garage
- Non-involved floors, some utility openings not sealed

Calculated Rate of Flow V/100
- Basement/sub basement = 2000 GPM
- 1st floor lobby = 4500 GPM
- 1st floor bank = 480 GPM
- 1st floor restaurant = 320 GPM
- 2nd floor lobby = 920 GPM
- Tower Floors w/open layout = 960 GPM
- Garage all levels = 760 GPM

Water Supply
- 20,000 GPM available from primary water supply
- Can be cross tied to suburban system adding 20,000 GPM
- Hydrants located on sides A, B, and C of building. Hydrant spacing 250'

Total Staffing/Apparatus Available

Alarm	Engine Companies	Truck Companies	Other
1st	1, 2, 3 and 4	1, 2, 3	Heavy Rescue 1, Bat. Chief 1
2nd	5, 6, 7, and 8	4 and 5	Bat. Chief 2, ALS 1
3rd	9, 10, 11, and 12	6	Fire Chief, 4 Asst. Chiefs, Heavy Rescue 2
4th	13, 14, 15, and 16	7	2 Bat. Chief Administrative Staff, ALS 2 and ALS 3
5th	17, 18, 19, and 20	8	Air Supply Truck, Command Van

Each fire company including Heavy Rescue staffed with an officer, driver and 2 fire fighters.
Department High-Rise SOP requires a full one alarm compliment be staged in Base when a working fire is declared.
Fire Department Has 40 Engine Companies, 18 Truck Companies, 3 heavy rescues, 12 ALS Units, 4 Assistant Chiefs, 5 on-duty Battalion Chiefs, 5 Administrative Battalion chiefs.
Approximately 700 off duty fire fighters can be recalled via a telephone relay system.

Fire Protection Systems
- Standpipe System with 1000 GPM pump
- FDC on Vine St. side of building
- 100 psi at each standpipe outlet, center stairs
- Field adjustable pressure-regulating valves, PRV tool in lock box.
- Dry Chemical hood system in kitchen area of restaurant on 1st floor

Property Conservation
- Computers and paper files - very susceptible to water damage

General Factors
Utilities (Water, Gas, Electric)
- Main Electric in Sub-basement (DO NOT ATTEMPT TO DE-ENERGIZE)
- Gas and Water Street Valves on Vine St.
- Gas and water equipment/meters in basement

FIGURE 12-24 *(Continued)*

FIGURE 12-25, **FIGURE 12-26**, and **FIGURE 12-27** are drawings showing floor layouts at various levels. Use the preplan information and scenario to guide your response. Assume the role of IC and retain that role, even though command may change.

Using the following information from the alarm card, call whatever assistance you need. Start by using on-duty resources, then request mutual aid and off-duty assistance as needed. Background information includes the following:

- Fire report from dispatch: "Smoke detector alarms were received from floors 19, 20, 21, 22, and 23, and the 27th floor equipment room. In addition, calls were received from the 20th floor reporting a fire in the office area. A call from 21st floor reported heavy smoke conditions. We are getting multiple calls on this one."
- Day/date: Tuesday, August 1
- Time: 1600 hours (4:00 PM)
- Weather conditions:
  - 90°F (32.2°C)
  - 90% humidity
  - Partly cloudy
  - Wind from the southwest at 1 mph (1.6 kmph)
- The report from Engine 1 is as follows:

"Engine 1 and Truck 1 are on the scene of a 26-story office building with a working fire. Fire is visible from several windows on an upper floor on the Central Parkway side. We are proceeding to the 18th floor using the elevator. Engine 1 is RGB command."

Engine 1 is carrying a high-rise pack including 2½-in. (64-mm) hose to the fire area. You arrive on the scene at the same time as Engine 2, Truck 2, and the heavy rescue unit. You notice that Engine 1 has a water

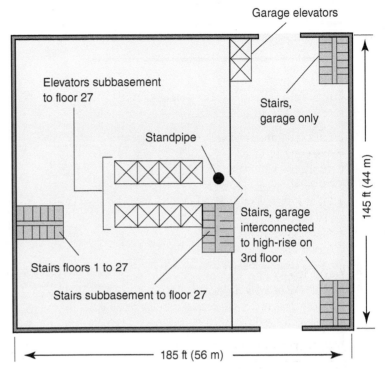

FIGURE 12-25 RGB high-rise typical floor plans—floors 1 to 3 of high-rise and floors 1 to 6 of garage.
© Jones & Bartlett Learning.

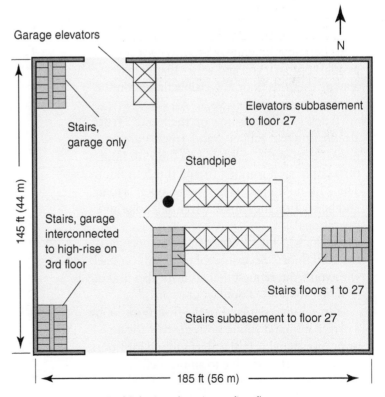

FIGURE 12-26 RGB high-rise plan view—first floor.
© Jones & Bartlett Learning.

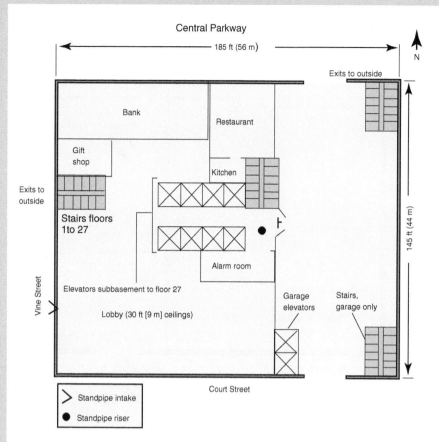

FIGURE 12-27 RGB high-rise plan view—upper levels.
© Jones & Bartlett Learning.

supply and the pump operator is at the apparatus hooking up to the hydrant. The security people approach you and verify that people leaving the 20th floor saw visible flames on the Central Parkway side.

Truck 1's officer is now on the radio: "Truck 1 reporting that Engine 1 and Truck 1 are hooking up to the standpipe on the 19th floor for a confirmed working fire on the 20th floor. The 20th floor appears to have an open layout and heavy fire. Elevator 2 is free of smoke and operational."

As the IC, you ask security, "Who occupies the 20th floor?" Security replies, "I will get back with you." You request that security also find out as much about the floor layout as possible.

You set up command in the first-floor lobby alarm/command room, noticing that the annunciator panel is in the alarm mode for floors 19 to the penthouse due to smoke detector activation.

It is now 4:08 P.M. You strike the second alarm, and Engines 3 and 4 as well as Truck 2 are approaching the command post. The heavy rescue unit is in the elevator lobby suggesting they assist the Engine 1, Truck 1 attack team on the 20th floor; you approve and they proceed toward the 20th floor with additional SCBA cylinders. All of the elevators have been recalled to the first floor. Truck 3 is assuming their normal role as the rapid intervention crew (RIC) and proceeding via an elevator to the 18th floor with extra SCBA cylinders and RIC equipment.

In accordance with SOPs, Truck 2 is setting up PPV in all three stairwells beginning with the evacuation stairway on the Vine Street side. They will continue as the vent team to improve conditions in the stairways and assist in evacuation in the stairways as needed.

Engine 1 is now reporting back to the command post via radio: "Engine 1 and Truck 1 are now on the 20th floor. We estimate the fire to cover the entire Central Parkway side of an open-layout floor plan. We have heavy smoke throughout the floor and we are not making progress on the fire. We need additional hose streams or a master stream up here."

Truck 1 is now reporting back to the command post via radio: "On the way up to the 20th floor we ran into at least 10 people evacuating via the stairway. The occupants looked okay but a few were coughing. The stairway has moderate smoke from the 19th floor to at least the 21st floor."

5. Consider the RGB scenario, but with the following changes to the situation and occupancy. Explain how these changes will affect the operations you described in the previous question:

- Place the fire in a residential occupancy for the elderly.
- Place the fire in a hotel.
- Change the situation as follows:
  - Elevators are out of service.
  - There is a 45-mi/h (72-km/h) wind from the north (Central Parkway side).

- The fire area is subdivided into small office areas.
- The fire occurs on Sunday afternoon in an office building.
- The building is protected by a working sprinkler system.
- There are heavy smoke conditions on the fire floor and all floors above.

REFERENCES

Ahrens, Marty. November 2016. "High-Rise Fire Buildings." *NFPA Fire Analysis & Research*; p. 2. https://www.nfpa.org/-/media/Files/News-and-Research/Fire-statistics-and-reports/Building-and-life-safety/oshighrise.pdf. Accessed March 9, 2020.

Averill, Jason D., Lori Moore-Merrell, Raymond T. Ranellone, Jr., et al. April 2013. *Report on High-Rise Fireground Field Experiments* [NIST Technical Note 1797]. Gaithersburg, MD: National Institute of Standards and Technology. https://nvlpubs.nist.gov/nistpubs/TechnicalNotes/NIST.TN.1797.pdf. Accessed March 9, 2020.

Best, Richard L., and David P. Demers. 1980. "MGM Grand Hotel Fire, Las Vegas, NV." *NFPA Fire Investigative Report*. Quincy, MA: National Fire Protection Association.

Brannigan, Francis L. 1998. "Paranoia May Save Your Life." *Fire Engineering*. July.

Demers, David P. 1982. *Investigative Fire Report on the Las Vegas Hilton Hotel Fire*. Quincy, MA: National Fire Protection Association.

Eisner, Harvey. 1991. "One Meridian Plaza Fire." *Fire Engineering* 8 (144). https://www.fireengineering.com/1991/08/01/215776/one-meridian-plaza-fire/#gref. Accessed March 11, 2020.

Independent Commission appointed by Cook County Board President John H. Stronger Jr. October 17, 2003. *Report of the Cook County Commission, Investigating the 69 West Washington Building Fire*. Chicago, IL: Independent Commission.

Isner, Michael S., and Thomas J. Klem. 1995. "World Trade Center Explosion and Fire." Fire Investigation Report. *NFPA Journal* 2: 59–67.

Kerber, Steve. March 2007. *Evaluating Positive Pressure Ventilation in Large Structures: High-Rise Pressure Experiments* [NISTIR 7412]. Gaithersburg, MD: National Institute of Standards and Technology.

Klem, Thomas J. 1988. "First Interstate Bank Building Fire, Los Angeles, CA." *Fire Investigation Report*. Quincy, MA: National Fire Protection Association.

Klem, Thomas J. 1992. "One Meridian Plaza, Three Fire Fighter Fatalities, Philadelphia, PA." *Fire Investigation Report*. Quincy, MA: National Fire Protection Association.

Madrzykowski, Daniel, and Stephen Kerber. January 2009. *Fire Fighting Tactics Under Wind Driven Conditions: Laboratory Experiments* [Technical Note 1618]. Gaithersburg, MD: National Institute of Standards and Technology.

Manning, William A. 1993. *The World Trade Center Bombing: Report and Analysis* [USFA-YR-076]. Emmitsburg, MD: U.S. Fire Administration.

National Institute of Standards and Technology. 2005. *NIST, NCSTAR 1 Federal Building and Fire Safety Investigation of the World Trade Center Disaster, Final Report*. Gaithersburg, MD: NIST.

National Fire Protection Association. 2003. *NFPA Fire Protection Handbook, Nineteenth Edition*. Quincy, MA: NFPA; pp. 12–119.

National Fire Protection Association. 2018. *NFPA 101: Life Safety Code*. Quincy, MA: NFPA.

National Fire Protection Association. 2020. *NFPA 1710: Standard for the Organization and Deployment of Fire Suppression Operations, Emergency Medical Operations, and Special Operations to the Public by Career Fire Departments*. Quincy, MA: NFPA.

National Institute of Standards and Technology. 2010. *Report on Residential Fireground Field Experiments* [Technical Note 1661]. Gaithersburg, MD: NIST.

NFPA Fire Investigations Department. 1995. "Residential High-Rise, North York, Ontario, Canada, January 6, 1995." *Fire Investigation Report*. Quincy, MA: National Fire Protection Association.

NFPA Fire Investigations Department. 1982. "The MGM Hotel Fire, Part 1." *Fire Service Today*. January: 18–23. Quincy, MA: National Fire Protection Association; Figure CS8-1.

Routley, J. Gordon. 1988. *Interstate Bank Building Fire, Los Angeles, California*. Emmitsburg, MD: U.S. Fire Administration.

Routley, J. Gordon, Charles Jennings, and Mark Chubb. 1991. *High-Rise Office Building Fire, One Meridian Plaza, Philadelphia, Pennsylvania*. Emmitsburg, MD: U.S. Fire Administration.

Appendix A
FESHE Correlation Guide

© Courtesy of David J. Jones, Cincinanti, Ohio.

Strategy and Tactics FESHE Course Outcomes	*Structural Firefighting: Strategy and Tactics, Fourth Edition* Chapter Correlation
1. Discuss fire behavior as it relates to strategies and tactics.	2, 3, 5, 6, 9, 12
2. Explain the main components of prefire planning, and identify steps needed for a prefire plan review.	2, 7, 8, 11, 12
3. Identify the basics of building construction and how they interrelate to prefire planning and strategy and tactics.	2, 5, 6
4. Describe the steps taken during size-up.	2, 12
5. Examine the significance of fire-ground communications.	1, 5, 12
6. Identify the roles of the National Incident Management System (NIMS) and Incident Management System (ICS) as it relates to strategy and tactics.	1, 8, 9, 12
7. Demonstrate the various roles and responsibilities in ICS/NIMS.	1, 8, 9, 12

Glossary

20-minute rule A general rule used for estimating the length of time until structural collapse occurs. It states that when a heavy volume of fire is burning out of control on two or more floors for 20 minutes or longer in a building of ordinary construction, structural collapse should be anticipated.

A

accountability system A system established on the fire ground to ensure that everyone entering the area has a specific assignment, to track all personnel at the scene, and to identify the location of any missing personnel if a catastrophic event should occur.

area of refuge A floor area with at least two rooms separated by smoke-resisting partitions in a building protected by a sprinkler system, or a space located in an egress path that is separated from other building spaces.

assembly occupancy An occupancy used for the gathering of 50 or more persons for deliberation, worship, entertainment, eating, drinking, amusement, awaiting transportation, or similar uses; or used as a special amusement building, regardless of occupant load.

attack pumper The first-arriving engine company that goes directly to the fire building without securing a water supply.

automatic mutual aid Resources dispatched on the initial alarm without special request from the authority having jurisdiction, and based on preexisting agreements between agencies.

B

backdraft A fire condition that occurs when oxygen (air) is introduced into a superheated, oxygen-deficient compartment charged with smoke and pyrolytic emissions, resulting in an explosive ignition.

backpressure Also known as elevation pressure, the pressure required to overcome the weight of water in a piping or hose system. Each vertical foot of water in a pipe, hose, or tank exerts a pressure of 0.434 psi (3 kPa) at the base.

balloon-frame construction An older type of wood-frame construction in which the wall studs extend vertically from the basement of the structure to the roof.

branches Immediately subordinate to section in NIMS hierarchy; these units are subordinate to the logistics, finance/administration, and planning sections. They are used to reduce the span of control at very large operations or to manage a particular function/agency.

business occupancy An occupancy used for account and record keeping or the transaction of business other than mercantile.

C

chain of command Organizational structure establishing a line of authority and responsibility along which orders and instructions are passed (e.g., IC to operations section, to branch director, to division supervisor, to company officer, to fire fighter).

Class A foam Foam for use on fires in Class A fuels such as vegetation, wood, cloth, paper, rubber, and some plastics.

cold zone An area where personal protective clothing is not required; the command post, rehabilitation, and medical treatment should be located in this zone.

collapse zone The area endangered by a potential building collapse; generally considered to be an area 1½ times the height of the involved building.

compartmentation Subdividing of a building into small areas (rooms) capable of limiting the spread of fire and the products of combustion.

complex preincident plan A plan used when a property has more than three buildings or when it is necessary to show the layout of the premises and relationship between buildings on the site.

convergence cluster phenomenon A reaction to fire conditions in which groups take shelter together to provide mutual support.

critical time The time available until the structure becomes untenable.

D

dead load The weight of a building; consists of the weight of all materials of construction incorporated into a building, including but not limited to floors, roofs, ceilings, stairways, built-in partitions, finishes, cladding, and other similarly incorporated architectural and structural items, as well as fixed service equipment.

decay phase The phase of fire development in which the fire has consumed either the available fuel or oxygen to a point that the fire begins to diminish in intensity.

defend-in-place A tactic utilized during a structure fire when it is very difficult to remove occupants from the building. Occupants are either protected at their present location or moved to a safe location within the building.

defensive A type of fire attack in which exterior fire suppression operations are directed at protecting exposures.

deluge system A sprinkler system in which all sprinklers or applicators are open. When an initiating device, such as a smoke detector or heat detector, is activated, the deluge valve opens and water discharges from all of the open sprinklers simultaneously.

detention and correctional occupancy An occupancy used to house four or more persons under varied degrees of restraint or security where such occupants are mostly incapable of self-preservation because of security measures not under the occupants' control.

direct attack Firefighting operations involving the application of extinguishing agents directly onto the burning fuel.

division Tactical-level management unit in charge of a geographic area.

division of labor principle The principle that an incident or task should be broken down into smaller, more manageable tasks and personnel assigned to complete those tasks (developing job skills in a concentrated area to allow for more productivity).

draft curtains Walls designed to limit horizontal spread of the fire that extend partially down (usually no more than 20% of the height of the compartment) from the underside of the roof.

drilling down Using a computer to navigate by pointing and clicking through a series of drop-down menus in a graphical user interface.

dry pipe sprinkler system A system in which the pipes are normally filled with compressed air or nitrogen. When a sprinkler is activated, it releases the air from the system, which opens a valve so the pipes can fill with water.

E

educational occupancy An occupancy used for educational purposes through the 12th grade by six or more persons for 4 or more hours per day or more than 12 hours per week.

extension Fire that moves into areas not originally involved, including walls, ceilings, and attic spaces; also, the movement of fire into uninvolved areas of a structure.

external exposures Buildings, vehicles, or other property threatened by fire that are external to the building, vehicle, or property where the fire originated.

F

finance/administration section The section that tracks and provides financial and administrative services required to compensate personnel or organizations providing goods and services at the incident scene.

fire perimeter A wide area beyond the hazard control zones, usually staffed by police to keep unauthorized people away from the scene.

flashover An oxygen-sufficient condition in which room temperatures reach the ignition temperature of the suspended pyrolytic emissions, causing all combustible contents to suddenly ignite.

flow path The movement of heat and smoke from the higher pressure fire area toward lower pressure areas on the interior and exterior of the structure.

foam house A fixed facility consisting of an enclosure housing a foam concentrate supply tank, a foam solution proportioning system, a pump, and sometimes an extra supply of foam concentrate that can be added to the proportioning system.

formal preincident plan A plan for a property with a substantial risk to life and/or property; includes a drawing of the property, specific floor layouts, and a narrative describing important features.

freelancing Performing tasks outside the incident organization structure.

frequency As related to fire fighter injuries, a measure of how often an injury occurs; for example, sprains and strains are the most frequent fire fighter injuries.

fuel load Fuels provided by a building's contents and combustible building materials; also called fire load.

fuel-controlled Refers to a fire in which the heat release rate and growth rate are controlled by the characteristics of the fuel, such as quantity, chemistry, and geometry, and in which adequate air for combustion is available.

fully developed phase The phase of fire development at which the fire is free-burning and consuming much of the fuel.

G

group The tactical-level management unit in charge of a function.

growth phase The phase of fire development at which the fire is spreading beyond the point of origin and beginning to involve other fuels in the immediate fire area.

H

hazard control zone The area in which emergency responders are working; it can be subdivided into no-entry, hot, warm, and cold zones.

healthcare occupancy An occupancy used for purposes of medical or other treatment or care of four or more persons, where such occupants are mostly incapable of self-preservation due to age, physical or mental disability, or security measures not under the occupants' control.

hot zone An operating area considered safe only for individuals wearing appropriate levels of personal protective clothing; established by the IC and safety officer.

hovering Accessing text or graphic items on a computer screen by placing the computer cursor over a specific area or by touching the screen in a touch screen environment.

I

ignition phase The phase of fire development at which the fire is limited to the immediate point of origin.

immediately dangerous to life and health (IDLH) Exposure to airborne contaminants that are likely to cause death or immediate or delayed permanent adverse health effects or prevent escape from such an environment. Examples include smoke or other poisonous gases at sufficiently high concentrations.

incident action plan The objectives for the overall incident strategy, tactics, risk management, and member safety that are developed by the incident commander and updated throughout the incident.

incident commander (IC) The person responsible for all incident activities, including the development of strategies and tactics and the ordering and release of resources.

incident management team (IMT) A team consisting of the incident commander and appropriate command and general staff personnel assisting the IC in managing an incident. May also consist of specially trained and credentialed members who, at the request of another jurisdiction, are called in to assist or manage an incident.

incident safety officer The command staff position assigned to monitor the scene for hazards or unsafe operations; enforces safety practices, establishes a safety plan, and coordinates with representatives from cooperating and assisting agencies.

indirect attack Firefighting operations involving the application of extinguishing agents to reduce the buildup of heat released from a fire without applying the agent directly onto the burning fuel; ventilation is kept to a minimum while a fog stream is directed at the ceiling. This approach is most useful in unoccupied, tightly enclosed spaces.

industrial occupancy An occupancy in which products are manufactured or in which processing, assembling, mixing, packaging, finishing, decorating, or repair operations are conducted.

initial rapid intervention crew (IRIC) A standby team of at least two fire fighters located outside the hazard area available to provide assistance to fire fighters operating within the hazard area until a formal RIT can be established.

intelligence/information section The section that provides information related to the prevention of criminal activity, including terrorism or the apprehension of perpetrators of criminal activity.

internal exposures Areas within the structure where the fire originates or within buildings directly connected to the building of origin that were not involved in the initial fire ignition.

L

liaison officer A member of the command staff who is the point of contact for agencies that are not assigned to operations functions.

live loads The weight of the building's contents, people, or anything that is not part of or permanently attached to the structure.

lobby control A high-rise assignment in which a crew is responsible for managing the stairways; elevators; heating, ventilation, and air-conditioning systems; and related duties.

logistics section The section that obtains needed supplies, equipment, and facilities.

M

mercantile occupancy An occupancy used for the display and sale of merchandise.

mixed multiple occupancy A multiple occupancy where the occupancies are intermingled.

multiple occupancy A building or structure in which two or more classes of occupancy exist.

N

National Fire Academy formula A rate of flow calculation that calculates the rate of flow as the area in square feet divided by 3 (A/3).

National Incident Management System (NIMS) A U.S. Department of Homeland Security system designed to enable federal, state, and local governments and private-sector and nongovernmental organizations to effectively and efficiently prepare for, prevent, respond to, and recover from domestic incidents, regardless of the cause, size, or complexity, including acts of catastrophic terrorism.

no-entry zone An area that is unsafe regardless of the level of personal protective equipment and that must be cleared of all personnel, including emergency response personnel.

notation A piece of information about the premises, such as damage to the building from a previous fire. This information may accompany a preincident plan or may be available when the building does not have a preincident plan.

O

occupant load The total number of persons that might occupy a building or portion thereof at any one time.

occupational injury An injury sustained during the duties, responsibilities, and functions of a fire department member. (NFPA 1500)

offensive A type of fire attack in which fire fighters advance into the fire building with hose lines or other extinguishing agents to overpower the fire.

operations section The section that manages all tactical units deployed at an incident scene.

overhaul Examination of all areas of the building and contents involved in a fire to ensure that the fire is completely extinguished.

P

planning section The section that gathers and evaluates information, assists the incident commander in developing the incident action plan, and tracks progress; also tracks resource status.

platform-frame construction A construction technique using separate components to build the frame of a structure (one floor at a time). Each floor has a top and bottom plate that act as fire-stops.

preaction sprinkler system A dry sprinkler system that uses a deluge valve instead of a dry pipe valve and requires activation of a secondary device before the pipes fill with water.

prefire conditions Factors that can contribute to a collapse in a building that is heavily involved in fire; these factors include construction type, weight, fuel load, damage, renovations, deterioration, support systems, and related factors such as lightweight truss ceilings and floors.

preincident plans Written documents resulting from the gathering of general and detailed information to be used by public emergency response agencies and private industry for determining the response to reasonable anticipated emergency incidents at a specific facility.

primary damage Damage caused by the products of combustion.

primary factors The most important factors, assessed during size-up, which change from incident to incident and depend on specific incident conditions.

public information officer (PIO) A member of the command staff responsible for interfacing with the public and media or with other agencies with incident-related information requirements.

pyrolysis The chemical decomposition of a compound into one or more other substances by heat alone; the process of heating solid materials until combustible vapors are emitted.

Q

quad A multifunction apparatus that is equipped to provide for the following four functions: water, pumps, hose, and ground ladders. In other words, this is a quint company minus the aerial ladder.

quint A multifunction apparatus that is equipped to perform both engine and truck company operations. It is equipped to provide the following five functions: water, pumps, hose, ground ladders, and an aerial ladder.

R

rapid intervention crew (RIC) A minimum of two fully equipped personnel onsite, for immediate rescue of injured or trapped fire fighters.

rate of flow The minimum water application rate required for extinguishment.

rational decision making (RDM) A form of decision making in which input is obtained from diverse sources and careful analyses of all options is considered.

recognition-primed decision making (RPD) A form of decision making in which the incident commander must decide on the proper course of action with limited information available in a relatively short period of time.

rehabilitation The process of providing rest, rehydration, nourishment, and medical evaluation to members involved in extended incident scene operations and/or extreme weather conditions.

residential board and care occupancy A building or portion thereof that is used for lodging and boarding of four or more residents, not related by blood or marriage to the owners or operators, for the purpose of providing personal care services.

residential occupancy An occupancy that provides sleeping accommodations for purposes other than health care or detention and correctional.

risk-versus-benefit analysis The process of weighing predicted risks to fire fighters against potential benefits for owners/occupants and making decisions based on the outcome of that analysis.

Royer/Nelson formula A rate of flow calculation that calculates the rate of flow as the volume in the fire area in cubic feet divided by 100 (V/100). This formula is based on the assumption that structure fires are primarily ventilation controlled.

S

secondary damage Damage caused by fire-ground activities or the operation of a fire suppression system.

secondary factors Less important factors at an incident, which change from incident to incident and depend on specific incident conditions.

separated multiple occupancy A multiple occupancy where the occupancies are separated by fire resistance-rated assemblies.

severity The extent of an injury's consequence, usually categorized as death, permanent disability, temporary disability, and minor.

single command A command structure in which one person is designated as the incident commander. This person is responsible for the development and implementation of the incident action plan. The incident commander can delegate staff and command positions as needed to assist in command and control functions.

smoke-proof tower A stairway that is designed to be separated from the building by a landing. This creates a separation that will limit the spread of smoke into the stairway and keep it clear for evacuation.

softening the target Cooling hot fire gases using rapid, short-duration application (usually 15 to 60 seconds) from a straight or smooth-bore fire stream aimed at a steep angle toward the ceiling from the safest effective location. When using this tactic, water often is applied from the exterior to the interior of a building.

span of control The number of people reporting to a supervisor. The span of control should not exceed seven people reporting to a single supervisor under emergency conditions. For example, a fire captain supervising an apparatus operator and three fire fighters would have a four-to-one span of control.

sprinkler system calculations Specific rate of flow calculations for sprinkler systems; based on the fuel load. These calculations can be found in various publications, including NFPA documents and Factory Mutual Data Sheets.

stack effect The vertical airflow within buildings caused by temperature differences between the building interior and the exterior; depending on conditions, stack effect could be positive or negative, causing smoke to move upward or downward.

standard operating procedures (SOPs) Written rules, policies, regulations, and procedures intended to organize operations in a predictable manner.

standpipe system A piping system with discharge outlets at various locations; in high-rise buildings an outlet will normally be located in the stairway on each floor level. Most are connected to a water source and the pressure is boosted by a fire pump.

storage occupancy An occupancy used primarily for the storage or sheltering of goods, merchandise, products, vehicles, or animals.

strike team A set number of resources of the same kind and type that have an established minimum number of personnel (e.g., five engine companies). These teams always have a leader (usually in a separate vehicle) and have common communications among resource elements.

Superfund Amendments and Reauthorization Act (SARA) A federal law enacted in 1986 and also known as the Emergency Planning and Community Right to Know Act. Title III of this law requires businesses that handle or store hazardous chemicals in quantities above specific limits to report the location, quantity, and hazards of those chemicals to the State Emergency Response Commission (SERC), Local Emergency Planning Committee, and local fire department. Some of this information has been classified since the attack on the World Trade Center on September 11, 2001.

T

task force Any combination of resources that can be temporarily assembled for a specific mission. These teams should be established to meet specific tactical needs and should be demobilized as single resources. All resource elements within this team must have common communications and a leader.

ten-codes Numeric codes used to communicate predefined situations or conditions. For example, "10-4" generally means "finished communicating." Their use is discouraged.

tenders NIMS term for vehicles that transport water from a water source to the fire scene; also referred to as a tanker or mobile water supply apparatus.

trench cut An opening in the roof that extends from bearing wall to bearing wall to prevent horizontal fire spread in a building.

two-in/two-out rule A rule mandating that fire fighters working inside the hazard area must work in crews of at least two people; these two people must be backed up by at least two people outside the hazard area who are properly equipped and immediately available to come to the aid of the inside crew. This rule applies to the first-arriving unit.

type I construction Construction method in which the structural members, including walls, columns, beams, girders, trusses, arches, floors, and roofs, are of approved noncombustible or limited-combustible materials and have the highest level of fire-resistance ratings.

type II construction Construction method in which the structural members, including walls, columns, beams, girders, trusses, arches, floors, and roofs, are of approved noncombustible or limited-combustible materials but the fire resistance rating does not meet the requirements for Type I construction.

type III construction Construction method in which exterior walls and structural members that are portions of exterior walls are of approved noncombustible or limited-combustible materials, and interior structural members, including walls, columns, beams, girders, trusses, arches, floors, and roofs, are entirely or partially of wood of smaller dimensions than those required for Type IV construction or of approved noncombustible, limited-combustible, or other approved combustible materials.

type IV construction Construction method in which exterior walls and structural members that are portions of exterior walls are of approved noncombustible or limited-combustible materials. Other interior structural members, including walls, columns, beams, girders, trusses, arches, floors, and roofs, are of solid or laminated wood without concealed spaces. Wood columns supporting floor loads are not less than 8 in. (20 cm) in any dimension; wood columns supporting roof loads only are not less than 6 in. (15 cm) in the smallest dimension and not less than 8 in. (20 cm) in depth. Wood beams and girders supporting floor loads are not less than 6 in. (15 cm) in width and not less than 10 in. (25 cm) in depth. Wood beams and girders and other roof framing supporting roof loads only are not less than 4 in. (10 cm) in width and not less than 6 in. (15 cm) in depth. Specifics for other structural members are required to be large-dimension lumber as well.

type V construction Construction method in which exterior walls, bearing walls, columns, beams, girders, trusses, arches, floors, and roofs are entirely or partially of wood or other approved combustible material smaller than material required for Type IV construction.

U

unified command An application of the National Incident Management System that allows all agencies with jurisdiction responsibility for an incident or planned event, either geographic or functional, to manage an emergency incident or planned event by establishing a common set of incident objectives and strategies; the role of incident commander is shared by representatives of various responding jurisdictions and/or agencies.

unity of command A pyramidal command system ensuring that no one reports to more than one supervisor.

V

ventilation controlled A fire within a compartment or building that is limited due to a lack of air (oxygen) even though sufficient vapor fuel is available to support continued burning.

W

warm zone An intermediate area between the hot and cold zones where personal protective equipment is required, but at a lower level than in the hot zone.

wet pipe sprinkler system A sprinkler system in which the pipes are normally filled with water.

wood truss An assembly made up of small-dimension lumber joined in a triangular configuration that can be used to support either roofs or floors.

Index

© Courtesy of David J. Jones, Cincinanti, Ohio.

Note: Locators followed by the letter '*f*' and '*t*' refer to figures and tables respectively.

A

accountability, fire-ground operations and, 183–184
aerial ladders, rescue and, 209
age of building, structural factors, 82
alarm information, 91. *See also* fire alarm system, response to
Alfred P. Murrah Federal Building bombing, 32
alphanumeric naming system, 28–30
ambulatory care facilities occupancies, 357
apartment buildings occupancies, 363–364
apparatus
    IC and, 88
    needs, offensive operation, 296–297
area of refuge, 75
arenas occupancies, 347–348
assembly occupancies, 343–349
    convention centers, 348–349
    eating and drinking establishments, 345–347
    sport arenas, 347–348
    theaters, 349
    worship, places of, 344–345
attack hose lines, offensive operation, 282, 284
attack hose size, offensive operation, 268–276, 269*f*, 270–271*t*
attack pumper, 130
automatic fire suppression equipment, 86
automatic mutual aid, 15

B

backdraft, 70
backpressure, 243
backup needs, offensive operation, 284
balloon-frame construction, 55
base, high-rise building fires and, 410–411
basement fires, 160–165
BASF plant fire, 26–27
big-box stores occupancies, 368–369
bowstring truss construction fire, 156
branches, defined, 25
Bremerton, Washington, apartment complex fire, 212–213, 364
building collapse. *See* collapse zone; structural factors
building communications systems, 32
building conservation. *See* property conservation
building construction
    balloon-frame, 55
    concealed spaces, 81–82
    construction type, 79–80
    fire extension, 166–168
    methods and materials, 153–156, 168*t*
    platform-frame construction, 55
    roof construction, 80
building design loads, risk management and, 152
buildings, occupancy and, 374–376
business occupancies, 369–370

C

calling for additional resources, 14–15
cell phones, communications and, 31–32
chain of command, 10
Chapel Hill fraternity fire, 212, 365
Charleston, South Carolina, furniture store fire, 169–170
chief officer, command by, 10–11
chief's aide, 21–22
Cincinnati residential fire, 264
Class A foam, 247
cold zone, 170
collapse zone, 79, 168
    defensive operations and, 155*f*, 310–312
college and university occupancies, 353
command modes, 10
command post, 12–13, 396–397, 397*f*
command staff, 17–19
    considerations, 18–19
    incident safety officer, 17
    liaison officer, 17–18
    positions, 17*f*
    public information officer, 18
command transfer process, 11–12
commercial buildings. *See* occupancy
commercial-type automatic sprinkler systems, 231
communications, 31–35
    advances in, 33
    building communication systems, 32
    cell phones and, 31–32
    computer systems, 32–33
    face-to-face, 31
    fire-ground operations and, 182
    hard-wire systems, 32
    high-rise building fires and, 408, 409*f*
    messengers, 31
    network, 33–35, 34*f*, 35*f*
    public address systems, 32
    radio designation for IC, 33
    satellite phones, 32

company constructions, types, 168t
company operations, 122–135
   coordination, 126–131
      apparatus positioning, 130–131
      quint and quad companies, 128–129
      ventilation, 129, 129t
   engine company tasks, 124–125
   introduction, 124
   truck company tasks, 126
company safety responsibilities, 188
compartmentation, 68
complexity and layout, structural factors, 82
complex preincident plan, 59, 60f, 61
computer systems, communications and, 32–33
concealed spaces, structural factors, 81–82
conflagrations, defensive operations and, 316, 317t, 318–320
conservation of property. *See* property conservation
construction methods and materials, 153–156, 168t
construction type, structural factors, during preincident plan, 79–80
convention centers occupancies, 348–349
convergence cluster phenomenon, 361
Cook County Administration Building fire, 395
critical time, 212–213

D

dead load, 81
death rates operating inside structure fire, 144f
decay phase, 175
defend-in-place concept, 207
defensive decision, 309
defensive operations, 101, 306–327
   apparatus needs, 315–316
   collapse zone, establishing, 310–312
   conflagrations, 316, 317t, 318–320
   defensive attack, classifying, 309–320
   direct attack, evaluating, 312–313
   exposures, evaluating, 312
   group fires, 316, 317t, 318–320
   introduction, 309
   master streams, estimating the number and type of, 313–314
   nonattack, classifying, 321
   staffing needs, 315
   water supply needs, estimating, 314–315
      nozzle type, selection of, 314–315
Delaware arson floor collapse, 162
delegation, 12
deluge system, 234
   working at property protected by, 240–241
      backing up the system, 241
      control valve and fire pump, checking of, 240
      interlocks, checking, 241
      operations, 240–241
demobilization, 15
deployment, 101–102, 105, 108, 110–111, 113–114, 116
detention and correctional occupancies, 359–360
direct attack, 278–282, 282f
   defensive operations and, 312–313
district resources and challenges, evaluating response, 50–55
   construction methods, 54–55
   response time, 51
   water supply, 51–54
division
   defined, 24
   of labor principle, 124

door marking system, 211f
dormitories occupancies, 364–366
draft curtains, 240
drilling down, 58
dry pipe sprinkler system, 233

E

educational occupancies, 350–353, 350t
   colleges and universities, 353
   elementary schools, 351–353
   middle and high schools, 353
electrical hazards, fire ground, 172–173
elementary schools occupancies, 351–353
elevated platforms, rescue and, 209
elevator rescues, 209
   high-rise building fires and, 388–392
emergency medical services (EMS), 205, 217–219
emergency voice/alarm communications system (EVACS), 394–395
EMS. *See* emergency medical services (EMS)
enclosures and fire separations, structural factors, 81
engine company tasks, 124–125
EVACS. *See* emergency voice/alarm communications system (EVACS)
evacuation status, 74–76
   awareness of occupants, 74
   familiarity with building of occupant, 74–75
   life safety and, 74–76, 210–213
      awareness of occupants, 74
   medical status of occupants, 76
   mobility of occupants, 74
   occupant proximity to fire, 74
   primary and alternative egress routes, 75–76
expected benefits, risk-versus-benefit analysis, 103, 106, 110, 113, 115
expected risk, risk-versus-benefit analysis, 103, 106, 110, 113, 115
exposures
   defensive operations and, 312
   evaluation of, 282, 283f
   external, 282
   offensive operation and, 282, 283f
extension probability, structural factors, 81
external exposures, 282
extinguishment, 83–86. *See also* manual extinguishment
   automatic fire suppression equipment, 86
   external exposures, 84
   high-rise building fires, 396
   internal exposures, 84
   life safety and, 204–205
   manual, 84–85
      apparatus pump capacity, 86
      fire suppression systems, 86
      fuel load, 84
      hose lines needed, 84
      rate of flow, calculation of, 84
      staffing, 84
      water supply, 84–85
   needs, estimation of, 68
   offensive/defensive/nonattack, 83
   probability of, 83
   ventilation status, 83–84

F

face-to-face communications, 31
finance/administration section, 19–20, 20f
fire alarm system, response to, 250
fire conditions, structural collapse, 166

fire department connection, 237–239, 242–243
fire escapes, rescue and, 208–209
fire extension, 166–168
fire fighter safety, 70–78, 136–199. *See also* life safety
    basement fires, 160–165
    company safety responsibilities, 188
    construction methods and materials, 153–156, 168t
    electrical hazards, fire ground, 172–173
    evacuation status, 74–76
        awareness of occupants, 74
        familiarity with building, of occupant, 74–75
        medical status of occupants, 76
        mobility of occupants, 74
        occupant proximity to fire, 74
        primary and alternative egress routes, 75–76
    fire conditions, structural collapse, 166
    fire extension, 166–168
    fire-ground operations, 182–188
        accountability, 183–184
        alternative egress, 184–185
        command and control, 182
        communications, 182
        fatality, 145f
        rapid intervention crews, 185–188
        safety officers, 184
    fire intensity, 174–177
    fire streams, opposing, 189–190
    floor construction, 159
    green construction, 159
    hazard control zone, 168–172, 172–173f
    incident safety officer, 139–140
    injuries and fatalities, 140–141t, 140–148, 142f, 144–146f
    introduction, 139
    mayday and, 188–189
    occupancy type, 73–74
    operational status, 76–78
    overhaul, 192–194
    personal protective clothing, 190–192, 191t
    prefire conditions, 165–166
    rehabilitation, 194–195
    risk management, 148–153
        building design loads, 152
        fire intensity, 150
        fuel load, 150–151
        structural stability, 152–153
    roof operations, 157–158
    smoke and fire conditions, 70–73
        direction of travel, 71–72
        location of fire, 71
    structural stability, 174–176
    tactical reserve, 180
    time, 174–177, 176t
        detection/transmission, 177
        dispatch, 177
        flashover and, 180–182
        setup, 178
        staffing, 178–180
        travel, 177–178
        turnout, 177
    ventilation status, 72–73
fire fighter safety, high-rise building fires and, 388
fire-ground operations, 182–188
    accountability, 183–184
    alternative egress, 184–185

    command and control, 182
    communications, 182
    rapid intervention crews, 185–188
    safety officers, 184
fire intensity, 174–177
    risk management and, 150
fire investigation, property conservation and, 335
fire perimeter, 170
fire protection systems, 228–253
    deluge system, working at property protected by, 240–241
        backing up the system, 241
        control valve and fire pump, checking of, 240
        interlocks, checking, 241
        operations, 240–241
    fire alarm system, response to, 250
    introduction, 231
    nonwater-based extinguishing systems, 246–248
        carbon dioxide systems, 247
        dry and wet chemical systems, 248
        foam systems, 246–247
        halon and other clean agents, 247–248
    sprinkler-protected building, working with evidence of fire showing from the exterior, 238–240
        backing up the system, 239
        fire department connection, supplying, 238–239
        gaining entry, 238
        main control valve, checking of, 238
        property conservation tasks, performing, 240
        ventilation, 240
    sprinkler-protected building, working with no signs of fire, 235–238
        fire department connection, supplying, 237–238
        fire pumps, checking of, 236
        fire/sprinkler operation, checking for, 237
        gaining entry, 235
        main control valve, checking of, 236
    sprinkler systems, 231–235
    standpipe system, working at building equipped with, 241–246
        discharge, connection to standpipe, 246
        equipment, providing of, 243–246
        fire department connection, supplying, 242–243
        fire pumps and main control valve, checking of, 242
    total flooding carbon dioxide or clean agent systems, working in, 248–249
        agent supply, checking of, 249
        final extinguishment and rescue, 249
        interlocks, checking of, 249
        manual activation, 249
        process, 248
        system restoration, 249
fire streams, opposing, 189–190
fire suppression water load, structural factors, 81
The First Interstate Bank Building fire, 413–415
fixed water spray system, 234f
flashover, 70
floor construction, fire fighter safety and, 159
floor layout and size, surveying, 214–216
flow path, 48
foam house, 247
formal preincident plan, 61–62, 61f
freelancing, 11
frequency, injury, 143

fuel-controlled fire, 259
fuel load, risk management and, 150–151
fully developed phase, 175

G

general staff, 19–23
   branches, divisions, and groups, 24–27
   sections, 19–23, 19f
   strike team, 28
   task force, 28
green buildings, 83
green construction, fire fighter safety and, 159
ground ladders, rescue and, 209
ground support, high-rise building fires and, 392–393
group, defined, 24
group fires, defensive operations and, 316, 317t, 318–320
growth phase, 175
gusset plates, 155f

H

hard-wire communication systems, 32
hazard control zone, 168–172, 172–173f
healthcare occupancies, 353–357, 354t
   ambulatory care facilities, 357
   hospitals, 354–355
   limited care facilities, 356–357
   nursing homes, 355–356
heat index chart, 195f
height and area, structural factors, 82
helicopter rescues, 210
high-rise building fires, 384–425
   case histories
      The First Interstate Bank Building fire, 413–415
      The One Meridian Plaza Fire, 412–415
      World Trade Center, terrorist attack in, 415–417
   command post location, 396–397
   elevators use, fire fighter, 388–392
   extinguishment, 395–396
   ground support, 392–393
   incident action plan, implementing, 404–408
      interior exposures, 406
      lead time, 406–407
      property conservation, 406
      stairways usage, 405–406
      ventilation, 406
      wide fire zone, establishing, 407–408
   introduction, 387–388
   life safety, 393
   NIMS, applying, 408–412
      base, 410–411
      communications, 408, 409f
      lobby control, 388, 412
      staging, 411–412
      tactical working sheets, 409, 410f
   operating procedures, 388
   preincident plans, building specific, 397–399
   rescuing and evacuating occupants, 393–395
      EVACS, 394–395
      helicopter rescues, 393–394
      partial or sequential evacuation, 394
   safety, fire fighter, 388
   size-up, analyzing situation through, 399–403
      heat of fire, 401

      smoke movement, 400–401
      stack effect, 401
      wind, 401–403
   wind and, 401–403
      application of PPV, 403
      controlling windows and doors, 402–403
      deployment of control devices, 403
      exterior fire streams, usage of, 403
high-rise decision tree, 401f
hose lines
   evaluation, 285
   offensive operation and, 285
hospitals occupancies, 354–355
hotels and motels occupancies, 366
Hotel Vendome fire, 15
hot zone, 170
Houston, Texas, 1979 conflagration, 319f
hovering, 58

I

IC. *See* incident commander (IC)
IDLH. *See* immediately dangerous to life and health (IDLH)
ignition phase, 175
immediately dangerous to life and health (IDLH), 100
IMT. *See* incident management team (IMT)
incident action plan, 96–120
   for church fire, 111–114
   defensive, developing, 101
   defined, 99
   deployment, 101–102, 105, 108, 110–111, 113–114, 116
   formulating, 101
   for fourteen-unit apartment building, 105–108
   for high-rise apartment building, 114–117
   high-rise building fires and, 404–408
      interior exposures, 406
      lead time, 406–407
      property conservation, 406
      stairways usage, 405–406
      ventilation, 406
      wide fire zone, establishing, 407–408
   implementation of, 87
   introduction, 99
   for large industrial complex, 108–111
   offensive, developing, 101
   resource capability and requirement, estimating, 100–101
   risk-versus-benefit analysis, 103–105, 106–108, 110, 113, 115
      expected risk, 103, 106, 110, 113, 115
      offensive/defensive decision, 104, 106, 110, 113, 115
      resource needs, 104–105, 106, 108, 110, 113, 115
      structural stability, 104, 106, 110, 113, 115
   for single-family detached dwelling, 102–105
   structural conditions, evaluating, 99–100
incident commander (IC), 6
incident management team (IMT), 23
incident safety officer, 17
   fire fighter safety and, 139–140
Indiana roof collapse, 171
indirect attack, 277, 277f
indirect damage, property conservation and, 331
industrial occupancies, 372–374
initial command, 8–10
initial rapid intervention crew (IRIC), 127
injuries and fatalities, fire fighter safety and, 140–141t, 140–148, 142f, 144–146f

injury frequency, 143
injury severity, 143
intelligence/information section, 22
interior exposures, high-rise building fires and, 406
interior stairs, rescue and, 208
interlocks, 241, 249
internal exposures, 282
IRIC. *See* initial rapid intervention crew (IRIC)

K
Kmart warehouse fire, 272

L
lead time, high-rise building fires and, 406–407
liaison officer, 17–18
life safety, 70–78, 202–227. *See also* fire fighter safety
    evacuation status, 74–76
        awareness of occupants, 74
        familiarity with building, occupant, 74–75
        medical status of occupants, 76
        mobility of occupants, 74
        occupant proximity to fire, 74
        primary and alternative egress routes, 75–76
    evacuation status, classifying, 210–213
    extinguishment, evaluation of probability, 204–205
    floor layout and size, surveying, 214–216
    high-rise building fires and, 393
    introduction, 204
    location/proximity, prioritizing by, 216–217
    mass-casualty flowchart, 220*f*
    medical status of victims, evaluating, 217–220
    needs, estimation of, 68
    occupancy type, 73–74
    operational status, 76–78
    rescue options, analyzing, 207–210
        aerial ladders, 209
        elevated platforms, 209
        elevator rescues, 209
        fire escapes, 208–209
        ground ladders, 209
        helicopter rescues, 210
        interior stairs, 208
        rope rescues, 209–210
    rescue priorities, 217*f*
    rescue versus fire attack, 205
    shelter, evaluation of need for, 220
    smoke and fire conditions, 70–73
        direction of travel, 71–72
        location of fire, 71
    staffing, 213, 220–221
    transport, medical priority and, 218–219
    triage, medical priority and, 218–219
    ventilation status, 72–73, 206–207
    victim evaluation in mass-casualty incidents, 218
lifestyle centers occupancies, 368
limited care facilities occupancies, 356–357
live and dead loads, structural factors, 81
live load, 81
lobby control, high-rise building fires and, 388, 412
location/proximity, prioritizing by, 216–217
logistics section, 20–21, 20*f*

M
manual extinguishment, 84–85. *See also* extinguishment
    apparatus pump capacity, 86
    fire suppression systems, 86
    fuel load, 84
    hose lines needed, 84
    rate of flow, calculation of, 84
    staffing, 84
    water supply, 84–85
mass-casualty flowchart, 220*f*
mayday, 188–189
medical status of victims, evaluating, 217–220
mercantile occupancies, 366–369, 366*t*
    big-box stores, 368–369
    enclosed shopping malls, 367–368
    lifestyle centers, 368
    multilevel department stores, 368
    shopping centers, 367
messengers, communications and, 31
MGM Grand fire, 417
mixed multiple occupancy, 374
mixed occupancy, 374
mixed-use industrial fire, New York, 151–152
mobile command, 33*f*
modular organizations, 16–17
multilevel department stores occupancies, 369
multiple occupancy, 374

N
naming system, 28–30, 29*f*
National Fire Academy formula, 259–260, 260*f*
National Fire Protection Association (NFPA)
    NFPA 1, 56, 63, 81
    NFPA 13, 231, 265–266
    NFPA 13D, 231
    NFPA 13E, 232
    NFPA 13R, 231
    NFPA 14, 241, 275
    NFPA 25, 233
    NFPA 101, 74, 343, 350, 353, 357–360, 366, 369–370, 372–374, 413
    NFPA 170, 63
    NFPA 220, 79
    NFPA 704, 63
    NFPA 1142, 53, 265
    NFPA 1221, 177
    NFPA 1410, 178
    NFPA 1500, 6, 49, 77, 139–140, 143, 148–149, 153, 178–179, 182, 186, 192, 309
    NFPA 1521, 17
    NFPA 1561, 6, 11, 28
    NFPA 1584, 194
    NFPA 1620, 59, 62
    NFPA 1710, 50, 77, 88, 100, 105, 177–180, 185, 289, 290, 295, 387
    NFPA 1971, 191
    NFPA 3000, 55, 351
    NFPA 5000, 79
National Incident Management System (NIMS), 6
    command and, 6–7
    evaluation of, 6–7
    hierarchy, 19*f*
    high-rise building fires and, 408–412
        base, 410–411

communications, 408, 409f
lobby control, 388, 412
staging, 411–412
tactical working sheets, 409, 410f
operational status and, 76
organizations and positions, 16–28
New Orleans warehouse fires, 372
NFPA. *See* National Fire Protection Association (NFPA)
NIMS. *See* National Incident Management System (NIMS)
no-entry zone, 170
nonattack strategy, defensive operations and, 321
nonwater-based extinguishing systems, 246–248
carbon dioxide systems, 247
dry and wet chemical systems, 248
foam systems, 246–247
halon and other clean agents, 247–248
North York, Ontario, fire in, 208
notation, 62
nozzle type, offensive operation and, 276
nursing homes occupancies, 355–356

O

occupancy, 338–382
assembly occupancies, 343–349
convention centers, 348–349
eating and drinking establishments, 345–347
sport arenas, 347–348
theaters, 349
worship, places of, 344–345
buildings (under construction, renovation or demolition), 374–376
business occupancies, 369–370
considerations of special occupancy fires, 376–377
detention and correctional occupancies, 359–360
educational occupancies, 350–353, 350t
college and universities, 353
elementary schools, 351–353
middle and high schools, 353
fire fighter safety and, 73–74
healthcare occupancies, 353–357, 354t
ambulatory care facilities, 357
hospitals, 354–355
limited care facilities, 356–357
nursing homes, 355–356
industrial occupancies, 372–374
introduction, 342
mercantile occupancies, 366–369, 366t
big-box stores, 368–369
enclosed shopping malls, 367–368
lifestyle centers, 368
multilevel department stores, 369
shopping centers, 367
multiple and mixed occupancy buildings, 374
residential board and care occupancies, 357–358, 357t
residential occupancies, 360–366, 361t
apartment buildings, 363–364
dormitories, 364–366
hotels and motels, 366
one- and two-family dwellings, 362–363
storage occupancies, 370–372
types, 73–74
classification of, 342–343
victims, estimation of, 377

occupant load, 342
occupational injury, 140
offensive attack, 309
property conservation and, 330–335
conservation needs, evaluating, 333–334
indirect damage, estimating, 331
overhaul needs, evaluating, 334, 335f
smoke damage, evaluating, 333
staffing needs, calculation of, 333
water damage, evaluating, 331–333
offensive/defensive decision, risk-versus-benefit analysis, 104, 106, 110, 113, 115
offensive operation, 101, 254–304
apparatus needs, determining, 296–297
attack hose lines, estimation of number of, 282, 284
attack hose size, selection of, 268–276, 269f, 270–271t
backup needs, estimation of, 284
exposures, evaluation of, 282, 283f
hose lines, evaluation of other, 285
introduction, 257
method of attack, choosing of, 276–282
direct attack, 278–282, 282f
indirect attack, 277, 277f
transitional, 281f
nozzle type, selecting, 276
rate of flow, calculation of, 257–265
choosing of best, 267–268
comparisons, 265–267, 266f
for first floor, 268
National Fire Academy formula, 259–260, 260f
percentage of area, estimating, 263–265
Royer/Nelson formula, 258–259, 259f
for second floor, 268
size of area, estimating, 261–263
for sprinkler, 260–261
for third floor, 268
staffing needs, calculating, 289–295, 291–296t
chief officer, 292–293
engine 1, 290–291
engine 2, 291–292
engine 3, 294
engine 4, 294
truck 1, 292
truck 2, 294–295
ventilation needs, estimating, 287–288, 287f
water supply needs, estimating, 285–287, 285f
Ohio basement fire, 161
Ohio residential fire flashover, 167
Oklahoma City bombing, 32
on-duty fire fighter deaths, 142f, 146f
The One Meridian Plaza Fire, 412–415
operational status, 76–78
fire fighter safety and, 76–78
high-rise building fires and, 76
operations section, 22–23, 22f
outside stem and yolk (OS&Y) valve, 236f
overhaul, 91
fire fighter safety and, 192–194
property conservation and, 334, 335f

P

paint manufacturing facility fire, 59
partial or sequential evacuation, 394

personal protective clothing, 170, 190–192, 191*t*
personal protective equipment (PPE), 144, 191*t*
Phoenix supermarket fire, 187
PIO. *See* public information officer (PIO)
planning section, 21, 21*f*
platform-frame construction, 55
post indicator valve, 236*f*
PPE. *See* personal protective equipment (PPE)
preaction sprinkler system, 233
prefire conditions, fire fighter safety and, 165–166
preincident plans, 47, 58–67, 79, 91
    checklist and drawings, 62–67, 66*f*
    high-rise building fires and, 397–399
    narrative, 64–65*f*
    by occupancy, 62
    safeguarding, 57
    types of, 59–62
preplanned structures, 67
preplanning *vs.* size-up, 68
primary damage, 331
primary factors, size-up and, 68
properties, evaluation of specific, 55–68
    extinguishment needs, estimation of, 68
    life-safety needs, estimation of, 68
    modification of SOPs, 67–68
    preincident plans, 58–67
    preplanned structures, 67
    preplanning *vs.* size-up, 68
    property conservation needs, estimation of, 68
    security concerns, 55–58
property conservation, 86–87, 328–337
    fire investigation and, 335
    high-rise building fires and, 406
    introduction, 330
    needs, estimation of, 68
    offensive attack, classification of, 330–335
        conservation needs, evaluating, 333–334
        indirect damage, estimating, 331
        overhaul needs, evaluating, 334, 335*f*
        smoke damage, evaluating, 333
        staffing needs, calculation of, 333
        water damage, evaluating, 331–333
    salvageable property, 86–87
    smoke damage, 87
    water damage, 87
public address systems, communications, 32
public information officer (PIO), 18
pyrolysis, 180, 280–281

Q
quint and quad companies, 128–129

R
rapid intervention crew (RIC), 77
rate of flow, 52, 257–265
    choosing of best, 267–268
    comparisons, 265–267, 266*f*
    for first floor, 268
    National Fire Academy formula, 259–260, 260*f*
    percentage of area, estimating, 263–265
    Royer/Nelson formula, 258–259, 259*f*
    for second floor, 268
    size of area, estimating, 261–263
    for sprinkler, 260–261
    for third floor, 268
rational decision making (RDM), 6
RDM. *See* rational decision making (RDM)
recognition-primed decision making (RPD), 6
rehabilitation, fire fighter safety and, 194–195
rescue options, life safety and, 207–210
    aerial ladders, 209
    elevated platforms, 209
    elevator rescues, 209
    fire escapes, 208–209
    ground ladders, 209
    helicopter rescues, 210
    interior stairs, 208
rescue priorities, 217*f*
rescue versus fire attack, 205
rescuing and evacuating occupants, high-rise building fires and, 393–395
    EVACS, 394–395
    helicopter rescues, 393–394
residential board and care occupancies, 357–358, 357*t*
residential occupancies, 360–366, 361*t*
    apartment buildings, 363–364
    dormitories, 364–366
    hotels and motels, 366
    one- and two-family dwellings, 362–363
residential sprinkler system, 231
resource needs, risk-versus-benefit analysis, 104–105, 106, 108, 110, 113, 115
Rhode Island, The Station Nightclub fire, 347
RIC. *See* rapid intervention crew (RIC)
risk management, fire fighter safety and, 148–153
    building design loads, 152
    fire intensity, 150
    fuel load, 150–151
    structural stability, 152–153
risk-versus-benefit analysis, 78, 100
    incident action plan and, 103–105, 106–108, 110, 113, 115
        expected benefits, 103, 106, 110, 113, 115
        expected risk, 103, 106, 110, 113, 115
        offensive/defensive decision, 104, 106, 110, 113, 115
        resource needs, 104–105, 106, 108, 110, 113, 115
        structural stability, 104, 106, 110, 113, 115
roof construction, structural factors, 80
roof operations, fire fighter safety and, 157–158
rope rescues, 209–210
Royer/Nelson formula, 258–259, 259*f*

S
safety. *See also* fire fighter safety
    fire fighter, high-rise building fires and, 388
safety officers, fire-ground operations and, 184
salvageable property, 86–87
satellite phones, communications, 32
schools occupancies, 353
search and rescue operations, 366
secondary damage, 331
secondary factors, size-up and, 68
security systems, 55–58
    collecting information and preparing preincident plans, 57–58
    lockboxes, 55–57
    safeguarding preincident plan information, 57

separated multiple occupancy, 374
severity, injury, 143
shelter, evaluation of need for, 220
shopping centers occupancies, 367
shopping mall fire in Altoona, Pennsylvania, 368
shopping malls occupancies, 367–368
single command approach, 7
size-ups
   analyzing situation through, 68–90
   checklist, 69f
   chronology, 90–91
   of evacuation status, 74
   fire fighter safety. *See* fire fighter safety
   life safety. *See* life safety
   preplanning *vs.*, 68
   smoke and fire conditions, 70–73
      direction of travel, 71–72
      location of fire, 71
smoke damage, property conservation and, 87, 333
smoke-proof tower, 399
softening the target, 49
SOPs. *See* standard operating procedures (SOPs)
span of control, 13–14, 14f
sprinkler-protected building, working with evidence of fire showing from the exterior, 238–240
   backing up the system, 239
   fire department connection, supplying, 238–239
   gaining entry, 238
   main control valve, checking of, 238
   property conservation tasks, performing, 240
   ventilation, 240
sprinkler-protected building, working with no signs of fire, 235–238
   fire department connection, supplying, 237–238
   fire pumps, checking of, 236
   fire/sprinkler operation, checking for, 237
   gaining entry, 235
   main control valve, checking of, 236
sprinkler systems, 231–235
   calculations, 260
St. Louis high-rise fire, 211
stack effect, 401
staffing
   defensive operations and, 315
   IC and, 87–88
   life safety and, 213, 220–221
   offensive operation and, 289–295, 291–296t
      chief officer, 292–293
      engine 1, 290–291
      engine 2, 291–292
      engine 3, 294
      engine 4, 294
      truck 1, 292
      truck 2, 294–295
   property conservation and, 333
staging, 15–16
   high-rise building fires and, 411–412
stairways usage, high-rise building fires and, 405–406
standard operating procedures (SOPs), 47
   areas covered in, 47–48t
   development of, 47–50
   guidelines controversies, 50
   modification of, evaluation of specific properties and, 67–68
   preincident plan and, 91
   purpose of, 48–49
   training and equipment, relationship of, 49–50
standpipe system, 241
   working at building equipped with, 241–246
      discharge, connection to standpipe, 246
      equipment, providing of, 243–246
      fire department connection, supplying, 242–243
      fire pumps and main control valve, checking of, 242
storage occupancies, 370–372
stream deflection, 396f
streams, defensive operations and, 313–314
strike team, 28
structural collapse, fire conditions, 166
structural factors, 78–83
   age of building, 82
   collapse, signs of, 78–79
   collapse zone, 79
   complexity and layout, 82
   concealed spaces, 81–82
   condition, 80
   construction type, 79–80
   enclosures and fire separations, 81
   extension probability, 81
   fire suppression water load, 81
   green buildings, 83
   height and area, 82
   live and dead loads, 81
   roof construction, 80
structural stability
   fire fighter safety and, 174–176
   risk management and, 152–153
   risk-versus-benefit analysis, 104, 106, 110, 113, 115
Superfund Amendments and Reauthorization Act (SARA), 55

T

tactical reserve, fire fighter safety and, 180
tactical worksheets, 101, 103f, 409, 410f
task force, 28
ten codes, 10
tenders, 53
Tennessee partial roof collapse, 181
Texas apartment building fire, 158
theaters occupancies, 349
time
   fire fighter safety and, 174–177, 176t
      detection/transmission, 177
      dispatch, 177
      flashover and, 180–182
      setup, 178
      staffing, 178–180
      travel, 177–178
      turnout, 177
   IC and, 89
total flooding carbon dioxide or clean agent systems, working in, 248–249
   agent supply, checking of, 249
   final extinguishment and rescue, 249
   interlocks, checking of, 249
   manual activation, 249
   process, 248
   system restoration, 249
transitional attack, 49, 281f
transport, medical priority and, 218–219
trench cut, 288

triage, medical priority and, 218–219
truck company tasks, 126
truss construction, 154
truss roof, 80
20-minute rule, 152
two-in/two-out rule, 24
type I construction, 79
type II construction, 79
type III construction, 79
type IV construction, 79
type V construction, 79

U
unified command approach, 7–8, 101
unity of command, 33
University of North Carolina, Chapel Hill Fraternity House fires, 365
utilities, IC and, 89–90

V
Vent Enter Isolate Search (VEIS) tactics, 205
ventilation
    and company operations, 129, 129t
    fire fighter safety and, 72–73
    high-rise building fires and, 406
    offensive operation and, 287–288, 287f
    status, 72–73
        life safety, 72–73, 206–207
ventilation controlled fire, 258
victim evaluation, mass-casualty incidents, 218
victims, occupancy and, 377

W
warm zone, 170
water damage, property conservation and, 87, 331–333
water relay, 52f
water supply needs
    defensive operations and, 314–315
    offensive operation and, 285–287, 285f
weather, IC and, 89–90
West Fertilizer explosion, 147
wet pipe sprinkler system, 231, 232f
wide fire zone, high-rise building fires and, 407–408
wind chill chart, 195f
wind-driven fires, 402–403
    application of PPV, 403
    controlling windows and doors, 402–403
    deployment of control devices, 403
    exterior fire streams, usage of, 403
wood truss, 54, 153–154, 376
Worcester, Massachusetts, warehouse fire, 149
World Trade Center, terrorist attack in, 415–417
worship occupancies, 344–345